P9-DNV-017

AN INTRODUCTION TO
BIO-INORGANIC
CHEMISTRY

UNH LIBRARY

3 4600 00227 6456

AN INTRODUCTION TO
BIO-INORGANIC
CHEMISTRY

Edited by

DAVID R. WILLIAMS

Chemistry Department
The University of St. Andrews
St. Andrews, Scotland

CHARLES C THOMAS · PUBLISHER
Springfield · Illinois · U.S.A.

Chem
QD
415
.I58

Published and Distributed Throughout the World by

CHARLES C THOMAS • PUBLISHER

Bannerstone House

301-327 East Lawrence Avenue, Springfield, Illinois, U.S.A.

This book is protected by copyright. No part of it
may be reproduced in any manner without written
permission from the publisher.

© *1976, by* CHARLES C THOMAS • PUBLISHER

ISBN 0-398-03422-2

Library of Congress Catalog Card Number: 75-8504

*With THOMAS BOOKS careful attention is given to all details of
manufacturing and design. It is the Publisher's desire to present books that are
satisfactory as to their physical qualities and artistic possibilities and
appropriate for their particular use. THOMAS BOOKS will be true to those
laws of quality that assure a good name and good will.*

Printed in the United States of America

P-4

Library of Congress Cataloging in Publication Data
Main entry under title:

An Introduction to bio-inorganic chemistry.

Includes bibliography and index.
1. Biological chemistry. 2. Chemistry, Inorganic. I. Williams, David Raymond. [DNLM: 1. Biochemistry. 2. Chemistry. QU4 I623]
QD415.I58   546   75-8504
ISBN 0-398-03422-2

CONTRIBUTORS

R. P. AGARWAL, Medical Chemistry Group, The John Curtin School of Medical Chemistry, Australian National University, Canberra, Australia

A. ALBERT, Department of Pharmacological Sciences, Health Sciences Center, State University of New York at Stony Brook, N.Y. 11794, USA

G. W. BATES, Texas Agricultural Experiment Station and Department of Biochemistry and Biophysics, Texas A & M University, College Station, Texas, USA

N. J. BLACKBURN, Department of Biochemistry, University of Oregon Medical School, Portland, Oregon, USA

F. L. BYGRAVE, Department of Biochemistry, Australian National University, Canberra, Australia

G. S. FELL, Department of Pathological Biochemistry, Royal Infirmary, University of Glasgow, Scotland

D. E. FENTON, Department of Chemistry, University of Sheffield, England

G. GLIDEWELL, Department of Chemistry, University of St. Andrews, Scotland

R. W. HAY, Department of Chemistry, University of Stirling, Scotland

H. M. N. H. IRVING, Department of Inorganic and Structural Chemistry, School of Chemistry, University of Leeds, England

R. F. JAMESON, Department of Chemistry, University of Dundee, Scotland

E. KÖRÖS, Institute of Inorganic and Analytical Chemistry, L. Eötvös University of Budapest, Hungary

A. D. B. MALCOLM, Department of Biochemistry, University of Glasgow, Scotland

R.-P. MARTIN, Laboratoire de Chimie Minerale, Universite Claude Bernard, Lyon, France

G. J. MOODY, Department of Chemistry, U.W.I.S.T., Cardiff, Wales

R. ÖSTERBERG, Department of Medical Biochemistry, University of Gothenburg, Sweden

D. D. PERRIN, Medical Chemistry Group, The John Curtin School of Medical Chemistry, Australian National University, Canberra, Australia

P. SALTMAN, Department of Biology, University of California at San Diego, La Jolla, California, USA

B. SARKAR, The Research Institute, The Hospital for Sick Children, Toronto, Canada

J. P. SCHARFF, Laboratoire de Chimie Minerale, Universite Claude Bernard, Lyon, France

H. SMITH, Department of Forensic Medicine, University of Glasgow, Scotland

J. D. R. THOMAS, Department of Chemistry, U.W.I.S.T., Cardiff, Wales

D. R. WILLIAMS, Department of Chemistry, University of St. Andrews, Scotland

PREFACE

THE IDEA OF A MULTIAUTHOR, interdisciplinary textbook of bio-inorganic chemistry was conceived in an Amsterdam bar by an *ad hoc* committee of bio-inorganic researchers delayed by a twenty-four-hour pilots strike: The book is designed to provide an introduction to graduates (and senior undergraduates) in inorganic chemistry, biochemistry and medicine to the field of metal ions in biological systems.

Recently there has been an avalanche of printed material concerning all aspects of the bio-metals. This book is designed as an introduction to scientists of any discipline who are interested in obtaining an overview of this emerging subject called bio-inorganic chemistry. Our aims are primarily to teach the *principles* of the subject but, in addition, we propose to draw the reader's attention to unanswered questions and to areas that have not yet been researched so that he or she may be tempted to tackle these problems.

Students and researchers are currently favoring topics that have altruistic overtones. This is leading them to interdisciplinary areas and therefore there is a concomitant need for intelligible communication across scientific boundaries—a need that will undoubtedly grow in the future. All biochemical, inorganic and medical students should have an understanding of bio-inorganic chemistry because much of our way of life depends upon bio-inorganic principles and the resulting technological applications of these concepts. Thus, it is not our intention to present streams of chemical facts unrelated to the world outside the lecture hall and the laboratory, but rather in compiling this book we have tried to portray bio-inorganic chemistry as a challenging field of research rather than as an established collection of factual knowledge. Creative science involves extrapolating beyond the information currently available.

In spite of the recent increase in cooperation between inorganic chemistry and biological chemistry, there is still a communications gap. Our aim is to collect the principles, and some subject matter, of this newly established discipline of bio-inorganic chemistry and to narrow this communications gap, chapter by chapter. The "biomass" (carbon, hydrogen, oxygen and nitrogen) is controlled and tantalized by the "trace elements"—a misnomer if it conveys the impression that the elements do not have an important role *in vivo*. Even if required as only one atom in ten million, life cannot exist without these essential trace elements. The literature on the subject is distributed throughout many journals from several disciplines and we have attempted a combination which forms a general picture of the growing interest in inorganic chemistry other than within the confines of reaction flasks and test tubes.

vii

The transition from an isolated laboratory system to the real multiphase multi-component phenomena of human life creates the possibility that this volume could contain an infinite number of chapters but we have limited our choice to sections describing the three areas, caused by two natural cleavage planes that divide the subject, I, of the general principles of bio-inorganic chemistry, II, the experimental methods used to produce the facts that gave rise to these principles, and, III, the applications of these principles to medicine. Within all three sections we have tried to take the reader to the most important of the frontiers of research within the constraints set by time lags in publishing schedules and by the limited size of this volume. We have not attempted to be encyclopaedic but rather to give a well-chosen sampling of material, the relative emphasis upon certain areas being dictated by a need to describe the majority of the principles carefully selected to comprise a coherent presentation of this very diffuse subject.

Workers at the interface between the disciplines of biological, inorganic and medical chemistry now talk in terms of atoms and nanometer units: It is hoped that our selection of the material for this book will persuade even more life scientists to devote their researches to the rational evolution of new drugs along the lines mentioned in Section III and to devote their teaching to new courses and practical training laboratories. (In this latter respect we must point out that, from students, bio-inorganic chemistry attracts added motivation since the subject offers ideals that can be realized.) During these last few years it has become apparent that academic growth either at pedagogic or research levels is no longer a sufficient goal in itself. The work has to have some meaning, some intellectual wealth, and some challenge. This text book initiates this process that leads to the more advanced reviews listed in the references and further reading. We hope that the interdisciplinary complexity of the subject stimulates rather than deters.

There will, of course, be opposition to the growth of this new subject—opposition from those intellectual Scrooges who react against the concept that a scientist trained in one discipline can ever do anything competent in another discipline. It is only human, for example, that inorganic chemists cling jealously to their prerogative to be the only teachers of inorganic chemistry. One can only hope that they too will soon appreciate (a) the fascination that interdisciplinary researches hold for men's minds, and (b) the many important achievements arising from multidisciplinary contributions.

I am indebted to a whole host of authors, advisers, critics, referees and publishers in helping me to edit this volume. We would also like to invite both students and teachers to advise us of constructive changes that will make an improved second edition. I hope that my prejudices and enthusiasms have not colored my presentation too much, and I hope that the readers will enjoy working with this book as much as I have enjoyed compiling it.

David R. Williams

CONTENTS

ix

AN INTRODUCTION TO
BIO-INORGANIC
CHEMISTRY

SECTION I

BIO-INORGANIC CHEMISTRY

CHAPTER 1

INTRODUCTION

David R. Williams

THIS BOOK HAS BEEN WRITTEN to teach the principles and uses of bio-inorganic chemistry. It is neither a reference book nor a compendium of information. Our aim has been to be clear rather than exhaustive and we have attempted to demonstrate to the student reader that bio-inorganic chemistry is a living, expanding subject that offers both interest and challenge.

We have also designed this book with a wider, nonspecialist, research audience in mind: Within the scientific community, cross fertilization of ideas and information is currently very essential because an increasing number of discoveries are being made by multidisciplinary research teams working in the areas between the traditional disciplines of chemistry, biology and medicine. This situation has produced a requirement for intelligible communication across scientific boundaries and it is hoped that this book will bridge some of the gaps involved in bio-inorganic chemistry.

Why has bio-inorganic chemistry been chosen? Metal ions play a vital role in numerous, widely differing biological processes, and, of late, new developments in instrumental techniques have accelerated studies involving the uncharted territory between inorganic chemistry and biological sciences thus producing many exciting developments to such an extent that the subject is now one of the most rapidly expanding areas of science. It is as incorrect to divide nature into the sharply defined areas of chemistry, biology and medicine as it was to divide chemistry into organic and inorganic from a "vital force" standpoint.

Thus, we endeavor to introduce the reader to the principles of the thousands of processes, each being a collection of many metal-dependent reactions, that occur in the human body. (In order to restrict this book to a manageable size the plant and lower animal kingdoms have been omitted.)[1-3]

What Is Bio-inorganic Chemistry?

A star belonging to our galaxy disappeared 4.6 to 4.7 thousand million years ago in a blinding supernova explosion, the fragments of which formed the solar system of which our earth is a part. The periodic table of elements comprising our planet are those which coalesced as products of the explosion. Gravity compacted them and temperature and solar energy reacted these elements into a primitive hydrosphere. This prebiotic soup then underwent very many further reactions to form simple molecules which through evolution (the survival of the fittest or perhaps more correctly the prevalence of the most reproducible) led to metabolism—the organization of series of reactions from a wide selection of random reactions—and this eventually gave cellular, plant, animal and human life as we know it today.

The majority of chemical elements can be found in minute quantities in the human body, their concentrations depending upon their concentrations in food, soil and the atmosphere. Some of these elements have been termed "essential" or "beneficial" in that homeostatic control mechanisms exist to govern their concentrations in various organs or body fluids. The twenty-five elements that are currently accepted as being necessary for healthy human life are shown in the Figure 1-1.[4] As expected, these elements follow the abundances of the elements in the earth's crust and in sea water since the process of natural selection has removed organisms dependent upon less readily available elements. Biochemistry is the study of compounds of these elements—their structures, reactions and mechanisms *in vivo*. However, traditional biochemistry has investigated the elements present in bulk *in vivo*—the non-

metals—but as our experimental abilities have improved to be capable of examining smaller concentrations, a new subject—bioinorganic chemistry—has evolved, a subject which is intent upon modernizing biochemistry by redressing the balance to consider metal-dependent reactions. Furthermore, more heavy element compounds are being used by our civilization and so our world is becoming more polluted. The interactions between pollutants and *in vivo* chemicals is also a part of bio-inorganic chemistry as is our evolution to future generations requiring essential elements in addition to the ones shown in Figure 1-1. Thus, *bio-inorganic chemistry is a branch of natural philosophy whose aims are to understand the chemistry of reactions involving the essential metals, and other trace elements, in vivo and to apply this knowledge altruistically.*

If the view that evolution and adaptation permit elements to traverse the scheme: *poisons → tolerable impurities → useful elements → essential elements* is accepted then one can understand how the seven elements printed in light type in Figure 1-1 have recently been shown to be essential for health because they occupy periodic table positions adjacent to the eighteen essential elements whose biochemical properties are well known—dietary deficiencies produce animal growth rates as low as two-thirds the normal growth rates.[5]

All three groups of metals *in vivo*—those essential for life, those essential for *healthy* life, and the polluting metals, have a cycle equivalent to the familiar nitrogen or carbon cycles. For example, Figure 1-2 shows the cycle recently composed for mercury in the biosphere.[6]

The basic elements of biochemistry, hydrogen, carbon, nitrogen and oxygen, comprise 99 percent of the atoms in the human body—chiefly as water, protein and fat.

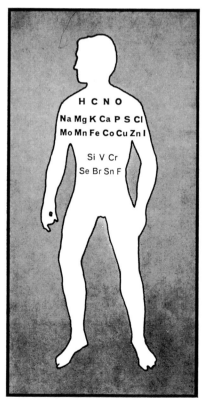

Figure 1-1. Elements essential for healthy human life.

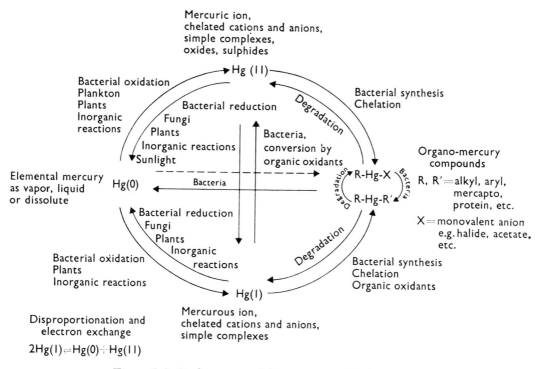

Figure 1-2. Cycle suggested for mercury in the biosphere.

Other *nonmetals* exist as anions in biological fluids. The total *metallic* component of our bodies is composed almost entirely of the main group metals including sodium, potassium, magnesium and calcium. The trace metals of the transition series—either as essential trace elements or as polluting ones—usually weigh less than 10 g *in toto*. Nevertheless, many volumes have already been published concerning the fascinating roles of these trace elements.

Historical Discoveries of Trace Metals

The history of the development of the idea that minute amounts of trace metals were indispensible for human life is intriguing. The term "trace" arose because early researchers, although aware of small amounts of mineral elements in living tissue, were unable to measure their precise concentrations. The adjective "essential" is only dubbed if the element is present in all healthy tissues, has a fairly constant concentration range between different animals, and when excluded from a body produces reproducible physiological abnormalities which are reversible upon readmittance of the element. Such definitions have excluded some two dozen trace elements which are ubiquitous to tissues but whose concentrations vary and whose physiological roles could not be determined. These must be assumed to be contaminants, some of which are toxic in small amounts (*all* trace elements are toxic in larger amounts).

Every element, essential, beneficial or polluting, has a spectrum of biological actions which depends upon the concentration of the element, or its compounds, in any particular organ or body fluid. Figure 1-3

Figure 1-3. The effects of varying the concentration of the trace element in the organ *in vivo*.

schematically shows how the state of health of an organ is dependent upon the concentration of an element in that organ.[7] As the concentration is increased from A to B the organ becomes progressively more normal in its reactions until a homeostatically buffered plateau B to C is reached. Clearly this "concentration for optimum health" plateau varies from essential element to essential element and from organ to organ. However, it would be hasty to assume that the only concentration of nonessential elements giving optimum health is zero: First, our bodies are skilled at tolerating quite high concentrations of polluting elements, especially if these elements are carried into the organs by the same mechanism that provides an essential element. For example, it is important to have sufficient zinc *in vivo* even if this means having to accept some cadmium as well. Secondly, regions D to G depict the property of a nonessential element to stimulate the body's defense mechanisms. Region C to D depicts the decline in health as excessive concentrations build up in an organ (for example, the siderosis effects associated with too much iron) until the curve reaches trough D. This is a passive protective intermediate concentration between C and E.

Some diseases, especially those caused by invading organisms such as viruses or bacteria, do not respond adequately to merely correcting elemental *in vivo* concentrations

to plateau B to C. In these instances there may be grounds for administering a higher concentration of a metal or of its complex. Region D to E shows the pharmacological effect of administering this element as doses of drugs. Such drugs stimulate the host's defense mechanisms. Naturally there is a limit to this process at plateau E to F. The presence of this plateau is fortunate in that it means an 80 kg man can be reasonably safely prescribed the same quantity of drug as a 50 kg girl. All therapies eventually change from excitation to a toxicological inhibition of the living process. F to G depicts this drug poisoning. Eventually, large doses of the element—as an essential element, as a polluting element, or as a medically administered compound of the element—cause an irreversible reaction, a complete decline in the living systems, leading to death.

Such curves differ from element to element, some having better homeostatic capacities than others and some having but a brief safety margin between optimum and toxic concentrations. Further, such *curves* ought really to be *surfaces* enclosed in three (or multi-) dimensional axes since healthy concentrations of elements are sometimes dependent upon the prevailing concentration of other elements. (For example, there are widely known mutual antagonisms between Fe and Co, Cu and Mn, Cu and Mo, Cu and Zn, and Ca and K concentrations.) Finally, we must remember that these curves or surfaces may have a varying amplitude according to prevailing circadian rhythms.

The first realization that trace elements were important *in vivo* came more than a century ago but it was not until the first quarter of this century that emission spectrophotography was applied to the study of iron and iodine in human health. Nuclear fission in World War II gave a boost to such studies from the viewpoint of assessing the maximum "safe" concentrations of radionuclides and their involvement in cycles such as the mercury example given. Then came the purified diet techniques for establishing "essentiality to health" criteria, such approaches helping to uncover many nutritional maladies in animals and humans and later chronic metal poisoning diseases were revealed. During the last two decades radiotracers and improved analytical techniques have helped to establish the metabolic movements and biochemical roles of these trace elements. Finally, the growing concern for human environmental health has boosted trace element biochemistry even further.

In summary: The absence of an attack upon living systems by the inorganic chemist on a scale comparable to that launched by the organic chemist has left us with a totally false impression of living things. Biochemistry is as much inorganic as organic.[8]

Outline of Chapter Contents

Music for the tone deaf can carry no message and so we have adopted the following three-part disciplining of our bio-inorganic receptors: Section I describes the principles of the subject and sufficient facts to illustrate them; Section II lists the instrumental techniques used to determine these facts; Section III puts the "bio-" into "bio-inorganic" by describing some of the useful products of the exercise—new drugs for Wilson's disease, for cancerous or polluted humans.

Before describing these sections more fully, an apology is in order since some aspects of bio-inorganic chemistry have been omitted from this book. This is usually because the topic has not yet reached the degree of maturity such that a simple survey at the level of the present book would

achieve anything.

Section I describes some specific examples to illustrate the general principles of bio-inorganic chemistry. From ancient times until the late nineteenth century, most great thinkers subscribed to the cult of Spontaneous Generation. Chapter 2 discusses modern twentieth century concepts of how we actually evolved from the elements distributed over the surface of the Earth and its environs. Although only a small fraction of this planet's atoms are involved in the fascinating set of processes we call "life," what this collection lacks in numbers it makes up for in activity and complexity. Yet, our evolution, like all chemical processes, could proceed no faster than the rate at which the scarcest elements became available.

Chapter 3 describes how the trace metals —once regarded simply as contaminants— are now known to be uniquely matched to ligand donor groups (and *vice versa*—the groups are matched to the available trace elements). Chapters 2 and 3 were not included for purely historical or academic interest: Man is both a part of, and a product of, the environment (a point reemphasized in Chapter 20) and so Chapters 2 and 3 are not only mere studies of how we came to be what we are, but also, they will help us to extrapolate to the future, to see the possible effects upon future generations of overpopulation, of increased material prosperity, of pollution and of agricultural practices which overstress natural systems.

Chapter 4 takes this specificity of metal ion-ligand bonding even deeper by studying a range of metalloenzymes—a description of the complete matching of a metal to its enzyme, to the enzyme substrate, and to the catalytic mechanism. Chapter 5 is also concerned with enzymology in general, and in particular, the inorganic chemistry of the activation of molecular oxygen and the description of the struggle involved as we human macroorganisms try to mimic the efficiency of microorganisms at nitrogen fixation.

In vivo we have learned to expect the unusual or the exceptional and this is found to be true for the bio-inorganic chemistry of *Homo sapiens* and bacteria in which the oxygen and nitrogen activating systems sometimes exhibit unexpected oxidation states and bond configurations. Much of Chapter 5 has been researched on model systems or on the simpler biological ligand (for example, amino acid) systems. Nucleic acids marshall the conjoining of these simple units into polymers and so it is appropriate that five chapters in Section I mention nucleic acid-metal ion interactions.

Man's ability to manipulate metal ion concentrations and bond configurations is currently limited to the use of *single* multidentate ligands as tools. Nature invokes *several* such ligands competing for a mutual metal ion. Chapter 6 describes our current understanding of these mixed ligand complexes. Each of Chapters 2 to 6 covers important subjects in their own rights: Too frequently students want to know the purpose of life before they have discovered what the chemistry of life is all about. Although it is possible to train young students in interdisciplinary work it is far more desirable to train them in specific topics (for example, those covered in Chapters 2 to 6) with the idea that they will eventually apply this expertise to other disciplines.

To illuminate the effectiveness of combining expertises from several disciplines to researches into one particular topic, Chapters 7 and 8 are overviews of our present state of knowledge of the reactions and roles of just three bio-metals—transition metal *iron,* and main group metals *calcium*

and *magnesium,* the latter emphasizing the essential interdependence of metal ion concentrations *in vivo.*

Section II stresses the *uses* of a variety of techniques rather than being a description of the exact instrumental approach in the belief that if a person is sufficiently motivated to apply a method he can learn the rigors of the method from one of the source books listed.

Although life is predominantly aqueous, there are no direct methods for determining the structures of complexes present in aqueous solution and so indirect (or as Tobe calls them, "sporting") methods are necessary. For example, extrapolations from crystal structures or of thermodynamic reasoning.

As in Section I, we have tried to concentrate upon patterns of behavior and to produce generalized views of the usefulness and applicabilities of the instrumental techniques described. Through these means we aim to bridge the gap between the standard inorganic or biochemical textbook and the research paper in another discipline. Understandably, there are techniques which attract only occasional glances in this book since, although their potentials may be immense, useful bio-inorganic achievements to date have been modest.

All life is based upon two important chemical reactions—the capture of solar energy for producing oxygen from carbon dioxide and water, and the reversion of this oxygen to carbon dioxide, water and muscular energy. Life has been defined as an organism continually reducing its entropy (and death is the reassertion of the claims of thermodynamic equilibria). Another unique feature of life is that organisms exist because of their kinetic stability—all living forms being thermodynamically unstable. Finally, life functions in cycles (circadian rhythms, heart beat, menstruation, etc.). Thus Section II commences with chapters on thermodynamics and kinetics. The entropy loss when life is created arises from the structural formation of peptides, membranes, and nucleic acids. Chapter 13 describes crystallographic means of determining the structures of living molecules.

Research is continually expanding the number of metabolic systems known to be influenced by trace elements and is revealing many hitherto unsuspected relationships between trace element concentrations and abnormal states of health. In the not too distant future, a determination of the concentration of trace elements *in vivo* will play a fundamental role in the diagnosis of illness and, of course, the manipulation of these concentrations will play an even greater role in its prevention. Thus, it is singularily appropriate that the best available methods of trace element concentration analysis—ion selective electrodes, neutron activation analysis and atomic absorption spectroscopy—are given a wide coverage (Chapters 12 and 14).

The "structure" of biological membranes and their temporal permeabilities to metals —two of life's most closely guarded secrets— are only just beginning to yield to research investigation using all available methods and so we conclude Section II with two chapters concerning multimethod approaches as tools for investigating membrane transport (Chapter 15), and for studying bio-oscillations (Chapter 16).

Section III, for many of us, provides the *raison d'etre* for continuing in bio-inorganic chemistry. We hope that students will also be more highly motivated after reading this section and so have their eventual goals in mind. The predictive, futuristic conclusions to most chapters ought also to supplement this motivation. We offer no excuses

to the purists for our "sugaring the pill" in Section III. At one time, Sir Henry Dale was warned against "selling his scientific birthright for a mess of commercial pottage" when he contemplated changing to pharmaceutical research. He did change and the "mess" led to his Nobel Prize!

Chapter 18 covers one disease (S. A. K. Wilson's disease) in relatively great depth from all three viewpoints (biochemical, inorganic and medical) and Chapter 19 reviews several dozen diseases associated with metal ions in far less detail than Chapter 18. Finally, Chapter 20 describes metal induced toxicity (caused by both the essential and nonessential *in vivo* elements) and the chelating ligands curently used to treat these conditions—therapeuticals which sorely need replacing by more specific ligands.

A century or so ago, natural philosophy could exactly define the atom and its electrons. The "uncertainty principle" smashed the absoluteness of their definitions and gave matter statistical rather than finite properties. Medicine and biology have also been sharply defined in terms of one vitamin or one potion to treat each specific disease, but, increasing knowledge of the interactions between *in vivo* chemicals also demands an integrated statistical approach. Thus, the time is opportune for researchers to integrate the whole of biochemistry, inorganic chemistry and medicine.

At the *student* level, science courses are changing rapidly in both structure and content. It is hoped that subsequent chapters offer new material for training students in several disciplines. The content, organization and coverage of each aspect are naturally governed by the present state of the subject (this is not a history book in which all the facts arc known and merely need assessing and presenting!). For clarity, only summary references have been chosen, original research papers being avoided. This is not to devalue the many original references upon whose data the principles in this book are firmly founded. We owe a great debt to these unmentioned authors.

REFERENCES

1. Eichhorn, G. L. (Ed.): Inorganic biochemistry. *Elsevier, 1 and 2,* 1973.
2. Williams, D. R.: *The Metals of Life.* London, Van Nostrand, 1971.
3. Lippard, S. J. (Ed.): Current research topics in bio-inorganic chemistry. *Prog in Inorg Chem, 18:*1973.
4. Williams, D. R.: *Education in Chemistry, 10:* 56, 1973.
5. Williams, D. R.: *Chem Revs,* 72:203, 1972.
6. N.B.S. Report: *Chemistry in Britain.* 1973, p. 49.
7. Venchikov, A. I.: *Vop pita, 6:*3, 1960.
8. Williams, R. J. P.: Bio-inorganic chemistry. American Chemistry Society Publication no. *100,* 1955 (1971).

CHAPTER 2

THE ORIGIN AND SPECIFICITY OF METAL IONS IN BIOLOGY[*]

R. Österberg

Introduction
The Elements of Life
Evolutionary Aspects of Metal Ion
Specificity
Concluding Remarks

INTRODUCTION

THE EARTH IS 4.7 billion years old, and life is known to have existed for at least 3.5 billion years—most probably preceded by a period of chemical evolution lasting about 500 million years. During this chemical evolution it is generally believed that all of the molecules essential for life were successively synthesized. Starting with pure inorganic compounds, a set of inorganic-organic reactions took place leading to the formation of the building stones for proteins and nucleic acids;[1, 5] these are amino acids and nucleotides. Similarly, during this chemical evolution period, a mixture of inorganic and organic polyphosphates was synthesized,[17–19] these being necessary to provide the primary chemical reserves of free energy for further evolution.[5, 17] As a consequence, polymerization and organization occurred; and, after a further period of development, the first cell evolved.

This cell, in all probability, was not as complicated as the present cells. For instance, in our present biological systems one single bacterium cell contains more than 3,000 different protein molecules; and an eukaryotic cell, which is much more complicated, contains more than 10^5 different protein molecules. However, if this very ancient cell functioned in a manner similar to our present cells, a minimum of about one hundred different protein molecules would have been required in order to maintain protein synthesis and anaerobic energy production.[2] Incorporated in this ancient cell were a series of metal ions. Some ions assumed functions required for the integrity of the cell, such as osmotic pressure; others (but initially very few) assumed functions of catalytic interest. The main metal ion catalyst at that time, perhaps the only one, must have been the magnesium ion; since, almost all of the reactions involved in fundamental cell reactions, such as protein biosynthesis and anaerobic energy production, require Mg^{2+} ions. Also, magnesium ions are known to catalyze several prebiotic condensation reactions.[17]

Further support for the assumption that magnesium ions were important early in bio-evolution arises from its presence in sea water and sea sediment,[6] an environment similar, it is assumed, to that from which our cell system once evolved. Table 2-I lists (a) the inorganic compositions of sea water and sea sediment, and (b) the

* The financial support of the *Swedish Natural Science Research Council* is greatly appreciated.

TABLE 2-I

COMPARISON BETWEEN THE AMOUNT OF IONS PRESENT IN SEA-WATER-SEDIMENT
AND MAMMALIAN INTRACELLULAR-EXTRACELLULAR SPACES

For the sea-water-sediment system the number of mmoles of the major components listed corresponds to 1 liter of sea water.

Ion	Sea Water (mM) (H)	Sediments (mmoles dissolved per 1 liter sea water) (S)	Extracell. (blood plasma) (mM) (E)	Intracell. (mmoles/ kg) (I)	Ratio (H/S)	Ratio (E/I)
Na	470	290	138	<10	2:1	14:1
Mg	50	480	1	20	1:10	1:20
Ca	10	550	3	($\leqslant 0.1$)*	1:55	—
K	10	400	4	110	1:40	1:28
Cl	55	—	100	—	>100:1	>100:1
HPO$_4$	0.001	20	1	100	1:20000	1:100
SO$_4$	28	40	1	5	1:2	1:5
Fe	0.0001	550	0.02	10	1:5500000	1:500
Zn	0.0001	1	0.02	1	1:10000	1:50
Cu	0.001	0.6	0.015	0.1	1:600	1:7
Co	$10^{-5.5}$	$\leqslant 0.001$	0.002			
Ni	10^{-6}	~ 0	~ 0			

* Solid phases in bone and teeth systems are not included.

inorganic compositions of a cell and its extracellular environment.[6, 20] Some interesting correlations can be made between the two systems. For instance, the same types of ions predominate both in the sea system and in living tissues: They are ions of sodium, potassium, calcium, magnesium and chloride. Also, the values of the trace metal ion concentrations are similar in the two systems: Zinc and copper ions are both present in dilute concentrations, and nickel and cobalt are present in extremely dilute concentrations.

A comparison between the extra- and intracellular spaces and between sea water and sediment indicates that there are high concentrations of potassium, magnesium and phosphate both in the sediment and within the mammalian cell as compared with the concentrations occurring in sea water and the extracellular space, whereas there is a high chloride concentration both in sea

water and in extracellular space compared with the amount of the chloride present in sediment and cells. These data indicate that a "solid crystalline" phase surrounded by its mother liquor might be regarded as a model for the cell and its extracellular fluid, and thus similar to the solid sea sediment surrounded by sea water. Further support for the assumption that sea water and its sediment might be regarded as a "model" system for the ionic distributions occurring in a cellular system is that the two concentration ratios, between sea water and sediment and between extracellular and intracellular spaces, show the same trend (Table 2-I); both series of quotients are either larger than unity or smaller than unity.

THE ELEMENTS OF LIFE

The elements of fundamental biological importance are all relatively light elements

	1a	2a	3a	4a	5a	6a	7a	8			1b	2b	3b	4b	5b	6b	7b	0
1	H																	
2		Be											C	N	O	F		
3	Na	Mg											Si	P	S	Cl		
4	K	Ca			V	Cr	Mn	Fe	Co		Cu	Zn		As	Se			
5			Y			Mo						Cd		Sn			I	
6																		
7																		

Figure 2-1. Elements essential for life in their periodic table positions. The eleven most abundant elements found in living organisms (H, O, C, N, Na, K, Ca, Mg, P, S, and Cl) are indicated in boldface lettering. The seven next most abundant elements are in Roman letters, and these are needed in trace amounts. The next important six elements are in italics, and are needed in ultratraces. The four elements shown as dotted letters are carcinogenic.

(indicated by boldface lettering in Figure 2-1). As a cofactor for biocatalysis, magnesium ions must have been the first metal ion of real importance. Later, when first photosynthesis and then aerobic energy production started, the magnesium ions were accompanied by the somewhat heavier trace elements (Figure 2-1). A common characteristic of the trace elements is that their ions form very strong metal complexes; for cobalt and zinc ions it is their particular metal complex forming property that is of biological importance. Copper and molybdenum also form strong metal complexes, but their main biological function depends upon their redox properties. Iron has at least two important characteristics in biology: Its complex forming ability is important for binding oxygen in hemoglobin and myoglobin, and its redox properties are important in the cytochromes and the iron-sulphur proteins. Here it should be noted that the reducing conditions supposed to prevail during the period of the chemical evolution and of early cellular life[1, 5] would favor the release of manganese(II) and iron(II) ions into prebiotic and early living systems from mineral and primeval sea sediments. Thus, manganese(II) appears to be one of those ions that acquired a specific biological function quite early. As an example, *de novo* synthesis of RNA by DNA polymerase requires Mn^{2+} but cannot proceed in the presence of Mg^{2+}. On the other hand, synthesis of DNA using a template requires either Mg^{2+} or Mn^{2+}.

In addition to the fundamental elements (H, C, N, O, . . .) and the trace elements (Mn, Fe, Co, . . .), which nowadays are required for almost all forms of life, there are also other additional essential elements[7, 8] (Figure 2-1). These other elements (V, Cr, Se, . . .) seem to be mainly involved in highly organized forms of life and have acquired their functions quite "recently." They are required for normal growth in the plant and animal kingdoms at ultratrace concentrations. It is the comprehensive studies by Schwarz and his collaborators on element-sterile rats that clearly showed that the elements V, Cr, Si, F,

Se, and Sn are essential.[7] In addition to these elements, boron is needed for the growth of certain plants. Of the ultratrace elements, fluorine is important for the proper development of dental and bone tissues, and chromium is important for the metabolism of glucose involving insulin. A deficiency of selenium in concentrations of less than 0.1 ppm (parts per million) results in liver necrosis and muscle dystrophy. In much higher concentrations, 10 ppm or larger, selenium becomes toxic, even carcinogenic.[8, 21]

Selenium also interferes and interacts with other elements. A relatively large selenium concentration, > 10 ppm, appears to decrease the *in vivo* concentrations of magnesium, copper, and especially manganese ions. In humans these effects were magnified when selenium was supplemented with cobalt.[21] The symptoms observed were similar to those described as "beer drinkers' cardiomyopathy" which caused many deaths in Omaha, Minneapolis, and other cities in 1965 and 1966. The toxicity was originally ascribed to the small quantities of cobalt ions added to commercial beer in order to stabilize its foam. At present, it seems more probable that cobalt might have potentiated the toxic action of naturally occurring selenium. It should also be noted that in small "physiological" concentrations, 0.1 ppm, there is a correlation between selenium and a low cancer frequency; in a larger concentration, > 10 ppm, selenium is carcinogenic. The mechanism is unknown.

Several elements have a carcinogenic action at concentrations far lower than those which produce other toxic effects (*cf.* [8, 21]); for instance, the beryllium and yttrium ions, which have the same closed-shell electronic structure as the physiologically important Na^+, K^+, Mg^{2+} and Ca^{2+} ions. But Be^{2+} has a smaller ionic radius than any of these ions, and Y^{3+} has a higher charge; thus, both Be^{2+} and Y^{3+} ions have a much higher charge density than the other biologically important ions. One explanation of their carcinogenic effects might be that the beryllium and yttrium ions simulate and probably block the action of Mg^{2+} and Ca^{2+} ions, respectively. Their higher charge densities make them more reactive and as a result they complex too firmly for the finely balanced, highly sophisticated, cell machinery.

Arsenic and selenium are also carcinogenic; these two elements are situated directly below N, P and O, S, respectively (Figure 2-1). In addition, cadmium belongs to this group of carcinogenic elements. It is tempting to speculate that the carcinogenic actions are due to cadmium simulating the action of the physiologically important zinc, arsenic simulating the action of phosphorus, and selenium simulating the action of sulphur.

In large concentrations, almost every metal ion of the transition series is toxic; this is especially true if the diet is deficient in essential nutrients. For example, many South African Bantus suffering from pellagra accumulate an excess of iron from their diet. Natives, who are heavy beer drinkers of Kaffir beer, have a high incidence of hemosiderosis. Kaffir beer contains 4 mg iron per 100 ml, and it is not uncommon for a Bantu to drink 4 litres of Kaffir beer in one session (about 160 mg of iron). The liver eventually accumulates enormous quantities of iron and, because of the accompanying irritation, the liver cells become necrosed and disappear. Carcinoma of the liver is a complication in about one-fifth of the cases.

EVOLUTIONARY ASPECTS OF METAL ION SPECIFICITY

In biological tissues the catalytic action of a metal ion is usually asserted when the metal ion or its metal complex reacts with a protein molecule. This is a general rule in biology: *The fundamental element by which a certain biological property is expressed is the protein molecule.* There are two types of metal-ion-protein complexes. One of these, the metalloprotein, usually contains *inertly* bound metal ion(s) and, after preparation, the metal ion(s) constitute(s) an integral part of the protein molecule. The other types of complexes are the *labile* metal-ion-protein complexes. It is indicated from structural studies that in the metalloproteins the metal ions are bound within the molecules and not on their surfaces. Thus, the coordination structure of a metalloprotein has a certain ligand geometry that fits only one or a few metal ions.

Our present biological systems evolved due to the selection of a series of random mutations.[9, 10] A mutation hits the informative system, i.e. the nucleic acids, but it is recognized by the environment and by the cell *via* a protein molecule. A single mutation may involve only one amino acid residue out of several hundreds. (There are about twenty different amino acids in a protein; an average protein with a molecular weight of 30,000 contains about 300 amino acid residues.) One or a few mutations may not only change the amino acid sequence but also the three-dimensional structure of the protein. As a result, a series of selected mutations makes it possible for a protein to develop a very specific metal ion binding site.

Abundance of Magnesium Ions

One explanation for the important role of magnesium ions in early life processes, such as prebiotic condensations, protein biosynthesis, anaerobic energy production and, later, photosynthesis, may be attributed to its abundance; at present its concentration (54 mM) in the oceans is higher than that of any other bivalent metal ion.[6]

In primeval seas the Mg^{2+} concentration may have been even higher, since NH_4^+, Fe^{2+}, and Mn^{2+} present under the reducing conditions could have replaced Mg^{2+} in the silicate phases.[20] Therefore, it is not surprising that Mg^{2+} most likely catalyzed a series of prebiotic condensation reactions, and that it is important for many fundamental steps in life processes; for example, DNA replication and protein biosynthesis. In these processes Mg^{2+} both stabilizes the structure of some cellular compounds, such as the tRNA:s and the ribosomes,[2] and acts as a cofactor for the enzymic reactions.

Magnesium also acts as a cofactor for several enzymatic steps in anaerobic energy production. In the enzymic reactions involving Mg^{2+}, magnesium ions usually react with a phosphorylated substrate rather than with the enzyme.[22] For instance, in its reaction with RNA and DNA polymerases, Mg^{2+} facilitates the favorable leaving group, magnesium pyrophosphate; and very few, if any, of the magnesium ions are bound to proteins in the form that is valid for the definition of a metalloprotein.

Another important biological process that requires magnesium is photosynthesis.[23] In photosynthesis, chlorophyll, a magnesium compound, captures light quanta and then utilizes this energy to fixate carbon dioxide and to evolve oxygen. This process might have evolved later than the early fundamental processes mentioned above; but once the porphyrins had been synthesized, the abundance of Mg^{2+} ions logically led to the formation of chlorophyll.

Availability of Fe(II) and Mn(II) Under Reducing Conditions

The atmosphere predominating during the time of chemical evolution and of early life is generally supposed to have been a reducing one characterized by a moderate pressure of methane (0.01 to 0.001 atm) and a very low oxygen pressure.[5] For such conditions, thermodynamic data[20] for the system air-sea-sediment indicate that at equilibria the redox potentials in the seas were as low as -325 mV (pH 8.1): Inorganic sulphur would then be in the form of the iron sulphide, $FeS_2(s)$ (pyrite), and perhaps partly as $Fe_{0.86}S(s)$ (pyrrhotite); the major excess of iron would be in the form of $Fe_3O_4(s)$ and there would be a certain amount in solution as Fe^{2+} ($\leqslant 0.2$ mM Fe^{2+}). Similarly, the Mn(II) concentration was much higher in primordial seas, perhaps as high as 50 mM. Due to these relatively high concentrations of Fe^{2+} and Mn^{2+} it is not surprising that manganese and iron were the most important elements in early redox processes.

Some of the first metalloproteins appear to have been the iron-sulphur proteins, and they formed the basis for photosynthesis. At this time the "cells" changed from being heterotrophs to autotrophs; that is, they "learned" how to make precursors for their own macromolecular synthesis. This was accomplished in part by using iron-sulphur proteins, which are generally strong reducing agents. Ferredoxins are one type of such iron-sulphur proteins; one of these, a bacterial ferredoxin, has recently been subjected to an X-ray crystal structure determination (cf.[11]), (Figure 2-2). There are two structural properties of iron-sulphur proteins that indicate that they must be very old proteins: one is due to the inorganic part of the protein, where iron is bound *via* sulphur atoms to the protein molecule,[24] and the other is due to the amino acid composition.[25]

To begin with, in iron-sulphur proteins the iron is bound to sulphur atoms—in the ferredoxins to both inorganic sulphide and to cysteine residues, and in rubredoxin only

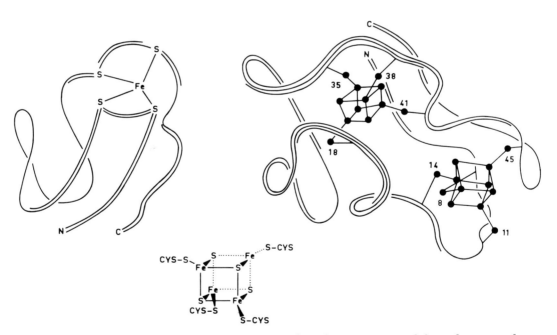

Figure 2-2. Schematic outline of the relationships between iron-sulphur clusters and protein chains in iron-sulphur proteins: (upper left) rubredoxin, (upper right) *M. aerogenes* ferredoxin, (below) iron-sulphur cluster of *M. aerogenes* ferredoxin. (From L. H. Jensen, *Trans Biochem Soc, 1*:27, 1973; cited ref. 11)

to cysteine residues (Figure 2-2). Also, crystals of the inorganic sulphides, $FeS_2(s)$ and $Fe_{0.86}S(s)$, are well-known semiconductors; they also have the capacity to transfer electrons. It has therefore been suggested[24] that inorganic electron carriers in the form of $FeS_2(s)$ and of $Fe_{0.86}S(s)$ were present prebiotically in the area where the first proteins evolved. When polypeptides containing cysteine residues formed, they reacted spontaneously with these iron sulphides and formed bonds similar to those present in ferredoxins. Under these conditions, iron in the form of Fe^{2+} would also react forming iron-sulphur proteins similar to rubredoxin. Thus, these data indicate that, very early, at least two different types of iron-sulphur proteins must have formed; and furthermore, that these proteins probably developed independently into two different classes of electron carriers, one of these corresponding to the low redox potential ferredoxins and the other to rubredoxin.[24]

Regarding the amino acid compositions of the iron-sulphur proteins, it should be noted that those of some bacterial ferredoxins are very similar to the ones present in the Murchison meteorite.[25] This meteorite fell in Australia in 1969. In the present era, small anaerobic organisms, such as the *Clostridium* bacteria, contain very simple ferredoxins—small proteins of fifty-five amino acid residues and a molecular weight of 6,000 with a primitive polypeptide structure. Significantly, the six aminoacids detected in the Murchison meteorite (glycine, alanine, valine, proline, glutamic acid and aspartic acid) constitute 64 percent of the total amino acid content of *C. butyricum;* the nine "common" amino acids (the six plus serine, cysteine, and *iso*leucine) comprise 91 percent of the ferredoxin molecule of *C. butyricum*. All nine of these amino acids have been synthesized abiogenically

in conditions which simulated the primitive Earth environment in the laboratory.[1] The amino acids methionine and histidine that are more difficult to synthesize under those laboratory conditions are absent from these ferredoxins. Thus, the primitive amino acid compositions of the ferredoxins further indicate that these proteins originated during a very early phase of life on this planet.

As well as the formation of iron-sulphur proteins, the relatively high concentration in the oceans of $Fe(II)$ and $Mn(II)$ must have facilitated reactions with other ligands, such as the porphyrins. This probably led to the formation of heme proteins and, in the case of manganese, to components required for photosynthesis in the primitive blue-green algae (some blue-green algae contain up to twelve atoms of manganese per reaction center).[23] All of these components along with chlorophyll were necessary for photosynthesis which led to the release of oxygen into the atmosphere.

When oxygen evolved, mainly *via* the photosynthesis of blue-green algae, it was at first extremely toxic for most living systems. The inorganic chemicals in these systems were in a reduced state, and thus they reacted strongly with oxygen. Initially, this was a serious disadvantage for the living systems. It has been suggested that the peroxisomes, nowadays present in cells of higher organisms, represent a primitive enzyme system that evolved to cope with oxygen and thus to protect the cell when oxygen first appeared in the atmosphere.[26] These enzymes react with oxygen first forming H_2O_2 and then H_2O. Among the enzymes present in these early cells were most probably primitive forms of superoxy dismutase, catalase, and peroxidase. The primitive superoxy dismutase might have been a manganese metalloenzyme. (A support for this assumption is that the super-oxy dismutase present in primitive cells (prokaryotic cells) contains manganese.[27]) Superoxy dismutase catalyzes the reaction:

$$2O_2^- + 2H^+ \rightarrow H_2O_2 + O_2(g)$$

Thus, it deactivates the reaction products of O_2. The abundance of O_2^- must have been higher during the more reducing conditions than in our present atmosphere. The other enzymes, catalase and peroxidase, are heme proteins and they contain iron in their active centers. They catalyze the trans-formation of hydrogen peroxide into water. It should be noted that later, in eukaryotic cells, another superoxy dismutase devel-oped that is a copper metalloenzyme. Un-der early reducing conditions, copper ions were most likely trapped in the extremely insoluble $Cu(I)$ sulphides (solubility prod-uct 10^{-50} (g ion)^2), and they could not be released until the air-sea-sediment system became more oxidized as a result of the evolving oxygen.

Thus, in the early phase of oxygen-evolv-ing photosynthesis magnesium, iron and manganese ions had become important for living systems. The new metalloenzymes that developed were required to inactivate oxygen, which must have been an extremely reactive gas under these reducing condi-tions. The low redox potential prevailing at that time facilitated the release of iron and manganese ions into proteins, but trapped copper in inert sulphide complexes.

Availability of Copper During Oxidizing Conditions

Once oxygen-evolving photosynthesis arose, the oxygen generated by increasing populations of photoautotrophs built up in the oceans and began to escape into the atmosphere. There the ultraviolet compo-nent of the sun's radiation dissociated some of the molecular oxygen into two highly reactive species—atomic oxygen and ozone. Atmospheric oxygen and its reactive deriva-tives then oxidized iron, sulphur, and other elements in the sediments which had been produced by the weathering of rocks, and this terminated the banded iron formations as an important sedimentary type and start-ed the extensive formation of continental red beds rich in ferric iron. The geological evidence supports this succession of events: Red beds are essentially restricted to rocks younger than approximately 1.8 billion years, while banded iron formation is found only in older rocks. Thus, for iron there was a change in the predominating phases from $FeS_2(s)$ and $Fe_3O_4(s)$ first to $Fe_3O_4(s)$ and then finally to $FeOOH(s)$; and for manganese, there was a change from Mn^{2+} and $Mn(II)$ compounds to $Mn_3O_4(s)$ and then to $MnO_2(s)$. At pres-ent the concentration of dissolved iron and manganese in sea water is less than $10^{-7}M$.

On the other hand, the copper which was trapped as extremely insoluble sulphides, during the reducing conditions, was re-leased when the air-sea-sediment system became more oxidized through the evolving oxygen. It can be estimated[24] that cop-per(I) complexes of compounds possessing thiol and amino groups could form at redox potentials of about +90 and +230 mV, respectively, and free Cu^{2+} ions could form at about + 250 mV. The situation may have been similar for other biological ions which form insoluble sulphides, such as Co^{2+} and Zn^{2+}. Nowadays, the concentration of dis-solved Cu^{2+} in sea water is about 1 μM. Hence, we may conclude that as a result of oxidation, iron and manganese became immobilized, while Cu^{2+}, and perhaps some other ions, was released into sea water and biological tissues.

An oxygen concentration of just 1 percent

of the present atmosphere is supposed to be adequate for the formation of sufficient ozone to screen out the most deleterious wavelengths of ultraviolet radiation.[5] This also happens to be the same level of oxygen at which Pasteur found that certain microorganisms switched over from an anaerobic type of metabolism to an oxidative one.[2] It has, therefore, been generally understood that this was the stage at which aerobic metabolism arose. The first unequivocal fossils of metazoan organisms are not older than 640 million years, but primitive eukaryotic cells (cells with nuclei and other organelles) have been identified as existing at least twice as long —about 1.3 billion years ago.[12] At that time at least some primitive forms of copper proteins had evolved (most likely, primitive forms of plastocyanin and cytochrome oxidase), and these were required for subsequent biological evolution requiring more efficient photosynthesis and aerobic energy production.

The Biological Specificity of Copper and Zinc Ions

Once copper ions were released by oxygen oxidation, they could in principle have been bound immediately to proteins. However, the formation of protein complexes must have been restricted by competition with low molecular weight ligands. Initially, for relatively reducing conditions, when Cu(I) species predominated, they were restricted by the Cu(I) complexes formed with ligands having thiol and amino groups. One may wonder whether some copper metalloproteins were already evolving at that time, at least in some primitive forms, since nowadays some copper proteins are best reconstituted by adding Cu(I) to the apoproteins. Later, when the air-sea-sediment system became even more oxidizing, the in-

corporation of copper into proteins must have been restricted by the powerful copper(II) complexes formed with amino acids.[13]

Reasonable quantities of amino acids are present in all cells, the concentration being of the order of 1 to 100 mM. Table 2-II lists the stability constants for a series of metal complexes of amino acids. It follows from these data that a mixed amino acid complex rather than a complex of only one kind of amino acid will predominate in an amino acid mixture (Table 2-II). Also, in the tissues there are relatively high concentrations of amino acids (1 to 100 mM) compared to the concentration of any single protein molecule ($\leqslant 100$ μM). Clearly, if a metal ion is to be bound to a protein molecule, the stability constant of the protein complex has to be significantly higher than that of the amino acid complexes. It is also clear that, for a protein complex, a binding site of a highly specific geometry is required in order to provide such a high stability. This geometry needs to be more specific as the stability of the corresponding amino acid complex increases. For example, copper(II) may require a far more sophisticated binding site than calcium(II),

TABLE 2-II

STABILITY CONSTANTS OF AMINO-ACID COMPLEXES[a]

Metal Ion	*Glycine*[a] Log β_1	Log β_2	*Histidine, Threonine (mixed complex)*[b] Log β_{1011}
Ca	1.4		
Mg	2.1	4	
Zn	5.2	9.5	10.5[c]
Cu(II)	8.4	15.3	17.6

[a] $M^{2+} + nA^- \rightleftharpoons MA_n^{(2-n)+}$, β_n.
[b] $M^{2+} + A^- + B^- \rightleftharpoons MAB$, β_{1011}.
[c] Estimated value.

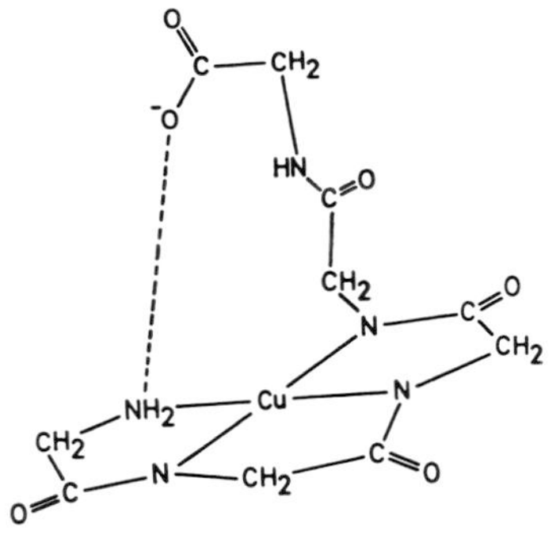

Figure 2-3. The structure of Cu(II)-penta-glycine.[15]

(*cf.* Table 2-II). A calcium ion can be bound, as indeed it is in some calcium metalloproteins, to just one or a few carboxylate groups, since the stabilities of its amino acid complexes at neutral pH are very low (*cf.*[3, 14]).

The following example will illustrate how the biological specificity might have been developed:[13] With respect to Cu^{2+} ions, the N-terminal part of a peptide chain is a relatively strong binding site, and model studies in the crystalline state indicate that the complex is formed *via* the *a*-amino group and two or three peptide bond amide groups[15] (Figure 2-3). Now, if a mutation should occur, such that a "neutral" amino acid in the N-terminal sequence is replaced by a histidine residue, and if this mutation hits position 2 or position 3, then there is a dramatic increase in stability. Using the ratio, $R = [CuProt]/[Cu(His)Thr]$, as a measure of the complex forming ability of the N-terminal site, we find, for the following three N-terminal sequences:
1) Gly-gly-gly-gly- . . . , 2) Gly-his-gly- . . . , 3) Asp-ala-his-lys- . . . , the following results: 1) log $R = -4.0$, 2) log $R = 1.9$, 3) log $R = 4.9$. Thus, these data indicate that the N-terminal part of a peptide or of a protein will bind copper(II) ions *in vivo* only if there is a histidine residue in position two or position three. A structural model for such a site is shown in Figure 2-4. On the other hand, the R-value for the first protein site, 0.0001, indicates that, *in vivo*, the free amino acids will protect the proteins and prevent copper ions from reacting with the proteins. In this example, site 3 corresponds to the site present in human serum albumin; these results confirm that the main fraction of labile copper in our blood plasma is bound to albumin.[13, 14]

Another example illustrates how a binding site specific for a given metal ion might have developed. We select the following ions: Ca^{2+}, Mg^{2+}, Cu^{2+}, and Zn^{2+}. And for these ions, we assume that the evolutionary restraint on the system involved the metal complexes formed by the amino acids. Under these circumstances, for efficient competition with the amino acids, the ratio $R = [MeProt]/[(amino acid complex)]$ has to be larger than unity, and so the binding site evolved has to form a stronger metal

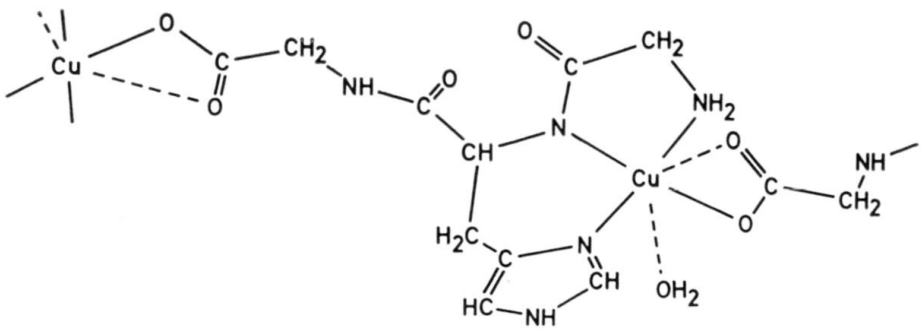

Figure 2-4. The structure of Cu^{II}(glycyl-L-histidylglycine)$NaClO_4 \cdot H_2O$.[29]

TABLE 2-III

STABILITY RATIO, *R*, BETWEEN THE
METAL COMPLEXES FORMED BY
PROTEIN BINDING SITES AND
THOSE FORMED BY
AMINO ACIDS

Metal Ion	Log R Type of Site (1)	(2)	(3)
Ca(II)	+0.6	—	—
Mg(II)	−1.5	—	—
Zn(II)	−1.7	+2.5	+1.6
Cu(II)	−6.7	−1.1	−2.2

complex than that of the amino acids (Me denotes the metal ion). If these assumptions are correct, we ought to be able to choose existing metal ion binding sites, and a calculation of *R* should indicate the true specificity. Table 2-III lists *R*-values calculated for the following three sites:[4]

1) four carboxylate groups (staphylococcal nuclease);
2) three imidazole groups (*cf.* carbonic anhydrase);
3) two imidazole and one carboxylate group (*cf.* carboxypeptidase and thermolysine).

(Details about these calculations, involving estimation of the stability constants for these sites from the data on the corresponding low molecular weight systems, are described in Ref. 13.) As shown by Table 2-III, site 1, which is a site in a calcium

metalloprotein, has the largest affinity for calcium. Sites 2 and 3, however, were selected from zinc metalloproteins, and are seen to preferably bind zinc ions (Table 2-III). It should be kept in mind that copper(II) ions do not appear to bind to these sites, in spite of the fact that their "copper complexes" have the highest stabilities. Rather, because the free amino acids form very stable copper complexes, less stable zinc complexes, and much less stable magnesium and calcium complexes, it is the calcium and zinc ions instead of copper ions that will be bound to the protein sites. Thus, copper ions appear to require a site that is different from those of the other ions. As yet, very little is known concerning the true nature of these copper binding sites, since as yet the coordination structure of a copper protein has not been determined. However, there are indications that some copper proteins have both mononuclear and binuclear copper binding sites;[13] a model structure for a binuclear site is shown in Figure 2-5.

Thus, these calculations have indicated that the presence of free amino acids imposes an important restraint on the mutants of metal binding proteins: The only protein complexes to become important are those forming more stable complexes than those formed by the amino acids.[13] Even nowadays this "amino acid selectivi-

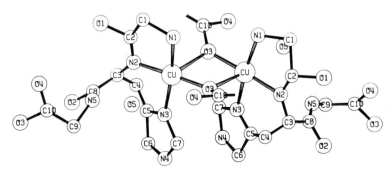

Figure 2-5. The structure of Cu^{II}(glycyl-L-histidylglycine) · xH_2O.[30]

ty" is still functional; the presence of amino acids in all forms of living tissues are important not only for the binding of each metal ion to just one specific protein out of a wide selection but also for the binding of one specific metal ion, out of a wide selection of metal ions, to its unique metal ion binding site.

The Role of Calcium, Potassium and Sodium Ions

From a quantitative viewpoint the concentrations of the transition elements in biological tissues are very low even though some of these elements, for example iron, constitute a large part of the Earth's crust. However, ions having a closed shell electronic structure, i.e. Na^+, K^+, Ca^{2+}, and Mg^{2+}, are far more abundant in the biosphere. One important function of these ions is the maintenance of the correct osmotic pressures. There are, however, additional and more specific biological functions that have evolved during a period of many millions of years.

The closed-shell electronic structure of the ions is important for their specific biological functions: Na^+ and Mg^{2+} ions have the same electronic configuration as that of neon, and K^+ and Ca^{2+} have the same configuration as that of argon. As a result, it is the electrostatic forces that are important for the reactions of these ions rather than quantum-mechanical forces; they form metal complexes as a consequence of their charges and ionic radii (Figure 2-6).

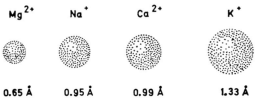

Figure 2-6. Ionic radii of Na^+, K^+, Mg^{2+} and Ca^{2+} ions.

Consequently, it is not surprising that the formation of the metal complexes of $Na(I)$, $K(I)$, $Mg(II)$, and $Ca(II)$ generally are favored by the entropy change rather than by the enthalpy change.[3] The main sources of this entropy are supposed to be the water molecules released from the ordered water structure surrounding the metal ion and the ligand. As an example, let us consider the following reaction:

$$M(H_2O)_x + L(H_2O)_y \rightleftharpoons ML(H_2O)_z + (x + y - z)H_2O$$

where M is the metal ion and L the ligand. Such equilibria have important implications for reactions occurring within the cells. For instance, the magnesium concentration is relatively high inside the cell, and many phosphates, such as the nucleotides, exist in the form of their magnesium complexes. This has the tremendous advantage that it maintains the highly ordered system that characterizes a cell. Complex formation involving two macromolecules, such as nucleic acids with other nucleic acids or nucleic acids with protein molecules, would lead to disorder within the cell (due to the release of a great number of water molecules), unless the nucleic acids were complexes by the Mg^{2+} ions.

It has been suggested previously that magnesium ions must have been important in very early life processes. *Calcium ions*, however, acquired their functions of specific importance much later—these specific Ca^{2+} functions involve processes that only characterize the higher forms of life, such as nerve transmission, muscle contraction and blood coagulation. The distributions of Mg^{2+} and Ca^{2+} ions in living tissues are also unlike. In blood plasma and in other biological liquids the Ca^{2+} ions predominate, and the concentration of Mg^{2+} is relatively low. However, within cells, Mg^{2+} is the most important bivalent ion, the concentration of Ca^{2+} being very low (Table

2-I). Thus, there is a pronounced gradient in the Ca^{2+} concentration across the cell membrane, and this is important for the biological action of Ca^{2+}. A further difference between the ions of Mg^{2+} and Ca^{2+} occurs at the molecular level. In enzymatic processes, such as protein biosynthesis and anaerobic energy production, it is the magnesium ions that react with the substrate rather than the enzyme. The larger Ca^{2+} ion having a higher coordination number (six, seven or eight) usually binds directly to the enzyme molecule; and, within the protein molecule it can induce a dramatic conformational change. An important example of the effect of Ca^{2+} on an enzyme is the Ca^{2+} activation of *staphylococcal* nuclease.[4]

The calcium ion has a larger ionic radius than the magnesium ion (Figure 2-6) and so the Ca^{2+} ion has a much lower charge density than that of a Mg^{2+}. The Ca^{2+} reactions are also characterized by kinetics that are much faster than those of Mg^{2+} (*cf.*[14]) and therefore, Ca^{2+} ions are not trapped into unimportant side reactions. These properties, together with the relatively high net charge, lead to the phenomenon that Ca^{2+} ions travel quite rapidly in living tissues. Therefore, it is not surprising that in higher organisms the calcium ion developed into some kind of a "messenger" that transfers signals between different cells; for example, the contraction of a muscle cell.

When a nerve impulse hits the membrane of the muscle cell, the permeability for Ca^{2+} increases and, because of the Ca^{2+} gradient across the cell membrane, calcium ions are then introduced into the muscle cell. Here, Ca^{2+} ions become specifically bound to a protein called troponin, and as a result troponin changes its conformation. This troponin conformational change initiates a series of reactions, which finally leads to the contraction of the muscle.[4] This cal-cium ion *trigger effect* seems to be more general: the action of a nerve impulse on a cell membrane increases the movement of Ca^{2+} ions into the cell and, as a result, hormones and transmitters (such as acetylcholine) are released. An important requirement for this effect is the capacity of Ca^{2+} ions to break the structure of water and of dispersed phases. Also, it has been suggested that these cells that are sensitive to Ca^{2+} may contain some microsystems similar to those of the muscle proteins. Thus, a "muscle contraction," on a microscale, may be responsible for the Ca^{2+} dependent release of a substance unique to certain cells.

We may conclude that, for bivalent ions, an increase in the ionic radius from 0.65 to 0.99Å leads to a dramatic change in their biological effects. As the ionic radius increases, the ion acquires a greater importance for processes characteristic of highly developed and sophisticated organisms rather than for those characteristic of primitive cells. For example, one specific function of Ca^{2+} in higher forms of life appears to be to act as some kind of a "messenger." Here, it is tempting to speculate, whether the fidelity of such a "messenger" can be improved any further. One possible candidate would be a trivalent ion of a low charge density, say the La^{3+} ion. However, the abundance of La^{3+} and other rare earth metal ions is very low. So, the question as to whether or not such a higher fidelity can exist, must await the exploration of life on other planets where rare earth ions may be more abundant.

Among univalent ions, the *potassium ion* is the one whose biological effect is somewhat reminiscent of Ca^{2+}. The potassium ion has the lowest charge density of all these four metal ions; therefore, it can move relatively freely, not just in tissues containing water but also in tissues containing lipids (*cf.* Fig. 2-6). In water solutions, K^+ forms interstitial solutions with the water molecules, there being very few ordered water molecules surrounding the K^+ ion. Thus, the K^+ ion has little or no water shell. This is a great advantage for its movement into other phases such as the lipid phase. A molecule that is specific for the transport of K^+ and that can neither transport Na^+, Ca^{2+} nor Mg^{2+}, should in general be unpolar and yet be just able to carry a single charge. One example is the so-called "football" ligands (Figure 2-7). And of the "football" ligands, those found in nature usually bind $K^+ > Rb^+$, but Na^+ and Li^+ very poorly.

During chemical evolution and during the first phase of bioevolution (for which it is generally assumed that the atmosphere was a reducing one) the NH_4^+ ions could have been much more abundant than they are at present. The NH_4^+ ion has almost the same size as the K^+ ion. Therefore, during the evolution, the potassium ion may have been substituted for NH_4^+ in many biological processes. Hence, some biological processes[16] that nowadays are activated by K^+,

such as some enzymatic reactions, can just as efficiently be activated by NH_4^+. Other processes however, such as the transmission of nerve impulses, seem to be more specific for the K^+ ion.

In the transmission of nerve signals not only K^+ ions but also Ca^{2+} and Na^+ ions are involved. The cells are poor in Na^+ and Ca^{2+} ions but rich in K^+ ions. As a phenomenon parallel to the stimulation of the nerve, potassium ions diffuse out from the nerve cell. Here, as in the case of Ca^{2+}, a concentration gradient is required across the cell membrane in order for the biological effect to occur. For example, the activity of the heart will cease if a very high intake of K^+ leads to the K^+ concentration becoming higher outside the cell than inside the cell. Potassium ions are able to move out from a resting nerve cell, and this generates a K^+ potential. On the other hand, when the membrane is activated it becomes permeable to sodium ions.

Sodium ions also are very important for maintaining membrane potentials and for transmitting nerve signals. But, Na^+ ions cannot penetrate a resting cell membrane.

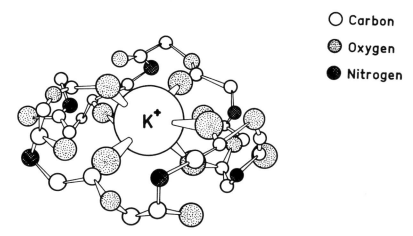

○ Carbon

◉ Oxygen

● Nitrogen

Figure 2-7. The structure of the K^+ complex of valinomycin,[31] a natural "football" ligand involved in biological K^+ transport.

Outside the cells, in blood plasma and other extracellular spaces, Na⁺ is the predominating cation (Table 2-I); and, the largest contribution to the extracellular osmotic pressure comes from these sodium ions. Apart from these general functions there are but few specific functions for Na⁺, mainly as a cofactor in enzymatic reactions.[16] The sodium ion is a univalent ion and, thus, insufficiently effective as a messenger. Its charge density is relatively high and Na⁺ is not as easily released as K⁺ from the water phase. As a rule, the concentrations of Na⁺ ions are low inside cells.

Nevertheless, within cells there is at least one specific function for Na⁺ ions: that is to activate the process called chromosome "puffing." The particular role of Na⁺ in this process is not known, but a suggestion of how Na⁺ ions may interact with nucleic acids may be obtained from the crystal structure of sodium-adenosyl-3′,5′-uridine phosphate[28] (ApU). In this structure, ApU forms a double helix in which the bases are hydrogen bonded to each other in the Watson-Crick manner.[2] Some sodium ions occupy positions adjacent to oxygen atoms of the diester phosphate and other sodium ions coordinate to the oxygen atoms of uridine. Hence, this structure is held together not only by hydrogen bonds between uridine and adenine but also through the sodium coordination of uridine molecules.[28]

CONCLUDING REMARKS

The preceding sections have illustrated that the more detailed knowledge about processes in biochemistry and molecular biology, as developed during the last twenty years, has led to a much better understanding of the role of metal ions in biology. Many of these ions are now known to have very specific biological functions, and they cannot be regarded *only* as passive passengers in life processes. These specific functions have evolved over a period of more than four billion years and the degree of perfection will be further illustrated in subsequent chapters.

REFERENCES

1. Miller, S., and Orgel, L. E.: *The Origins of Life*. London, Chapman & Hall, 1973.
2. Watson, J. D.: *The Molecular Biology of the Gene*. Benjamin Co., New York, 2nd ed., 1970.
3. Sillén, L. G., and Martell, A. E.: *Stability Constants*, 2nd ed. Chem Soc Spec Publ no 17 (1964); Suppl No 1 Chem Soc Spec Publ no 25 (1971).
4. *Cold Spring Harbor Symp Quant Biol*, 1972, p. 36.
5. Ponnamperuma, C.: *Quart Rev Biophys*, 4:77, 1971.
6. Sillén, L. G.: *Oceanography*, 459, 1959.
7. Schwarz, K.: In Mills, C. F., Livingstone, E., and Livingstone, S. (Eds.): *Trace Element Metabolism in Animals*, 1970.
8. Frieden, E.: *Scientific American*, 52, July 1972.
9. Eigen: M.: *Q Rev Biophys*, 4:149, 1971.
11. Orme Johnson, W. H.: *Ann Rev Biochem*, 42:159, 1973.
12. Cloud, P., and Gibor, A.: *Sci Am*, 111, Sept. 1970.
13. Österberg, R.: *Coord Chem Rev*, 12:309, 1974.
14. Österberg, R.: In Sigel, H. (Ed.): *Metal Ions in Biological Systems*. New York, Marcel Dekker, 1974, p. 45.
15. Freeman, H. C.: *Adv Protein Chem*, 22:257, 1967.
16. Evans, H. J., and Sorger, G. J.: *Ann Rev Plant Physiol*, 17:47, 1966.
17. Lohrmann, R., and Orgel, L. E.: *Nature*, 244: 418, 1973.
18. Österberg, R., and Orgel, L. E.: *J Molec Evol*, 1:241, 1972.
19. Österberg, R., Orgel, L. E., and Lohrmann, R.: *J Molec Evol*, 2:231, 1973.
20. Sillén, L. G.: *Arkiv Kemi*, 25:159, 1966.
21. Maugh, T. H.: *Science*, 181:253, 1973.
22. Bryant, T. N., Watson, H. C., and Wendel, P. L.: *Nature*, 247:14, 1974.
23. Olson, J. M.: *Science*, 168:438, 1970.

24. Österberg, R.: *Nature, 249:*382, 1974.
25. Hall, D. O., Cammack, R., and Rao, K. K.: *Nature,* 233:136, 1971.
26. De Duve, C.: *Science, 182:*85, 1973.
27. Keele, B. B., Jr., McCord, J. M., and Frido-wich, I.: *J Biol Chem, 245:*6176, 1970; *246:* 2875, 1971.
28. Rosenberg, J. M., Seeman, N. C., Kim, J. J. P.,

Suddath, F. L., Nicholas, H. B., and Rich, A.: *Nature, 243:*150, 1973.
29. Österberg, R., Sjöberg, B., and Söderquist, R.: *Acta Chem Scand, 26:*4184, 1972.
30. Österberg, R., Sjöberg, B., and Söderquist, R.: *Chem Commun,* 983, 1972.
31. Duax, W. L., Hauptman, H., Weeks, C. M., and Norton, D. A.: *Science, 176:*911, 1972.

CHAPTER 3

SELECTIVITY IN METAL COMPLEX FORMATION

R. F. JAMESON

Introduction
Thermodynamic Functions
The Effect of the Cation
Hard and Soft Acids and Bases
The Effect of the Ligand
Oxidation States and Innocent
  Ligands

INTRODUCTION

Selectivity or Specificity?

THE STUDY OF METAL ion-ligand inter-actions in media that are essentially aqueous is of importance to an immensely diverse range of chemistry ranging from inorganic chemical analysis to what might best be termed the "life sciences." In all of this work the *selectivity* of these metal-ion interactions plays a considerable part. For example, it might either enable one to test for the presence of a specific metal ion in the presence of others, or help to explain the role of a particular metal ion in a bio-system.

But note the use of the word *"selectivity,"* rather than *"specificity"*: In actual practice specificity for a single metal ion by a ligand is rare, even in biological systems. For example, the metal ions at the active site of many metalloenzymes can often be replaced *(in vitro)* by a whole range of metal ions, some of which are considerably more cata-lytically active (cobalt(II) replacing zinc(II) in carboxypeptidase). In fact, there are three general observations that result in

good reasons for suspecting a lack of speci-ficity:

1. The binding of ions such as Na^+, K^+, Ca^{2+}, etc. depend largely on charge/size re-lationships (as we will discuss in detail lat-er) and thus, in complex-forming situations, it would in general be very difficult to dis-tinguish between Tl^+ and K^+ or even Ca^{2+} and Eu^{3+}.

2. In both analytical chemistry and in bio-systems a method of prior separation of cations into "Groups" followed by a "chro-matographic" method of further separation can be used in the former and might well be expected in the latter: In other words, the carboxypeptidase mentioned above would never be "offered" a cobalt(II) ion (in any case cobalt(III) would more likely be present) and even cations of very similar properties can be transported to different sites because they must first interact with a whole range of complexing ligands before arriving at their destination.

3. Nature seems to make diverse use of any one group of what are basically the same macromolecules. For example, the iron atom in the haem groups and the mag-nesium in chlorophyls share the same square-planar ligand thus making *specificity downright undesirable!* This of course, might well be a direct result of the evolu-tionary aspect of life (haemoglobin perhaps evolving from an early symbiotic, plant-like, oxygen-carrying discrete organism).

However, before commencing a study of

29

those factors that influence complex formation, an attempt will be made to outline the main points concerning the use of the thermodynamic functions ΔG, ΔH and ΔS that enable a study of the interactions between chemical species to be put on a sound quantitative footing. The influence of the cation and of the ligand will then be treated as separate problems, although, of course, such a separation of roles is, strictly speaking, an impossibility. We will, however, be partially unifying the treatment by means of a discussion of the Hard and Soft Acids and Bases (HSAB) concept of Pearson.[1, 2] The problems of oxidation states and the "innocence" of cations and ligands will conclude this chapter.

THERMODYNAMIC FUNCTIONS

Equilibrium Quotients

In order to measure the extent of interaction between metal ions (or protons) and ligands it is customary to write the reaction in the form

$$M + nL \rightleftharpoons ML_n \qquad (1)$$

where any charges associated with the species have been omitted in the interests of generality and clarity. Note particularly the absence of solvent (H_2O) molecules in formulating (1) although, for example, they certainly act as ligands towards M: We will return to this point several times.

The next step towards the quantification of our theory is the definition of a *"standard state"* in order that free energy *changes* might meaningfully be discussed (*cf.* the role of the "ideal gas" when discussing changes in the free energy of gases). This choice is to a large extent arbitrary and although pure solvent is invariably chosen as the standard state to which one can refer changes in the (Gibbs) free energy of the *solvent*, when dealing with electrolytes the properties of the pure *solute* differ so

much from those of the solution that it would be misleading to choose it for our standard. Instead we define a *hypothetical solution* to have the required properties at *unit concentration* (and at the temperature and pressure of the solution). Very often the properties of this hypothetical solution are obtained by *extrapolating the properties of the actual solution* to zero concentration (often called "infinite dilution") because other-than-ideal interactions can then be ignored. Note very carefully, however, that such an infinitely dilute solution is *not the standard state itself* (we have already defined it, above, as a solution at a hypothetical mean *unit* concentration)—one has only to compare the effect of putting the concentration terms equal to either unity or to zero in equations (2) below. Or, perhaps better, note how although the behavior of a real gas approaches that of the standard ideal gas as the pressure approaches zero, we do not go on to *equate* an ideal gas with an infinitely expanded real gas.[†]

It is thus possible to define an *"activity"* of a solute X *according to the concentration scale to be used*, namely

$$\begin{aligned}
\overline{G}_x &= \overline{G}_x^{0(m)} + RT\ln a_x(m) & \text{for the molal scale} \\
&= \overline{G}_x^{0(c)} + RT\ln a_x(c) & \text{for the molar scale} \\
&= \overline{G}_x^{0(N)} + RT\ln a_x(N) & \text{for the mole fraction} \\
& & \text{scale}
\end{aligned}$$

$$(2)$$

where the $\overline{G}_x^0$ refer to the standard state. Note particularly that, although for a given solution at fixed temperature and pressure, the chemical potential of X, namely $\overline{G}_x$, is fixed, both a_x and $\overline{G}_x^0$ *depend on the scale*

[†] Note that the *practical* definition of the standard chemical potential is, of course

$$\overline{G}^\circ = \lim_{c \to 0} [\overline{G} - RT \ln c]$$

where c is the concentration on the appropriate scale, which leads directly to the misapprehension as to what constitutes the standard *state*.

chosen. This is usually the *molar* scale in this work although especially accurate data for related studies (such as the ionic product for water, K_w) are often quoted on the *molal* scale.

Having defined activities by the appropriate equation (2) we now require to relate this to the measured concentrations and in order to do this define an *"activity coefficient"* by the relationship:

$$\text{activity} = \text{concentration} \times \text{activity coefficient} \quad (3)$$

Thus for an equilibrium between a metal M and a ligand L we have

$$M + L \rightleftharpoons ML \quad (4)$$

for which we can write an *"equilibrium constant,"* usually referred to as a *"stability constant"* or a *"formation constant."*

$$K_1{}^M = \frac{a_{ML}}{a_M a_L} \quad (5)$$

$$= \frac{[ML]}{[M][L]} \cdot \frac{f_{ML}}{f_M f_L} \quad (6)$$

where the f_X are the (dimensionless) activity coefficients of X as defined by equation (3).

Equation (6) is then rewritten as

$$K_1{}^M = K_1{}^{Mc} f_{ML} / f_M f_L \quad (7)$$

where $K_1{}^{Mc}$ is the concentration quotient applicable to equilibrium (4) and is, of course, *the quantity that we usually measure.* It is extremely important to note that $K_1{}^{Mc}$ depends both on the medium chosen and in general on the concentrations of M, L and ML. This is so often forgotten and pK_a values for amino acids, for example in 0.1 M KNO_3 solution, are used without reservations when discussing "pH" effects in the cells of biological systems.

Values of $K_1{}^M$ can then be obtained either by extrapolation of $K_1{}^{Mc}$ to "infinite dilution," when the f_X are *by definition*

unity, or by calculation of the f_X, e.g. by equations such as those derived by Debye and Hückel.[3] Either method is rather tedious, however, and usually abandoned in favor of the *"inert background electrolyte"* method that we will discuss next.

Background Electrolytes

For *dilute* solutions (usually <0.1 M) the activity coefficients, f_X, have been found to depend on the total ionic strength of the medium rather than on the ions actually concerned. In this context, the *"ionic strength,"* μ, of a solution is defined by the relationship

$$\mu = \tfrac{1}{2} \Sigma \; C_i z_i{}^2 \quad (8)$$

where C_i is the concentration of ions of charge z_i (in electron units). In other words, if the equilibrium measurements are carried out in a solution made up to constant ionic strength by ions that can be considered inert with respect to M, L, ing the use of an inert electrolyte should and providing that the reacting species are of relatively low concentrations compared with those of the inert species, then the values of f_M, f_L, and f_{ML} will depend only on the total ionic strength of the solution. Thus defining the standard state as the solution of mean unit concentration in *that particular medium* (say 0.1 M $NaClO_4$ solution) rather than in pure solvent (water) makes $K_1{}^{Mc}$ thermodynamically significant because (1) we have defined the activities as unity in that medium and (2) what is experimentally more important, we can assume to a considerable degree of accuracy that the activity coefficients are unaffected by alteration in the concentrations of the species under examination.

The use of background electrolytes to maintain constant ionic strength in this way is now widespread but unfortunately neither particular concentrations nor even particular salts are in general use. Different workers report results in concentrations ranging from 0.1 M to 3.0 M and the salts used are mainly KCl, KNO_3 and $NaClO_4$. Two widespread misapprehensions concerning the use of an inert electrolyte should be emphasized here: namely that *either* the concentration need not be accurate, as it is only the "background" *or* that the higher the concentration the better. First, for dilute electrolytes (*cf.* 0.1 M), the activity coefficients vary rapidly with concentration, and secondly, in 3 M $NaClO_4$ solution, the interionic distances are only about 6Å— just compare this with the dimensions of many organic ligands or even with the 5.5Å diameter of the hydrated sodium ion!

Finally, we reiterate that concentration quotients obtained at a given temperature in a given medium are only valid for that temperature and for that medium.

Stepwise and Overall Stability Constants

The reaction between a metal ion and n ligands may be divided into steps in two ways, leading to related but different formulations of the equilibrium constants, *viz:*

where the K's are called the "*stepwise stability constants*" and the β's are called the "*overall stability constants*." The nomenclature should be self-evident.

There are three main reasons for expecting a steady decrease in the K_i values as the number of ligands, i, increases and three main reasons for experimental violations of this.

One would expect a steady decrease in K_i on (1) a statistical argument: Usually n ligands are replacing n solvent molecules and so the first entering ligand has a choice of n sites, the second $(n-1)$ sites and so on; (2) a Coulombic factor especially pronounced if the ligands are charged; and (3) a steric factor if the ligands are larger than the solvent molecules that they are replacing. This argument could fail if (1) the coordination number changes during coordination, e.g. for zinc(II)-amine complexes were K_4 relates to tetrahedrally coordinated zinc whereas K_5 and K_6 relate to octahedral complexes, (2) a change in electronic structure with the entry of another ligand, e.g. iron(II) complexes with one or two 2,2'-bipyridine ligands are para-

$$\text{I} \quad \text{M} + \text{L} \rightleftharpoons \text{ML} \qquad\qquad K_1{}^M = \frac{[\text{ML}]}{[\text{M}][\text{L}]}$$

$$\text{ML} + \text{L} \rightleftharpoons \text{ML}_2 \qquad\qquad K_2{}^M = \frac{[\text{ML}_2]}{[\text{ML}][\text{L}]}$$

$$\cdots\cdots\cdots\cdots\cdots \tag{9}$$

$$\text{ML}_{(n-1)} + \text{L} \rightleftharpoons \text{ML}_n \qquad\qquad K_n{}^M = \frac{[\text{ML}_n]}{[\text{ML}_{(n-1)}][\text{L}]}$$

$$\text{II} \quad \text{M} + \text{L} \rightleftharpoons \text{ML} \qquad\qquad \beta_1{}^M = \frac{[\text{ML}]}{[\text{M}][\text{L}]} = K_1$$

$$\text{M} + 2\text{L} \rightleftharpoons \text{ML}_2 \qquad\qquad \beta_2{}^M = \frac{[\text{ML}_2]}{[\text{M}][\text{L}]^2} = K_1 K_2 \tag{10}$$

$$\cdots\cdots\cdots\cdots\cdots$$

$$\text{M} + n\text{L} \rightleftharpoons \text{ML}_n \qquad\qquad \beta_n{}^M = \frac{[\text{ML}_n]}{[\text{M}][\text{L}]^n} = K_1 K_2 K_3 \ldots K_n$$

magnetic whereas the complex with three ligands is diamagnetic. (The change in spin-state when haemoglobin coordinates O_2 is also relevant here.) And finally, (3) especially large steric effects might only become operative at certain stages of complex formation, e.g. 6,6′-dimethyl-2,2′-bipyridine will often not form *tris* complexes, or even *bis* complexes in some cases, with ions that readily form *tris* complexes with the parent 2,2′-bipyridine ligand.

Free Energy, Enthalpy and Entropy Changes

The thermodynamic relationships

$$-\Delta G^\oplus = RT\ln K$$
$$\text{and} \quad \Delta G^\oplus = \Delta H^\oplus - T\Delta S^\oplus \quad (11)$$

are probably the most useful and yet most misused relationships in the study of complex equilibria. We will commence by looking at the astonishingly widespread misinterpretation of the so-called "*chelate effect.*" This effect can be illustrated as follows:

Consider the two equilibria

(i) $Ni^{2+}_{(aq)} + 6NH_{3(aq)} \rightleftharpoons [Ni(NH_3)_6]^{2+}_{aq}$ $\log \beta_6 = 8.61$

(ii) $Ni^{2+}_{(aq)} + 3\,en(aq) \rightleftharpoons [Ni(en)_3]^{2+}_{aq}$ $\log \beta_3 = 18.28$

$$(12)$$

where *en* is 1,2-diaminoethane ("ethylenediamine") and is a chelating ligand occupying two coordination positions:

$$(12a)$$

Now *disregarding* for a moment the log β values in equation (12), look at the compositions of equilibrium mixtures illustrated

in Figure 3-1. These show clearly that *en* is a much more efficient complexing agent than is NH_3, i.e. for a given concentration of ligand considerably more metal ion in a given solution is complexed. Figure 3-1 also shows the dramatic effect of increasing the number of chelate rings: *This is the* "*chelate effect.*"

But what do the relative sizes of the log β values in (12) show us? Unfortunately, very little. Consider a change in units of concentration from moles l^{-1} to moles ml^{-1}, whereupon we can write

(concentration in mole l^{-1}) = 10^{-3} (concentration in mole ml^{-1})

which leads to

$$\begin{array}{l} \log \beta_6 \text{ in } 12(\text{i}) \rightarrow 26.61 \\ \log \beta_3 \text{ in } 12(\text{ii}) \rightarrow 27.28 \end{array} \quad (13)$$

In which the "effect" has nearly vanished and indeed the relative values of $\beta_6^{(\text{i})}$ and $\beta_3^{(\text{ii})}$ could readily be *reversed* by choosing an even smaller unit of volume. In other words, *because $\beta_6^{(\text{i})}$ and $\beta_3^{(\text{ii})}$ have different units a direct comparison is invalid.* This point might be made even clearer if one imagines the figures on the pL scale in Figure 3-1 to be increased by $3(= -\log 10^{-3})$: *The* "*chelate effect*" *has obviously not altered although summing the log K values* from the curves has the effect noted in equations (13). Thus, any substitution in equations (11) leading to the result that the chelate effect is an extremely large *entropy effect* is *nonsense* as are calculations by this method of entropy effects on the basis of β_1 and β_2 values for the same metal-ligand system.

Such comparisons *are* possible, however, if one takes into account the role of the solvent: the "concentration" of the water

---

* When n = 0.5 then $K_n = 1/L$, i.e. log K_n = pL (to show this, it is necessary to start with n = 1).

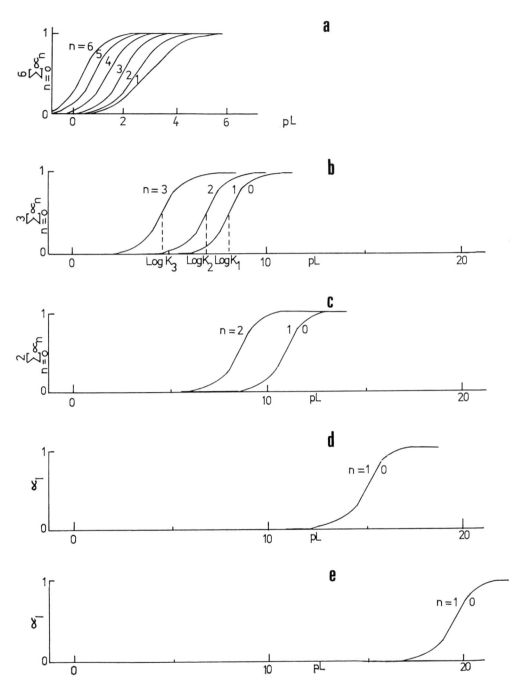

Figure 3-1. The "chelate effect" illustrated for nickel (II) complexes with some nitrogen donor ligands: *(a)* Ammonia *(monodentate)*—note different scale to curves *(b)* to *(e)*; *(b)* Ethylenediamine *(bidentate)*; *(c)* Diethylenetriamine *(tridentate)*; *(d)* $\beta'\beta''\beta'''$triamino-triethylamine *(tetradentate)*; *(e)* NNN'N'tetra (2-aminoethyl)ethylenediamine *(hexaden-tate)*. $\alpha_n = [\mathrm{NiL_n}]/[\mathrm{Ni}]_{\mathrm{Total}}$, and $\mathrm{pL} = -\log_{10}[\mathrm{L}]$.

is approximately 55.5 molar and thus all concentration terms applicable to the equilibrium can be expressed as (dimensionless) mole fractions, for example from 12(ii) we obtain

$$\beta_3^{(N)} = \frac{\dfrac{[ML_3]}{55.5}}{\dfrac{[M]}{55.5}\left\{\dfrac{[L]}{55.5}\right\}^3} \quad (14)$$

i.e. $\log \beta_3^{(N)} = \log \beta_3 + 3 \log 55.5$ (15)

Note *(a)* that these $\beta_n^{(N)}$ values are now dimensionless making comparisons valid and *(b)* that when used in conjunction with equation (11) the entropy contribution to the chelate effect becomes very small to zero (according to the metal and ligand system concerned), but does afford useful information. For further details see the papers by Prue[4] and Adamson.[5] But if we are to make sensible comments on the relative stabilities of complex species in order to look at selectivity effects we must be aware of this problem of units. (Note that ΔH, being a unitary quantity, is immune from this problem.)

Even so, we would not be correct in assuming that there is *no* entropy effect linked with the formation of chelated complexes, only that it is not as large as many naive calculations that ignore units would have us believe: After all, *one* bidentate chelating ligand displaces *two* monodentate ligands (or solvent molecules) into the bulk of the solution.

However, assuming that we make proper use of equations (11) in this way, what information is obtainable from these correctly evaluated ΔG ΔH and ΔS values for the interactions of metal ions with ligands? We can only deal briefly here with some of the ways in which information may be obtained from a knowledge of these thermodynamic functions or we would be going beyond the scope of this chapter, but further informa-

tion is available from a variety of sources.[6]

Before the advent of direct calorimetry, ΔG values (derived from equilibrium data) were the only reliable data available and so most workers have confined themselves to relationships based on these, and indeed for our purposes, that is the study of selectivity, the equilibrium constants themselves are often sufficient although giving little information concerning the type of bonding (covalent or electrostatic). But various correlations based on free energy changes are very useful and are based on the following type of argument:

Consider the equilibrium

$$M + HL \rightleftharpoons ML + H; \quad (16)$$

(in which the ligand, L, exists in a protonated form, HL) for which we can write:

$$K = \frac{[ML][H]}{[M][HL]} = K^{ML}/K^{HL} \quad (17)$$

(where the K^{ML} and K^{HL} are the *stability* constants for both metal and proton complexes respectively of L).

The thermodynamic changes accompanying complex formation are then as follows: For protonation of L:

$$\Delta G^\circ(HL) = -RT\ln K^{HL} = G^\circ(HL) - G^\circ(L) - G^\circ(H) \quad (18)$$

For formation of metal complex:

$$\Delta G^\circ(ML) = -RT\ln K^{ML} = G^\circ(ML) - G^\circ(L) - G^\circ(M) \quad (19)$$

and subtraction of (19) from (18) leads on rearrangement to:

$$\log K^{ML} = \log K^{HL} + \{[G^\circ(M) - G^\circ(H)] + [G^\circ(HL) - G^\circ(ML)]\}/2.303 \text{ RT} \quad (20)$$

For a given metal ion $[G^\circ(M) - G^\circ(H)]$ is a constant and experimentally (Figure 3-2) it is frequently found that a plot of $\log K^{ML}$ against $\log K^{HL}$ for a given ligand is a straight line often of unit slope. For

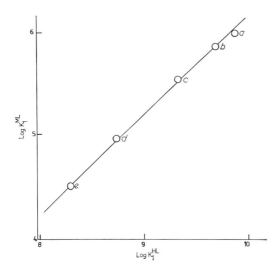

Figure 3-2. Correlation of log K_1^{ML} with log K_1^{HL} for nickel(II) amino acid complexes ($\mu = 0.1M$; 25°C). Key: a = alanine, b = glycine, c = phenylalanine, d = DOPA, e = glycylglycine.

unit slope this implies that the term $[G^{\ominus}(HL) - G^{\ominus}(ML)]$ is either constant or negligible with respect to the first term in square brackets (Equation (20)). Nonunit slope means that $[G^{\ominus}(HL) - G^{\ominus}(ML)]$ must also be a linear function of log K^{HL}.

Since $\Delta G = \Delta H - T\Delta S$, deviations from linearity are usually ascribed to ΔH and or ΔS variations. For example ΔH changes on protonation are directly related to σ-bonding whereas π-bonding may also be important in metal complex formation.

Similar relationships to (20) also hold for two *related* ligands with varying metal ions, i.e. log K^{ML} v. log $K^{ML'}$, illustrated in Figure 3-3, and further relationships connecting $\Delta G^{\ominus}$ with ionic radii are mentioned in later sections.

The measurement of $\Delta G^{\ominus}$, or more accurately log K^M values, for metal-complex formation is also the initial basis for classifying cations according to the extent of their interaction with different donor atoms.

Thus some cations form the strongest complexes with fluoride and the weakest with iodide, e.g. calcium(II) and others such as silver(I) have the reverse order: We extend this argument when we discuss the role of the cation in this chapter.

But if we are interested in the strength of bonding or *bond-type* then we must use $\Delta H^{\ominus}$ rather than $\Delta G^{\ominus}$ as our criterion. For metal-ligand interactions *in water* it can be shown[4] that because the dielectric constant for water varies inversely with temperature the contribution to $\Delta H^{\ominus}$ due to the formation of ion-pairs, i.e. *purely* Coulombic bonding, is *always positive*. In other words, a large electrostatic contribution to the bonding between metal and ligand is indicated if $\Delta H^{\ominus}$ for formation is positive, i.e. endothermic, or is very small, while a large negative $\Delta H^{\ominus}$ value, i.e. exothermic, is indicative of a large contribution from covalent bonding.

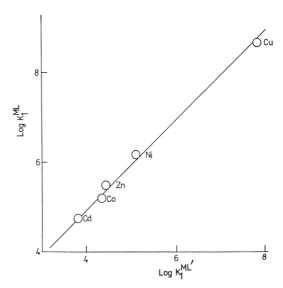

Figure 3-3. Correlation of log K_1^{ML} for glycine complexes with log $K_1^{ML'}$ for β-alanine complexes of several divalent cations ($\mu = 0.1M$; 25°C). The line is drawn with slope = 1 and intercept = log $K^{HL}/K^{HL'}$.

Although for the formation of *ion-pairs*, ΔH^Θ may be positive, the *stability constant can still be large* because of the entropy contribution (Equations (11)). In such a reaction involving a large K^M the ΔS^Θ value will be large because of the liberation of many water molecules from initially charged species on the formation of a species of at least lower, possibly zero- or even negative-charge. Note here that to a close approximation the $T\Delta S^\Theta$ contribution is *independent of the type of bonding*[†] which makes very questionable the argument that a large ΔS^Θ is the "driving force" (whatever that means) behind the formation of strong yet essentially ionic complexes. And this is very important when a change of solvent is envisaged—indeed if the solvent is omitted altogether, the stability constant will be very much larger with no thanks due to the $T\Delta S^\Theta$ term! So once again, we see the importance of the solvent when we write an equation purporting to represent the interaction of a metal cation with a ligand.

To summarize briefly, ΔG^Θ values are a measure of the *extent* of complex formation, ΔH^Θ sometimes gives us an insight as to the *typing of bonding* and ΔS^Θ often gives us an idea of changes in *solvation* and/or changes in the *structure* of the ligand.

THE EFFECT OF THE CATION

A-, B-, and C-Metal Cations

Perhaps the obvious place to begin looking for effects giving rise to specific or at least selective formation of complexes is in the Periodic Table and hence at the electronic structure of the metal cations. We begin by classifying cations as follows:[7]

1. *Class A-cations.* This class comprises all cations having a "rare gas" (octet) structure. For example all Group IA and IIA ions such as K^+ and Ca^{2+}, ions such as B^{3+}, Al^{3+}, Si^{4+} and even "ions" such as Mn^{7+}. The ions of the lanthanides are also included in this series as the *f*-orbitals are deep-seated and do not influence complex formation to any extent.

2. *Class B-cations.* These are cations with a "pseudo rare gas" (d^{10}) structure. Ions such as Cu^+, Ag^+, Zn^{2+}, In^{3+}, Sn^{4+}, Sb^{5+}, etc. fall into this classification.

3. *Class C-cations.* More usually called the transition metal ions and for our present purposes these are defined as those ions having an incomplete *d*-shell, e.g. Cu^{2+}, Ni^{2+}, Fe^{3+}, etc.

a. *Complex formation by class A-cations.* It is easily seen that these ions form a very coherent class indeed: In aqueous solution they virtually will form only complexes with F^- and O-donor ligands, i.e. alcohols, ethers, carboxylic acids, etc. Fajans' Rules are strictly adhered to in that charge-size relationships are very important. Thus we have the well-known *"diagonal relationships"* that express the close similarity between the cations Li^+, Mg^{2+} and, even more marked, the pair Be^{2+}, Al^{3+}. In fact, wherever the first member of a pair differs from the rest of its Group in properties, it will nearly always resemble the second member of the next Group. For example, Be^{2+} so resembles Al^{3+} in its properties that for some time after its discovery it was assumed to be trivalent. This dependence on size and charge also indicates why Tl^+ (ionic radius 1.40 Å) can often be used as a probe to study the reactions of K^+ (1.33 Å) in biological systems by making use of either the $6s^2 \rightarrow 6s^1p^1$ fluorescence at 215 nm or of the nuclear spin of one-half of ^{215}Tl. In the same way, Eu^{3+} (radius 1.12 Å) can be used to study the behavior of

---

[†] Although a large ΔG^Θ is *never the result of both a large ΔH^Θ and a large $T\Delta S^\Theta$* term; one or the other dominates.

Ca^{2+} (radius 0.99 Å) because the increase in size is offset by the increase in charge.

ΔH_1° for complex formation of these ions is always small and often positive (endothermic) with large ΔS_1° values when the equilibrium constant is large and this of course lines up with the above observations by indicating that the bonding is largely electrostatic in nature. Ligands such as NH_3, CN^-, and S-donors which would normally form strong *covalent* bonds with any cation that they complex with merely act as Brønsted bases and result in hydroxo-complexes or precipitates, e.g. the reaction of ammonia with aluminium solutions.

An important exception, as we shall see later, to these generalizations concerning size/charge factors is found when we encounter ligand species with many (> 4) donor groups such as edta and especially the cyclic polyethers, for example:

(21)

dicyclohexyl-18-crown-6

In compounds such as these, the *size* of the hole becomes of paramount importance leading in fact in the case of the "crown" ligands to a considerable degree of specificity, the example above (21) being remarkably specific for K^+.

b. *Complex formation by class B-cations.* This class of cations is very poorly defined when we look at the stability of the complexes. Starting with the ions of lowest charge, namely Cu^+, Ag^+, and Au^+ we find completely different behavior with respect to donor atoms. Complexes with fluoride and oxygen donors are now weak, the strongest complexes now being with I^-

($> Br^- > Cl^- >> F^-$) and sulphide ligands. Ammonia and cyanide likewise are stronger and in nearly all cases ΔH_1° is large and negative (exothermic) and dominates ΔG_1°.

Such behavior is slightly less apparent when we come to the divalent cations namely Zn^{2+}, Cd^{2+}, and Hg^{2+} and the trivalent ions (such as In^{3+}) and tetravalent ions, e.g. Sn^{4+} parallel class A-behavior very closely. (For example, in failing to complex with ammonia at all, instead precipitating the hydroxo-species from solution, and in complexing more strongly with fluoride as the charge is increased).

In other words, as the charge *increases*, so *electrovalency increases* which might at first sight be rather surprising until one realizes that as the d^{10}-shell therefore contracts it more closely begins to resemble the true rare gas configuration. In fact the reverse also applies in that as the rare gas shell gets farther from the nucleus, it also begins to take part in more covalent bonding as is exemplified by the formation of the xenon fluorides, and by the fact that Cs^+ is by no means clearly class-A.

c. *The transition-metal cations.* Complexes formed by these ions, namely those with an incomplete d-shell, have been more widely studied and for a longer time than those of any other group of cations. This is due in part the varied colors and paramagnetism associated with the unfilled d-shell have made complex formation an obvious phenomenon rather than because these complexes are essentially different to those found in other parts of the Periodic Table. Thus, spectroscopic and magnetic properties enable many features of the complexes to be monitored in a way impossible for other cations (and hence the use of, for example, Co^{2+} as a replacement probe for the properties of Zn^{2+} in many metallo-enzymes).

Complex formation is dominated here by the *d*-orbital occupancy and the resulting crystal field stabilization energy (CFSE). The Irving-Williams order[8] of stabilities for the divalent cations is

$$Mn^{2+} < Fe^{2+} < Co^{2+} < Ni^{2+} < Cu^{2+} > Zn^{2+} \quad (22)$$

and this is illustrated in Figure 3-4. This diagram also shows clearly that electrostatic interaction (oxalato-complexes) gives rise to a smaller *range* of values than does covalent interaction (1, 2-diaminoethane in the diagram) once again emphasizing that the latter is less dependent on changes in ionic radii that in this series are relatively small (Figure 3-5 shows the change in ionic radii obtained from the solid fluorides). In Figure 3-5 the d^0 and d^{10} cations are included to illustrate the absence of any CFSE effects in the d^0, d^5, and d^{10} ions (symmetrically occupied, or empty, *d*-shell).

The Irving-Williams rule is obeyed by an astonishing range of ligands except those that have special stereochemical require-

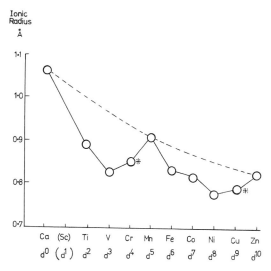

Figure 3-5. Ionic radii of divalent transition metal ions obtained assuming octahedral coordination in the fluorides ⧺ = distorted ocahedra (Jahn Teller effect).

ments (noting, for example, that Ni^{2+} accepts regular octahedral sites more readily than does Cu^{2+}—the Jahn-Teller[19] distortion being operative in the latter) or those that cause spin-pairing effects, e.g. $[Fe(CN)_6]^{4-}$ is *diamagnetic* with three *pairs* of d-electrons.

To conclude then, we have, based on electronic structure alone, defined one very coherent class of cations namely class-A but left ourselves with too other less well defined classes. Hence, in the next section of this chapter, we see what happens if we use a *purely chemical* method of classification.

The (a)- and (b)-Cation Classification

This method of classification,[10] due to Ahrland and Chatt,[11] divides all metal cations in *aqueous solution* into two groups according to their behavior towards complexing atoms as follows:

Class (a). These cations form their strongest complexes, as measured by $\Delta G^{\ominus}$

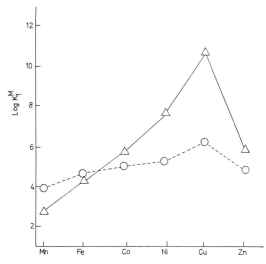

Figure 3-4. Irving-Williams order for stability of complexes with divalent metal ions: O = oxalate and Δ = 1,2-diaminoethane.

($\log K^M$) values with the first atom of each Periodic Group, namely N, O, F⁻.

Class (b). Form their strongest complexes with donor atoms with the second or subsequent atom of the Periodic Groups, i.e. P, S, I⁻.

The actual sequences are:

Donor Group	(a) cations	(b) cations
VB	F >> Cl > As > Sb	N << P > As > Sb
VIB	N >> P > Se > Te	O << S ~ Se ~ Te
VIIB	O >> S > Br > I	F << Cl < Br < I

$$(23)$$

In practice, however, a good working definition of these classes is obtained by considering the order of stability of the halides (MX_n) in *water* since for class (a) this is $MF_n \gg MCl_n > MBr_n > MI_n$ whereas for class (b) it is $MI_n > MBr_n > MCl_n \gg MF_n$, i.e. exactly the reverse order.

Unfortunately this then leads to very few members of the (b)-class namely:

		Cu	Zn			
	Pd	Ag	Cd	In		
Os	Ir	Pt	Au	Hg	Tl	Pb	Bi

$$(24)$$

but a brief discussion of the thermodynamic factors involved in making this group separate will prove helpful: We begin with the properties of the *gaseous* complexes.

We first note that the *enthalpy* of formation of a metal halide from its constituent ions in the gas phase is a measure of the strength of the bond (which we can compare later with the interaction of the aquated *ions* in water) and then the surprising fact is that *all metal cations in the gas phase are of class (a) type*, i.e. fluoride is always bound more strongly than iodide. However, two important ways of distinguishing at least in partial agreement with our aqueous solution interactions, the class (b) cations are (1) $\Delta H_I - \Delta H_F$ is markedly smaller for class (b) cations, e.g. for calcium iodide and fluoride the difference is 418 kilojoule mole⁻¹ whereas for mercury iodide and fluoride the difference is only 180 kilojoule mole⁻¹ and (2) class (b) ions have a generally larger ΔH for the formation of the gaseous halides from gaseous ions if compounds of the same internuclear distance for a given halide are compared.

The reason for leaving aqueous solutions for a moment should now become clear. First, we can avoid talking about π-bonding to the iodide ion in class (b)-metals—it is obviously not true because it must disappear in the gas phase! Second, $\Delta H^\ominus$ of formation of the complex is the important factor despite the fact that we used $\Delta G^\ominus$ values for our initial separation into classes. Fortunately, we succeeded because $\Delta S^\ominus$ changes *always* favor class (a) behavior which leads us to the third and final point namely that the class (b) cations owe their separate behavior to an increase in covalent bonding as we have already discussed in the section on Free Energy. Thus, the iodide ion transfers *more* charge to the class (b) metal-cation rather than the opposite effect predicted on the π-bonding model (already ruled out above).

The fact that the iodide ion is highly *polarizable* in comparison with the fluoride ion leads one to associate class (b) behavior with *anion* polarizability and hence to the term "*soft*." Unfortunately, however, a highly polarizable *cation*, e.g. Cs⁺ associated with a small anion, e.g. F⁻ leads to greater effective charges on the ions and hence to *enhanced class (a)* behavior! We will examine this point a little further in the next section.

HARD AND SOFT ACIDS AND BASES

Complex Formation as Acid-Base Interaction

A *Lewis acid* is defined as an electron-pair acceptor and a *Lewis base* as an electron pair donor and is thus identical with a *Brønsted base* as shown by the first of the following reactions.

$$H^+ + :NH_3 \rightleftharpoons NH_4^+$$
$$Cu^{2+} + :NH_3 \rightleftharpoons [Cu(NH_3)]^{2+} \quad (25)$$

The second reaction in (25) indicates clearly that metal-ligand interactions also fall into this pattern and hence may be treated as (Lewis) acid-base reactions.

The "Hard" and "Soft" Nomenclature

Furthermore, if we follow Pearson[1] we will describe all bases that are of high electronegativity, are hard to oxidize, and are of low polarizability "*hard*" bases, e.g. donors such as F^-, N, O, etc. Conversely, all donors that are of low electronegativity, readily oxidized, and are highly polarizable will be "*soft*." We have then generated the donor atom sequences used to define class (a) and (b) cations in sequence (23).

We then describe the acids, i.e. cations in the *same language* namely instead of (b)-cations we refer to "*soft acids*" and (a)-cations become "*hard acids.*" Note, however, from what we said at the conclusion of our discussion on (a)/(b) behavior, that to extend *polarization* arguments to the *acids* is extremely dubious and misleading. However, the HSAB (hard and soft acids and bases) concept is an extremely useful one. Even on the basis of terminology we now have a "label" for *both* cation and ligand that corresponds, and leads automatically to the basic rule of the concept that, "hard prefers hard and soft prefers soft." It also has the advantage that terms like "very hard," "moderately soft" and "borderline" come easier to the tongue than terms like "highly (a)" or "some (b)-characteristics." Table 3-I lists some common cations and ligands according to these ideas.

The Applications of HSAB

Although, as we have indicated, it is very difficult to give a reliable theoretical interpretation of the "hard" and "soft" phenomena, and indeed is even impossible to grade each group uniquely (the order of the acids being too dependent on the base chosen and *vice versa*), and provided that we use it as an empirical rule, it is very useful indeed. For example, the coordination end of an ambidentate ligand such as

TABLE 3-I

CLASSIFICATION OF LEWIS ACIDS AND BASES

	Hard	*Borderline*	*Soft*
(a) Acids:			
	H^+, Na^+, K^+	Fe^{2+}, Co^{2+}, Ni^{2+}	Cu^+, Ag^+, Au^+
	Be^{2+}, Mg^{2+}, Ca^{2+}, Mn^{2+}	Zn^{2+}, Pb^{2+}, Ru^{2+}	Pd^{2+}, Cd^{2+}, Hg^{2+}
	Al^{3+}, Sc^{3+}, Cr^{3+}, Fe^{3+}	$[R_3C^+]$	Pt^{4+}, Te^{4+}
(b) Bases:			
	H_2O, HO^-, F^-	Br^-, NO_2^-, N_3^-	I^-, SCN^-, RS^-
	CH_3COO^-, Cl^-, NO_3^-	RNH_2	CN^-
	SO_4^{2-}, CO_3^{2-}		R_2S, R_3P
	NH_3, N_2H_4		RNC, CO, C_2H_4

CN^- or SCN^- can often be correctly predicted, most soft cations coordinating through the S end of thiocyanate, for example. An intriguing example at first sight is the existence of compounds of the type

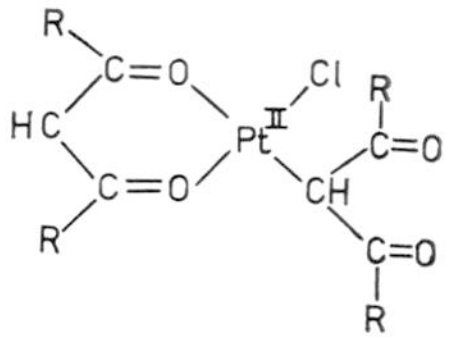

$$\begin{array}{c}\text{(structure 26)}\end{array} \qquad (26)$$

in which an acetyl acetone is coordinated via the *carbon* atom. But this is not too surprising on reflection because Pt^{II} is a soft cation and soft cations may be expected to use the C-atom for coordination (as in CN^- complexes).

In general, we may also expect a *symbiotic* effect exemplified by the rule "hard bases tend to flock together and soft bases tend to flock together." This is essentially because *hard* bases tend to *increase* the effective charge on the cation, thus making it *harder* and *mutatis mutandis*. Thus BH_3 will coordinate the soft CO ligand, while BF_3CO is unknown (although BF_3 does, as expected, form *etherates*). Unfortunately however, strongly π-bonding ligands, e.g. triphenylphosphine, often lead to a reversal of this rule by virtue of the fact that they may *remove* more charge *via* the π-orbitals than they donate *via* the σ-bond. A good example of this "antisymbiotic" effect is given by the pair of complexes:

$[Pt(NH_3)_2(SCN)_2]$ but $[Pt(P\emptyset_3)_2(NCS)_2]$
S-bonded; expected   N-bonded; not
                         expected   (27)

Related to this is the reversal of the *trans*-effect when soft ligands are involved: In square-planar complexes the *trans*-ligands share the same orbitals for bonding (as do the ligands in linear complexes) and hence, compounds such as $[Au(CH_3)_2 (H_2O)_2]^+$ always have the soft, i.e. CH_3

ligands *cis* to each other, where the *symbiotic* effect is still observable.

Finally, the HASB principle is very useful in predicting the $\Delta H^\ominus$ value expected on complex formation; hard-hard interactions leading to low or positive $\Delta H^\ominus$ values, and this of course, coupled with the need to choose the solvent correctly can help in preparative work.

The HSAB concept has also been invoked to study kinetic effects and in the study of *organic* reactions, but a discussion of these is excluded by the nature of this chapter.

THE EFFECT OF THE LIGAND

Availability of Lone Pairs

Although much of what is to be said in this section must be understood to be limited with respect to cation effects as outlined in previous sections, there are still quite a few generalizations pertinent to ligand type that can be made. Since we have already indicated that a ligand is essentially an electron-pair donor, then one important feature we must examine is the "availability" of lone pairs.

Consider the sequence of molecules

$$H_2O > ROH > ROR > RCHO \qquad (28)$$

when it should come as no surprise that we have essentially listed these (a) in order of "looseness of binding" of the lone pairs and (b) in order of effectiveness of the molecules as ligands.

But this is by no means the whole story; Many ligands, such as CO, form strong complexes, usually with very soft cations or neutral atoms and yet they do not have very strong Lewis base properties. In CO, for example, the "lone pair" is located largely on the C-atom in a molecular orbital that is antibonding with respect to the O-atom as illustrated in Figure 3-6. But here the important feature is to be found

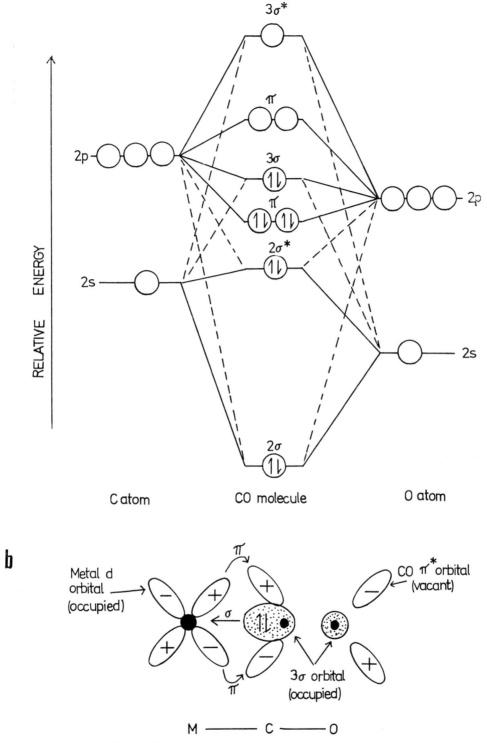

Figure 3-6. M. O. Scheme (a) for carbon monoxide, and (b) for the δ and π components of bonding orbitals in metal carbonyls: Full lines indicate main "parentage": Note that the 3σ orbital is almost nonbonding and located mainly on the C-atom effectively giving rise to a "lone pair."

in the empty π^* *(antibonding)* degenerate pair of orbitals, which have the correct symmetry to combine with and accept electron density from the occupied orbitals of transition metals with nearly filled d-orbitals, e.g. Ni°. This would result, if it really occurred in this way, in a buildup of charge on the CO molecule which in turn would result in a strengthening in the σ-bond when the charge "returned" to the metal atom *via* this route; i.e. a symbiotic "effect" involving the σ- and the π-orbitals results in a strong multiple bond. Even in a ligand such as NH_3, two of the group orbitals associated with the three protons have π-symmetry and so might interact with the central metal atom d-orbitals. Witness the effect of the $-CH_3$ group in toluene where the esr spectrum for the toluene anion clearly proves the existence of the "hyperconjugation" effect and incidentally explains the *"electron releasing"* properties for $-CH_3$ claimed by the organic chemists *contrary* to the order of ligands in (28).

The Chelate Effect

We have the *interpretation* of this effect in an earlier section of this chapter, but in practical terms it is very important. Consider the three potential ligands in (29):

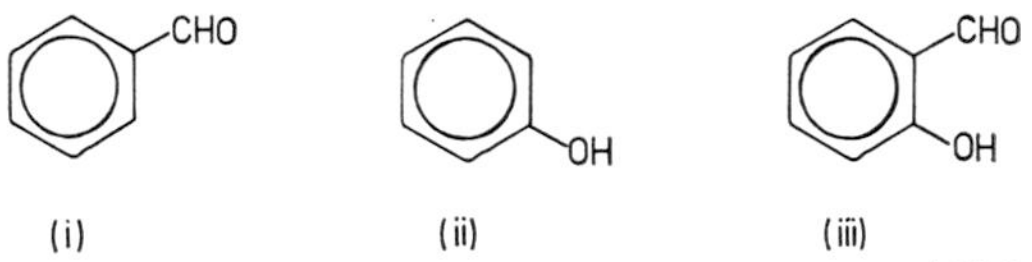

(29)

Of the first two, benzaldehyde is a very poor ligand, and phenol is only moderately effective whereas the combination of these two functional groups in (iii) results in a very good chelating ligand:

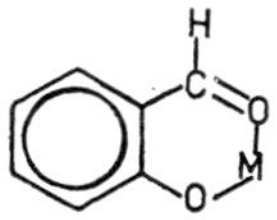

(30)

(metal chelate with salicylaldehyde)

The *size* of the chelate ring is important; in general 5-membered rings are the most stable, followed by 6-membered whereas longer chains between the donor atoms often result in the use of the cation of only one end of the ligand.

As illustrated in Figure 3-1, the number of donor groups is also important and this is emphasized by the fact that multidentate ligands can often force unusual coordination stereochemistry onto the central atom, for example $\beta'\beta''\beta''$-triaminotriethylamine $(N(CH_2CH_2NH_2)_3)$ forces *tetrahedral* geometry onto Pt^{II}.

Once again π-bonding can be of considerable importance, a simple example being the acetylacetone complexes of transition metal ions: In the case of copper(II) there is esr evidence for almost complete delocalization of the π-system:

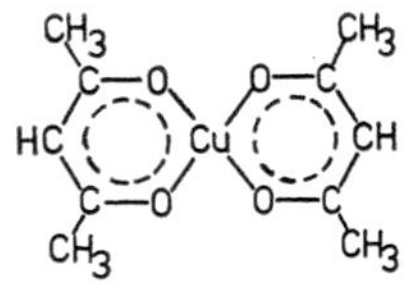

(31)

and similarly in planar ligands such as the porphyrin molecule:

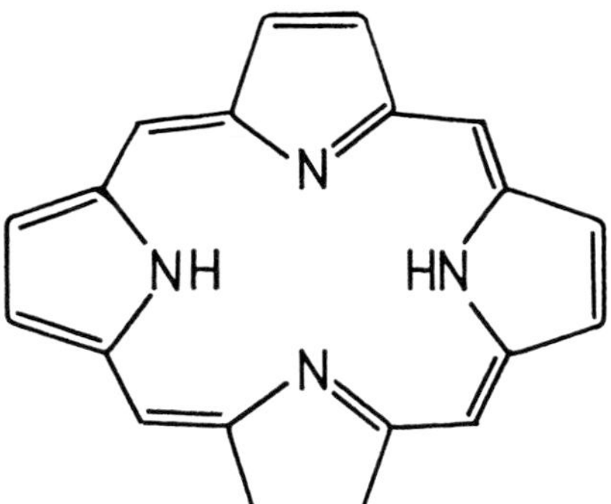

(32)

esr evidence again shows considerable delocalization of the d-electrons over the N's, e.g. nitrogen hyperfine structure due to four equivalent N's is superposed on each of the four Cu^{2+} peaks.

Selectivity by Means of Polydentate Ligands

We have in effect already discussed general selectivity on the basis of hard and soft interactions and we now wish to see how this is enhanced by the use of polydentate ligands. Essentially two effects are of importance, namely (1) stereochemical effects due to coordination number and geometry and (2) effects due to the size of the "hole" available to house the metal cation. Essentially, the first group are important when *covalent* interactions are of major importance and the second when *Coulombic* interactions predominate. We will therefore, as far as possible, treat these phenomena separately.

a. *Coordination geometry and number.* The shape of many polydentate ligands often favors one coordination geometry over another. For example consider the two tridentate ligands:

$$(33)$$

(I)   (II)

Both can complex with a cation of octahedral geometry, the ligand (I) complexing in one plane whereas ligand (II) requires one donor N out of the plane, *viz:*

$$(34)$$

(I_M) (II_M)

Thus complexes of type (II_M) would be expected to be weaker than type (I_M), but this *difference* would be considerably en-

hanced if the metal cation coordinated ligands in an elongated tetragonal form (quasi-square planar) as does copper(II). This is readily confirmed by comparing the appropriate log K^M values for nickel and copper:

	Ni^{II}	Cu^{II}
For ligand (I): log K^M	10.7	16.0
For ligand (II): log K^M	9.4	11.1

$$(35)$$

Furthermore, the out-of-plane N in type (II_M) is fairly readily protonated (leaving one chelate ring),* rather more readily for copper than for nickel complexes although too much must not be read into this because electronegativity effects have not been taken into account.

In other words, matching the stereochemistry of the ligand to the stereochemistry of the cation will always enhance stability. An especially interesting example of this is the ligand diethylenetriaminepentaacetic acid (dtpa):

$$(36)$$

which has *eight* potential coordinating position, all usable by an ion of coordination number eight such as Th^{4+} which, therefore, forms very strong complexes with it (log $K_1^M = 28.8$).

Of course, further selectivity is achieved if the ligand atoms are chosen in accord with our HSAB rules. For example, complexes such as the "crown" polyethers (21) do not bind Ag^+ very strongly whereas compounds in which O is replaced by S (or N) bind silver much more firmly (but hard

---

* Protonation need not always involve the removal of the complexing atom concerned, e.g. catechol complexes of iron(III) remain chelated even when both O-atoms are protonated, c.f. also OH^- and OH_2 complexes of Al(III).

cations such as K[+] much *less* readily, of course).

Introducing noncomplexing substituents into a chelating agent nearly always lowers the stability of the complex, but this can often lead to considerable changes in selectivity. To take an example from analytical chemistry; 8 hydroxyquinoline (37)(a) complexes with both Al^{3+} and Mg^{2+} but 2-methyl-8-hydroxyquinoline (37)(b) will precipitate out the Mg^{2+} complex even in the presence of large amounts of Al^{3+}.

![Chemical structures of 8-hydroxyquinoline (a) and 2-methyl-8-hydroxyquinoline (b)] (37)

a   **b**

Finally, it is of considerable interest that optically active ligands sometimes chelate stereoselectivity. For example, ($\pm$) histidine and some ($\pm$) α-hydroxyamidines complex with copper(II) to yield optically pure species, i.e. [M(+)L(+)L] or [M(−)L(−)L] rather than the [M(+)L(−)L] 2:1 complexes that might have been expected. Thus, in a sense, the ML species initially formed is acting selectively in a manner analogous to the use of optically active acids or bases in organic chemistry to effect separation of isomers. There is, in fact, quite a lot of evidence that although stereoselective effects are not absolute, in general the most symmetrical diastereoisomer is the most stable. Another related example (known since 1909) is the reaction of $[CoCl_2(-pn)_2]$ with $+pn$ (where pn represents propylenediamine) which is almost exclusively as follows:

$$3[CoCl_2(-pn)_2] + 3(+pn) \rightarrow$$
$$2(-)[Co(-pn)_3]^{3+} + (+)[Co(+pn)_3]^{3+}$$
$$+ 2Cl^- \quad (38)$$

b. *Selectivity by size of site.* Because of the considerable amount of thermodynamic data published by Pedersen[12] and by Frensdorf[13] we will confine our studies in this section to the "crown" polyethers. The principles can readily be extended to other types of macrocyclic ligands such as the antibiotic valinomycin, or to bicyclic or tricyclic polyethers, etc.

The "crown" terminology is in fact due to Pedersen and the name *n*-crown-*m* means a cyclic polyether with a total of *n* atoms in the ring with *m* donor atoms. For example:

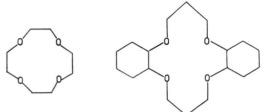

12-crown-4   dicyclohexyl-14-crown-4
(39)

We begin by noting a few "hole" diameters as estimated by Frensdorf

Ligand	Diameter/Å
14-crown-4	1.2-1.5
15-crown-5	1.7-2.2
18-crown-6	2.6-3.2
21-crown-7	3.4-4.3

(40)

and then refer to Figure 3-7 when the importance of the diameter of the hole becomes obvious with respect to the *unhydrated* cation diameter (given in Figure 3-7 together with hydrated diameters).

Figure 3-8 shows how the stability of the complexes varies with increase *above* the optimum hole size of 18-crown-6 for K[+] (hole diameter = 2.6 − 3.2Å, cation diameter = 2.88Å). The sudden increase with the 30-crown-10 ligand might be surprising until one looks at a tennis ball and sees how the seam is arranged i.e. enormous *conformational changes* are possible with large

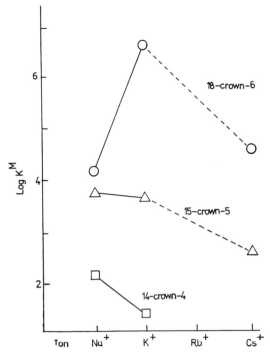

Figure 3-7. Stability of some "crown" complexes of the alkali metal cations. $\bigcirc$ = 18-crown 6; $\triangle$ = 15-crown 5; $\square$ = 14-crown 4, The ionic diameters — crystallographic, and *hydrated* are Na$^+$ 2.24, *5.52*; K$^+$ 2.88, *4.64*; Rb$^+$ 3.16, *4.56*; and Cs$^+$ 3.68, *4.56* Å respectively.

rings in order to accommodate small *(unhydrated)* cations.

This large conformational change is also present when K$^+$ (or Na$^+$) are complexed by large cyclic antibiotics such as valinomycin or nonactin and it should be fairly apparent that the *complex* will present the *hydrophobic groups* to the outside. Immediately we see a possible mechanism of transport of alkali-metal ions across an essentially nonaqueous membrane which can be written

$$K_{aq}^+ + L_{aq} \longrightarrow (KL) \xrightarrow[\substack{\text{"non aqueous"} \\ \text{medium}}]{\text{through}}$$

$$(KL) \longrightarrow K_{aq}^+ + L_{aq} \qquad (41)$$

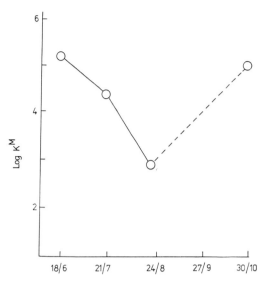

Figure 3-8. Some crown complexes of K$^+$; the effect of hole size. Dibenze-*n*-crown-*m* expressed as *n/m*.

Increasing the number of rings can also have a dramatic effect and ligands of the type:

$$(42)$$

(often referred to as a "football" ligand) have almost invariably high stability constants, e.g. K^M for ligand (42) reacting with K$^+$ is about 10^4 times that for K$^+$ and valinomycin.

Changes in solvent will also change the stability constants, of course, and very often also change the degree of selectivity between cations, showing that the diameters of cation and hole are not the only factors although they certainly dominate.

Finally, as before, changes in the donor

atoms can be used to effect very large changes according to our HSAB principle. For example, changing the oxygens for sulphurs in the 15-crown-5 ligand changes log K^M for Ag^+ from ~ 1 to 5.2 and for K^+ from 0.75 to no detectable complex formation.

OXIDATION STATES AND INNOCENT LIGANDS

The Assignment of Oxidation States

In assigning oxidation states to atoms in a molecule, a lot of chemical judgement as well as the application of rules must come into it. For example, the carbon in CH_4 should be $-IV$, in CH_2Cl_2O, and in CCl_4 it should be $+IV$ if the rules were rigidly applied; instead we define organic reactions in general as *substitution* rather than *redox* reactions and it is, of course, only in the latter that the use of formal oxidation states is of value. Note that they tell us very little about either valency (compare the oxidation state of the S in

with that of the central O in

which is an allotropic form of an element) or even the *electron density* about the central atom in a complex.

A set of working rules for the assignment of oxidation states would be as follows:

1. The oxidation states of free elements are zero, e.g. H_2, O_3, etc.
2. The oxidation state for a simple ion is its charge, e.g. Na^+ is I; Cl^- is $-I$; Ca^{2+} is II. (Note use of Roman Numerals.)
3. The oxidation state of covalently

bound H is $+I$, e.g. in H_2O, HCl, PH_3, but with some difficulty in carbon chemistry.
4. Covalently bound halides (except for interhalogens) are taken as $-I$, e.g. in SF_6, $[FeCl_4]^-$.
5. Covalently bound O is taken as $-II$ excepting for peroxides and related compounds, e.g. as in CO_2, SO_4^{2-}, it is $-II$.
6. The sum of oxidation states for a species is equal to the change on that species, e.g. for XeF_6 $Xe = +6$, $F = -1$; total $= 0$ or $[Co(NH_3)_6]^{3+}$ $Co = +3$, $NH_3 = 0$ (neutral species); total $= +3$.

The Existence of Noninnocent Ligands and Cations

Consider the two reactions

$$(1)\ [Fe(CN)_6]^{3-}_{(aq)} + CO(g) \rightleftharpoons [Fe(CN)_5CO]^{2-}_{(aq)} + CN^-_{(aq)}$$

$$(2)\ [Fe(CN)_6]^{3-}_{(aq)} + NO(g) \rightleftharpoons [Fe(CN)_5NO]^{2-}_{(aq)} + CN^-_{(aq)}$$

$$(43)$$

On the face of it they are completely parallel, and thus, application of the rules given in the previous section should lead invariably in both cases to an oxidation state of $+III$ for the iron. But, $[Fe(CN)_5 NO]^{2-}$ is *diamagnetic* with no unpaired electrons and thus the iron is best termed iron(II), i.e. with three pairs of d-electrons coordinated to 5 CN^- and 1 NO^+ ligand.

The oxidation state of the *ligand* is also variable and this behavior is called "noninnocent" behavior.[14] However, in the case above we had other evidence for referring to the oxidation state of the iron as $+II$, namely magnetic, but this is not always so, and thus we cannot unambiguously assign an oxidation state to the central cation. A very good example of a *ligand* with important redox properties is one of the type:

$$(44)$$

and complexes of the type $[NiL_2]^{2-}$ are almost certainly $Ni(II)$ complexes but what about the oxidized complexes $[NiL_2]$? Formally they *could* contain $Ni(IV)$, but almost certainly the ligand is not innocent.

Even in simple sulphides, the sulphur atom is not innocent; for example, GdS has a magnetic moment corresponding to $Gd(III)$ whereas EuS has the expected value for $Eu(II)$ and hence arithmetic will not aid the determination of oxidation states.

But what sort of ligands are noninnocent? Essentially two main sources of noninnocence can be separated. The first being summarized by noting that the softer the ligand the more we should suspect noninnocent behavior with cations of possible variable oxidation state. The second group is very common among biologically important ligands with large conjugated systems involving hetereoatoms. For example, ferric haem can be oxidized by chlorine gas to give what might seem to contain Fe^V, namely $Fe(haem)Cl_2$, but in this case esr evidence is that at least one electron is lost from the macrocyclic ligand rather than from the iron atom.

But up to now we have assumed that nonintegral oxidation states are ruled out. However, the internal redox reaction between a noninnocent pair (acid-base) is important, there is no reason to believe that it always results in *unit* transfer of charge. For example, if we look at iron(III) —catechol interactions, the *charge transference* band is intense and at such a *low energy* that the colors of these complexes are

diagnostic of the catechol group. In fact, it was on the basis of the existence of these low-energy charge transfer bands that Jørgenson originally defined a noninnocent ligand system.[14]

Some Problems in Biological Chemistry

In recent years there has been considerable use of metal-ion substitution techniques to study the roles of metal ions in bio-systems. In particular, the use of Tl^+ to monitor K^+ has been much advocated: However, it should be used with caution since not only is Tl^+ quite considerably softer than K^+ but acts noninnocently with sulphur containing ligand sites, i.e. to a first approximation (whole-number oxidation states) we can often write

$$Tl^+ + SR \rightleftharpoons [Tl^{III}SR] \qquad (45)$$

and hence find what is *externally* Tl^I bound strongly to sites that K^+ would not interact with at all.

Another problem that arises is the binding of O_2 to the iron atom in heme-like ligands—a great deal of time is wasted wondering what the oxidation state of the iron is before and after O_2 coordination. As in the $[Fe(CN)_5NO]^{2-}$ system, spin-pairing arguments lead to a *formal* and invariant oxidation state of II (iron III heme does not bind O_2 anyway) and the confusion seems to arise from trying to equate oxidation state with a measure of electron density, even in the case of $[Fe(CN)_6]^{3-}$ vs. $[Fe(CN)_6]^{4-}$, Mössbauer spectroscopy shows little difference in the *electron density* at the iron atoms.

But it does pose a problem in trying to predict the behavior of some noninnocent cations with conjugated systems on the basis of HSAB arguments. This is because the *cation and ligand* can undergo internal but *reversible* redox reactions on coordination. (Note the distinction between the reaction type (45) where the Tl^I is regenerated in solution and the reaction $Cu^{II} + 2I^- \rightarrow CuI + \frac{1}{2}I_2$ where Cu^{II} is merely reduced permanently to Cu^I.)

REFERENCES AND GENERAL BIBLIOGRAPHY

1. Pearson, R. G.: *J Am Chem Soc*, 85:3533, 1963.
2. Pearson, R. G. (Ed.): *Hard and Soft Acids and Bases*, Stroudsburg, Penn., Dowden, Hutchinson and Ross, 1973. [A selection of reprinted articles, with comments by the Editor, on the HSAB principles.]
3. *See, for example,* Glasstone, S., *Textbook of Physical Chemistry*, 2nd Ed. London, Macmillan, 1955, p. 966.
4. Prue, J. E.: *J Chem Educ, 46:12, 1969.*
5. Adamson, A. W.: *J Am Chem Soc, 76:1578, 1969.*
6. Beck, M. T.: *Chemistry of Complex Equilibria.* London, Van Nostrand, 1970.
   Jameson, R. F.: In Skinner, H. A. (Ed.): MTP International Review of Science—Physical Chemistry. London, Butterworths, 1972, Pages,   Series One, vol. 10.
   Nancollas, G. H.: *Co-ord Chem Revs,* 5:379, 1970.
   Nancollas, G. H.: *Interactions in Electrolyte Solutions.* London, Elsevier, 1966.
   Rossotti, F. J. C.: In Lewis, J., and Wilkins, R. G. (Eds.): *Modern Co-ordination Chemistry.* New York, Interscience, 1960.
7. Schwartzenbach, G.: *Anal Chem,* 32:6, 1960.
8. *See, for example,* Cotton, F. A. and Wilkinson, G.: *Advanced Inorganic Chemistry,* 3rd ed. London, Interscience, 1972, p. 596.
9. *See, for example,* Karplus, M. and Porter, R. N.: *Atoms and Molecules.* New York, Benjamin, 1970, p. 280.
10. Schwartzenbach, G.: *Advances in Inorganic Chemistry and Radiochemistry,* Academic Press, 3:257, 1961.
11. Ahrland, S., Chatt, J. and Davies, N. R.: *Quarterly Revs.,* London, 12:265, 1958.
12. Pedersen, C. J.: *J Am Chem Soc,* 89:7017, 1967.
13. Frensdorf, H. K.: *J Am Chem Soc,* 93:600, 1971.
14. Jørgenson, C. K.: *Structure and Bonding,* Berlin, 1:234-248, 1966.
15. *Structure and Bonding.* Berlin, Heidelberg, New York, Springer-Verlag. [This Journal often contains articles of considerable value to the study of metal-complex phenomena.]
16. Cotton, F. A. and Wilkinson, G.: *Advanced Inorganic Chemistry.* London, Interscience, 3rd ed., 1972. [An outstanding general coverage of inorganic chemistry.]
17. Sillén, L. G. and Martell, A. E. (Compilers): *Stability Constants of Metal-ion Complexes.* London, The Chemical Society, 1964 and Supplement, 1971. [An almost exhaustive compilation of stability constant and related data up until the end of 1968.]
18. a. Ashcroft, S. J. and Mortimer, C. T.: *Thermochemistry of Transition-metal Complexes.* London, Academic Press, 1970.
    b. Christensen, J. J. and Izatt, R. M.: *Handbook of Metal-Ligand Heats and Related Thermodynamic Quantities.* New York, Marcel Dekker, 1970. [Sources of thermodynamic, mainly calorimetric, data.]

CHAPTER 4

METAL ION CATALYSIS AND METALLOENZYMES

ROBERT W. HAY

INTRODUCTION

WITH THE DEVELOPMENT of sophisticated analytical techniques applicable to biological systems, a large number of enzymatic reactions have been found to require metal ions as essential components. Of the 840 enzymes known in 1964, 27 percent have either metals built into their structures, require added metals for activity, or are further activated by metal ions. The metal ions most frequently encountered in biological systems fall into two fairly distinctive categories,

(a) The alkali metals Na(I) and K(I) and the alkaline earths Ca(II) and Mg(II).

(b) Metals of the second half of the first transition series, Mn, Fe, Co, Ni, Cu and Zn [zinc(II) although a d^{10} ion is commonly grouped with metals of the first transition series].

Nickel(II) is not widely involved in enzymatic catalysis, although it is found in procarboxypeptidase A. It has been suggested that the limited occurrence of nickel(II) in biological systems is due to the fact that it binds strongly to soil silicates. However, it has recently been observed that the element occurs at relatively high levels in a plant (Hybanthus Floribundas) growing in the nickel-ore bearing regions of Western Australia. Nickel is accumulated by the plant at a concentration much higher than in the local soil and this suggests that this species may have an essential requirement for nickel.

Only three of the twenty-five elements known to be essential for animal life have an atomic number in excess of 35. All three are needed only in trace amounts; molybdenum (Z = 42) tin (Z = 50) and iodine (Z = 53). Another biologically important metal is vanadium but it occurs under rather restricted conditions. It is found in the vanodocytes of the sea squirt in the presence of 0.9M H_2SO_4 to maintain the element in a reduced oxidation state. Molybdenum occurs in a small group of metalloflavoproteins. Iron, of course, is found in the cytochromes, peroxidases and catalases and such compounds were early recognized due to their intense visible absorption spectra. Both Frieden[3] and Williams[17] have recently reviewed the occurrence and distribution of metal ions in biological systems, and metals such as tin and chromium are now known to be essential to some organisms. Table 4-I lists some typical metalloenzymes. Extensive surveys of metalloenzymes have recently been given by Coleman[18] and by Mildvan.[10] In 1955, Vallee clearly defined metalloenzymes as enzymes with tightly

TABLE 4-1

REPRESENTATIVE METALLOENZYMES

Metal	Enzyme	Biological Function
Iron	Ferredoxin	Photosynthesis
	Succinate Dehydrogenase	Aerobic oxidation of carbohydrates
Iron in Heme	Aldehyde Oxidase	Aldehyde oxidation
	Cytochromes	Electron transfer
	Catalase	Protection against hydrogen peroxide
	(Hemoglobin)	Oxygen transport
Copper	Ceruloplasmin	Iron utilization
	Cytochrome Oxidase	Principal terminal oxidase
	Lysine Oxidase	Elasticity of aortic walls
	Tyrosinase	Skin pigmentation
	Plastocyanin	Photosynthesis
	(Hemocyanin)	Oxygen transport in invertebrates
Zinc	Carbonic Anhydrase	CO_2 formation; regulation of acidity
	Carboxypeptidase	Protein digestion
	Alcohol Dehydrogenase	Alcohol metabolism
Manganese	Arginase	Urea formation
	Pyruvate Carboxylase	Pyruvate metabolism
Cobalt	Ribonucleotide Reductase	DNA biosynthesis
	Glutamate Mutase	Amino acid metabolism
Molybdenum	Xanthine Oxidase	Purine metabolism
	Nitrate Reductase	Nitrate utilization
Calcium	Lipases	Lipid digestion
Magnesium	Hexokinase	Phosphate transfer

bound metals which are retained on purification. The major difference between metalloenzymes (which have "built in" metal ions) and metal-activated enzymes (to which metals must be added for activity) is not clear-cut. In the latter enzymes, the affinity for the metal ion is generally rather low, but it might well be that the role of the metal ion is the same as in the metalloenzyme.

Properties of Metal Ions Relevant to Catalysis

Metal ions, like protons, are Lewis acids and act as electron pair acceptors. Westheimer first suggested that metal ions could be regarded as "superacid catalysts in neutral solution." Metal ions can exist in oxidation states in excess of +1 and can provide Lewis acid catalysis at pH values close to neutrality where a proton is ineffective. The term "superacid" is somewhat misleading since the activity of metal ions as catalysts is generally very much less than that of the proton, primarily as a result of the small size of the proton. The polarizing power of a cation, to a first approximation, will depend on the charge/size ratio. Obviously, small ions with a large charge should be the most effective catalysts.

In contrast to protons, metal ions can function as three dimensional templates for the binding and orientation of Lewis bases either as monodentate or polydentate ligands. An excellent example of a template effect is the reaction of acetone with $Ni(en)_3^{2+}$ (en = 1,2-diaminoethane) to give a macrocyclic complex (46a) by an internal aldol-type reaction,

$$(46)$$

As a result of their filled d-orbitals, transition metal ions are much larger and more polarizable than protons and can donate electrons to form π-bonds in addition to accepting electrons to form π-bonds, as in Figure 4-1.

The polarizability of metal ions which correlates with their ability to donate electrons for π-bonding, has been referred to as "softness." A thorough discussion of the factors operating in hard and soft interactions is given in Chapter 3, but it may be noted that the hard species, both acids, i.e. metal ions and bases (ligands) tend to be small slightly polarizable species, and soft acids and bases tend to be larger and more polarizable. Hard acids prefer to bind to hard bases and soft acids prefer to bind to soft bases.

In addition to their ability to act as Lewis acid catalysts many metal ions can exist in a number of different oxidation states, i.e. Fe^{II}/Fe^{III} and or, Cu^{I}/Cu^{II}. As a result, metal ions are involved in many catalytic processes involving electron transfer.

For example, ascorbic acid oxidase is a blue-green copper(II) protein obtained from summer crookneck squash (*Curcubita pepo condensa*). The enzyme has a molecular weight of 150,000 and contains about 0.26 percent copper, corresponding to six copper atoms per molecule. The enzyme catalyzes the oxidation of L-ascorbic acid (47a) to dehydroascorbic acid (47b).

$$(47)$$

When a small amount of ascorbic acid is added to a blue solution of ascorbic acid oxidase in 0.1M acetate buffer at pH 5.6 the color is rapidly bleached to a light yellow ($Cu^{II} \rightarrow Cu^{I}$). Admission of oxygen to the system slowly restores the color. The visible absorption spectrum of the enzyme (AAO) has an absorption maximum at 606nm and a shoulder at 412nm; when ascorbic acid (AH_2) is added, the 606nm band disappears but the shoulder is not affected, as in Figure 4-2.

The oxidation of ascorbic acid is also catalyzed by copper(II) ions, however, the enzyme catalyzed oxidation of L-ascorbic acid exhibits a different oxygen stoichiometry to the copper(II) catalyzed reaction. In the latter reaction twice as much oxygen is absorbed, and hydrogen peroxide is readily detected as a final product in addition to dehydroascorbic acid. In the enzymic reaction no hydrogen peroxide is de-

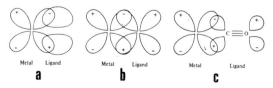

Figure 4-1. Pi bonds between a metal *d* orbital and ligand (a) *p* orbitals, (b) *d* orbitals, and (c) π^* antibonding orbitals.

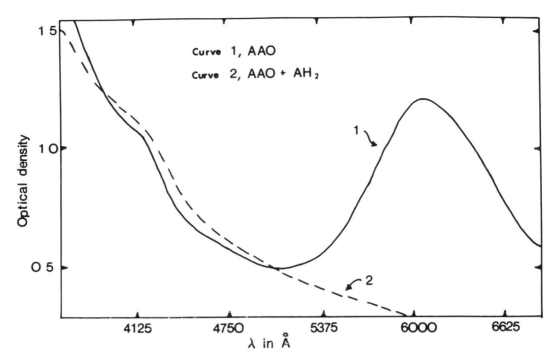

Figure 4-2. The visible absorption spectra of ascorbic acid oxidase (AAO), and in the presence of substrate. [From Dawson, C. R.: *Ann NY Acad Sci*, 88: *Art 2*, 353, 1960. Reproduced with permission.]

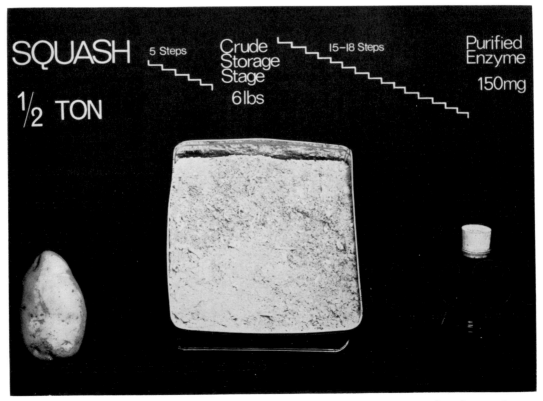

Figure 4-3. The purification of ascorbic acid oxidase from summer crookneck squash. [From Dawson, C. R.: *Ann NY Acad Sci*, 88: *Art 2*, 353, 1960. Reproduced with permission.]

tectable. On an equivalent copper basis the enzyme is about 10^3 more active than the aquo-copper ion. The enzyme also catalyzes the aerobic oxidation of L-ascorbic acid more rapidly than D-ascorbic acid, the copper(II) ion does not exhibit such stereospecificity being equally effective towards both D- and L-ascorbic acid.

Figure 4-3 illustrates the problems involved in the preparation of relatively pure samples of some metalloenzymes.

METAL PROTEIN EQUILIBRIA

Although simple hydrated metal ions catalyze a large number of organic reactions in solution, their catalytic activity is usually several orders of magnitude lower than the analogous reactions catalyzed by metalloenzymes.

Enzymic reactions are usually at least 10^9 times faster than the uncatalyzed reactions. Many metal ion catalyzed reactions are known with rate accelerations of $ca 10^6$ and rate accelerations of 10^7 to 10^9 have been reported for the hydrolysis of esters and nitriles, however, such large rate accelerations are relatively uncommon.

All enzymes are proteins, and the catalytic properties of metal ions in biological systems are intimately connected with the properties of the metal ion after coordinate bond formation with the protein (apoenzyme).

$$M^{n+} + \text{apoenzyme} \rightleftharpoons \text{metalloenzyme}$$

Most proteins are multidentate ligands, and in many cases there is a simple 1:1 stoichiometry, i.e. one protein molecule binds to a single metal ion.

The *in vitro* preparation of the apoenzymes by removal of the native metal ion from metalloenzymes by various chelating agents has been an important part of the study of these catalysts. Obviously, the preparation of a catalytically inactive metal

free apoenzyme which can be completely reactivated by the addition of an appropriate metal ion confirms the metal ion requirements of the enzyme system. For the three zinc(II) metalloenzymes, carboxypeptidase A, carbonic anhydrase, and alkaline phosphatase, stable apoenzymes have been prepared and their properties studied in detail. In the case of carbonic anhydrase, for example, it is possible to introduce other metal ions into the apoenzyme, and cobalt(II) carbonic anhydrase has a not too dissimilar catalytic activity to the native zinc(II) enzyme. Such metal substitution reactions can provide very useful information. As zinc(II) is a d^{10} ion, it has no visible absorption spectrum, the d^7 cobalt(II) ion has a visible spectrum markedly dependent on its stereochemistry (octahedral or tetrahedral) so that information about the metal binding site can be obtained.

The simplest possible example of the binding of a metal ion to a protein could be represented by the equilibrium

$$M^{n+} + \text{Protein} \overset{K}{\rightleftharpoons} M - \text{Protein}^{n+}$$
$$K = [\text{M-Protein}^{n+}]/[M^{n+}] \, [\text{Protein}]$$

The constant K would be the formation constant for the complexation reaction. The formation constants for the complexes of bovine carboxypeptidase A with Mn(II), Ni(II), Cu(II), Zn(II), Cd(II) and Hg(II) were the first to be determined, Table 4-II. The apoenzyme is stable, and one divalent metal ion at the active site readily exchanges with another in the series.

In contrast to carboxypeptidase A, the central Zn(II) ion in carbonic anhydrase will not exchange with metal ions in solution at a measureable rate. The zinc(II) ion can, however, be removed using suitable chelating agents in the pH range 5 to 6. The resulting apoenzyme is stable. By using various chelating agents and apocar-

TABLE 4-II

FORMATION CONSTANTS OF
METALLOCARBOXYPEPTIDASES*

Metal	Log K Apparent pH 8.0	Log K Corrected pH 8.0 for Cl⁻ and Tris
Mnᴵᴵ	5.6	5.6
Coᴵᴵ	5.8	7.0
Niᴵᴵ	5.7	8.2
Cuᴵᴵ	5.1	10.6
Znᴵᴵ	8.3	10.5
Cdᴵᴵ	7.9	10.8
Hgᴵᴵ	6.7	21.0

* Coleman, J. E. and Vallee, B. L.: *J Biol Chem*, 236:2244, 1961.

bonic anhydrase as competing ligands for various metal ions it has been possible to obtain values for the formation constants for the Mn(II), Co(II), Ni(II), Cu(II), Zn(II), Cd(II) and Hg(II), carbonic anhydrases, Table 4-III. The formation constants were calculated from the final equilibrium concentrations of the two metal complexes, the free metal ion concentration and the known formation constants of the metal complexes of the small chelating ligands. The ligands employed were 1,10-phenanthroline, 8-hydroxyquinoline-sulphonic acid and EDTA.

It has usually been necessary to determine the formation constants for metalloenzymes at a single pH, because of possible conformational changes in the protein as the pH varies which might alter the nature of the metal binding site. Knowledge of the state of protonation of the ligand donor atoms is generally not known and thus the constants are not pH independent, since a distinction cannot be made between protonated and unprotonated ligands. The apparent stability constants would be expected to increase with pH, thus with

TABLE 4-III

FORMATION CONSTANTS FOR
METALLOCARBONIC ANHYDRASES*

Metal	Log K, pH 5.5
Mnᴵᴵ	3.8
Coᴵᴵ	7.2
Niᴵᴵ	9.5
Cuᴵᴵ	11.6
Znᴵᴵ	10.5
Cdᴵᴵ	9.2
Hgᴵᴵ	21.5

* Lindskog, S. and Nyman, P. O.: *Biochim Biophys Acta*, 85:462, 1964.

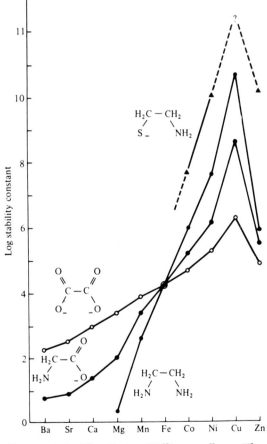

Figure 4-4. The Irving-Williams effect. The stability increases in the series Ba-Cu, decreased with Zn. [From Sigel, H. and McCormick, D. B.: *Acc Chem Res*, 3:201, 1970. Reproduced with permission.]

zinc(II)—carbonic anhydrase log K = 7 at pH4.5, rising to log K = 15 at pH 10 when log K becomes essentially independent at pH.

Oxygen, nitrogen and sulphur are the donor atoms of greatest biological interest. The logarithms of the formation constants of the 1:1 complexes of a number of divalent metal ions of the first transition series with ligands containing these donors are plotted in Figure 4-4. Also included for comparison are the constants for Ba^{II}, Ca^{II}, Sr^{II}, and Mg^{II}. The well-known Irving-Williams stability order $Mn^{II} < Fe^{II} < Co^{II} < Ni^{II} < Cu^{II} > Zn^{II}$ is attributable to a gradual decrease in ionic size as the series is traversed (small ions form more stable complexes) and an irregular crystal field stabilization energy. Thus for spin free octahedral complexes the CFSE's are Mn^{II}(zero) Fe^{II}(0.4Δ_o), Co^{II}(1.4Δ_o), Ni^{II}(1.8Δ_o), Cu^{II}-(0.6Δ_o) and Zn^{II}(zero), (Δ_o is the crystal field

in octahedral stereochemistry).

Manganese(II) complexes generally have the lowest formation constants except when the donor atoms are exclusively oxygen. In the metalloenzymes the same general trends are apparent, however, the zinc(II) metalloenzymes appear to be rather more thermodynamically stable than might be expected.

Kinetics of Metal Binding

In recent years the rates of many metal complexation reactions have been studied in detail as a result of the development of relaxation methods by Eigen and his co-workers. Figure 4-5 shows the half-lives for the exchange of solvent water with a variety of hydrated metal ions in solution

$$[M(H_2O)_x]^{n+} + XH_2\bullet \rightleftharpoons$$
$$[M(H_2\bullet)_x]^{n+} + XH_2O$$

For octahedral complexes it is possible to rationalize this data by considering the

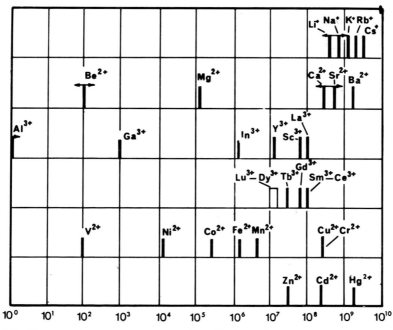

Figure 4-5. Characteristic rate constants (sec^{-1}), for substitution of inner sphere H_2O of various aquo ions.

charge and size of the metal ions. Small highly charged ions form the most thermodynamically stable complexes. Complexes having central atoms with small ionic radii react more slowly than those having larger central ions, for example,

$$[Mg(H_2O)_6]^{2+} < [Ca(H_2O)_6]^{2+} < [Sr(H_2O)_6]^{2+}$$

For a series of octahedral metal complexes containing the same ligands, the complexes having metal ions with the largest charge-to-radius ratios will react at the slowest rate.

Of the first-row-transition elements in Figure 4-5, the slowest water exchange rate is found with $[Ni(H_2O)_6]^{2+}$, a d^8 system. Such a result is predicted by crystal field theory. (The hydrated M^{2+} ions of the first row transitions elements are all high spin complexes). The rapid exchange rate for $[Cu(H_2O)_6]^{2+}$ has been attributed to the exchange of water molecules above and below the square plane of the tetragonally distorted octahedral complex. The four water molecules in the equatorial plane appear to react at a much slower rate. In general, four coordinate complexes (both tetrahedral and planar) react more rapidly than analogous six coordinate systems. For octahedral complexes the enthalpy of activation is strongly influenced by the breaking of metal ligand bonds (dissociative mechanism), as a result a large positive charge on the central ion retards the loss of a ligand. In four coordinate systems the formation of new metal ligand bonds is of increased importance (associative mechanism) and is favored by a large positive charge on the metal ion. Therefore, the guidelines of charge and ionic size which can be used to predict the rate behavior of six coordinate systems will often not apply to complexes having lower coordination numbers.

Rather limited information is presently available on metalloproteins. Metal ions exchange readily for each other in carboxypeptidase A. The replacement of $_{27}{}^{60}Co$ in carboxypeptidase A by zinc has been studied, as has the reverse reaction the replacement of $_{30}{}^{65}Zn$ by cobalt. Larger concentrations of cobalt(II) are required to replace zinc(II) than the reverse. Such exchange rates are presumably largely dependent on the dissociation rate of the metal-protein complex. The dissociation rate of the cobalt(II) carboxypeptidase is presumably more rapid than that of the zinc(II) enzyme.

The most thorough and complete study of the kinetics of formation of a metalloprotein has been carried out by Henkens and Sturtevant[19] (1968) on bovine carbonic anhydrase. The reaction is first order in zinc(II) and first order in apoenzyme with k = 10^4 M^{-1} s^{-1} at $25°$. Although the reaction of zinc(II) with the protein is rapid, the rate constant is at least 10^2 lower than for the reaction of zinc(II) with small chelating ligands where rate constants of 10^6 to 10^8 M^{-1} s^{-1} are commonly observed. The activation energy for the formation of the metalloprotein is *ca* 88 kJ mol^{-1} this large value being partially compensated by a large positive entropy of activation ($\Delta S^{\ddagger} = +502$ J mol^{-1} K^{-1}). The corresponding thermodynamic parameters for the reaction of zinc(II) with small ligands are $E_a = 29\text{-}33$ kJ mol^{-1} and $\Delta S^{\ddagger} = -17$ to -33 J mol^{-1} K^{-1}. By using the formation constant previously determined for zinc(II) carbonic anhydrase, the dissociation rate of the metalloprotein can be calculated to be *ca* 1.5 $\times$ $10^{-9}s^{-1}$. The dissociation rate for zinc(II) carboxypeptidase A has been estimated to be *ca* 2.5 $\times$ $10^{-5}s^{-1}$, some 10^4 to 10^5 times faster than the dissociation rate of $Zn(II)$ carbonic anhydrase. For this latter enzyme the half-life for the dissociation of zinc(II) at pH 8.0, I = 0.1M is 5 to 6 years.

TABLE 4-IV

ABSORPTION SPECTRA OF COBALT(II) AND COPPER(II) METALLOENZYMES[a]

Co(II) Metalloenzymes	*Band Position, nm*		*Intensity, $\epsilon\ (M^{-1}\ cm^{-1})$*
Co(II) carboxypeptidase	500		—
	555		160
	572		160
	940		~25
Co(II) carbonic anhydrase	520		205
	555		340
	615		230
	640		240
	900		~25
	1,250		95
Co(II) carbonic anhydrase +CN⁻		310[a]	
		345	
		450	
	520	545	350
	570	585	650
Co(II) carbonic anhydrase + acetazolamide		465	
		515	
	520	550	350
	570	570	550
	600	590	500
Co(II) alkaline phosphatase	640		260
	605		220
	555		378
	510		335
Co(II) alkaline phosphatase + HPO_4^{2-}	640		120
	535		350
	480		260
Co(II) alkaline phosphatase + $HAsO_4^{2-}$	500		~240
	550		~260
Cu(II) carboxypeptidase	790		< 100
Cu(II) carbonic anhydrase	590		50
	750		100
	900		75
Cu(II) carbonic anhydrase + CN⁻	700		130
	900		80
Cu(II) alkaline phosphatase	~750		~100
Copper Oxidases			
Laccase [Cu(II)]	730		~500
	615		1,400
	532		~300
Azurin (psuedomonas blue protein) [Cu(II)]	806		~600
	621		2,800-3,500
	521		~300
	467		~400
Ascorbic acid oxidase [Cu(II)]	606		770
	412		~500
Ceruloplasmin [Cu(II)]	605		1,200
	370		~500

[a] Figures in this column determined from band positions in the C.D. spectra. Data from ref. 18.

Physical Methods, Illustrative Examples

A number of physical techniques can be used to study metalloenzymes, and some illustrative examples will be discussed in this section.

Electronic Absorption Spectra and Optical Activity

In addition to the usual absorption bands present in proteins (at *ca* 280nm due to the aromatic amino acid residues L-β-phenylalanine and L-tyrosine), metalloproteins may have additional absorption bands arising from the d-d transitions of the metal ion, and charge transfer bands resulting from electronic transitions between the metal ion and the ligand.

The d-d transitions arising in transition metal ions with unfilled d-orbitals are usually of low intensity ($\epsilon \ll 1,000$ M^{-1} cm^{-1}) since they are forbidden under the usual selection rules for electronic transitions. The charge transfer bands associated with metal complexes are more intense ($\epsilon > 10^3$ M^{-1} cm^{-1}) and usually occur in the ultraviolet region.

The absorption spectrum of ascorbic acid oxidase is shown in Figure 4-2. Two bands are observed, one at 606nm ($\epsilon = 770$ M^{-1} cm^{-1}) and a shoulder at 412nm ($\epsilon \sim 500$ M^{-1} cm^{-1}). Typical absorption spectra of metalloenzymes are listed in Table 4-IV.

Although cobalt(II) metalloproteins are not common, it is possible to substitute cobalt(II) at the active site of some zinc(II) metalloenzymes with retention of catalytic activity. Figure 4-6 shows the visible absorption spectra of cobalt(II), nickel(II) and copper(II) carbonic anhydrases prepared from the native zinc(II) enzyme. Also included are the spectra observed in the presence of some inhibitors of the copper and cobalt enzymes.

The spectrum of the cobalt(II) enzyme

is similar to that expected for a tetrahedral complex (ϵ of several hundred), as in Figure 4-7, rather than an octahedral complex ($\epsilon \sim 10$ M^{-1} cm^{-1}). The 2Å resolution X-ray structure[20] of the active site of the zinc enzyme indicates the presence of three protein ligands (histidine residues) and a fourth site probably occupied by a coordinated water molecule or hydroxide ion. Both the X-ray data and the visible spectrum are consistent with a distorted tetrahedral geometry about the metal ion. Cyanide and acetazolamide are both believed to displace the coordinated water molecule. The shift to a narrower more intense spec-

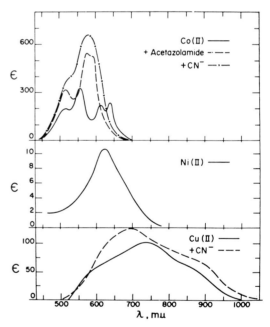

Figure 4-6. Visible absorption spectra of Co(II), Ni(II) and Cu(II) human carbonic anhydrases B and inhibitor complexes. [From Coleman, J. E.: *Progress in Bioorganic Chemistry*, Vol. 1. Kaiser, E. T. and Kézdy, F. J. (Eds.). Wiley, Interscience, 1971. Reproduced with permission.] (Note. Acetazolamide is a sulphonamide which binds to the enzyme by coordination at the metal site, a point which has been confirmed by x-ray studies.)

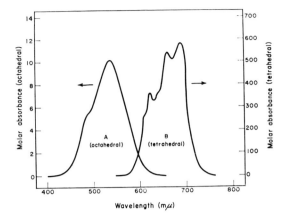

Figure 4-7. The visible spectra of $[Co(H_2O)_6]^{2+}$ (curve A) and $[CoCl_4]^{2-}$ (curve B). The molar absorbance scale at the left applied to curve A, and that at the right to curve B. [From Cotton, F. A. and Wilkinson, G.: *Advanced Inorganic Chemistry*. Wiley, Interscience. Reproduced with permission.]

trum on inhibitor binding has been interpreted as a shift to a more regular tetrahedral geometry. The visible absorption spectrum of cobalt(II) human carbonic anhydrase B is pH-dependent, as in Figure 4-8.

The change in the spectrum at 640nm follows a titration curve with a $pK_a \sim 7.5$, very similar to the pH rate profile observed for carbonic anhydrase. A good deal of evidence now indicates that this ionization is due to the process $Co - OH_2 \rightleftharpoons Co - OH + H^+$ and that attack on a substrate carbonyl group by the coordinated hydroxide ion occurs.

Optical rotary dispersion and circular dichroism measurements have also proved useful in studying metalloenzymes. Circular dichroism in general proves to be much

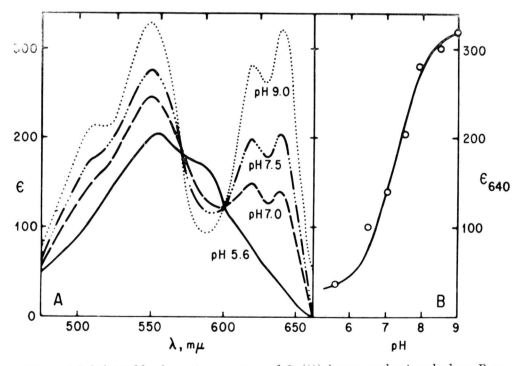

Figure 4-8 (A) Visible absorption spectrum of Co(II) human carbonic anhydrase B as a function of pH. (B) Molar extinction coefficient at 640 nm as a function of pH. [From Lindskog, S. and Nyman, P. O.: *Biochim Biophys Acta*, 85:462, 1964.]

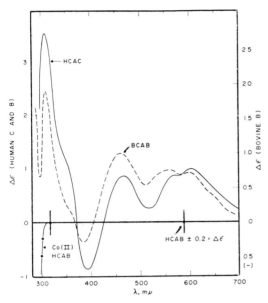

Figure 4-9. Visible CD of three isozyme and species variants of Co(II) carbonic anhydrase. (——) human isozyme C; (- - - -) bovine isozyme B; (•——•) human isozyme B. [From Coleman, J. E.: *Progress in Bioorganic Chemistry*, Vol. 1. Kaiser, F. T. and Kézdy, F. J. (Eds.): Wiley, Interscience, 1971. Reproduced with permission.]

more sensitive to small changes in dissymmetry at the coordination site than the absorption spectrum. Thus, while the absorption spectra of the cobalt(II) derivative of the human B, human C and bovine carbonic anhydrases are almost identical, only the human C and bovine enzymes show pronounced circular dichroism associated with the d-d bands of cobalt(II), Figure 4-9. There are, therefore, pronounced differences in asymmetry at the active sites of these variants of carbonic anhydrase.

Much information has been obtained from esr measurements, and perhaps more copper(II) proteins have been examined by esr than any other type of metalloprotein. Copper(II) complexes (d^9) with a single unpaired electron, give an easily re-

solved esr signal at g-values only slightly higher than that of a free electron (g = 2.0023). As copper(I) is d^{10} with no unpaired electrons, esr measurements can establish that copper(II) is reduced to copper(I) when a substrate is added. Thus the blue copper oxidase, mushroom laccase is reduced on the addition of the substrate catechol, as in Figure 4-10.

There are two major nuclear hyperfine intereactions which have proved useful in studies of copper(II) binding. The copper(II) signal is split into four lines by the copper nuclear hyperfine interaction (I = 3/2), and if nitrogen is one of the ligand donor atoms, hyperfine lines appear from the $^{14}_{7}$N nuclei (I = 1). The number of nitrogen hyperfine lines observed is therefore a function of the number of nitrogen donors. The room temperature esr spectrum of copper(II) triglycine (48) is shown in Figure 4-11.

$$(48)$$

In this complex there are three nitrogen donors and one oxygen donor. Seven hyperfine lines are observed on the high field side of the spectrum as a result of the three nitrogen donors in addition to the copper hyperfine splitting. (Each esr signal of an electronic system which interacts with a group of n equivalent nuclei of spin I is split into $(2nI + 1)$ lines, in this case n = 3 and for $^{14}_{7}$N, I = 1, so that $(2nI + 1)$ = 7.) Nitrogen hyperfine splittings have been observed in a number of copper(II) complexes, such as the copper(II)-conalbumins, copper(II)-transferrins, copper

(II)-carboxypeptidase and copper(II)-insulin).

MODEL SYSTEMS

Simple hydrated metal ions catalyze a large number of organic reactions in solution. Many of the reactions catalyzed by metalloenzymes are also catalyzed by aquometal cations and these reactions can be regarded as simple models for the much more complex enzymatic reactions. The decarboxylation of oxaloacetic acid is a typical reaction of this type,

$$HO_2CCOCH_2CO_2H \rightarrow$$
$$HO_2CCOCH_3 + CO_2$$

In biological systems the reaction is catalyzed by the enzyme oxaloacetate decarboxylase. This enzyme has recently been crystallized by Kosicki[21] and found to require manganese(II) for activity. The decarboxylation is catalyzed by cations of the first transition series (and by lanthanide ions). The catalytically active species in the reaction is the keto complex (49b) which is in equilibrium with the catalytically inactive enolic complex (49a).

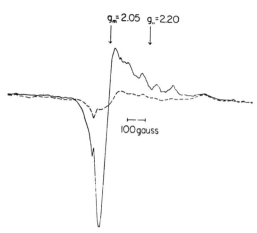

$g_m = 2.05$ $g_\parallel = 2.20$

$\vdash\dashv$
100 gauss

Figure 4-10. Electron spin resonance signal of mushroom laccase (——) before and (- - - -) after treatment with 10 mM catechol. Temperature 77°K, field modulation 6.6 gauss. [From Nakamura, T. and Ogura, Y., in *Magnetic Resonance in Biological Systems*, Ehrenberg, A., Malmström, B. G. and Vanngard, T. (Eds.): New York, Pergamon Press, 1967.]

Polarization of the carbonyl group by the metal ion in the keto complex (49b) assists the transfer of electrons from the carbon-carbon bond undergoing cleavage. The

(49)

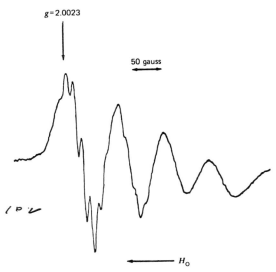

Figure 4-11. Room temperature spectrum of [Cu(II) (glycylglycylglycine)]. [From Wiersema, A. K. and Windle, J. J.: *J Phys Chem*, 86:2316, 1964.]

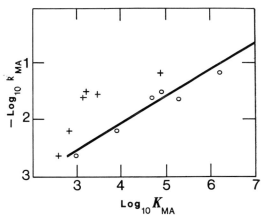

Figure 4-12. Plots of log K_{MA} for oxaloacetates (+) and oxalates (O) versus the rate constants k_{MA} for the decarboxylation of the metal oxaloacetates. After Gelles, E. and Salama, A.: *J Chem Soc*, 3689, 1958.

catalytic activity of a variety of transition metal ions on the reaction has been studied and the formation constants of the 1:1 complexes obtained, where A^{2-} is the dianion of oxaloacetic acid

$$M^{2+} + A^{2-} \overset{K_{MA}}{\rightleftharpoons} MA \overset{k_{MA}}{\to} CO_2 + \text{Products}$$
$$K_{MA} = [MA]/[M^{2+}][A^{2-}]$$

Values of K_{MA} and k_{MA} are listed in Table 4-V.

TABLE 4-V

RATE CONSTANTS AND FORMATION CONSTANTS FOR METAL OXALOACETATES AT 37° AND I = 0.1M

Species	$10^3 k\ (min^{-1})$	Log K_{MA}
H_2A	0.345	—
HA^-	15.4	—
A^{2-}	4.2	—
CaA	144	2.6
MnA	390	2.8
CoA	1,440	3.1
ZnA	1,860	3.2
NiA	1,380	3.5
CuA	3,960	4.9

A linear free energy relationship might be expected to exist between the logarithms of the appropriate rate constants and the formation constants, as the formation constants give a measure of the strength of the metal-ligand binding. Such a relationship is not observed, Figure 4-12, however, the rate constants do correlate quite well with the formation constants of the corresponding metal oxalates. It has been suggested that the interaction of the metal ion with the transition state (50a) which closely resembles the oxalate structure (50b) is more important than interactions with the ground state.

$$\text{a} \qquad\qquad \text{b} \qquad (50)$$

A somewhat similar transition state interaction has been invoked in the nickel(II) catalyzed hydrolysis of 2-cyanophenanthroline to give the corresponding carboxamide. The base hydrolysis of the 1:1 complex of nickel(II) with 2-cyanophenanthroline is some 10^7 times faster than base hydrolysis of 2-cyanophenanthroline (51).

$$(51)$$

The rate acceleration of 10^7 is due exclusively to a change in $\Delta S^{\ddagger}$. For straightforward base hydrolysis $\Delta S^{\ddagger} = -84$ J mol^{-1}K^{-1} while for the nickel(II)-catalyzed reaction $\Delta S^{\ddagger} = +59$ J mol^{-1}K^{-1}. The unusually large positive entropy of activation in the catalyzed reaction is unexpected for a bimolecular process. It has therefore been suggested that the transition state must involve bonding of the developing imino anion to the metal ion, perhaps displacing a water molecule coordinated to the metal ion (in addition to the liberation of the solvent shell around the nucleophile), and thus a more positive entropy of activation. The metal ion thus effectively "solvates" charges developed in the transition state.

$$(52)$$

The hydrolysis of amino acid esters is catalyzed by transition metal ions. Amino acid esters can act as mono-dentate or bidentate ligands.

$$(53)$$

Formation of a bidentate species leads to significant polarization of the ester carbonyl group by the metal ion and thus

ready attack by a nucleophile such as hydroxide ion in solution. Bidentate metal complexes have been isolated from solution and their infrared spectra studied. The vCO (ester) frequency at *ca* 1745 cm^{-1} in the free ligand is reduced to *ca* 1600 cm^{-1} in the bidentate copper(II) complex indicating significant polarization of the carbonyl group by the metal ion. Base hydrolysis can, therefore, occur by the following mechanism,

$$(54)$$

The copper(II) catalyzed, or more correctly, promoted reactions are *ca* 10^5 to 10^6 times faster than base hydrolysis of the uncomplexed ligand.

The hydrolysis of peptides is also catalyzed by metal ions, for example, copper(II) catalyzes the hydrolysis of glycylglycine within the pH range 4 to 5. At these pH values the predominant complex in solution is the 1:1 carbonyl bonded species (55a). As the pH is raised deprotonation of the amide proton occurs with a concomitant molecular

$$(55)$$

rearrangement to give the complex (55b). Only the carbonyl bonded complex is susceptible to hydrolysis by water and hydroxide ion. The deprotonated complex is apparently inert to nucleophilic attack. In addition to copper(II), other metal ions such as cobalt(II) and nickel(II) give deprotonated peptide complexes. These reactions illustrate a further important aspect of complexed ligands. The pK_a for the deprotonation of free glycylglycine can be estimated to be approximately 14.5;

$$NH_2CH_2CONHCH_2CO_2^- \rightleftharpoons$$
$$NH_2CH_2CON^-CH_2CO_2^- + H^+$$

so that complexing to the metal ion leads to an increase of 10^{10} in the acidity of the ligand.

If kinetically labile complexes, e.g. copper(II), zinc(II), cobalt(II) etc., are used in studies of metal ion promoted reactions, it is often difficult to identify the catalytically active complexes and so define the various mechanistic pathways. Therefore, in recent years there has been considerable interest in employing inert cobalt(III) complexes which can provide more precise mechanistic data.

Alexander and Busch studied the Hg(II) promoted hydrolysis of the *cis*-[Co(en)₂X (glyOR)]²⁺ ion in acid solution (X = Cl,Br; R = $CH_3C_2H_5$, i-C_3H_7, gly = NH_2CH_2 CO).

$$[Co(en)_2X(glyOR)]^{2+} + Hg^{2+} + H_2O \rightarrow$$
$$[Co(en)_2gly]^{2+} + ROH + H^+ + HgX^+$$

and proposed that a chelated ester species [Co(en)₂(glyOR)]³⁺ (56) was the reactive intermediate in the reaction. Evidence for such a species

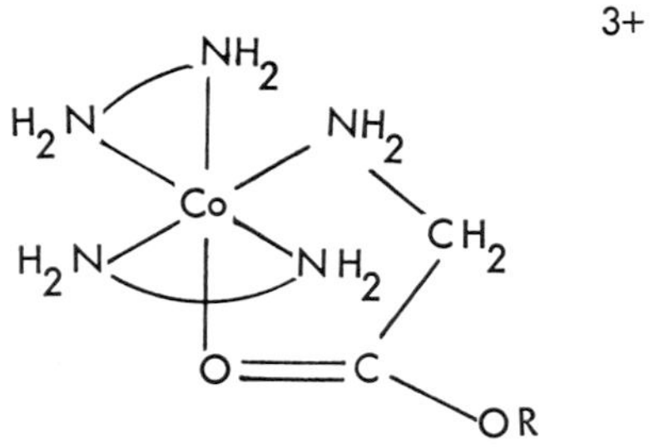

(56)

was obtained primarily from the change in the CO stretching frequency as the monodentate ester (1735 cm⁻¹) was first chelated (1610 cm⁻¹) and them hydrolyzed to [Co(en)₂gly]²⁺ (1640 cm⁻¹). The mercury(II) ion assists removal of the coordinated halide ion to give a 5-coordinate species and the ester carbonyl oxygen competes so effectively with solvent water for the vacant coordination site, that the chelated ester complex is formed exclusively,

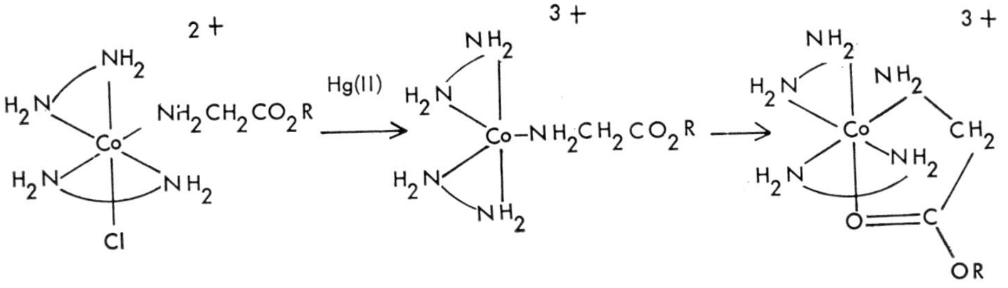

(57)

The perchlorate salts of the chelated ester $[Co(en)_2(glyOR)](ClO_4)_3$ were subsequently isolated by Buckingham et al., and used for the synthesis of peptide esters. Treatment of the chelated ester complex with amino acid or peptide esters in anhydrous sulpholane or dimethyl sulphoxide or acetone solutions gave the $[Co(en)_2(peptide-OR)]^{3+}$ ion (58).

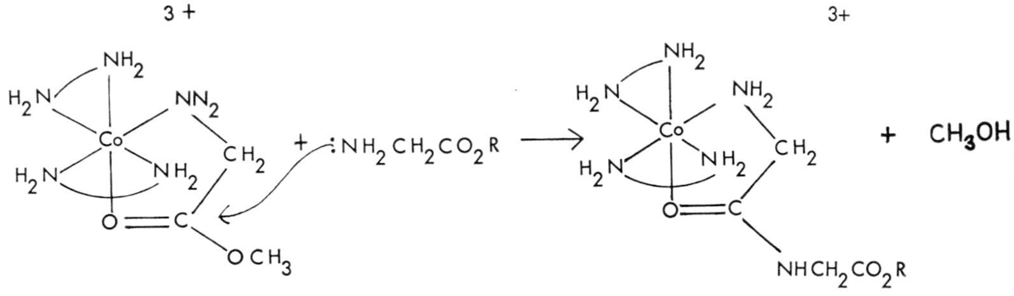

has been shown that base hydrolysis is some 10^5 times faster than for the uncomplexed dipeptide.

In recent years there has been considerable discussion on the nucleophilicity of hydroxide ions coordinated to metal ions. Such "bound nucleophiles" have often been proposed in enzymatic reactions, but their importance has been somewhat question-

(58)

Base hydrolysis of the chelated ester complex (57) is some 10^6 times faster than for uncomplexed glycine esters. Using dipeptide complexes such as (58; R = H) it

able. Recently, it has been shown that hydroxide ions attached to cobalt(III) are extremely power nucleophiles. The two reactions,

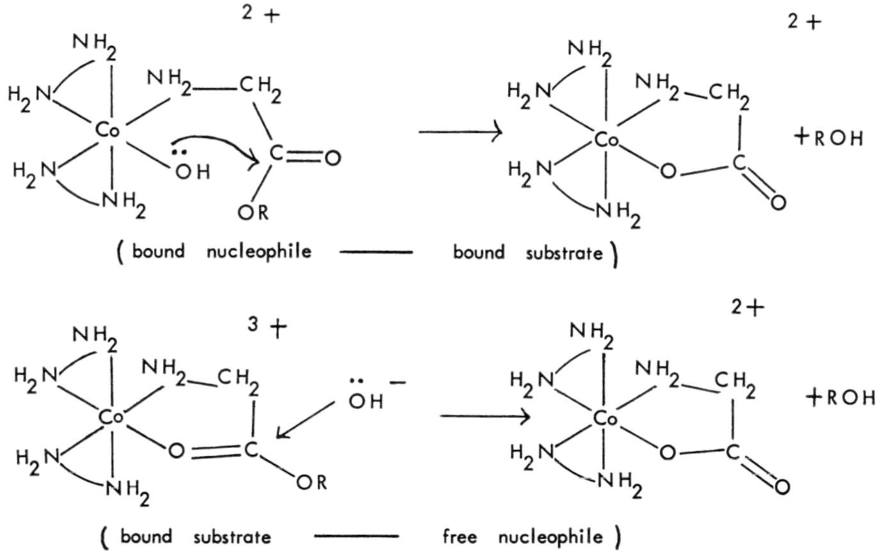

(bound nucleophile ——— bound substrate)

(59)

(bound substrate ——— free nucleophile)

have been investigated in some detail. The rate acceleration for the first reaction is *ca* 10^9, while for the second reaction it is *ca* 10^6. A number of complexes of the type $[CoN_4(OH)(OH_2)]^{2+}$ (N_4 = a system of four nitrogen donors as occurs with 2en (en = 1,2-diaminoethane) or trien (60a) or tren (60b) stoichiometrically cleave the N-terminal amino acid

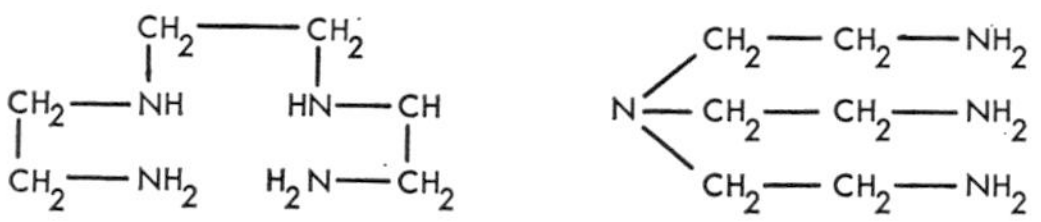

(a) "trien"   (b) "tren"

(60)

from di- or tri-peptides. Although the reaction is stoichiometric rather than catalytic, it provides an interesting model for the amino-peptidases (enzymes cleaving the N-terminal amino acid from a peptide chain). A necessary prerequisite for hydrolytic activity in the cobalt complex, is the availability of at least two *cis*-sites in the octahedral coordination sphere, the remaining sites being filled by nonlabile ligands.

It has often been suggested that in metalloenzymes the metal ion acts as a bridge between the apoenzyme and the substrate. There has, therefore, been considerable interest in studying ternary metal complexes as models for such systems. Sigman and Jorgenson[22] have (1972) found that zinc(II) catalyzes the transesterification reaction between N-(β-hydroxyethyl) ethylenediamine and p-nitrophenyl picolinate. The reaction is considered to involve a reactive tertiary complex in which the zinc(II) ion perturbs the pK_a of the hydroxyethyl group of N-(β-hydroxyethyl) ethylenediamine to provide a high effective concentration of the nucleophile. Nucleophilic attack then occurs at the carbonyl group of p-nitrophenyl picolinate (62)

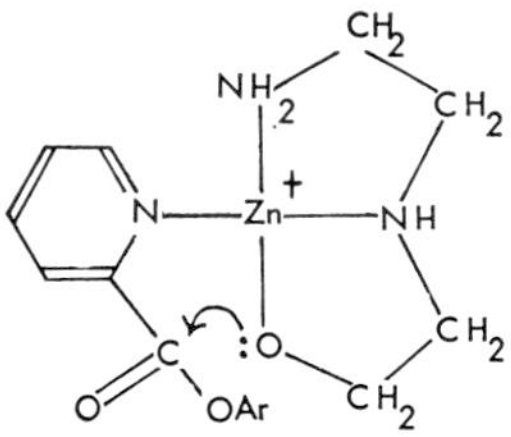

In addition to ester hydrolysis, peptide

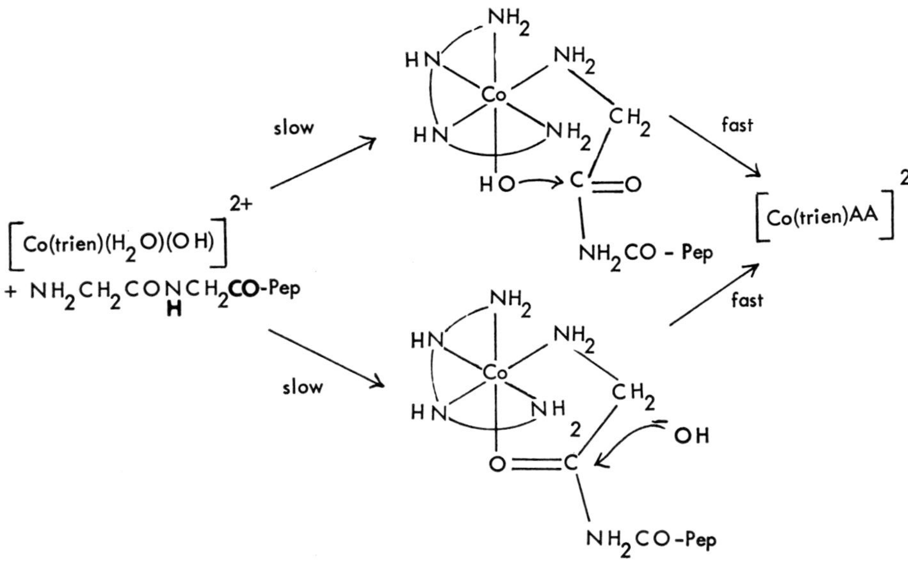

Proposed mechanism for peptide hydrolysis

hydrolysis, and decarboxylation reactions many other organic reactions are subject to metal ion catalysis and some of these are summarized in Table 4-VI.

The hydrolysis of the Schiff's base formed by condensation of 1,2-diaminoethane with thiophene-2-aldehyde (63a) is subject to metal ion catalysis as is the Schiff's base salicylideneaniline (63b)

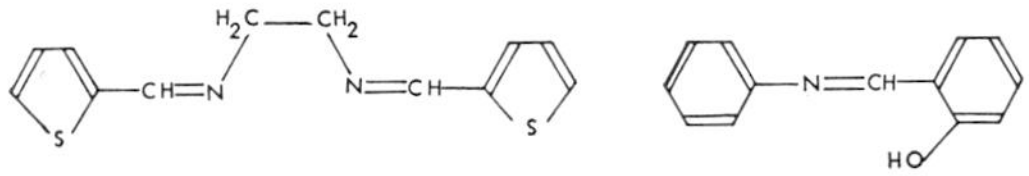

a   **b**

(63)

With salicylideneaniline it appears that the

metal complex (64) is formed.[23]

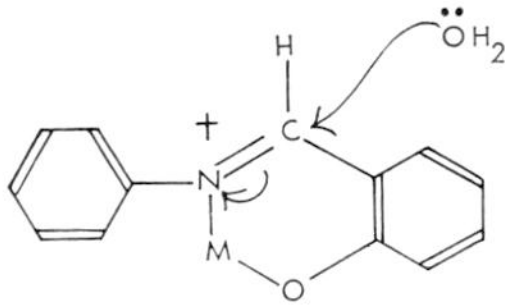

(64)

Polarization of the $-CH = N-$ linkage by the metal ion assists nucleophilic attack by a water molecule. In the case of the metal complexes of the Schiff's base obtained by condensation of salicyladehyde with glycine (65), the coordinate bonds involve donor groups from both constituents of the Schiff's base and the hydrolysis is retarded by metal ions due to the stabilizing influence of the chelate rings.

TABLE 4-VI

Some metal ion catalysed (or promoted) reactions in solution

Reaction	Example
Decarboxylation	$HO_2C\,CO\,CH_2\,CO_2H \longrightarrow HO_2C\,CO\,CH_3 + CO_2$
Schiff base hydrolysis	
Phosphate ester hydrolysis	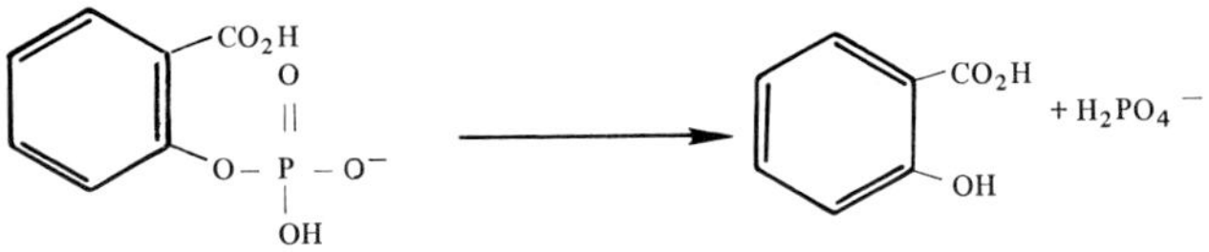 Salicyl phosphate
Carbonyl hydration	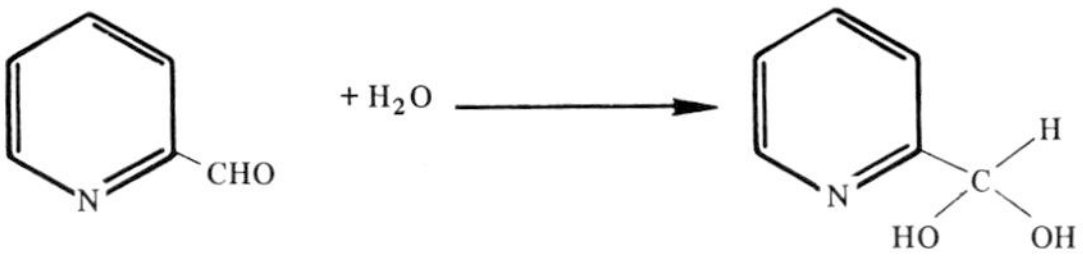
Glycoside hydrolysis	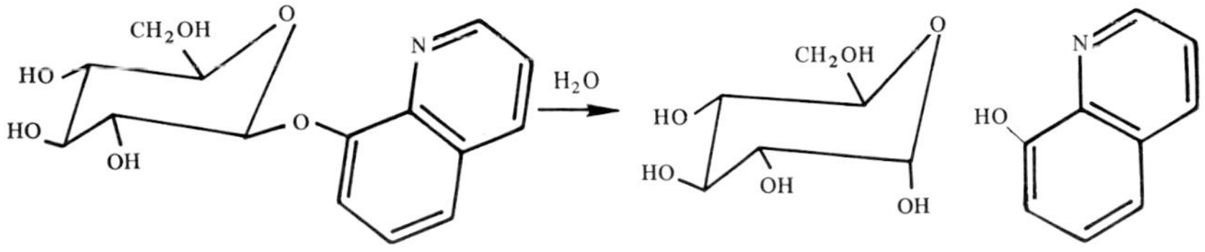

TABLE 4-VI Cont.

Hydrogen exchange

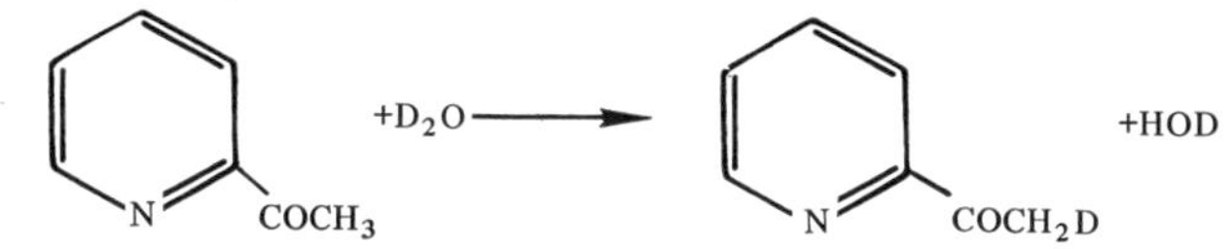

$+D_2O \longrightarrow$ $+HOD$

Sulphate ester hydrolysis

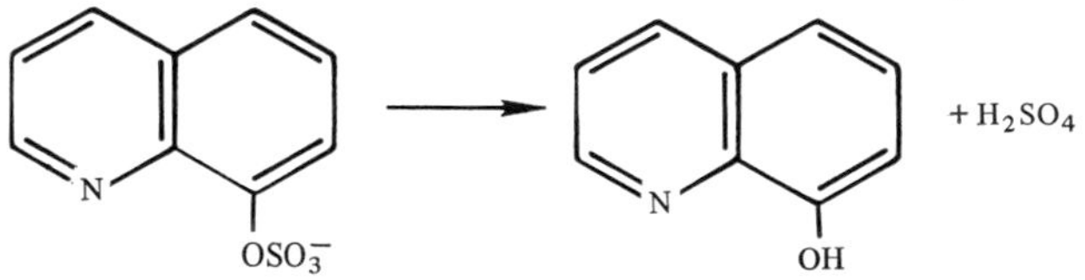

$\longrightarrow$ $+ H_2SO_4$

Carboxylic ester hydrolysis

$$NH_2CH_2CO_2R + OH^- \longrightarrow NH_2CH_2CO_2^- + ROH$$

Fluorophosphate hydrolysis

sarin $+H_2O \longrightarrow$ $+HF$

Nitrile hydrolysis

$+OH^- \longrightarrow$

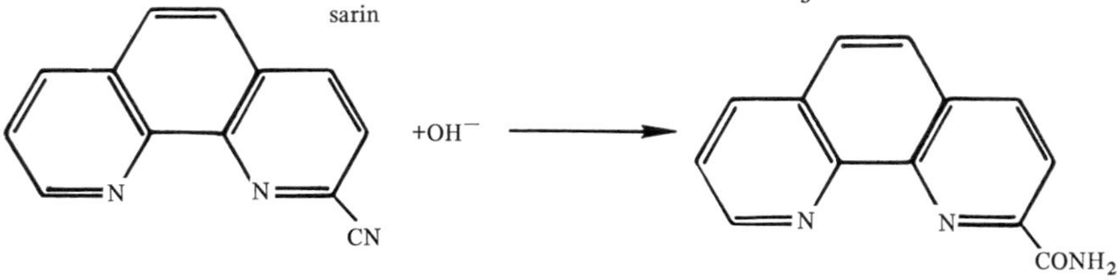

Amide hydrolysis

$$NH_2CH_2CONH_2 + OH^- \longrightarrow NH_2CH_2CO_2^- + NH_3$$

Peptide bond formation

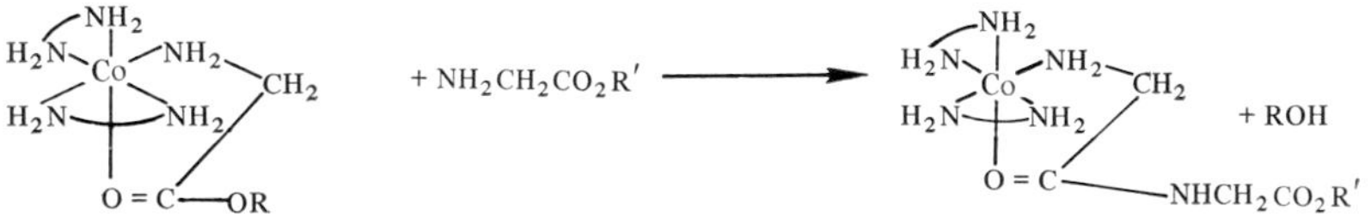

$+ NH_2CH_2CO_2R' \longrightarrow$ $+ ROH$

Peptide bond hydrolysis

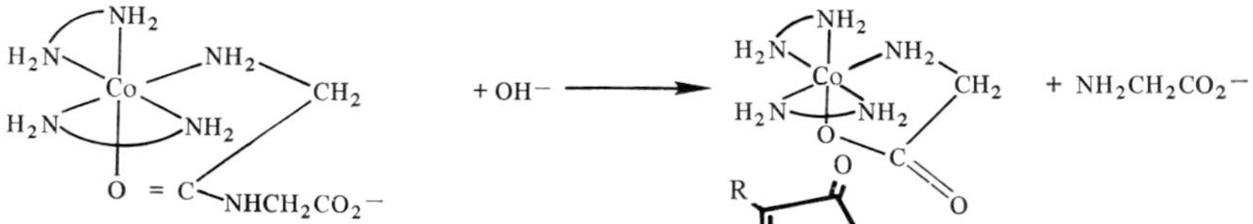

$+ OH^- \longrightarrow$ $+ NH_2CH_2CO_2^-$

Schiff base formation

$+ RCOCO_2^- \longrightarrow$

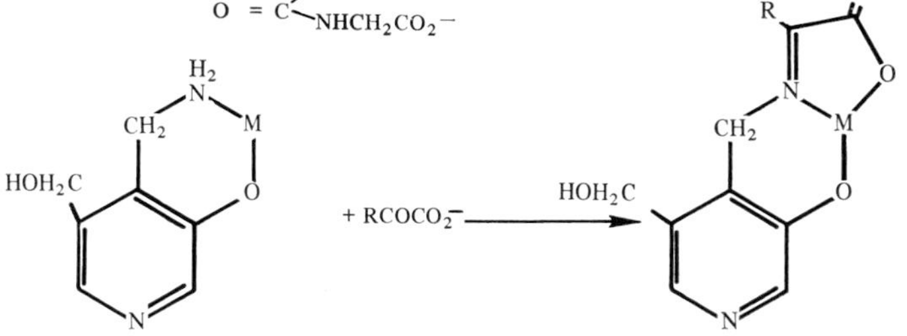

Transphosphorylation

$$ATP + Be^{II} + CH_3CO_2^- \longrightarrow ADP + CH_3COOPO_3^- + Be^{II}$$

Carboxylation

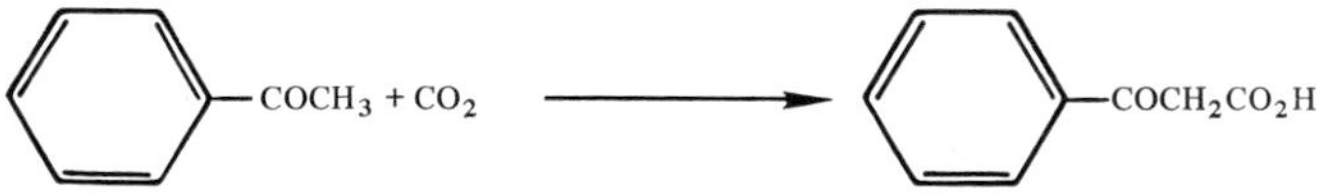

$-COCH_3 + CO_2 \longrightarrow$ $-COCH_2CO_2H$

Thiol ester hydrolysis

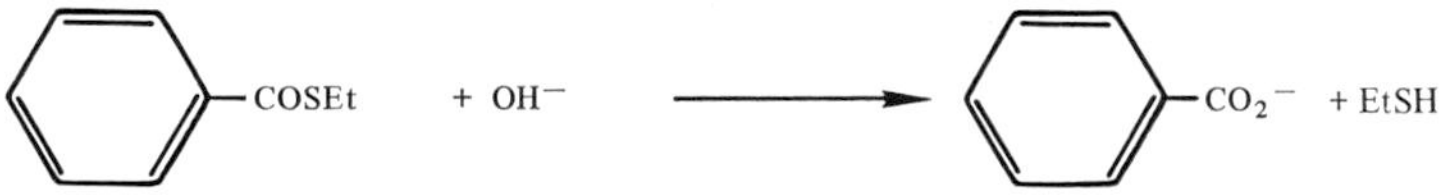

$-COSEt \quad + OH^- \longrightarrow$ $-CO_2^- + EtSH$

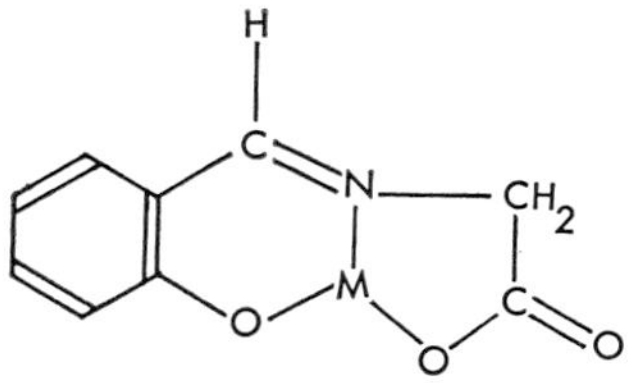

$$(65)$$

A rather similar type of reaction is the racemization of optically active amino acid esters when incorporated into metal complexes as their Schiff's base derivatives. Such reactions are similar to those of vitamin B_6 and were first studied by Pfeiffer and his collaborators.[24]

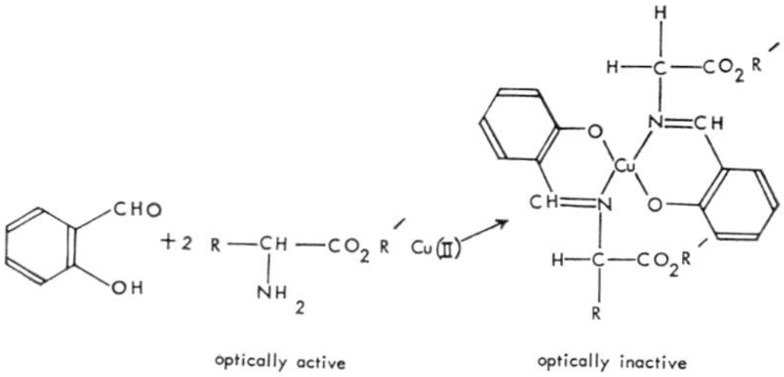

The formation of the Schiff's base also facilitates the oxidative deamination of amino acid esters and transesterification with alcohols in solution. The transesterification reactions, have been studied in detail.

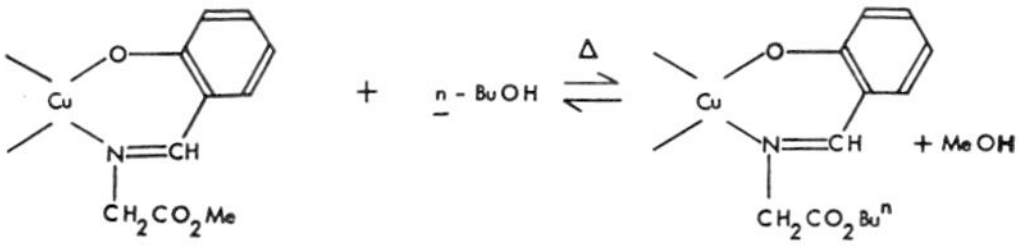

$$+ \quad n\text{-}BuOH \quad \xrightarrow{\Delta} \quad + \; MeOH \quad (67)$$

Vitamin B_6 is 2-methyl-3-hydroxy-4,5-(dihydroxymethyl)pyridine.

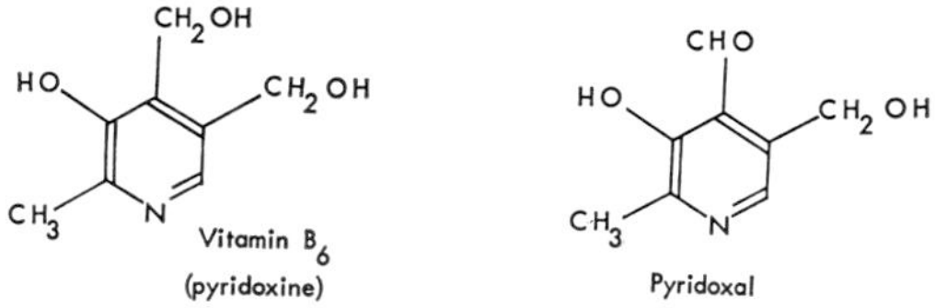

Vitamin B_6 (pyridoxine)

Pyridoxal

$$(68)$$

An example of a transamination reaction involving pyridoxal, an L-amino acid and copper(II), is the following,

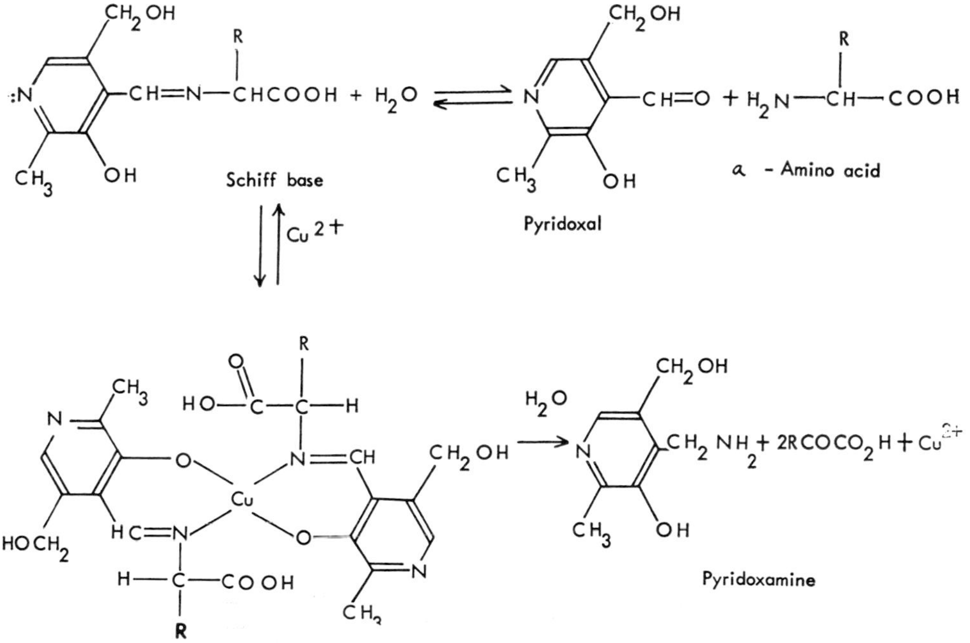

optically active   optically inactive

$$(66)$$

Schiff base

Pyridoxal

α – Amino acid

$$(69)$$

Pyridoxamine

The metal complexes of Schiff bases derived from pyridoxal provide a suitable intermediate for several types of reaction, depending upon the position of bond cleavage. The possible reactions can be understood from the structure of the intermediate

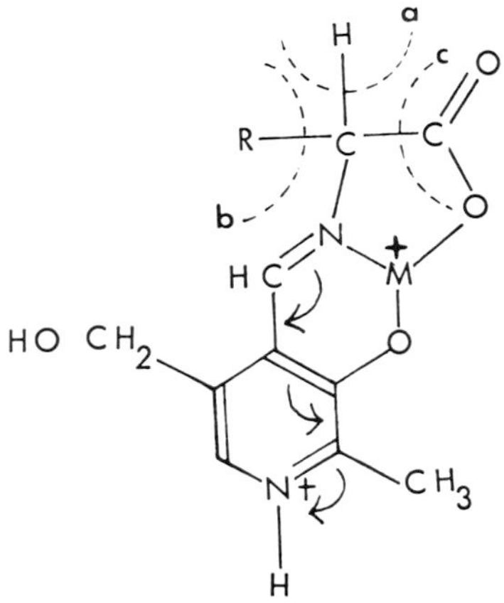

(70)

Cleavage at *a* leads to loss of a proton. Protonation of the intermediate gives a racemic amino acid which can be regenerated by hydrolysis of the Schiff base. An alternative pathway for racemization would involve proton loss at *a*, followed by protonation at the formyl carbon of pyridoxal to give the intermediate (71)

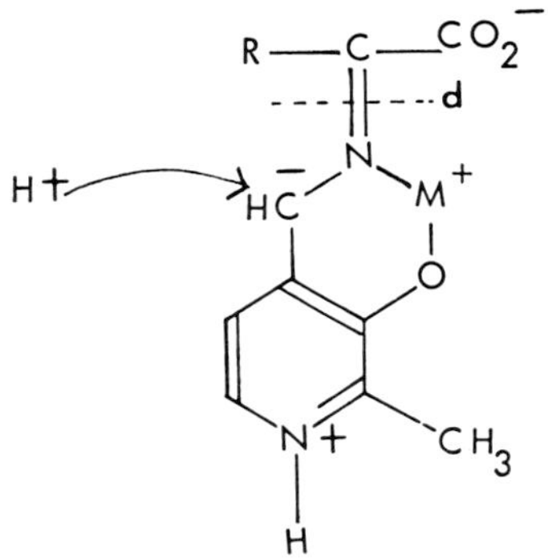

(71)

which can undergo hydrolysis at *d*. The products of this reaction are an α-keto acid and pyridoxamine.

Cleavage at *b* with loss of the carbonium ion R⁺ provides a route for the synthesis of glycine from other L-amino acids. Cleav-

age at *c* leads to decarboxylation.

Carbonyl hydration is a very interesting reaction thus, the enzyme carbonic anhydrase in addition to catalyzing the hydration of carbon dioxide also catalyzes the hydration of aldehydes,

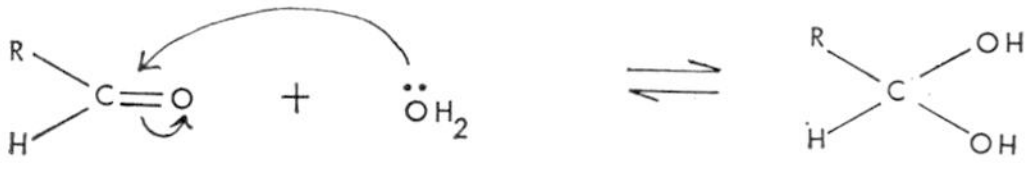

(72)

If a substrate can provide a further binding site for a metal ion, in addition to the carbonyl group, very substantial catalysis of carbonyl hydration occurs. Thus the hydration of pyridine-2-aldehyde[25] (73a) is subject to metal ion catalysis while that of pyridine-4-aldehyde (73b) is not so catalyzed, as in the latter compound chelate ring formation is not possible.

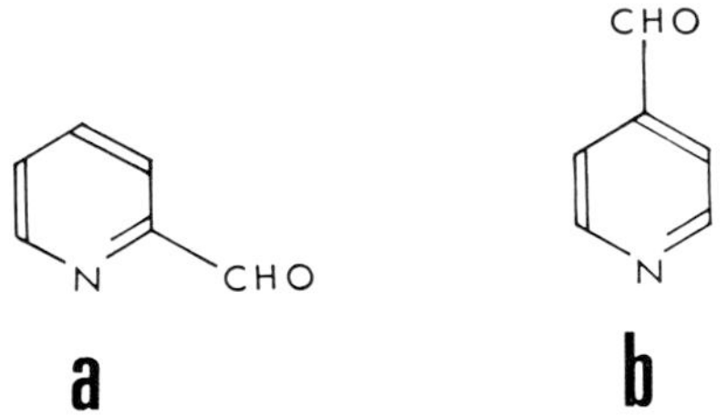

(73)

a **b**

The effectiveness of metal ions in catalyzing the reaction follows the order $Cu^{2+} > Co^{2+} > Ni^{2+} > Zn^{2+} > Cd^{2+} > Mn^{2+} > Mg^{2+} \simeq Ca^{2+}$, Table 4-VII. The hydration

TABLE 4-VII

CATALYTIC RATE CONSTANTS IN THE
REVERSIBLE HYDRATION OF
PYRIDINE-2-ALDEHYDE*

Catalyst	$k_{cat}/M^{-1}min^{-1}$	k_{cat}/k_{H2O}
H_2O	0.0043	1
Zn^{2+}	27,900	6.5×10^6
Co^{2+}	32,500	7.6×10^6
BCA	365,000	8.5×10^7
OH^-	2,600,000	6.0×10^8

BCA = Bovine carbonic anhydrase.
* Data from Y. Pocker and J. E. Meany, *J Am Chem Soc*, 89:631, 1967.

of acetaldehyde is also catalyzed by zinc(II) ions and catalysis is enhanced by certain anions.[26]

The hydrolysis of a large number of phosphate esters and acylphosphates is catalyzed by metal ions, e.g. acetyl phosphate (74a), 3-hydroxy-2-pyridylmethylphosphate (74b), salicylphosphate (74c), 8-quinolyl phosphate (74d) and 3-pyridylphosphate (74e)

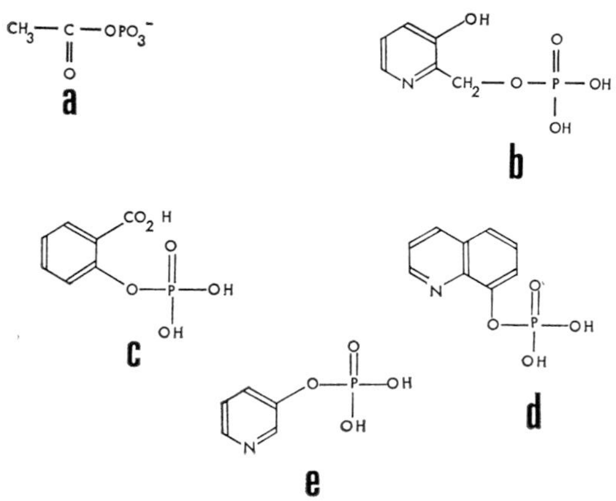

(74)

Murakami[27] has proposed that for catalytic activity to be observed by metal ions, the following requirements should be met, (1) the metal ion should form a chelate of significant stability with the organic phosphate prior to hydrolysis (2) the metal ion should interact with the ester oxygen in the transition state in such a way as to form a chelate ring involving the ester oxygen atom.

In view of the mechanistic similarities between phosphate and sulphate ester hydrolysis it would be expected that suitable sulphate esters would also be subject to metal ion catalysis, and it has been found that the hydrolysis of 8-quinolyl sulphate (75) is catalyzed by copper(II).[28]

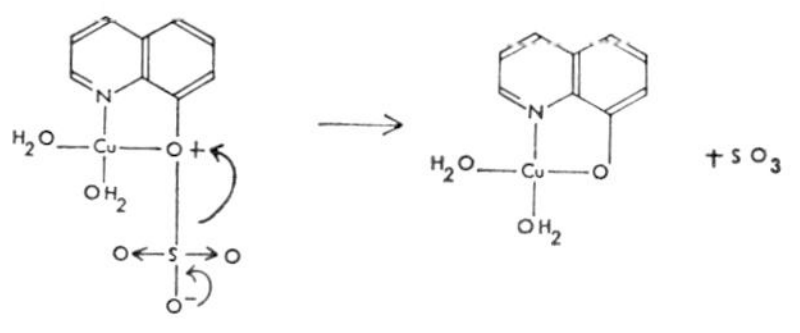

(75)

Polarization of the ester-oxygen atom by the copper(II) ion in the 5-membered chelate ring assists the transfer of electrons from the sulphur-oxygen bond undergoing cleavage.

In conclusion, it can be said that metal ions will catalyze all reactions which are subject to specific acid catalysis, with the proviso that the substrate must carry additional functional groups to allow chelate ring formation and a direct interaction between the metal ion and the labile bond. These ideas can be illustrated by a typical example.

The hydrolysis of glycosides is subject to general acid catalysis, the intermediate carbonium ion, and hence probably the transition state is stabilized by resonance with the ring oxygen,

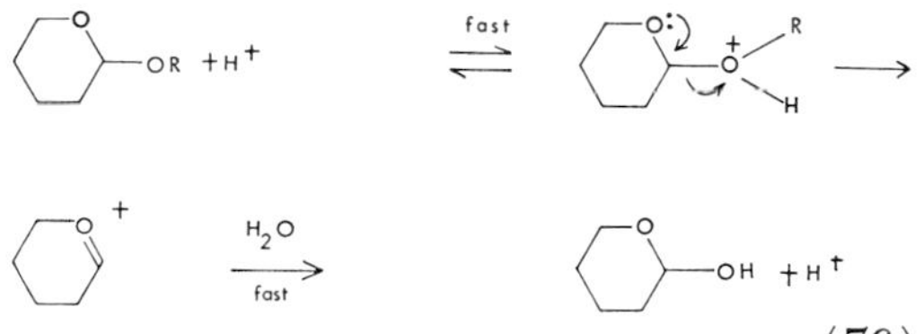

(76)

The hydrolysis of simple glycosides is not catalyzed by metal ions, however the hydrolysis of 8-hydroxyquinoline-β-D-glucopyranoside (77) is subject to metal ion catalysis by copper(II)[29]

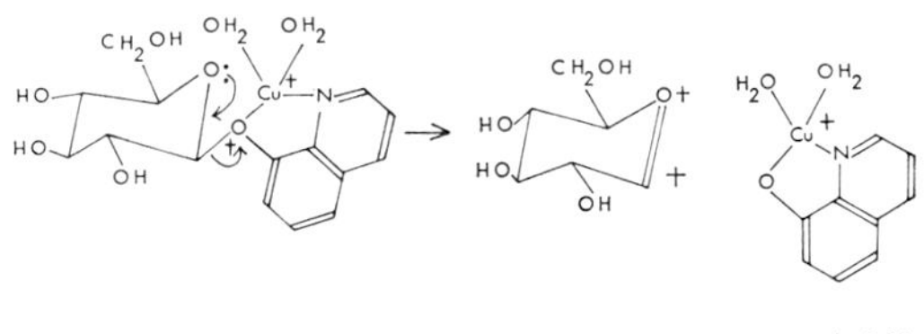

(77)

The aglycone 8-hydroxyquinoline provides a suitable binding site for the metal ion, so that it can interact with the glycosidic oxygen, the resulting polarization leads to more facile cleavage of the glycosidic bond. In the enzymatic reactions the apoenzyme provides the binding site for the metal ion so that only monodentate, rather than the polydentate interactions with the substrate are required.

In most of the metal ion catalyzed reactions which have been subjected to detailed study, it has been found that the rate-accelerations between the catalyzed and uncatalyzed reaction is due primarily to a more positive $\Delta S^{\ddagger}$. The enthalpies of activation ($\Delta H^{\ddagger}$) are often very similar to the values observed in the uncatalyzed reactions. In this connection, it is interesting to note the effect of an entropy change on the rate of reactions, Table 4-VIII. Although at first the entropy changes have a relatively small effect on the rate constant, the effect is exponential and becomes very large with larger entropy changes. Bimolecular reactions between uncharged reactants normally have $\Delta S^{\ddagger} = -10$ e.u., however, if the species carry a charge, as in the reaction $BrCH_2CO_2^- + S_2O_3^{2-} \rightarrow {}^-O_3S_2CH_2$ $CO_2^- + Br^-$ where $\Delta S^{\ddagger} = -28$ e.u., quite large negative entropies of activation are observed, leading to quite slow reactions. Hydroxide ions are involved in many bimolecular reactions of biological importance, and large negative entropies of activation occur with negatively charged substrates. If the substrate is complexed with a metal ion, more positive entropies of activation are possible leading to faster reactions.

TABLE 4-VIII

EFFECT OF ENTROPY OF ACTIVATION CHANGES ON RATE CONSTANTS AT A CONSTANT ENTHALPY OF ACTIVATION

$\Delta(\Delta S^{\ddagger})$ e.u.*	k/k_0
1	1.64
2	2.72
5	12.2
10	148
20	21,900
30	3,240,000

* 1 e.u. = 4.184 J. mol^{-1} K^{-1}.

In aqueous solution the electrostatic effect of charge on the reactants in a bimolecular reaction can be estimated from the equation, $\Delta(\Delta S^{\ddagger}) \sim -10Z_A B_B$ where Z_A and Z_B are the charges on the reactants. Thus $\Delta(\Delta S^{\ddagger}) = -10$ e.u. for a reaction between two negatively charged species, so that $\Delta S^{\ddagger} = -20$ e.u. since $\Delta S^{\ddagger}$ for a bimolecular reaction $= -10$ e.u.

REPRESENTATIVE ENZYMES

In this section, some of the representative enzymes in which the metal ion functions, as a Lewis acid catalyst are discussed.

Carboxypeptidase A

The enzyme carboxypeptidase A catalyses the hydrolysis of the C-terminal amino acid residue from a peptide or protein chain

$$\text{mwwww-NH—CH—CO} \mid \text{NH—CH—CO}_2^-$$

exopeptidase cleavage

(78)

Carboxypeptidase is an example of an *exopeptidase* which cleaves only terminal peptide bonds. *Endopeptidases* catalyze the hydrolysis of peptide bonds which are nonterminal. The carboxypeptidases cleave the C-(carboxyl)-terminal peptide bond in a substrate while the amino-peptidases catalyze the hydrolysis of the peptide bond associated with the amino-terminal residue.

The absolute requirements for carboxypeptidase A are that the C-terminal residue must have the L-configuration and that its carboxyl group must be free. Substrates in which the side chain R is aromatic are favored, although carboxypeptidase A has a fairly broad specificity in this respect; peptides possessing almost any C-terminal residue except proline will be hydrolyzed.

Not all the features of structure (78) are necessary in a substrate, and the terminal residue may be an α-hydroxy-acid. Benzoylglycyl-L-phenyl-lactic acid is a good substrate for the enzyme.

Carboxypeptidase A consists of a single polypeptide chain of 307 amino acid residues with a molecular weight of 34,600.[34] The amino acid sequence has now been completely established. The enzyme contains one zinc(II) ion per molecule which is essential for activity. The zinc(II) ion may be removed either by dialysis at low pH or by dialysis at neutral pH against buffers containing the chelating agent 1,10-phenanthroline. The loss of activity from samples treated in this manner exactly parallels the loss of zinc(II). The removal of zinc(II) appears to have little effect on the overall structure of the protein. The optical rotations of the native and metal free enzymes and their behavior on sedimentation are identical. Addition of zinc(II) to the apoenzyme restores the catalytic activity, the activity of the reconstituted enzyme being proportional to the amount of zinc added up to one atom per mole of protein.

Certain other transition metal ions may replace zinc(II) and regenerate peptidase activity in the apoenzyme, for example Fe(II), Mn(II), Co(II) and Ni(II). These metal ions occupy the site formerly occupied by the Zn(II) ion. The cobalt(II) enzyme has been reported to be a better catalyst in the hydrolysis of benzyloxycarbonylglycylphenylalanine than the zinc(II)-derivatives, Table 4-IX.

Lipscombe[30] has published electron density maps at 2.0Å resolution for carboxypeptidase A and a complex of the enzyme with the dipeptide glycyl-L-tyrosine (a poor substrate). The complete primary amino acid sequence was reported in 1969

TABLE 4-IX

RELATIVE APPARENT ACTIVITIES OF SOME METALLOCARBOXYPEPTIDASES*

Metal	Peptidase Activity†	Esterase Activity‡
apo (no metal)	0	0
Zn	1.0	1.0
Co	1.6	0.95
Ni	1.1	0.87
Mn	0.08	0.35
Cd	0	1.50
Hg	0	1.16
Pb	0	0.52
Ca	0	0

* From data of Coleman and Vallee. *J Biol Chem*, 236:2244, 1961.

† Based on rate assays using the dipeptide substrate, carbobenzylxyglycyl-L-phenylalanine (CGP). Carried out at 0°C at pH 7.5 in a 20mM EDTA buffer containing 0.1M NaCl. Initial substrate concentration 20mM.

‡ Based on rate assays using the ester substrate, hippuryl-DL-β-phenyllactate (HPLA). Carried out at 25°C at pH 7.5 in a 5mM tris buffer containing 0.1M NaCl. Initial substrate concentration 10mM.

by Neurath et al.[31] Carboxypeptidase A thus became the first metalloenzyme for which the high resolution structure and sequence were known.

From the X-ray and chemical sequence studies, it is known that there are three amino acid ligands in the enzyme which bind the metal ion, His-69, Glu-72 and His-169 in addition to a water ligand, Figure 4-13. The most important structural information from the mechanistic point of view was obtained from the 2.0Å resolution X-ray investigation of the carboxypeptidase A glycyl-L-tyrosine complex. This study showed that the only parts of carboxypeptidase A which are sufficiently close to the peptide bond to be involved directly in catalysis are the zinc(II) ion, Glu-270 and Tyr-248. Significantly the carbonyl group of the peptide bond has displaced the water ligand from the zinc ion

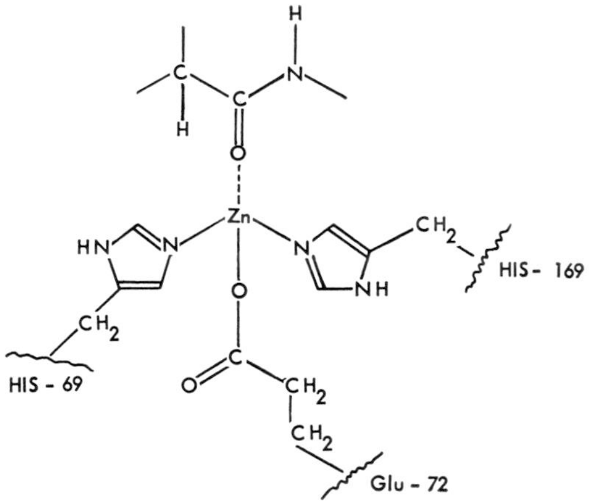

Figure 4-13. The coordination of the Zinc(II) ion at the active site of carboxypeptidase A.

The mechanism proposed for the action of carboxypeptidase A on esters and peptides is shown in Figure 4-14.

According to this mechanism, the carboxylate group of Glu-270 acts as a nucleophile, attacking the carbonyl group of the peptide. The zinc ion serves to orientate and polarize the carbonyl group of the peptide so making it more susceptible to attack by a nucleophile. The tetrahedral intermediate so formed breaks down to give an anhydride species with concomitant formation of the amino acid from the C-terminal residue of the peptide. Finally, the anhydride hydrolyses, regenerating the free

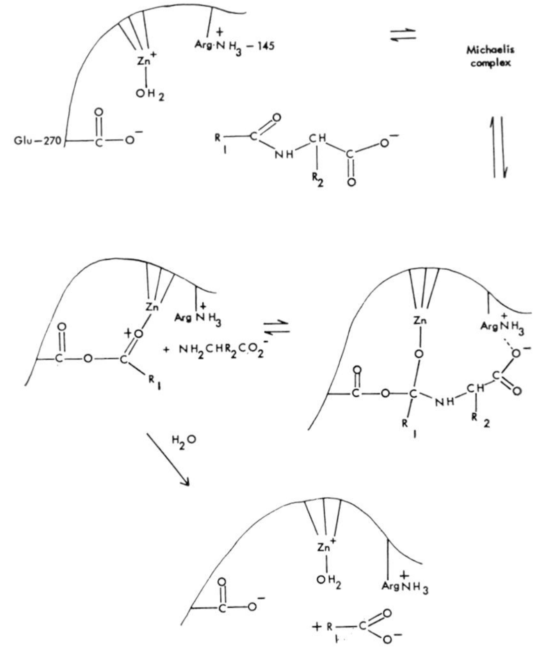

in the complex. The only other group of the protein within 3Å of a functional group of the substrate is Arg-145 which binds the terminal carboxyl group present in the peptide.

The enzyme is rugby-football-shaped, $45 \times 45 \times 55$Å in diameter, with a depression at one end and a groove running around the side of the molecule from this depression. The depression is the active site and the polypeptide substrate binds in the groove with the carboxyl end fitting into the active site.

Clearly, one gets only a static picture of an enzyme or enzyme-substrate complex from structural determinations of the type described. It is dangerous to draw mechanistic conclusions based on reaction dynamics on such a static picture alone. For example, glycyl-L-tyrosine is a very poor substrate for carboxypeptidase A, and the enzyme-substrate complex observed by X-ray diffraction is likely to be nonproductive. However, it is believed that the arrangement of the zinc ion, Glu-270, Tyr-248 and Arg-145 in the complexes of carboxypeptidase with reactive peptide substrates is essentially the same as in the case of glycyl-L-tyrosine.

Figure 4-14. Proposed mechanism for the action of carboxypeptidase A as an esterase and a peptidase. The NH group in a peptide is replaced by an O atom in an ester. After Kaiser, E. T. and Kaiser, B. L.: *Acc Chem Res,* 5:219, 1972. Glu-270 acts as a nucleophilic catalyst giving rise to an anhydride intermediate which undergoes subsequent hydrolysis. Alternative mechanisms in which Glu-270 acts as a general base catalyst are possible.

enzyme. It should be stressed that such mechanisms are only working hypotheses at the present time.

Carbonic Anhydrase

Carbonic anhydrase is a hydrolytic enzyme which catalyzes the hydration and hydrolysis of many carbonyl derivatives. A number of slightly different enzymes, carbonic anhydrase A, B, and C occur in different organisms. The enzyme catalyses a reaction of great physiological importance, the hydration of carbon dioxide or the reverse reaction, the dehydration of carbonate,

$$CO_2 + H_2O \rightleftharpoons HCO_3^- + H^+$$
$$CO_2 + OH^- \rightleftharpoons HCO_3^-$$

In addition, the enzyme also catalyzes the hydration of many aldehydes

$$CH_3CHO + H_2O \rightleftharpoons CH_3CH(OH)_2$$

the hydrolysis of esters

$$CH_3COOAr + OH^- \rightarrow$$
$$CH_3CO_2H + ArOH$$

and the hydrolysis of certain sultones and of 2,4-dinitrofluorobenzene.

The turnover number (under saturation conditions) for CO_2 hydration is *ca* $10^6 s^{-1}$. The uncatalyzed hydration rate is $7.0 \times 10^{-4} s^{-1}$, so that the rate acceleration with carbonic anhydrase is *ca* 10^9. The second order rate constant for the hydroxide reaction is 8×10^3 $M^{-1}s^{-1}$ at $25°$. The enzyme has a molecular weight of *ca* 30,000 and its amino acid composition has been reported by Nyman and Lindskog.[32]

In 1972 a Swedish group[33] reported the three-dimensional crystal structure of carbonic anhydrase C to 2.0Å resolution. Chemical work has shown that carbonic anhydrase contains one zinc ion per molecule, the removal of which leads to loss of enzymic activity. The X-ray studies have established that the molecule is roughly ellipsoidal, $40 \times 45 \times 55$Å and that the zinc(II) ion lies near the center of the molecule in a cleft. The zinc ion is ligated by three histidine residues (his-117, his-93 and his-95) in approximately tetrahedral stereochemistry, with the fourth site open to the surrounding medium. The active site region is relatively hydrophobic. The enzyme is inhibited by sulphonamides and the X-ray data has shown that sulphonamides project into the cleft and bind onto, or close to, the metal ion.

Lindskog[33] first succeeded in removing the very tightly bound zinc under conditions mild enough to avoid denaturation. The equilibrium

$$\text{Enzyme-Zn} + \text{n}(o\text{-phen}) + \text{H}^+ \rightleftharpoons$$
$$\text{Enzyme-H} + \text{Zn}(o\text{-phen})_n^{2+}$$

was driven to the right by removing the $Zn - (o\text{-phen})^{2+}$ complex by dialysis. The apoenzyme was almost completely catalytically inactive, and most of the reactivity was restored by the addition of zinc(II) in the molar ration 1:1. Sedimentation and ord data have shown that the apoenzyme and the native enzyme have the same gross tertiary structure, suggesting that the function of the zinc is not to stabilize the tertiary structure but to take part in the catalytic activity of the enzyme.

As discussed previously, new derivatives of carbonic anhydrase can be prepared by adding metal ions to the apoenzyme. Only the Zn(II) and Co(II) derivatives show marked catalytic activity.

Complex formation often involves the displacement of protons from basic donors if complexation occurs at pH values below the pK of these groups, e.g. $R - N^+H_3 + M^{2+} \rightleftharpoons RNH_2 - M^{2+} + H^+$. Lindskog has shown that a quite different situation occurs with carbonic anhydrase, thus at pH6 (CA = carbonic anhydrase)

$$Zn^{2+} + \text{apoCA} \rightarrow \text{ZnCA} + (1.3 \pm 0.15)H^+$$

while at pH9

$$Zn^{2+} + \text{apoCA} \rightarrow \text{ZnCA} + (2.2 \pm 0.1)H^+$$

The zinc(II) ion is, therefore, inducing a further ionization on binding. Such an ionization should effect the ligand field round the zinc ion as the pH increases from 6 to 9. The effect cannot be observed in the d^{10} ion, but it should be seen in the d-d spectrum of the cobalt derivative. The visible spectrum of cobalt carbonic anhydrase is indeed pH dependent in the pH range 6 to 9, Figure 4-8. The two tight isosbestic points provide strong evidence for only two species in equilibrium related by a pK_a of *ca* 7.1.

It seems very likely that the pK_a of 7 is due to ionization of a water molecule bound to zinc. The evidence that CO_2 is bound to close to, but not on to zinc, suggests a mechanistic scheme of the type.

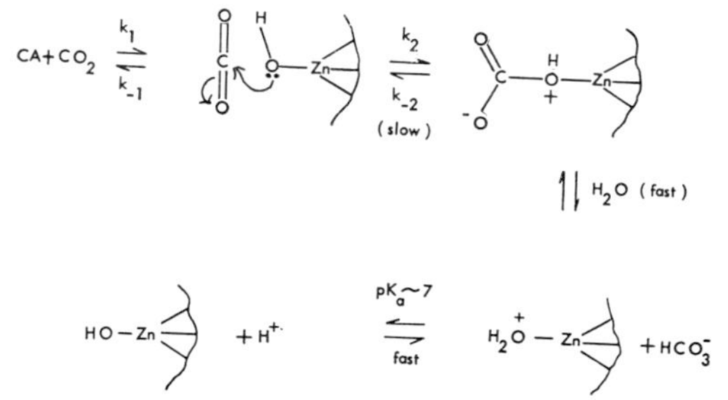

$$(79)$$

For the alkaline form of bovine carbonic anhydrase A the rate constant k_2 for the hydration of carbon dioxide is $8 \times 10^5 s^{-1}$ and $K_M = (k_{-1} + k_2)/k_1 = 1.24 \times 10^{-2}$ M^{-1}. Since k_2 is pH dependent while K_M is not, it is reasonable to assume that $k_{-1} >> k_2$. Assuming a lower limit for $k_{-1} = 10k_2$, the lower limit for k_1 is ca 6×10^8 M^{-1}s^{-1}. This rate constant is close to the diffusion-controlled limit, and therefore requires a fairly open and accessible active site, which is consistent with the observed binding of larger aromatic substrates and inhibitors.

The pK_a for the equilibrium,

$$[Zn(H_2O)_6]^{2+} \rightleftharpoons [Zn(H_2O)_5(OH)]^+ + H^+$$

is 10.5, while the value for carbonic anhydrase hydrase is 7.1, a substantial change. One apparent difficulty with the mechanistic scheme suggested for carbonic anhydrase is that the zinc hydroxo complex loses hydroxide to the substrate and regains water, which then ionizes. In the CO_2 hydration reaction (turnover $10^6 s^{-1}$) the zinc bound water must ionize at least as fast. However, if the pK_a of the coordinated water molecule is 7.0 and the maximum diffusion controlled protonation rate is 10^{10} M^{-1}s^{-1}, then the predicted maximum deprotonation rate is $10^{10}/10^7 = 10^3 s^{-1}$.

In addition to the "zinc-hydroxide" mechanism discussed above, a "zinc-carbonyl" mechanism could also be considered. Normally metal ions are envisaged as coordinated to the oxygen atom of the carbonyl group, e.g. in carboxypeptidase A. Such a mechanism could be written

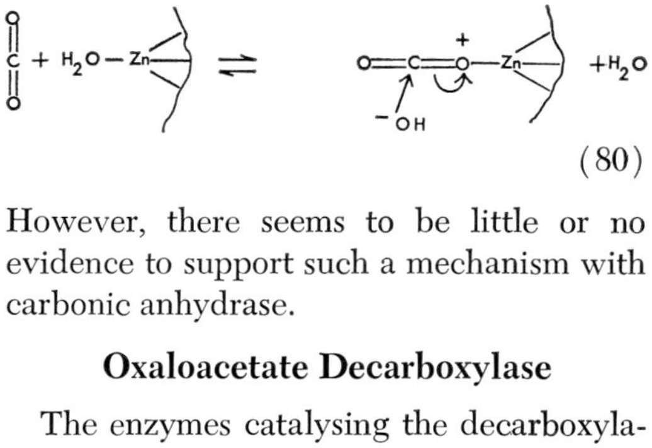

$$(80)$$

However, there seems to be little or no evidence to support such a mechanism with carbonic anhydrase.

Oxaloacetate Decarboxylase

The enzymes catalysing the decarboxylation of β-oxo-acids (β-keto-acids) are of two types, (a) the "active lysine" enzymes such as acetoacetate decarboxylase which are not dependent on metal ions for activity and (b) the metal dependent enzymes such as oxaloacetate decarboxylase. The intermediate in the active lysine enzymes is a Schiff's base formed by condensation of the ϵ-amino group of lysine ($NH_2CH_2CH_2$ $CH_2CH_2CH(NH_2)CO_2H$) with the carbonyl group, of the keto acid. The protonated Schiff's base acts as an "electron sink" so assisting the transfer of electrons from the carbon-carbon bond undergoing cleavage,

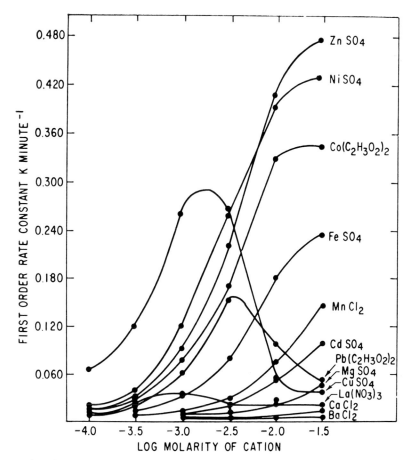

$$(81)$$

Sodium borohydride inactivates acetoacetate decarboxylase in the presence of acetoacetate, but not in its absence. Hydrolysis of the inactivated enzyme gives N-ε-isopropyl-lysine as a component of the hydrolysate. The oxaloacetate decarboxylases from *Microdoccus lysodeikticus*, from cod and pigeon liver require metal ions for activation, Figure 4-15.

The oxaloacetate decarboxylase from cod has been crystallized by Kosicki[21] and has been shown to depend on manganese(II) for activity. The enzyme catalyses the decarboxylation of oxaloacetate to pyruvate, a reaction important in the Krebs cycle,

$$^-O_2CCOCH_2CO_2^- \rightarrow {}^-O_2CCOCH_3 + CO_2$$

Sodium borohydride has no effect on the activity of the enzyme regardless as to the

Figure 4-15. Effect of metal ions on the enzymatic decarboxylation of oxaloacetic acid. After Speck, J. F.: *J Biol Chem, 176:*997, 1948.

presence or absence of the substrate or of the product (pyruvate), however, the reduction of pyruvate by borohydride in the presence of manganese(II) and the enzyme leads to the formation of an excess of D-(-)-lactate,[35]

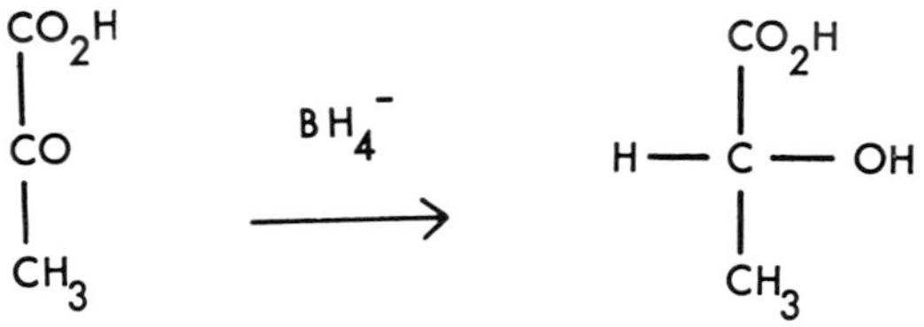

D-(-)-Lactic Acid

(82)

The formation of an excess of D-(-)-lactic acid is consistent with the view that a mixed ligand complex of the enzyme and oxaloacetate with manganese(II) is involved in both reduction and decarboxylation. As the enzyme is not activated by borohydride in the presence of substrate, a Schiff's base mechanism such as occurs with acetoacetate decarboxylase, is excluded, and a possible mechanism due to Kosicki and Westheimer[35] is shown in (83), tracing the enzymatic decarboxylation of oxaloacetic acid.

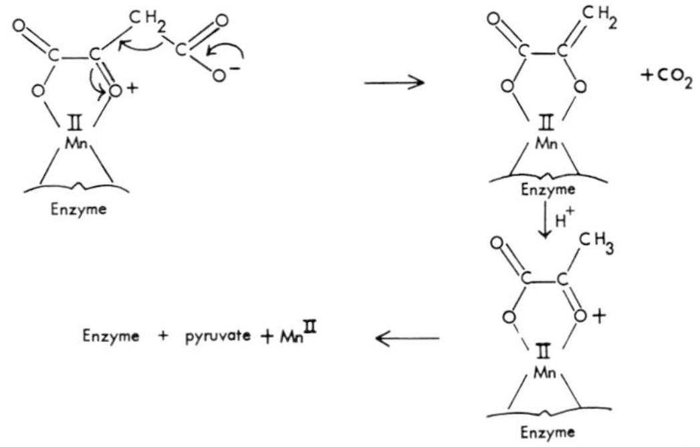

(83)

Since a metal ion is required for the enzyme catalyzed borohydride reduction, pyruvate is presumably bound for the reduction as it is as a product of the decarboxylation, as seen in (84). Formation of optically active lactic acid confirms that the reduction is enzymatic. The proposed mechanism for the enzymatic reaction is similar to that suggested for the metal ion catalyzed reaction. It has been observed[36] that the manganese-αα-dipyridyl complex is some sixteen times more active as a catalyst in the nonenzymatic reaction than the aquo-manganese ion $Mn(H_2O)_6^{2+}$. It has been suggested that a mixed ligand complex (as is 85a) is formed, the enhanced activity of the mixed ligand complex being due

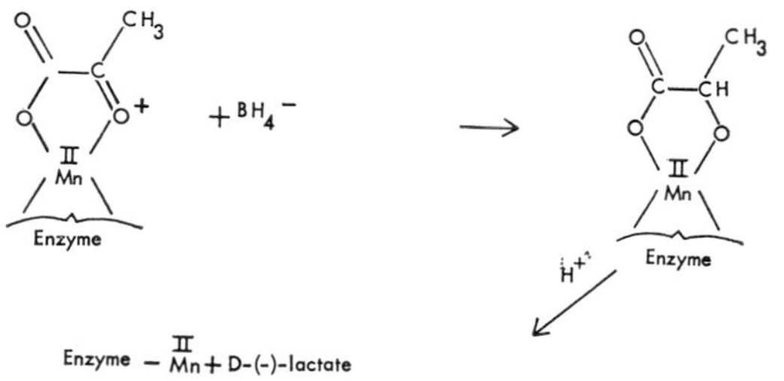

(84)

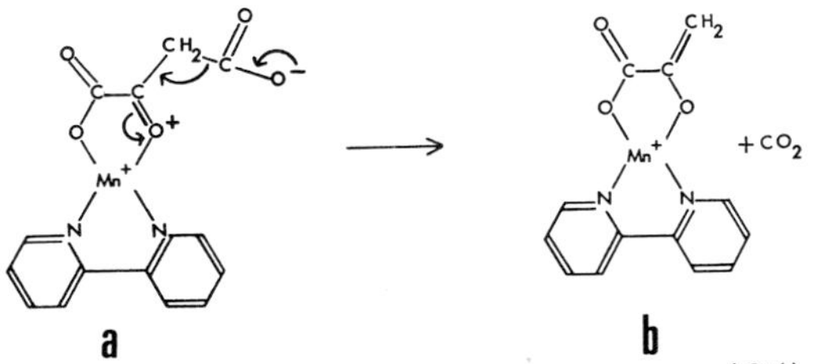

a   b

(85)

to d_π-p_π bonding, in which there is back-donation of electron density from the filled d-orbitals of the metal ion into the antibonding orbitals of the unsaturated ligand. The effect of such back-donating would be to increase the effective positive charge on the manganese and so increase its catalytic activity.

Alkaline Phosphatase

Alkaline phosphatase is a zinc containing enzyme obtained from *E. coli* with a molecular weight of about 86,000, which catalyzes the hydrolysis of a wide range of phosphate monoesters. Its designation as *alkaline* phosphatase arises from the fact that its optimum pH is near 8 in contrast to another type of phosphomonoesterase *(acid phosphatase)* which functions optimally at lower pH values.

It has been established that alkaline phosphatase from various sources (mammalian and bacterial) acts by a two-step process involving the intermediate formation of an enzyme-phosphate,

strate complete via nucleophilic attack by the hydroxyl group of serine (*cf.* Chymotrypsin). The metal ion is envisaged as providing a binding site for the dianionic phosphate species, and in addition, the formation of the metal ion-phosphate complex favors nucleophilic attack on the phosphorous atom with concomitant P-O cleavage.

$$\text{Enzyme} + \text{HO}-\underset{\underset{O^-}{|}}{\overset{\overset{O}{\|}}{P}}-\text{OR} \quad \underset{\text{Hydrolysis}}{\overset{\text{Phosphorylation}}{\rightleftharpoons}} \quad \begin{array}{c}\text{Enzyme - phosphate + Et O H} \\ \text{("phosphoryl enzyme")}\end{array} \tag{86}$$

$$\text{Enzyme - phosphate} + \text{H}_2\text{O} \quad \rightleftharpoons \quad \text{Enzyme} + \text{HPO}_4^{=}$$

Hydrolysis of a phosphate monoester is the presence of alkaline phosphatase takes place with P-O cleavage rather than C-O cleavage

$$\text{HO}-\underset{\underset{O^-}{|}}{\overset{\overset{O}{\|}}{P}}\overbrace{-O-}^{P-O}\underbrace{CH_2R}_{C-O} \tag{87}$$

Complete hydrolysis of the phosporyl-enzyme gives O-phophorylserine, and the amino acid sequence around the "active site" serine has been shown to be Thr, Asp, Ser, Ala, Ala.

Schwartz[37] has suggested a possible mechanism for alkaline phosphatase based on the available information, as seen in (88). (Alkaline phosphatase from *E coli* is a zinc(II) metalloenzyme.)

Hydrolysis occurs in the enzyme-sub-

METALLOENZYMES IN REDOX PROCESSES

The metal ions involved in biological oxidations are primarily iron and copper, but cobalt and molybdenum are also important. In these systems the catalytic activity is associated with a change in the oxidation state of the metal ion. It is a characteristic property of the transition metals that they have several stable oxidation states, thus copper(II) is a very effective catalyst for one-equivalent redox reactions as it is easily reduced to copper(I). A purely inorganic example of copper(II) catalysis is seen in the reaction

$$\text{Fe}^{III} + \text{V}^{III} \rightarrow \text{Fe}^{II} + \text{V}^{IV}$$

If copper(II) is added to the reagents at a concentration roughly equal to that of the reactants, the reaction rate is increased. The mechanism involves the intermediate

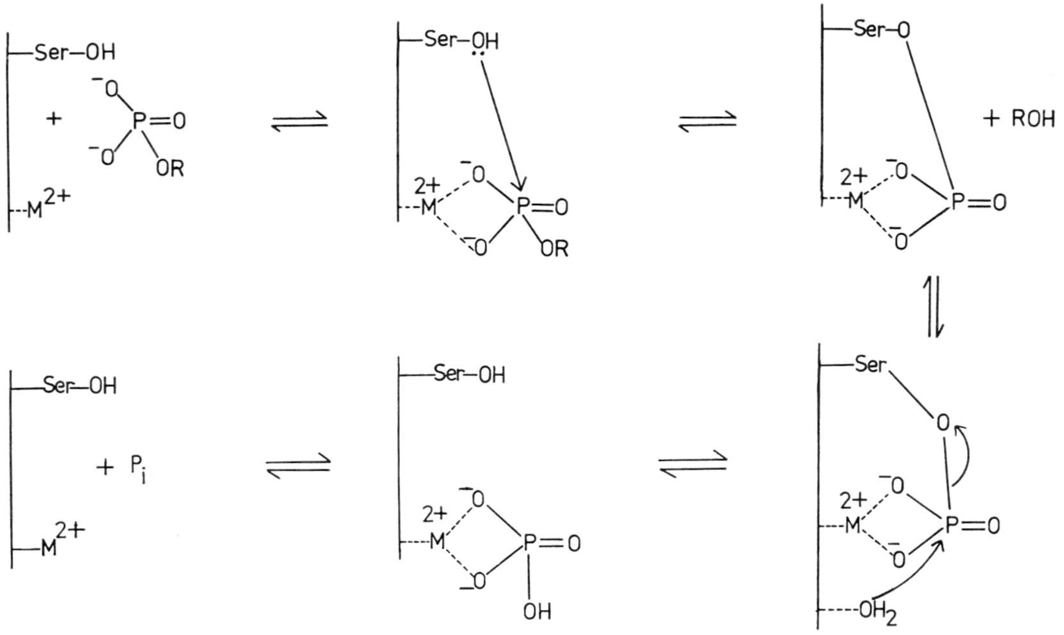

—Ser—OH

$+$

$P=O$, OR

$\rightleftharpoons$

—Ser—OH

$2+$ M --- P=O, OR

$\rightleftharpoons$

—Ser—O

$+$ ROH

$2+$ M --- P=O

$\updownarrow$

—Ser—OH

$+ P_i$

$\rightleftharpoons$

—Ser—OH

$2+$ M --- P=O, OH

$\rightleftharpoons$

—Ser

$2+$ M --- P=O

---OH$_2$

Proc Nat Acad Sci, 49:871, 1963.　　　　　　　　　　　　　　　　(88)

copper(I) which is formed in the slow rate-determining step, followed by the rapid oxidation of copper(I) by iron(III).

$$V^{III} + Cu^{II} \xrightarrow{slow} V^{IV} + Cu^I$$
$$Cu^I + Fe^{III} \xrightarrow{fast} Cu^{II} + Fe^{II}$$

If given oxidizing and reducing agents have the correct redox potentials (or standard free energies) for reaction to occur, i.e. the reactions are thermodynamically feasible, the reactions may still be slow due to kinetic factors. Such a situation frequently occurs with organic reducing agents. In these cases a metal ion having a number of different oxidation states may greatly accelerate the reaction by providing a different reaction pathway.

Copper and iron as well as other metals having two or more stable oxidation states, can "transport" electrons in redox reactions by a chain mechanism in which the catalytic metal is successively oxidized and reduced. It should also be noted that the non-transition elements zinc and magnesium are also associated with certain redox systems, however, their function is obviously different from that of the transition metals.

Most of the metalloenzymes are concerned with the oxidation of organic substrates with oxygen as the final electron acceptor being reduced to water or hydrogen peroxide. Biological oxidations may occur by a number of pathways. Thus if S = oxidized substrate and SH$_2$ = reduced substrate, the following reactions can occur

(a) electron transfer with oxygen (or another species) as the terminal electron acceptor

(b) hydrogen atom abstraction

$$2SH_2 + O_2 \rightarrow 2S + 2H_2O$$

or

$$SH_2 + O_2 \rightarrow S + H_2O_2$$

(c) hydride ion transfer

(d) oxygen atom incorporation, and

(e) hydroxyl group incorporation.

The oxygenases are enzymes catalyzing the incorporation of both atoms of molecular oxygen directly into the substrate,

$$(89)$$

The hydroxylases are similar to the oxygenases except that only one atom of molecular oxygen is introduced into the substrate. The other oxygen atom is incorporated into water by reaction with a reduced acceptor which therefore acts as a cosubstrate, i.e.

$$(90)$$

The reactions listed under (b) represent the overall stoichiometry of the process. In practice, the reaction may proceed by a series of intermediate carriers with oxygen as the terminal hydrogen acceptor, i.e.

$$(91)$$

A further important example of electron transfer is that of linked reactions. Here, one carrier intervenes between two separate dehydrogenases (if the acceptor is not molecular oxygen, the enzymes are known as dehydrogenases) using two different substrates SH_2 and $S'H_2$.

$$(92)$$

In this case, instead of the reduced carrier being oxidized by a second carrier, it is reoxidized by a molecule of a second substrate.

Many metalloprotein systems are often involved in these electron transfer chains, usually being arranged in sequence according to the values of their redox potentials. Thus cytochromes, nonheme iron proteins and copper proteins may intervene between flavin coenzyme and oxygen.

The value of the redox potential for any transition metal couple in a metalloenzyme system will determine the role it may play in electron transport or, for example, whether a metalloprotein will carry oxygen or be oxidized by it. In a brief discussion of this type, it is obviously impossible to deal with these systems in any detail, and thus a brief account of some copper metalloenzymes will only be given.

Copper

The relative thermodynamic stabilities of the copper(I) and copper(II) oxidation states are shown by the following electrode potentials

$$Cu^+ + e \rightarrow Cu \qquad E^\circ = 0.52v$$
$$Cu^{2+} + e \rightarrow Cu^+ \qquad E^\circ = 0.153v$$

As a result an oxidizing agent strong enough to oxidize copper to copper(I), is more than strong enough to oxidize copper(I) to copper(II).

Put alternatively, copper(I) in aqueous solution disproportionates to copper(II) and metallic copper.

$$2Cu^+ \rightleftharpoons Cu + Cu^{2+} \qquad E^\circ = 0.37v$$

and $K = \dfrac{[Cu^{2+}] \, [Cu]}{[Cu^+]^2} \sim 10^6$

The relative thermodynamic stabilities of Cu^I and Cu^{II} in aqueous solution is strongly dependent on the nature of the ligands present. In aqueous solution only low

TABLE 4-X

ELECTRODE POTENTIALS
$Cu^{2+} + e \rightarrow Cu^+$

System	$E_o{}'$ (volts)
Cu^{2+}/Cu^+	0.16
$Cu(imidazole)^{2+}$—$Cu(imidazole)^+$	0.255
$Cu(imidazole)^{2+}$—$Cu(imidazole)_2{}^+$	0.345
$Cu(glycine)_2{}^{2+}$—$Cu(glycine)_2{}^+$	−0.160
$Cu(alanine)_2{}^{2+}$—$Cu(alanine)_2{}^+$	−0.130
Copper blue proteins	0.4
Ceruloplasmin	0.39
$CuL_2{}^{2+}$—$CuL_2{}^+$ (L = 2-methylthioethylamine)	0.243
$Cu(pyridine)_2{}^{2+}$—$Cu(pyridine)_2{}^+$	0.270
$Cu(en)_2{}^{2+}$—$Cu(en)_2{}^+$	−0.38
Laccase. Cu^{2+}—Cu^+	0.415

equilibrium concentrations of copper(I) can exist.

Copper(I) is much softer than copper(II) thus sulphur donor ligands are bound more strongly by copper(I) than copper(II), as are all unsaturated ligands such as o-phenanthroline and 2,2′bipyridyl. The data in Table 4-X shows the effect on the electrode potentials of replacing aquo ligands. The replacement of one and then two water molecules by nitrogen donors results in a more positive potential. The aliphatic α-amino-acids and ethylenediamine results in negative electrode potentials. As $\Delta G^o = -nFE$, the negative values give a positive free energy change for the reduction which is therefore not favored. Such ligands prefer to bind to copper(II) and will in fact, cause disproportionation of copper(I) to copper metal and copper(II).

Copper(I) and copper(II) have very different stereochemistries. Copper(I) is often linear, two coordinate, but may increase its coordination number to four in which case a tetrahedral stereochemistry is preferred. On the other hand, copper(II) prefers a tetragonally distorted octahedral structure. One important principle that affects electron transfer reactions is the Franck-Condon principle which states that

there must be no movement of nuclei during the time of the electronic transition. As a result, the geometry of the two species after reaction must be identical to that existing before the actual electron transfer occurred.

The copper enzymes particularly involved in electron transfer reactions are the blue copper proteins where the copper is in the copper(II) state. One of the most unusual features of the blue copper proteins is the intensity of the blue color. The extinction coefficients on a protein basis for the band in the 600 nm region vary from 3,500 M^{-1} cm^{-1} in azurin to 11,300 M^{-1} cm^{-1} for ceruloplasmin. However, ceruloplasmin contains two "blue" copper(II) ions per mole, hence the extinction coefficient per gram atom of "blue" copper(II) is 5,600 M^{-1} cm^{-1}.

Extinction coefficients due to d-d transitions in tetragonal copper(II) are generally less than 100 M^{-1} cm^{-1}. The high values in the blue copper proteins have been attributed to distorted metal site stereochemistries. The irregular stereochemistry imposed on the copper ion in the blue copper proteins by the protein ligand is presumably one that lies between the two favored structures for the two oxidation states, thus implying rapid electron transfer and unusual redox potentials. Williams[15] has suggested that the active metal site of a metalloenzyme is in a geometry approaching that of the transition state of the appropriate reaction and as such, is uniquely fitted for catalytic action. Williams has used the term "entatic state" to describe this situation.

Copper Oxidases

Typical copper oxidases are tyrosinase, laccase (p-diphenol oxidase) and ascorbic acid oxidase. The properties of these, and some other enzymes are listed in Table 4-XI.

TABLE 4-XI

PROPERTIES OF SOME COPPER OXIDASES

Enzyme	Source	MW	Reaction Catalyzed
Tyrosinase	Mushroom	120,000 4 Cu	Oxidation of monohydric phenols (e.g Cresol) to the *o*-dihydric compound and oxidation of the *o*-dihydric phenols (catechol) to the *o*-quinone
Laccase	Latex of the lac tree	120,000 4 Cu	*p*-Diphenols $+ O_2 \rightarrow$ *p*-Quinones $+ H_2O$
Ascorbic acid oxidase	Squash Cucumber	146,000 6 Cu	L-Ascorbate $+ O_2 \rightarrow$ dehydroascorbate $+ H_2O$
Ceruloplasmin	Human plasma	151,000 8 Cu	Oxidizes *p*-phenylene-diamine, ascorbic acid and some *o*- and *p*-dihydroxy phenols
Pseudomones blue proteins (Azurins)	*Pseudomonas aeruginese*	16,400 1 Cu	Respiratory chain protein
Diamine oxidase	Pea seedlings	96,000 1 Cu 1 pyridoxal phosphate	$H_2N(CH_2)NH_2 + H_2O + O_2$ $\rightarrow NH_2(CH_2^n)_{n-1} CHO + NH_3 + H_2O$

Ascorbic Acid Oxidase

Ascorbic acid oxidase is a blue, copper-containing protein, having eight atoms of copper per molecule. Ultracentrifuge studies have established its molecular weight to be 146,000 ($\pm$15,000). Below pH 4.0 the copper dissociates from the enzyme, causing inactivation. The enzyme catalyzes the oxidation of ascorbic acid (47a) to dehydroascorbic acid (47b). The reaction involves a four-electron-transfer dehydrogenation of ascorbic acid and concomitant reduction of O_2 to H_2O. Hydrogen peroxide has not been detected in the reaction products. There are apparently no titratable sulphydryl (SH) groups in the native enzyme. However, in 8M urea or in 0.2M sodium dodecyl sulphate the copper is liberated and from ten to twelve SH groups are titratable. Complete reduction of the copper-free proteins (apoenzyme) with mercapto-ethanol gives sixteen to twenty SH groups, indicating between three and five disulphide cross-links per molecule in addition

to the copper bound sulphydryl groups. Addition of copper(I) to a solution of the apoenzyme leads to complete restoration of activity when 7.5 to 8 atoms of copper(I) are added per mole of the enzyme. Chelating agents capable of removing the copper inactivate the enzyme. The enzyme is capable of oxidizing 4×10^5 moles of ascorbic acid per mole of the enzyme in one minute at 37°.

The oxidation state of copper in this enzyme is apparently not uniform, both copper(II) and copper(I) being present. The oxidation of ascorbic acid is also catalyzed by copper(II) ions in aqueous solution and the reaction may be presented,

$$O_2 + Cu^{II} + \underset{monoanion}{ascorbic\ acid} \longrightarrow dehydroascorbate + H_2O_2 + Cu^{I}$$
$$Cu^{II} \longleftarrow O_2$$

(93)

Hydrogen peroxide is the reduction product of oxygen in the copper(II) catalyzed reaction while only H_2O occurs in the enzymic reaction. A possible mechanism for the enzyme catalyzed reaction (due to Hamilton[38]) is shown in (94).

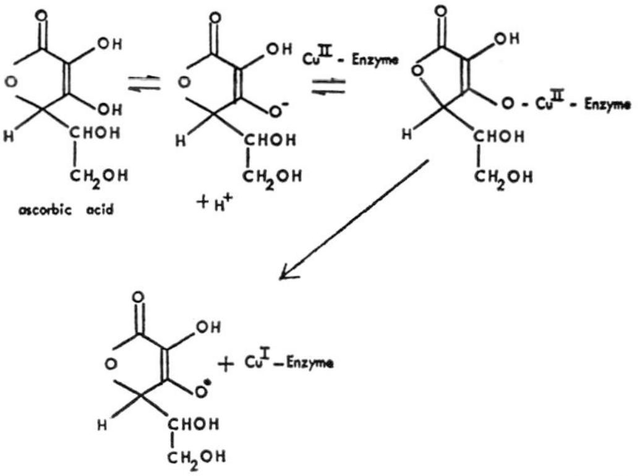

(94)

The above reaction scheme is not complete, however, since in the ascorbic acid oxidase reaction a mole of oxygen is converted to two moles of H_2O. The process discussed above results in the transfer of one electron to one of the eight Cu^{II} ions on the enzyme to give Cu^I. As there is no evidence for the generation of an enzyme-bound radical, Hamilton[39] has suggested that regeneration of the Cu^{II}-enzyme occurs after production of the quadruply reduced structure shown in (96) by reaction with four ascorbate molecules.

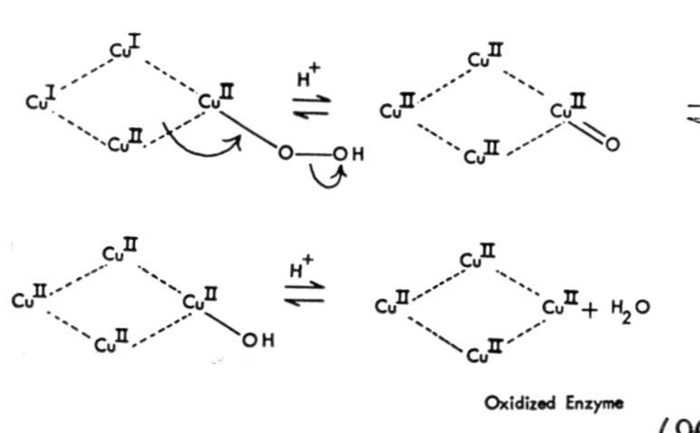

(96)

In order for the above reactions to occur the four copper(I) ions must be electronically connected since in the oxidation of copper(I) only one electron is involved while both steps in the reduction of oxygen are two-electron transfers.

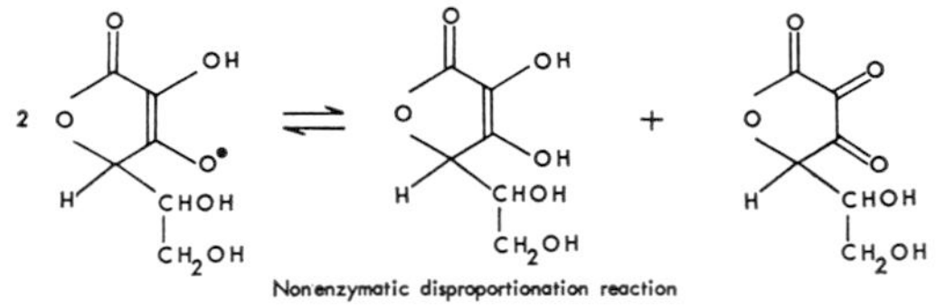

Nonenzymatic disproportionation reaction

(95)

Yamazaki and Piette[40] have shown by esr measurements that the ascorbic radical is produced by the enzyme during the oxidation of ascorbic acid. Their data suggest that a Cu^{II}-enzyme is reduced to Cu^I-enzyme when the radical is generated. Apparently the ascorbic radical is then released into solution and subsequently reacts with another ascorbic radical in a nonenzymic disproportionation reaction (95).

Tyrosinase

In addition to hydroxylase activity (equation 1), this enzyme also acts as a dehydrogenase (equation 2).

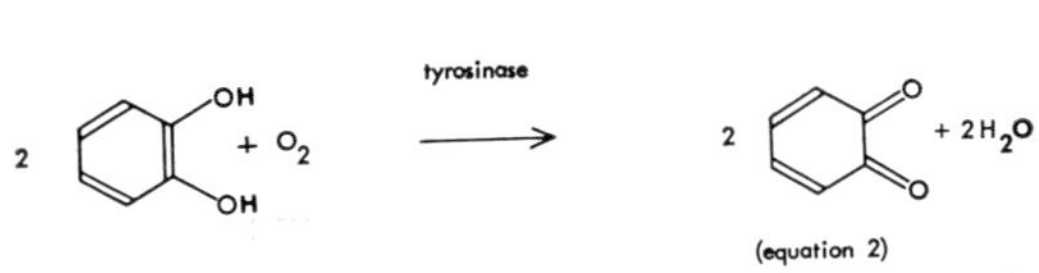

(equation 1)

(equation 2)

(97)

In the hydroxylase reaction a substituted phenol is oxidized to the corresponding catechol. The required co-substrate is a substituted catechol which is oxidized to the *o*-quinone. In many cases the substituent at the site of oxygen substitution in the substrate (labelled X) migrates to the adjacent position in the molecule during the reaction (NIH shift).

As a result of its bifunctional nature, there is much confusion regarding the nomenclature of this enzyme, for example the names catechol oxidase, dopa oxidase, phenolase complex, polyphenol oxidase and tyrosinase are all frequently encountered.

The richest source of the enzyme is the mushroom, where it occurs to the extent of 40 μg per gram of plant tissue (0.004%). The enzyme has a molecular weight of 130,000 and a copper content of 0.2 to 0.3 percent suggesting approximately five copper atoms per molecule. The copper occurs only as Cu^I and the oxidation state does not alter during catalysis. The copper may be removed by treatment with 0.01M HCN followed by exhaustive dialysis against neutral phosphate buffer. The resulting apoenzyme is completely inactive but the catalytic activity is restored by the addition of the appropriate amount of copper(I). Many investigations have confirmed that there are no SH groups in the apoenzyme, so that the metal binding sites must be different from those in ascorbic acid oxidase.

In the conversion of an *o*-dihydroxybenzene to the *o*-quinone, tyrosinase acts as a four-electron transferase converting one mole of oxygen to two moles of water during the oxidation of 2 moles of substrate. Mason and his coworkers[41] have studied the reaction of tyrosinase with catechol using electron paramagnetic resonance techniques and have concluded that radicals do not occur in the enzymic process. Hamilton[38] has suggested that for this type of nonradical process involving the loss of two electrons and two protons from each substrate molecule in a stepwise fashion, the mechanism shown in (98) applies. The copper(I) ion brings

(98)

the reactants catechol and oxygen together to form a pseudo-aromatic complex where electron transfer may readily occur, particularly when assisted by general acid or base catalysis from the protein molecule.

In conclusion, there remains much to be done on the two enzymes discussed in this section, the mechanisms suggested are simply plausible suggestions, and at present no clear function for the protein component of either enzyme has been demonstrated. In spite of marked similarities in physical properties (molecular weight, copper content, etc.) the mechanisms of action of the two enzymes are obviously quite different.

REFERENCES

1. Bender, M. L. and Brubacher, L. J.: *Catalysis and Enzyme Action.* McGraw-Hill, 1973. (An introductory account of enzyme catalysis.)
2. Eichhorn, G. L.: *Inorganic Biochemistry.* Amsterdam, Elsevier, 1973, vols. 1 and 2. (Contains accounts of various aspects of bio-inorganic chemistry.)
3. Frieden, E.: The chemical elements of life. *Sci Am,* 227:52, 1972.
4. Gould, R. F. (Ed.): Bio-inorganic chemistry. *Adv Chem Ser,* 100:1973. American Chemical Society, Washington. (A discussion of various aspects of bio-inorganic chemistry containing articles on metalloenzymes.)
5. Gray, C. J.: *Enzyme Catalysed Reactions.* New York, Van N-Rein, 1971. (A general account of enzyme catalysis.)
6. Hughes, M. N.: *The Inorganic Chemistry of*

Biological Processes. Salt Lake City, Wiley, 1972. (An introductory account of bio-inorganic chemistry.)

7. Jones, M. M.: *Ligand Reactivity and Catalysis.* New York, Academic Press, 1968. (A general discussion of catalysis and the reactions of coordinated ligands.)

8. Kaiser, E. T. and Kaiser, B. L.: Carboxypeptidase A, a mechanistic analysis. *Acc Chem Res, 5:*219, 1972.

9. Lippard, S. J. (Ed.): Current research topics in bio-inorganic chemistry. *Progr Inorg Chem, 18:*1973. (A very detailed account of metalloprotein redox reactions is given by L. E. Bennett.)

10. Mildvan, A. S.: *The Enzymes: Kinetics and Mechanisms.* Boyer, P. D. (Ed.): Metals in enzyme catalysis. New York, Academic Press, 1970, 3rd edition, vol. 2, p. 445.

11. Sigel, H., and McCormick, D. B.: On the discriminating behaviour of metal ions and ligands with regard to their biological significance. *Acc Chem Resh, 3:*201, 1970.

12. Vallee, B. L. and Wacker, W. E. C.: Metalloproteins. In Neurath, H. (Ed.): *The Proteins.* New York, Academic Press, 1970.

13. Vallee, B. L. and Williams, R. J. P.: Enzyme action: Views derived from metallo-enzyme studies. *Chem Brit,* 397.

14. Williams, R. J. P.: Role of transition metal ions in biological processes. *R I C Reviews, 1:*13, 1968.

15. Williams, R. J. P.: Catalysis by metallo-enzymes: The entatic state. *Inorg Chem Acta Rev, 5:*137, 1971.

16. Williams, R. J. P. and Dennard, A. E.: Transition metal ions as reagents in metallo-enzymes. *Transition Metal Chemistry.* London, R. L. Carlin, Edward Arnold, 1966, vol. 2.

17. Williams, D. R.: *Educ in Chemistry, 10:*56, 1973; *11:*124, 166, 1974.

18. Coleman, J. E.: "Metal Ions in Enzymatic Catalysis" in Progress in Bioorganic Chemistry, Ed. Kaiser, E. T. and Kézdy, F. J. Vol. 1. Wiley Interscience N.Y. 1971.

19. Henkins, R. W. and Sturtevant, J. M.: *J Amer Chem Soc, 90:*2669, 1968.

20. Liljas, A., Kannan, K. K., Bergstén, P.-C., Waara, I., Fridborg, K., Strandberg, B., Carlbom, U., Jarup, L., Lövgren, S., and Petef, M.: *Nature New Biology, 235:*131, 1972.

21. Kosicki, G. W.: *Biochemistry, 7:*4299, 1968.

22. Sigman, D. S., and Jorgenson, C. T.: *J Amer Chem Soc, 94:*1724, 1972.

23. Dash, A. C., and Nanda, R. K.: *J Amer Chem Soc, 91:*6944, 1969.

24. Pfeiffer, P., Offerman, W., and Werner, H.: *J prakt Chem, 159:*313, 1942.

25. Pocker, Y., and Meany, J. E.: *J Amer Chem Soc, 89:*631, 1967.

26. Prince, R. H., and Woolley, P. R.: *JCS Dalton,* 1548, 1972.

27. Murakami, Y., and Sunamoto, J.: *Bull Chem Soc Japan, 44:*1827, 1971.

28. Hay, R. W., Clark, C. R., Edmonds, J. A. G.: *JCS Dalton,* 9, 1974.

29. Clark, C. R., and Hay, R. W.: *JCS (Perkin II),* 1943, 1973.

30. Lipscombe, W. N.: *Chem Soc Rev, 1:*319, 1972.

31. See Lipscombe, W. N., *Acc Chem Res, 3:*81, 1970.

32. Nyman, P.-O., and Lindskog, S.: *Biochim Biophys* Acta, *85:*141, 1964.

33. Lindskog, S., and Malmstrom, B. G.: *J Biol Chem, 237:*1129, 1962.

34. Smith, E. L., and Stockell, A.: *J Biol Chem, 207:*501, 1954; Lipscombe, W. N., *et al. Proc Nat Acad Sci U.S.A., 58:*2220, 1967.

35. Kosicki, G. W., and Westheimer, F. H.: *Biochemistry, 7:*4303, 1968.

36. Hay, R. W., and Leong, K. N.: *J Chem Soc* (A), 3639, 1971; see also Rund, J. V., Plane, R. A., *J Amer Chem Soc, 86:*367, 1964.

37. Schwartz, J. H.: *Proc Nat Acad Sci U.S.A., 49:*871, 1963.

38. Hamilton, G. A.: *Adv Enzymol, 32:*55, 1969.

39. Hamilton, G. A.: Progress in Bioorganic Chemistry, Eds. Kaiser, E. T., and Kezdy, F. J., Wiley Interscience New York, N.Y., 1971, pp. 83-157.

40. Yamazaki, I., and Piette, L. H.: Biochim Biophys Acta, *50:*62, 1961.

41. Mason, H. S., Spencer, E., and Yamazaki, I.: *Biochem Biophys Res Commun, 4:*236, 1961.

Some Additional Reading References

42. Khan, M. M. T., and Martell, A. E.: Homogeneous Catalysis by Metal Complexes, Vol. 1, Academic Press, New York 1974 (Contains articles on the activation of dihydrogen, dioxygen, dinitrogen, carbon monoxide and nitric oxide.)

43. Sigel, H. (Ed.) Metal Ions in Biological Sys-

tems Vol I-V, Marcel Dekker New York, N.Y. 1975. (Volumes deal with such topics as the reactivity, synthesis, stability, structure and formation of biological compounds of low and high molecular weight containing metal ions; the metabolism and transport of metal ions and their complexes; and new models of complex natural structures and processes.)

44. Dhar, S. K. (Ed.), "Metal Ions in Biological Systems (Studies of Some Biochemical and Environmental Problems)" Plenum Press, New York, N.Y., 1973.

45. McAuliffe, C. A. (Ed.), "Techniques and Topics in Bioinorganic Chemistry" Macmillan, London, 1974. (The first volume contains articles on structural and electronic aspects of metal ions in proteins, the principles of cataly-

sis by metalloenzymes, the biochemical function of molybdenum, polynuclear proteins containing iron(III), and metal ions as probes in nuclear magnetic resonance studies in biochemistry.)

46. "Oxidases and Related Redox Systems," King, T. E., Mason, H. S., and Morrison, M., (Eds.) Wiley, New York 1965.

47. The Biochemistry of Sodium, Potassium, Magnesium and Calcium, Williams, R. J. P., *Quart Rev, 24*:331, 1970.

48. Vallee, B. L., and Williams, R. J. P.: *Chem in Britain, 4*:397, 1968.

49. "Chemical and Biological Aspects of Pyridoxal Catalysis," Eds. Snell, E. E., Fasella, F. M., Braunstein, A. E., and Faneli, A. R., Pergamon Press, London, 1963.

CHAPTER 5

ACTIVATION OF OXYGEN AND NITROGEN IN BIOLOGICAL SYSTEMS

R. F. JAMESON AND N. J. BLACKBURN

INTRODUCTION

O F ALL THE CHEMICAL ELEMENTS and compounds necessary to support life on our planet, oxygen is perhaps the most important. The consequences of oxygen starvation require little elaboration, less than three minutes being sufficient to cause severe, if not fatal, brain damage. The fact that such irreparable damage to the tissues occurs after so short a period without oxygen indicates that the element is utilized very quickly and efficiently within living systems, and is hence highly reactive chemically; yet atmospheric oxygen (which constitutes 20% of the surrounding air) does not appear to manifest the same high degree of reactivity: iron rusts only slowly, while petrol, wood and other inflammable materials do not spontaneously burst into flames. When fire *does* break out, however, there is no longer any doubt as to the re-activity of molecular oxygen.

Molecular nitrogen, the other major component of the atmosphere, is utilized by certain bacteria (for example, *Azobacter* and *Clostridium pasteurianum*) and some blue-green algae, being converted into ammonia in the process known as nitrogen fixation. The ammonia can then be used in the synthesis of amino acids and proteins by higher organisms, so that atmospheric nitrogen may eventually find its way (via the food chain) into the tissues of the highest species, man. Yet molecular nitrogen is widely used by chemists as an inert atmosphere under which reactions susceptible to atmospheric oxidation may be conducted.

It would appear then that some kind of "activation" of both molecular oxygen and molecular nitrogen is necessary before they may be efficiently utilized in biological systems, and it is the origin of this activation and subsequent reactivity which forms the subject of this chapter. Many of the ideas developed will be purely inorganic, rather than biochemical (if such a distinction exists), one of the ultimate goals of all "life scientists" being surely to elucidate the mechanism of physiological processes in terms of the basic chemistry of their constituents. The approach used will, therefore, be to discuss the inorganic chemistry of oxygen and nitrogen, and to discover how fully the more complex phenomena observed in biochemical and physiological processes may be explained in terms of such basic inorganic ideas. It must, however, always be borne in mind that by definition "model systems" are only approximations to the actual process under consideration, and that any conclusions derived from the study of model systems must be treated with caution. In short, oversimplification may easily lead to the wrong conclusion.

Biological oxidations by molecular oxygen can roughly be divided into two rather diffuse categories: (a) Dehydrogenations where the oxygen serves to remove two hydrogens from the oxidizable substrate, and (b) Oxygenations where the O_2 is partially or totally incorporated into the oxidation product. Such reactions generally require an enzymic catalyst which almost invariably contains a transition metal ion, e.g. iron, copper, molybdenum, and it is well established that in most cases this metal ion participates in the catalytic mechanism, the enzyme being termed an oxidase in class (a) oxidations, and an "oxygenase" in class (b) oxidations. Examples of dehydrogenations are provided by the oxidation of ascorbate: catalyzed by the copper containing enzyme ascorbic acid oxidase (Figure 5-1), the oxidation of catechols, glycols and sugars, etc., while the oxidation of tryptophan (Figure 5-2) and certain amines serve as examples of oxygenations. These reactions occur in the body tissues and consequently a third class of proteins termed "oxygen carriers" are necessary to transport the oxygen to the sites within the tissues where the oxidations occur. The

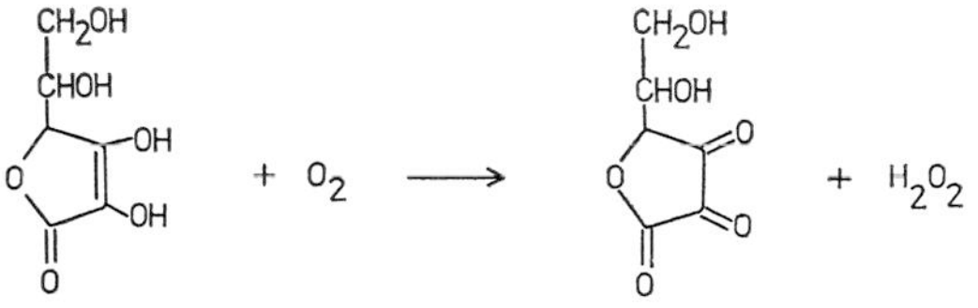

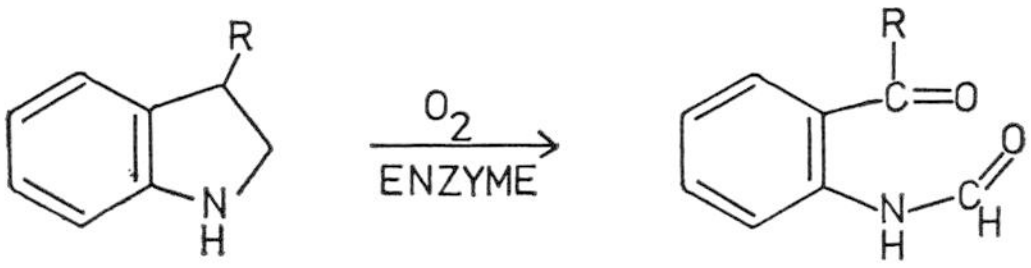

Figure 5-2. Oxidation of tryptophan by molecular oxygen and tryptophan oxygenase.

most common oxygen carrying proteins are the well-known iron containing haemoglobin, the hemerythrins (iron), and the haemocyanins (copper).

Since both oxidases and carriers interact with the oxygen molecule, an understanding of the inorganic chemistry of O_2 may allow some parallels to be drawn between the two types of interactions, and help formulate criteria for the oxidase activity or carrying properties found in a particular protein. Indeed, it is a surprising feature of these proteins that there appears to be great structural similarities between the two groups: both the cytochromes (proteins which form an electron transfer chain terminating in the transfer of four electrons to O_2, the latter reaction being catalyzed by cytochrome oxidase), and the haemoglobins contain the haem prosthetic group which can bind the oxygen molecule. Similarly, the haemocyanins (the copper containing oxygen carriers of mollusks and crustaceans) appear to have structural similarities to copper oxidases, and it seems likely that in each case multicenter sites with nitrogen or sulphur containing ligands are involved in the oxygen binding site. This could be due to evolutionary factors in that quite small variations of the binding site may be sufficient to change its function from that of an oxygen carrier to that of an oxidase.

Before examining what is known about the chemistry of these macromolecular systems however, it is worthwhile studying the chemistry of oxygen in model systems.

Figure 5-1. Oxidation of ascorbate by molecular oxygen giving dehydroascorbate.

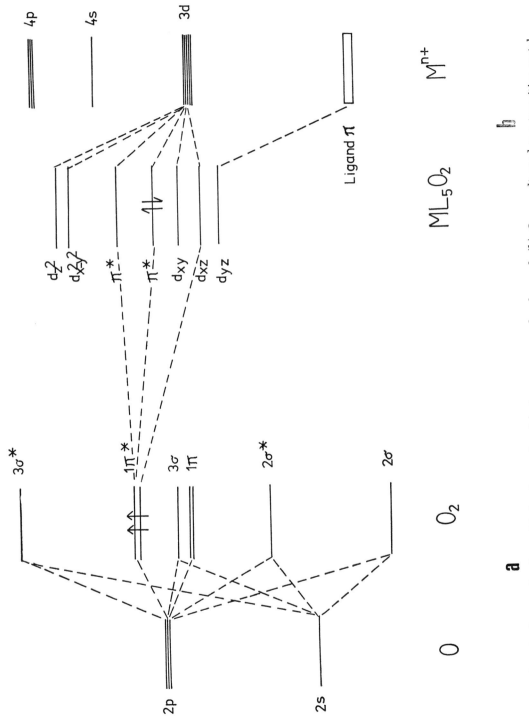

Figure 5-3. Outline molecular orbital scheme for (a) the oxygen molecule, and (b) O_2 coordinated to a transition metal.

THE ELECTRONIC STRUCTURE OF THE OXYGEN MOLECULE

The molecular orbital diagram for the oxygen molecule is shown in Figure 5-3a.

Filling in the twelve electrons from the 2s and 2p levels of the two individual atoms, we find that the highest occupied molecular orbital is the π^* level which is doubly degenerate. Hence the electrons in this orbital must have their spins parallel giving the oxygen molecule a triplet ground state (two unpaired electrons). Reduction to hydrogen peroxide corresponds to the transfer of two electrons into this half-filled orbital (Figure 5-3), and since the latter is antibonding, corresponding to a repulsion between the two oxygen nuclei, a decrease in the oxygen-oxygen electronic interaction, and hence a lengthening of the O-O bond length will result. Further reduction to water corresponds to the transfer of two more electrons into the $3\sigma^*$ level followed by schism of the O-O bond (the bond order of the O_2 molecule being reduced to zero).

There are three possible mechanisms for the reduction of oxygen to water:

that in case (a) the first one-electron transfer to form the peroxyl ($HO_2\cdot$) radical is thermodynamically very unfavorable; even though the subsequent three steps more than compensate for this initial energy barrier there will still be a high activation energy in the one-electron transfer mechanism.

Cases (b) and (c) do not have this restriction and hence any mechanism which will favor two-electron transfer is likely to be preferred. At the same time it must be noted that two electron steps are rare, e.g. it has been established that Sn^{III} is a common intermediate in the oxidation $Sn^{II} \rightarrow Sn^{IV}$.

There is another consequence of the triplet character, i.e. two electrons with "parallel" spins, of the ground state of molecular oxygen. Most oxidizable organic substrates are singlet molecules, i.e. they contain no unpaired electrons in their oxidized or reduced forms. Hydrogen peroxide and water are both singlet molecules and a reaction such as

Triplet + Singlet → Singlet + Singlet

is forbidden by the rules of spin conserva-

(a) a series of four one-electron transfer steps:

	E^o potentials (Volts)
(1) $H^+ + O_2 + e^- \rightarrow HO_2\cdot$	-0.32
(2) $H^+ + HO_2\cdot + e^- \rightarrow H_2O_2$	$+1.68$
(3) $H_2O_2 + e^- \rightarrow OH\cdot + OH^-$ (bond breaking step)	$+0.80$
(4) $H^+ + OH\cdot + e^- \rightarrow H_2O$	$+2.74$

(b) two two-electron steps:

(1) $2H^+ + O_2 + 2e^- \rightarrow H_2O_2$
(2) $H_2O_2 + 2H^+ + 2e^- \rightarrow 2H_2O$
(bond breaking step)

and finally (c) 1 four-electron step:

(1) $4H^+ + O_2 + 4e^- \rightarrow 2H_2O$ 1.23V
(simultaneous bond breaking)

From the values for the standard redox potentials of the individual steps we see

tion. In order to obey the spin conservation laws, the oxygen molecule would have to be promoted to its first excited state ($^1\Delta_g$) which corresponds to the pairing of two electrons in one of the π^* levels, leaving the other degenerate orbital empty. This state lies 23 kcals above the ground state and hence a high activation energy will be called for.

In fact, oxygen does react with singlet molecules in the *gas phase,* but the mechanism involves *hydrogen atom extraction* rather than *electron transfer.* These reactions are usually chain reactions involving complex free radical intermediates.

The foregoing discussion explains the apparent nonreactivity of molecular oxygen in the atmosphere around us. In order to make it reactive, we must decrease the high activation energies which occur, a process termed catalysis or activation.

TRANSITION-METAL—O₂ COMPLEXES

Consider the oxygen molecule bound to a transition metal ion with the oxidizable substrate also bound as a ligand. We can imagine the bonding of O_2 to the central metal as depicted schematically in very simple terms in Figure 5-4. The oxygen can form a σ-bond with the d_{z^2} orbital on the central metal while the empty π^* orbitals can interact with the metal d_{xz} and d_{yz}. (Remember that there is another π^* orbital at right angles to the one shown, i.e. perpendicular to the plane of the page). So long as these metal orbitals are filled, then electron density can flow back from the metal d orbitals into the π^* levels on the oxygen, which is what reduction of O_2 entails. Since the reducing agent is also bound as a ligand, and so long as the latter has orbitals of the right symmetry to interact with these same metal d orbitals, thus forming an extended molecular orbital over the substrate-metal-O_2 complex, then a very facile mechanism of charge transfer has

been achieved whereby electron density can flow from reducing agent to oxidizing agent.

Figure 5-5 shows the three basic modes of coordination possible for an oxygen molecule bound to a transition metal complex, (a) perpendicular, (b) linear, (c) angular. Type (a) will include all those orientations where the metal oxygen bond does not lie along the principal axis of symmetry of the complex, in this case the z axis. For those familiar with group theory, the symmetry of type (a) (perpendicular bonding) is C_{2v}, type (b) is C_{4v}, and type (c) is C_s, and a brief examination of the character tables for these groups indicates that in (a) the x and y

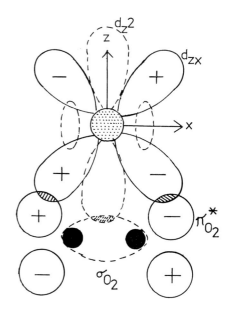

 Metal ion

 Oxygen atoms

Figure 5-4. Bonding in a transition metal complex (dotted lines represent σ bonding and solid lines are π bonding).

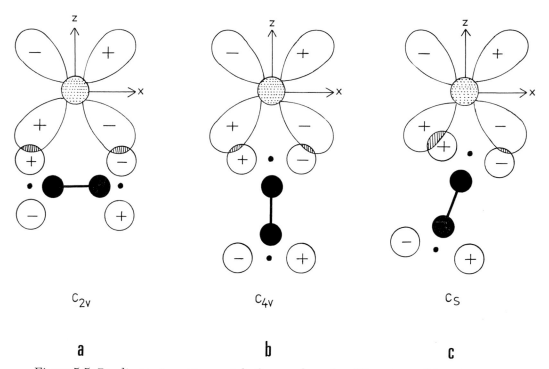

C_{2v} C_{4v} C_s

a   **b**   **c**

Figure 5-5. Bonding in transition metal $-O_2$ complexes for different possible orientations of the O_2 molecule (a) perpendicular, C_{2v}, (b) linear, C_{4v} (c) angular, C_s.

directions are nondegenerate, while in the other two cases the degeneracy remains unsplit. This means that in case (a) degeneracy of the π^* levels on the oxygen molecule (formely π_x and π_y) must be lifted by the symmetry of the ligand field, while in the other two cases they remain degenerate. For those who are not familiar with this kind of argument, the effect can be shown pictorially (Figure 5-5). When the oxygen is bound perpendicularly we imagine that a π-bonding orbital can form a σ-bond with the d_{z^2} orbital on the central metal. The π_x^* orbital on the oxygen can interact with the metal d_{xz} orbital but since the π_y^* orbital is perpendicular to the plane of the page, with two of its lobes projecting upwards in the positions shown as black dots, and the other two projecting downwards into the page, we see that over-

lap with the d_{yz} orbital is now greatly reduced, as the latter projects out of the plane of the page but along the z axis. Hence the π_y^* orbital will go up in energy, while the π_x^* orbital will go down in energy, while similar arguments can be used to show that the π^* levels remain degenerate in the other two cases. Figure 5-3b depicts the outline MO scheme for the perpendicularly bound metal-oxygen-complex showing the splitting of the π^* levels, this situation applying to all cases where the metal oxygen bond does not lie along the axis of symmetry of the molecule.

In the majority of known stable di-oxygen complexes of transition metals the oxygen molecule is in fact bonded in this perpendicular or skewed perpendicular fashion. The removal of the degeneracy of the π^* levels serves to remove the spin restrictions on any ensuing reaction, and furthermore it favors the situation whereby electrons can be transferred to the oxygen antibonding orbitals in pairs, i.e. in a two-electron mechanism already described as being more favorable thermodynamically.

By forming a transition metal O_2 complex we have thus achieved our aim, namely a facile mechanism for transfer of electrons in pairs into the oxygen antibonding orbitals via an extended molecular orbital system. This is not to say that every time an oxygen is bound to a transition metal complex reduction will ensue. If this were so, oxygen carriers would not exist. Indeed, as we shall shortly see, there are relatively few transition metal complexes which will bind oxygen at all, the binding powers being very dependent on both the nature of the metal and on the nature of the ligands. Di-oxygen complexes could well be described as being "pseudo activated complexes" in which there is fractional charge transfer from ligand to metal to oxygen.

We can set up a "model" which will apply both to oxygen carriers and to complexes showing "oxidase" activity. It is reasonable to assume that in both cases an initial metal-oxygen complex is formed in an equilibrium step and is either stabilized by its surrounding ligands or reacts further to form oxidation products and H_2O_2 or water:

$$L_nM + O_2 \overset{K}{\rightleftharpoons} L_nMO_2 \overset{k}{\rightarrow}$$
$$L_{n-1}M + L_{oxidized} + H_2O_2$$

Oxidase activity will occur if k is large, while oxygen carrying properties are favored by large K and small k.

Although the metal ion is acting as a catalyst and can therefore be recovered unchanged at the end of the reaction, the "pseudo activated complex" has been assumed to back donate electrons into the π^*

levels of the oxygen molecule. Complete transfer of two electrons would result in an increase of two in the formal oxidation state of the metal which is a highly unlikely situation, and in fact never happens in the stable di-oxygen complexes so far characterized. In almost every case we cannot assign a formal oxidation number to the central metal ion as the ligands surrounding it also interact with the metal-d and oxygen-π^* orbitals, through the delocalized MOs to which we have already alluded, and these serve to decrease the positive charge formally present. An examination of the properties of some stable di-oxygen complexes may help to clarify the situation, as well as indicate the origin of their stability; a greater understanding of oxidase activity may hopefully be achieved by extrapolation to cases where such effects are absent.

MODEL COMPOUNDS FOR OXYGEN CARRIERS

Table 5-I shows the oxygen-oxygen bond lengths together with the geometry and reversibility of some representative dioxygen complexes of transition metals, the bond lengths found in the various oxidation states of the oxygen molecule being shown for reference. The first point to notice is that in most cases the O-O bond length is in fact close to or greater than that found in peroxide indicating that even in these simple stable complexes charge equivalent to approximately two electrons has been transferred to oxygen. This means that the stability is conferred by factors other than the extent of charge transfer, i.e. by properties associated with either the ligands or the metal.

The Effect of the Metal

Metals at the beginning of the transition series will react with hydrogen peroxide to form peroxy compounds but since the start-

TABLE 5-I

O-O BOND LENGTHS IN SOME DIOXYGEN COMPLEXES*

Complex	Bond Length (Å)	Geometry	Reversibility
O_2^+	1.123		
O_2	1.207		
O_2^-	1.28		
O_2^{2-}	1.49		
$[Cr(O_2)_4]^{3-}$	1.405		
$[UO_2(O_2)_3]^{4-}$	1.51		
$[K_2Ti(O_2)C_7H_3O_4N]_2 \cdot 0.5H_2O$	1.45		
$[(NH_3)_5Co(O_2)Co(NH_3)_5]^{4+}$	1.65	Angular	Rev.
$IrCl(CO)(PPh_2Et)_2(O_2)$	1.573	Perpendicular	Rev.
$IrCl(CO)(PPh_3)_2(O_2)$	1.30	Perpendicular	Rev.
$IrI(CO)(PPh_3)_2(O_2)$	1.51	Perpendicular	Irrev.
$[Ir(dp)_2O_2]^+$	1.625	Perpendicular	Irrev.
$[Rh(dp)_2(O_2)]^+$	1.418	Perpendicular	Rev.
$Pt(Ph_3P)(O_2)1.5C_6H_6$	1.45	Perpendicular	Irrev.
$Fe(DMG)_2(base)_2(O_2)$	—	—	Rev.

* Choy and O'Connor: *Coord Chem Revs,* 9:145, 1972-1973.

ing product, H_2O_2 or O_2^{2-} has filled π^* orbitals, the flow of charge is from O_2^{2-} to metal. In the "hard and soft" notation introduced in Chapter 3, O_2^{2-} is a "hard-base" which interacts with the "hard-acid" metal ion. At the beginning of the transition series, higher oxidation states tend to be stabilized and this charge donation from peroxide reduces the positive charge on the metal. Examples in Table 5-I are $[Cr(O_2)_4]^{3-}$, $K_2Ti(O_2)[C_7H_3O_4N]_20.5$ H_2O. Clearly, the direction of charge transfer is opposite to that expected in O_2 complexes and is of little further interest.

Towards the end of the transition series, however, we find a large number of complexes which will bind O_2 reversibly, i.e. show O_2 carrying properties, cobalt(II) and the Os, Re, Ir, Rh, Pt, Pd, group of metals being particularly good examples. The latter are all "*b*" class or soft π donors with a large number of d electrons; they have low oxidation numbers, and interact quite strongly with the soft π-acceptor molecular oxygen. X-ray studies have shown the oxygen to be bonded perpendicularly,

and it can be thought of as occupying either 1 or 2 coordination positions as shown in Figure 5-6. If the metal-oxygen interaction becomes strong enough then the O-O bond order will decrease sufficiently to approximate to two separately coordinated oxygen atoms, but this is *not* the same as oxidase activity since the oxygen remains firmly bound.

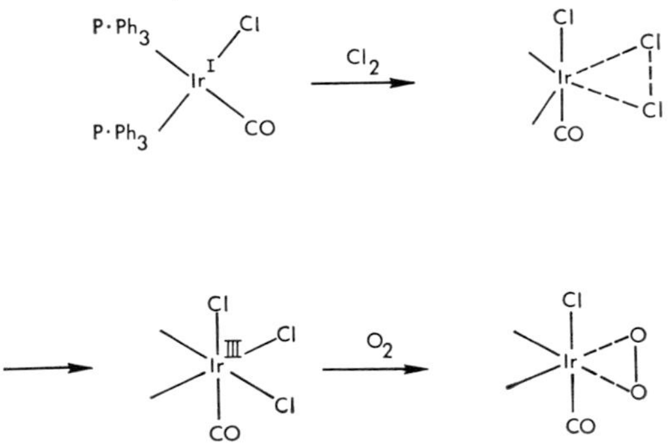

Figure 5-6. Formation of adducts with chlorocarbonyl-bis(triphenyl phosphinato)-Iridium(I). In the dioxygen complex iridium has a nonintegral oxidation state and may be considered either five or six coordinate. The trichloro complex is define to contain iridium (III).

Some representative cobalt complexes with molecular oxygen are shown in Table 5-II. With Cobalt(II) there is a tendency to form dinuclear complexes, e.g. $(NH_3)_5$-Co-O_2-Co$(NH_3)_5$ and X-ray work shows the oxygen to be bound in a bent fashion between the two metal ions, the resulting complex being diamagnetic. This would suggest that the degeneracy of the π^* orbitals on the oxygen has been lifted, but it is likely that the MOs extend over the whole complex, so that the π^* levels can no longer be considered as individual orbitals.

The effect of the metal is more general than specific. Low oxidation states are preferred, but there must be a higher oxidation state quite close in energy since on oxygen binding electron transfer from metal to oxygen takes place. Again, no attempt can be made to assign an integral oxidation state to the metal in the complex. $[(NH_3)_5$-Co-O_2-Co$(NH_3)_5]^{4+}$ is formally cobalt(III) and O_2^{2-} as shown by the O_2 bond length (Table 5-I). One electron oxidation of this compound leads to $[(NH_3)_5Co$-O_2-Co$(NH_3)_5]^{5+}$ which is paramagnetic (Table 5-II and has been shown by esr to contain one unpaired electron delocalized over both cobalt nuclei. This is indeed proof of the concept of a nonintegral oxidation state somewhere between +3 and +4.

Some other rather direct evidence of the above effect has been produced by Vaska. He took the compound chloro-carbonyl-bis (triphenylphosphinato) Iridium (I) shown in Figure 5-6 and reacted it with various small gaseous molecules (O_2, H_2, HCl, SO_2, etc.) to form adducts. The parent compound is square planar and addition of the molecule Cl_2, for example, can be thought of in terms of initial bonding of Cl_2 to form a five co-ordinate complex followed by complete transfer of two electrons from Ir(I) to the chlorine, causing the Cl-Cl bond to break (strengthening the Ir-Cl interaction) and formally oxidizing the iridium to Ir(III) (Figure 5-7). With less strongly oxidizing molecules oxidation is likely to be less complete.

The infrared stretching frequency of the C = O group can be used as a measure of the "oxidation state" of the central metal ion. This is because the CO stretching frequency depends on the electric dipole within the C = O unit. Obviously, if charge is back donated from iridium to, say, oxygen in the oxygen adduct then the increased positive charge on the central metal will

TABLE 5-II

SOME DIOXYGEN COMPLEXES OF Co(II)

Complex	Comments
$[(NH_3)_5Co - O_2 - Co(NH_3)_5]^{4+}$	Diamagnetic, reversible
$[(NH_3)_5Co - O_2 - Co(NH_3)_5]^{5+}$	Paramag, irrev., esr shows unpaired electron delocalized over both Co atoms
$[en_2 \cdot Co - O_2 - Co \cdot en_2]$	Diamag., rev.
Co $\cdot$ EDTA	No uptake
	No uptake
	Weak association
	Inner sphere complex

tend to pull electrons towards itself result-ing in the transfer of charge from the car-bonyl group to the metal with a correspond-ing decrease in bond order and hence in I.R. stretching frequency.

Table 5-III shows the stretching frequen-cies for a number of adducts. If the parent compound is taken to be iridium(I) and the trichloro compound described above is assumed to be iridium(III) then we can draw up a correlation between infrared stretching frequency and oxidation state as shown in Figure 5-7 and can then read off the oxidation states corresponding to the frequencies obtained for the other adducts. The results in the table show that the oxi-dation state of the metal changes slowly and in a nonintegral way as the oxidizing power of the particular small molecule in-creases, i.e. as the tendency towards an in-

TABLE 5-III

CARBONYL STRETCHING FREQUENCIES (ν_{CO}) AND RELATIVE OXIDATION STATES OF IRIDIUM IN THE ADDUCTS OF [IrCl(CO)((C₆H₅)₃P)₂] WITH SOME COVALENT MOLECULES*

Covalent Molecule	ν_{CO} (cm⁻¹)	Relative Oxidation State
Parent	1,967	1.00†
O₂	2,015	1.89
SO₂	2,021	2.00
D-D	2,034	2.24
H-Cl	2,046	2.46
CH₃I	2,047	2.48
C₂F₄	2,052	2.57
C₂(CN)₄	2,057	2.67
BF₃	2,067	2.85
I-I	2,067	2.85
Br-Br	2,072	2.95
Cl-I	2,074	2.98
Cl-Cl	2,075	3.00†

* Vaska, L.: *Acc Chem Res*, 1:335, 1968.
† Defined.

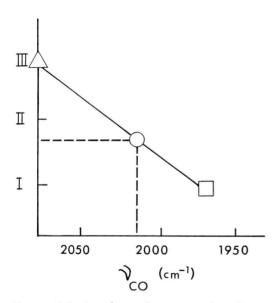

Figure 5-7. Correlation between infrared car-bonyl stretching frequency (ν_{CO}) and relative oxidation state of iridium. The position of O₂ is shown as an example (intercepts = oxidation state 1.89 and ν = 2015 cm⁻¹). Key: △ = Cl₂ adduct, □ = parent and ○ = oxygen ad-duct.

ternal redox reaction increases. This is really an extremely important result be-cause it indicates once more that the ligand orbitals are interacting with the orbitals on the central metal and on the adduct, pro-viding a mechanism for the delocalization of electron density over the whole complex. The extent of charge transfer or of internal oxidation reduction is thus determined to a large extent by the properties of these lig-ands.

Let us now go back and examine the ef-fect of the ligands in greater detail. In the class "b" metal dioxygen complexes in Ta-ble 5-I we see that the other ligands are al-ways those with delocalized π systems (for example, triphenyl phosphine, and dipy-ridyl), having orbitals of π symmetry which can interact with metal d and oxygen π and π* orbitals to form delocalized M.O.s. This interaction helps to reduce the increased positive charge on the central metal, or put in another way the electrons which are transferred to the π^* orbitals on oxygen come from the metal ligand complex as a whole which is very electron rich. Such ligands are termed "soft" and in extreme cases "non innocent," changing their own oxidation state when complexed to a metal and thus producing an electron distribution within the complex very different from that of free ligand or metal alone. This effect is largely responsible for the stability of the O_2 adducts and has been termed the "symbiotic" effect: "Soft" ligands tend to flock together. (O_2 is itself a "soft" acid since it has orbitals of π symmetry which can interact with metal.) Once one or two of these "soft" ligands are complexed the strong interaction with the metal ion so produced tends to reduce its effective positive charge and thereby increases its own "softness," and favors the coordination of more "soft" ligands. This is very loose terminology, but does serve to explain the effect quite well.

Some other ligand effects are shown in Table 5-II. $[Co(NH_3)_5-O_2-Co(NH_3)_5]^{4+}$ binds O_2 reversibly as does $[Co(en)_2-O_2 Co(en)_2]^{4+}$. Cobalt (EDTA) however takes up no oxygen at all and this illustrates the stabilising effect of nitrogen containing bases via a σ donor effect (but the π and σ metal ligand interactions are difficult to separate, and are best thought of in terms of a single covalent interaction). The bulky substituted cyclohexyl ligand shown at the bottom of the table does not react with oxygen either, probably due to steric hindrance; there is not enough room for the oxygen molecule to occupy a position in the inner coordination sphere of the complex indicating that O_2 is in fact less strongly bound than the nitrogen containing bases.

Iron and copper form very few dioxygen complexes, but these metals are generally the best catalysts for oxidations involving molecular oxygen. This is very significant

from the biological point of view since they are precisely the metals which the biosystem uses as oxygen binding sites in both O_2 carriers and in oxidases. A few model Fe(II) complexes will reversibly bind O_2, for example, Fe(II) dimethylglyoxime and certain Fe(II) porphyrins with strongly covalent bonding ligands such as histidine and pyridine coordinated in the axial positions. The ability to bind O_2 is much less widespread than among cobalt or the heavy metals discussed above, and the requirement for very soft or even noninnocent ligands is very stringent. Copper forms few truly reversible O_2 adducts. On the other hand, both iron and copper act as very efficient redox catalysts for many oxidations involving O_2. One mechanism generally proposed for these oxidations is the initial formation of a substrate-metal-oxygen complex followed by internal electron transfer to O_2 coupled with two electron oxidation of the substrate, i.e. the two step mechanism proposed at the beginning of the discussion to illustrate the relationship between carrying and oxidase properties.

$$L_nM + O_2 \overset{K}{\rightleftharpoons} L_nMO_2 \overset{k}{\to} L_{n-1}M + L_{oxidized} + H_2O_2$$

Copper and iron are good catalysts for these reactions because the peroxy complex ("pseudo activated complex") formed after initial internal charge transfer is generally unstable and decomposes releasing peroxide, O_2^{2-}. However, in the absence of an oxidizable group and with skillful choice of ligands we have seen that such a complex can be stabilized by electron delocalization. Cu and Fe will thus be well suited to form the O_2 binding sites, both in O_2 carrying and storage proteins, and in oxidases and electron transfer proteins, since suitable choice of electron-rich ligands will first of all help to bind the O_2 (a necessary requirement of both types of proteins), and secondly very slight modifications of these ligands may destabilize the peroxide so formed allowing its dissociation and the binding of another O_2 to the site, thus completing the catalytic cycle. In the dinuclear cobalt-O_2 complexes shown in Table 5-II, one electron oxidation can take place without dissociation of the peroxide due to the inertness of cobalt(III) while the heavy metal dioxygen complexes tend to irreversibly break the O-O bond leading to

type structures. Neither type of complex would thus seem very suitable for use in biological systems.

Model systems, however, fail to imitate very subtle and important effects produced by the presence of the macromolecular structure which we shall briefly discuss before going on to O_2 carriers and the oxidases in greater detail.

PROTEIN CONFORMATION

The protein conformation relates to the actual stereochemistry of the amino acids and peptide links. Polar groups on a polypeptide chain are often linked to one another via (a) hydrogen bonding or (b) salt linkages, producing the characteristic helices present in many protein chains, while nonpolar groups tend to gather together in areas of low dielectric constant on the inside of the molecule, giving rise to globular-type structures. A complete discussion of protein conformation and tertiary structure is outside the scope of this chapter, but is amply treated elsewhere, and so only those aspects relevant to metal site and metal protein interactions will be mentioned here:

The metal ion is usually bound to the main protein chain via some or all of its ligands, these often being amino acid residues. For example, in haemoglobin the axial ligand is the imidazole part of a histidine residue on the polypeptide chain, while the porphyrin ring interacts with the protein via Van der Waals forces. When oxygen (or in fact any small molecule) binds to a metal ion that is attached to a protein, as we have seen in model compounds, the coordination number of that metal ion changes, and thus its stereochemistry will change also. Electron transfer increases the net positive charge on the metal altering its own ionic radius as well as the bond length in the bound molecule. (We have seen that transfer of two electrons into the π^* levels on O_2 increases its bond length.) When oxygen binds to a metal ion in a site, we can then expect quite large changes in the stereochemistry of the site, and since the metal ion is also bound via its ligands to the polypeptide chain this change of stereochemistry will be transmitted along the polypeptide chain and produce conformational changes at points either far from the metal ion or in its immediate vicinity.

A schematic example of this "allosteric" effect is shown in Figure 5-8. When an O_2 binds to the metal site charge is partially transferred to oxygen. This produces an increased positive charge on the metal ion, and to compensate, ligands A and B, which will probably be strongly covalently bond-

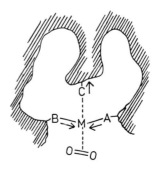

Figure 5-8. Schematic allosteric effect.

ing, will increase their interactions with the metal and move towards it as shown. This may cause ligand C to move away from M slightly due to conformational changes transmitted through the protein chain. If C were also responsible for the stabilization of the O_2 complex formed, this moving away might cause enough destabilization of the peroxide complex to make it dissociate to O_2^{2-}. This is highly speculative but does serve to illustrate the type of allosteric interactions that could be involved in oxidase mechanisms. Conformational changes could also be triggered by the lengthening of the oxygen-oxygen bond length which occurs on charge transfer.

A protein is capable of having pockets of low dielectic constant where nonpolar groups tend to cluster. If the metal ion is present in such a site separation of charge on complexing the oxygen will be hindered and thus will hinder the internal redox reaction by opposing internal electron transfer. This effect is thought to be partly responsible for stabilization of the bound oxygen in oxyhaemoglobin since the haem group does in fact sit in a hydrophobic site.

To summarize very briefly, we expect an oxygen binding site in which there are strongly covalently bonding or noninnocent ligands surrounding the metal both in oxidases and in carriers. We expect the O_2 to bind in a perpendicular manner (perhaps skewed) in which the MO bond does not lie along the principal axis of symmetry, such that degeneracy of the π^* levels is lifted. Very small changes in the ligands surrounding the metal are all that is required to promote oxidase activity over carrying properties or vice versa and such modifications could easily be brought about by conformational changes occurring on oxygen binding or in the structure of the metal site itself. We shall now examine the natural oxygen carriers and oxidases in

greater detail to see how fully our qualitative ideas lead to a greater understanding of the structure and function of these proteins.

HAEMOGLOBIN AND MYOGLOBIN

Haemoglobin is the oxygen carrier of mammalian species. In fact, a number of variations of this protein are found but they all contain the protoheme IX group shown in Figure 5-9, and have a molecular weight of around 65,000. The protein consists of two subunits of two different types called a and β, each subunit containing one polypeptide chain and one protoheme IX group. The sequence of amino acids for a number of haemoglobins has been determined by X-ray crystallography. Myoglobin has only one subunit with one haem group, but its structure is similar to the individual subunits of haemoglobin. Its function is an oxygen storage protein. The haem group is located in a hydrophobic pocket in the protein with the four in-plane coordination positions occupied by the N's of the por-

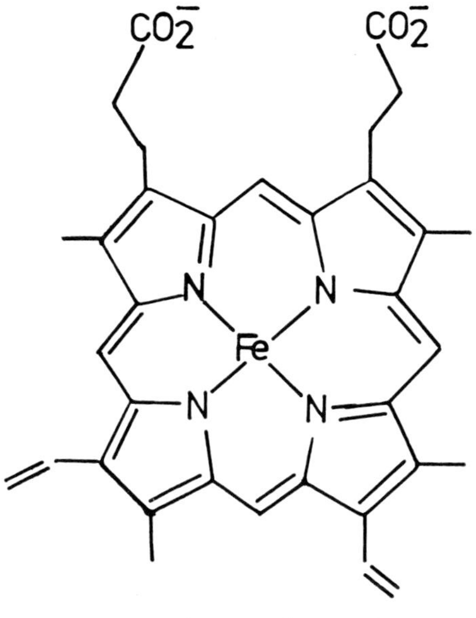

Figure 5-9. Protoheme IX.

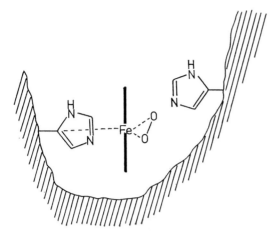

Figure 5-10. Representative diagram of haem complex in its site showing the proximal and distal imidazole groups.

phyrin ring while the fifth axial ligand is supplied by the imidazole group of a histidine. The sixth coordination position is left open allowing small π bonding ligands such as O_2, CO, CN, NO, etc. to coordinate. There is a more distant imidazole group as shown in Figure 5-10, but its function is not clear at the present time. In the de-

oxygenated form of the protein the iron is five-coordinate and adopts a square pyramidal stereochemistry such that the Fe atom lies about 0.9Å out of the plane of the porphyrin ring towards the imidazole nitrogen, and the spectral and magnetic properties indicate that the iron is in a high-spin Fe(II) state. When oxygen binds to haemoglobin the latter moves into the plane of the porphyrin ring, the oxygen-haem complex adopting octahedral stereochemistry; at the same time the spin-state changes to a low-spin configuration and the complex becomes diamagnetic. Figure 5-11 shows the relationship between the high- and low-spin cases for the d^6 Fe(II) ion and the d^5 Fe(III) ion. (It should be emphasized that the haem-oxygen complex does not have truly octahedral *symmetry*, the true symmetry probably being nearer C_{2v} depending on the orientation of the O_2 to the Fe atom, but for the purpose of the discussion, the deviations from octahedral symmetry may be treated as a perturbation and are not great enough to decouple the spins under the strong ligand field of O_2.)

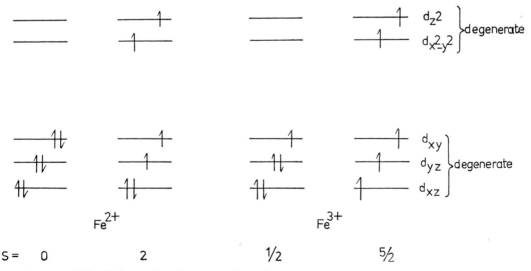

Figure 5-11. Relationship between low (S = 0 or ½) and high (S = 2 or ⁵⁄₂) spin Fe²⁺ and Fe³⁺ in an octahedral environment.

In polar environments the heme group is easily oxidized to Fe(III) and forms a series of ferri- or met-haemoglobin derivatives with ligands such as CN^-, F^-, H_2O, NH_3, azide, etc. bound in the axial position. The met-complexes are usually low spin.

In spite of the large amount of structural knowledge there are still unresolved problems associated with haemoglobin chemistry. One such is the actual mode of binding of O_2 to the iron atom and the diamagnetic nature of the complex so produced since the O_2 is itself paramagnetic; another is the co-operative effect, i.e. that the affinity of the protein for oxygen increases as each haem group in the four subunits is progressively oxygenated. The sigmoidal curves obtained for O_2 binding are shown in Figure 5-12 which also shows that more oxygen is bound at lower partial pressures of CO_2. Parallel to the effect of CO_2 is that of pH where oxygen affinity decreases with increasing hydrogen ion concentration since protons are released on O_2

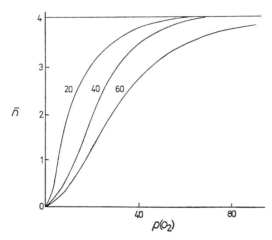

Figure 5-12. Oxygenation curve for haemoglobin at different partial pressures of CO_2 (as numbered on curves). $\bar{n}$ represents the average number of oxygen molecules bound per molecule of haemoglobin. All pressures are in mm Hg.

binding, this latter property being termed the "Bohr effect." The above properties make haemoglobin a very efficient oxygen carrier, the cooperative binding ensuring that the maximum amount of O_2 will be carried, the effect of CO_2 partial pressure ensuring that the O_2 will be released in the tissues where the partial pressure of CO_2 is relatively high and consequently the pH lower. Figure 5-12 also shows that maximum uptake of O_2 by mammalian forms of haemoglobin occurs at approximately the partial pressure of O_2 in the atmosphere. Furthermore, maximum uptake by fetal haemoglobin occurs at lower partial pressures, allowing transfer of oxygen to take place between mother and fetus via the placenta. Such properties serve to emphasize the remarkable efficiency of haemoglobins as oxygen carriers, and are almost certainly attributable to evolutionary factors.

The technique of Mössbauer spectroscopy has shed some light on the problems of the mode of oxygen binding and of the valence state of the iron in oxy- and deoxyhaemoglobin. In Mössbauer spectroscopy we observe transitions between two nuclear energy levels of the ^{57}Fe nucleus, a less abundant isotope of iron, these levels being designated $I = 1/2$ and $I = 3/2$ referring to the magnetic moment of the nuclear spin. Transitions can be observed between these nuclear energy levels corresponding to absorption of gamma radiation from an appropriate gamma ray source. (Figure 5-13) The energy separation, ΔE, depends on both the electronic charge density and the asymmetry of charge at the nucleus which in turn depend on the electronic state of the iron, i.e. the occupancy of d orbitals, but the precise nature of the interactions between electronic properties and nuclear energy levels is outside the scope of the discussion. Suffice it to say that char-

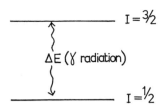

Figure 5-13. Nuclear energy levels and absorption of γ-radiation in Mössbauer spectroscopy.

acteristic spectra are obtained for different electronic and spin states of the iron, Fe(II) usually giving very different spectra from Fe(III). Mössbauer studies on oxyhaemoglobin give spectra which more closely resemble the Fe(III) valence state but show no interactions due to the unpaired electrons which must be present in low spin Fe(III) compounds. A similar situation is found in the Mössbauer spectra of ferri and ferrocyanide, $Fe(III)(CN)_6^{3-}$, $Fe(II)(CN)_6^{4-}$, the spectra of both ions being similar, the explanation being that there is a large degree of covalent bonding due to the strongly π bonding CN^- ligands, resulting in delocalization of charge over the whole complex. Exactly analogous behavior seems to exist in oxyhaemoglobin, the covalent interaction due to the strong π bonding porphyrin group causing electron delocalization. This is just the result that we would expect on the basis of the discussion in the first part of this chapter. We should not, in fact, have tried to assign an integral oxidation state to the Fe, its true value lying between +2 and +3 owing to the noninnocent character of the ligands. This covalency will help to stabilize the O_2 complex with respect to reduction. We could think of the bonding in terms of electron donation from the nitrogen-containing ligands to Fe (serving to decrease the positive charge on the metal) together with back bonding from the Fe to oxygen, the porphyrin ring and the imidazole group

via orbitals of π symmetry. This π back bonding is the basis of the so-called "symbiotic" effect alluded to before, and should serve to strengthen the metal oxygen bond.

As outlined above, the diamagnetism of the oxygen complex can be explained if we postulate some form of perpendicular bonding for O_2 since the π^* levels on oxygen will then be split. This model allows for considerable charge transfer to the oxygen molecule without its complete oxidation to Fe(III) since the other ligands can donate electron density to the metal; the latter need not, therefore, have unpaired electron density centered on it as is the case in the Fe(III) complexes. In addition, this model approximates to a seven coordinate complex and will help to promote the conformational changes responsible for the cooperative effect discussed below (Figure 5-14).

Proton nmr and Mössbauer spectra show no signs of deviation of the nonoxygenated Fe(II) ions in *partially oxygenated* haemoglobin from their initial high-spin Fe(II) state. Furthermore, myoglobin, which has only one subunit does not exhibit co-operativity. These two factors indicate that the latter is an allosteric effect due to interactions between the subunits and changes in the tertiary and quaternary structure of the protein, rather than a result of the electronic properties of the haem groups themselves. The subunits are joined together by

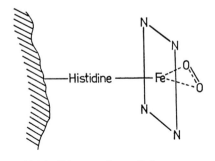

Figure 5-14. "7 co-ordinate" haem complex.

salt linkages, i.e. electrostatic interactions between positive and negative ions or dipoles situated on different polypeptide chains. A mechanism for the cooperative binding of O_2 has been proposed by Perutz in terms of the restrictions which these linkages impose on the protein conformation, and is shown schematically in Figure 5-15. The figure shows the four subunits of the haemoglobin molecule represented by the blocked out sections and constrained in the deoxy conformation (Figure 5-15a) by the association of the salt linkages, these being absent in the oxy-form. In the former conformation, as previously mentioned, the haem groups lie in pockets within the protein, the Fe atom lying 0.9Å out of the plane of the porphyrin ring, but there is another feature of this deoxy form, namely that a tyrosine residue is located in a nearby pocket, and it is to the side chain of this tyrosine that the polar groups which form the salt linkages are attached. On oxygenation the Fe atom moves into the plane of the porphyrin ring necessitating the corresponding movement of its axial imidazole ligand (which is attached to the peptide chain), thus triggering a series of conformational changes along the peptide chain which eventually result in the expulsion of the tyrosine from its cleft and the consequent rupturing of a salt linkage (Figure 5-15b). The initial breaking of one salt linkage will be thermodynamically more difficult than the breaking of the following ones, since it destroys the deoxyconformation and introduces steric strain among the subunits within the molecule. Hence, as subsequent haem groups become progressively oxygenated, each tyrosine can move out of its cleft and break its constraining salt linkages more easily than the last, and a model for cooperativity has been achieved. (Figure 5-15 c-f).

The Bohr effect is also partially explicable if we remember that the ionic and dipolar interaction between the subunits will be a function of pH since the proton can compete with the positively charged groups for the negatively charged ones. Thus as the H^+ concentration increases more salt linkages break owing to protonation of negatively charged groups, and consequently it will be easier for the protein to take up the oxy conformation.

The discussion with regard to haemoglobin shows that most of our ideas are at least qualitatively correct. The diamagnetism of the oxygenated protein can be explained by postulating nondegenerate oxygen π^* levels in C_{2v} symmetry. The ligands surrounding the metal are of the expected very soft π bonding type, while the oxidation state of the Fe is probably nonintegral indicating strong delocalization and fractional charge transfer. The fact that the dielectric constant of the environment of the oxygen-haem complex is low, irreversible oxidation to Fe(III) occurring in more polar media, points to stabilization caused by the unfavorability of charge separation in a hydrophobic environment. We must not be too smug however because even though the overall picture is quite convincing at a fairly elementary level, there are many interesting and subtle questions left unanswered. For example, what is the effect of the structural changes within the haem groups of the cytochromes and why do such changes produce electron transfer properties rather than O_2 binding? The answer must lie in subtle differences in electronic structure of the haem group, the dielectric constant of the environment or in more dramatic changes of protein structure in the vicinity of the active site.

Hemerythrin

Hemerythrin is a nonheme iron protein which serves as an oxygen carrier in some

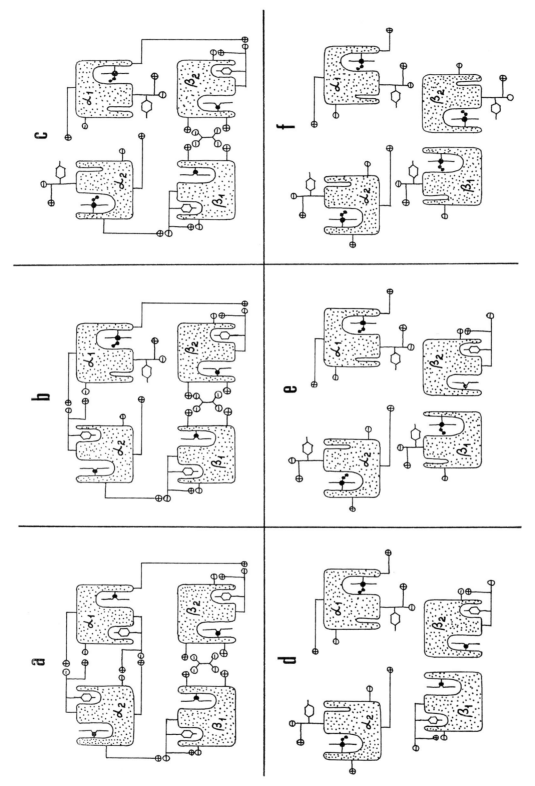

Figure 5-15. Possible mechanism for oxygenation of haemoglobin. (Perutz, M. F., *Nature*, 228:726, 1970.)

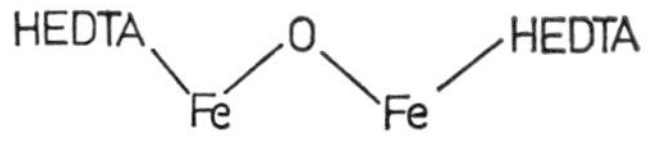

Figure 5-16. Oxo-bridged dinuclear complex of iron and hydroxyethylene-diaminetriacetate (HEDTA).

lower species such as branchiopods and spinculids. It can be dissociated into eight subunits of molecular weight 13,500 each containing two Fe atoms, which can reversibly bind one mole of oxygen. Because of the lack of X-ray data on the detailed structure of the protein, and the Fe-binding sites to which the O_2 molecule is attached, techniques such as electronic, Mössbauer and esr spectroscopy have largely been used in the study of the active site or sites. The results obtained from the different methods will be presented followed by a discussion of the probable mode of O_2 binding.

(a) Electronic Spectroscopy

The optical spectra of both oxyhemerythrin and methemerythrin (the oxidized form of the protein) are very similar to that of a model complex containing two Fe(III) ions linked via an oxo bridge. (Figure 5-16) The spectrum of both oxy- and met-hemerythrin, and the model complex show certain bands exhibiting increased intensity which is thought to be due to the coupling together of the "electronic states" of the two iron atoms via the oxo bridge. This makes some transitions allowed which would be forbidden in the absence of such spin-coupling. Other transitions present in both protein and models may be attributed to charge transfer between Fe(III) and bridging O, suggesting that the oxygenated form and the oxidized met-derivatives both contain a site with a pair of spin-coupled high-spin Fe(III) ions linked through an oxo bridge. The deoxy form, however, shows no such coupling and the spectrum resembles that of a high-spin Fe(II) ion.

(b) Magnetic Susceptibility and esr

None of the forms of hemerythrin shows an esr spectrum but magnetic susceptibility measurements at room temperature indicate that the deoxy form has a magnetic moment corresponding to four unpaired electrons (high-spin Fe(II)), while oxy- and met-hemerythrin have low susceptibilities which decrease with temperature and approach the values of S = 0 and S = ½ respectively at 4.2°K. This behavior is typical of strong spin-coupling between the two Fes in the site and is consistent with the lack of esr spectrum and high intensity spin coupled bands in the optical spectrum. It therefore supports the oxo bridge dimer suggested earlier.*

(c) Mössbauer Spectroscopy

Mössbauer spectroscopy shows that deoxyhemerythrin contains high-spin Fe(II) while methemerythrin spectra are similar to those of methemoglobin and metmyoglobin, i.e. the iron is basically Fe(III), and also show large distortions of the Fe(III) configuration which by analogy with haemoglobin are probably due to strong covalently bonding ligands. The oxyhemerythrin spectrum on the other hand, shows unambiguously that on oxygenation the Fe sites become nonequivalent. The results are summarized in Table 5-IV.

A model for O_2 binding can now be proposed (Figure 5-17). The deoxy form is high-spin Fe(II) which may have two aquo Fe(II) ions in close proximity. On oxygenation the water molecule can form an oxo bridged dimer while the oxygen

---

* The presence of high-spin Fe(II) but no esr spectra in deoxyhemerythrin means that although the Fes have unpaired electrons they must be sufficiently close together to influence the electronic relaxation time and provide a mechanism for relaxation via dipole-dipole interactions.

TABLE 5-IV

COMPARISON OF SOME PROPERTIES OF DEOXY-, OXY-, AND MET-HAEMERYTHRINS

	Oxidation No.	Spin State	Structure	Equivalences of Fe's
Deoxy-	II	High	Not bridged	Equivalent
Oxy-	Distorted III	Coupled	Bridged	Nonequivalent
Met-	Distorted III	Coupled	Bridged	Nonequivalent

may either bind as a peroxy bridge (*c.f.* cobalt-O_2 dimers) or to one or other of the iron atoms. Two possibilities for the oxygenated product are shown, both of which would result in nonequivalent Fe atoms. The conformational change which would have to occur when the oxo bridge is formed is particularly interesting in view of the fact that the kinetics of oxygen binding to hemerythrin show that the mechanism is a two-step process (having two relaxation times as measured by temperature jump methods) and such a mechanism is very easily explained in terms of initial O_2 binding followed by a conformational change or vice versa.

Hemerythrin shows no cooperative effect on its own, but if the perchlorate ion is bound to the protein initial cooperativity followed by inhibition is found (Figure 5-18).

When one or two out of the total eight sites is occupied in the presence of perchlorate there is a marked cooperative effect while six or seven oxygenated sites inhibit further uptake of O_2. Perchlorate also strongly affects the binding properties of methemerythrin towards ligands such as azide, thiocyanate, Br^-, Cl^-, CN^-, etc. which can coordinate to the oxidized form of the protein. The enhancement of the affinity of the site for oxygen by electrostatic forces cannot in itself be responsible for cooperativity since O_2 is uncharged. The ClO_4^- must influence the groups adjacent to the ion, i.e. change the pKs of acid or basic

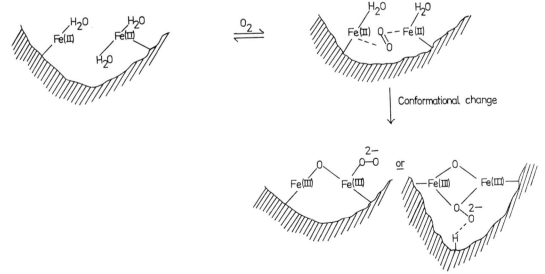

Figure 5-17. Model for oxygen binding to hemerythrin.

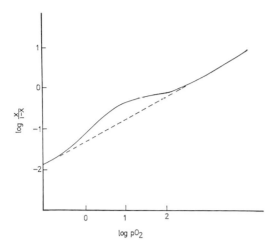

Figure 5-18. Hill plot for O_2 binding to hemerythrin at 25°C, pH = 7.13. — — — — = absence of perchlorate, ——— = presence of perchlorate, x = mole fraction of oxygen bound. (DePhilips, H. A., *Arch Biochem and Biophys, 144:122, 1971.*)

groups by modification of their degree of protonation. (This could cause conformational changes through cleavage of H-bonded structures, and indeed one suggested structure for the oxygenated hemerythrin evolves an O_2 . . . H hydrogen bonded structure of the type that could be affected by this type of interaction). It could, however, alter the polarity of the site, which we have seen in the case of haemoglobin to have a marked effect on the uptake of O_2. Conformational changes, which in themselves could bring about the formation of the oxo-bridged dimeric structure, would obviously increase the degree of oxygen binding lending more weight to this patricular model.

Hemerythrin then, shows some very different chemistry from haemoglobin, but there are certain similarities which do fit the general picture: (1) the complex is again diamagnetic indicating that the O_2 π^* levels are nondegenerate or at least spin-coupled, thus facilitating two electron

transfer, (2) Mössbauer studies show that the Fe(III) forms of the protein do indeed have delocalized electron density suggesting that symbiotic stabilization of the O_2 adduct may again be present, and (3) once more, the importance of the macromolecular conformational change in the stabilization of the O_2 complex is apparent.

Due to the large distortions in the electron clouds shown to be present by Mössbauer spectroscopy in the oxy- and methemerythrins we should not be too hasty to assign an oxidation state of +3 to the iron atoms and −2 to the oxygen. By analogy to our model systems and haemoglobin, noninnocent ligands are again expected to be present and charge transfer may well be only partial. An example of a molecule which exhibits similar type of behavior is the nitro-prusside ion $[Fe(CN)_5NO]^{2-}$ which is formally Fe(III). The complex is diamagnetic which seems to suggest Fe^{2+} − NO^+, but there is both physical and chemical evidence for a structure such as Fe^{3+} − NO. There is convincing but not unambiguous evidence for Fe(III) in oxyhemerythrin from the similarities in the optical spectra to Fe(III) model complexes, and from the spin-coupling of the two iron atoms, but Mössbauer evidence is far from conclusive. It is probable that the oxidation state of the iron is again nonintegral (but nearer +3) with incomplete charge transfer to O_2.

COPPER PROTEINS

Many copper containing proteins are found in nature, having quite diverse functions. We shall be concerned primarily with four, haemocyanin (the oxygen carrier found in mollusks and crustaceans), and laccase, ascorbic acid oxidase and ceruloplasmin, proteins which have oxidase activity and catalyze the oxidation of substrates such as ascorbate, noradrenalin, and cer-

TABLE 5-V

SOME DATA ON REPRESENTATIVE COPPER PROTEINS

Function	Protein	No. of Cu Atoms Cu^{2+}	$Cu\oplus$	λ_1 (nm)	λ_2 (nm)	λ_3 (nm)	λ_4 (nm)
Oxygen carriers	Deoxy-haemocyanin	0	>2	—	—	—	—
	Oxy-haemocyanin	>2	0	347	440	570	700
Iron metabolism, some oxidase activity	Ceruloplasmin	8	0	332	459	610	794
Oxidases	Polyporous laccase	4	0	330	450	614	788
	Ascorbic acid oxidase	~8	0	330	425	607	760
Electron carriers	Stellacyanin	1	0	—	448	604	845
	Plastocyanin	2	0	—	460	597	770
Mixed function oxidase	Deoxytyrosinase	0	1	—	—	—	—

tain catechols among others, with associated reduction of molecular oxygen to water:

$$2AH_2 + O_2 \rightarrow 2H_2O + 2A$$

The purpose of the discussion will be first, to learn what we can about oxygen binding and reduction, and gain an overall picture of oxidase enzyme mechanism, and in particular, the mechanism of charge transfer, then to compare the oxygen carrier haemocyanin with the copper oxidases. Table 5-V shows the absorption maxima, number of copper atoms in various oxidation states and the function of some representative copper proteins. One is immediately struck by the amazing similarities between the bands in proteins with very diverse roles and one is tempted to argue that the structures of the chromophores responsible for these bands must be very similar. There is one difference, however, and that is that only those proteins with multicopper sites and which interact with

molecular oxygen either as carriers or oxidases show the band at $\sim$ 330 nm.

On the basis of the above spectra, esr spectroscopy and oxidation-reduction titrations, the copper ions in enzymes may be classified into four types shown in Table 5-VI, three involving copper(II) and one involving copper(I).

Type I copper is thought to be responsible for the absorption maximum at approximately 610 nm, and hence would also be responsible for the blue color of the enzymes and of oxygenated haemocyanin. The extinction coefficient for this band, however, is very much larger than expected for an absorption arising from transitions within the d-shell of an octahedral Cu^{2+} ion (d-d spectra) since these transitions are forbidden and therefore, very weak. The increased intensity could arise if the copper ion were bound in a highly nonsymmetrical site such as a distorted tetrahedron, and this is one currently held view.

TABLE 5-VI

TYPES OF COPPER IONS IN PROTEINS

TYPE I	Cu^{2+}	Distorted site, paramagnetic, $\lambda = 610$ nm
TYPE II	Cu^{2+}	Normal site, paramagnetic
TYPE III	$Cu^{2+} - Cu^{2+}$	Spin paired dimer, diamagnetic, $\lambda = 330$ nm
TYPE IV	Cu^+	Cuprous

Type II copper more closely resembles Cu^{2+} in low molecular weight complexes of nearly octahedral symmetry and does not show any particularly unusual properties. It does, however, appear to bind anions such as F^-, N_3^- and CN^- quite strongly.

Type III copper is perhaps the most interesting in that it appears to consist of two Cu^{2+} ions which form a spin-paired dimer. It is not detected by either esr or magnetic susceptibility measurements and is therefore diamagnetic.

This classification was primarily based on extensive studies of the multicopper-containing enzymes laccase, ceruloplasmin and ascorbic acid oxidase (often called the "blue" oxidases) performed mainly by Malmstrom and his team in Sweden, and it is both interesting and instructive to consider these experiments in some detail. The enzymes act as redox catalysts, i.e. electrons can be transferred from substrate to oxygen by a very facile mechanism involving higher or lower oxidation states of the metal. For copper:

$$Substrate + 2Cu^{2+} \rightarrow$$
$$substrate\ (oxidized) + 2Cu^+$$
$$2Cu^+ + O_2 \rightarrow 2Cu^{2+} + O_2^{2-}$$

This scheme is only illustrative and does not attempt to indicate the actual electron transfer mechanism, but the Cu^{2+}/Cu^+ redox couple is presumed to be involved in reactions catalyzed by the blue oxidases. They can be reduced from the oxidized Cu^{2+} state to the reduced Cu^+ state by suitable reducing agents such as ascorbate and the reduction can be performed as a titration in which the exact number of electron-reducing equivalents can be measured. Malmstrom's team carried out redox titrations by measuring the absorbancies at 610 and 330 nm as a function of the total number of reducing equivalents added. They

found that (1) with polyporous laccase, both these absorbancies decreased linearly and at the same rate consuming approximately three electron equivalents of reducing agent. Their interpretation of this was that there are three electron accepting sites represented by the 610 and 330 nm maxima and that these sites were acting in a cooperative manner, the redox potential for the protein as a whole increasing with the number of electrons added; (2) sodium fluoride was added to the oxidized enzymes and the experiment repeated. One electron equivalent was taken up by the site corresponding to the 610 nm maxima, and only after this absorption had fallen to zero did the absorbance at 330 nm decrease, taking up in all two electron equivalents. This indicated that the two bands represented two different electron accepting sites, one of which was a one electron acceptor, the other a two electron acceptor.

The kinetics of anaerobic reduction of laccase were followed by measuring the changes in the 610 nm and 330 nm absorptions as the reaction proceeded. After an initial very fast decrease in the 610 nm absorbance, reduction of the site proceeded more slowly, and during this latter stage, reduction of the 330 nm band commenced. Such results indicated that initially the one-electron site was reduced by substrate, but the electrons then reorganized themselves within the enzyme thereby reducing the two-electron site. This electron "reshuffle" within the enzyme added weight to the idea of cooperativity among the sites, since after the initial one-electron site reduction the electron distribution then came to equilibrium. Under steady-state conditions, the 610 nm band was found to be almost fully reduced, while the 330 nm band was almost fully oxidized.

These results led to (1) the assignment of the 610 nm band to type I copper, a one-electron accepting site associated with the substrate binding site, and (2) the 330 nm band as a spin-paired diamagnetic Cu^{2+} — Cu^{2+} dimer, a two-electron accepting site most likely to be the site of O_2 binding. Thus two electrons could be donated to the oxygen in a thermodynamically favorable two-electron step.

The cooperativity between the sites suggests that they are connected via, for example, delocalized π systems through which electrons can be transferred from one site to the next. Indeed there is evidence that the substrate may not be bound to type I Cu^{2+} at all. If the enzyme is reduced with the very powerful reducing agent the hydrated electron, the absorption due to e^- disappears about one hundred times faster than that due to type I copper, indicating that the electron may go into a group on the surface of the protein and then be transferred through a delocalized π system to the Cu ions and finally to O_2. This shows that the properties of the metal ion and its specialized environment are not the only factors governing oxidase activity, and that organic electron transfer systems are extremely important in producing an efficient redox enzyme. The role of the metal is, therefore, considerably more complex, and the study of these electron transfer systems and their interaction with the metal ion will be essential for a complete description of the mechanism.

In view of the fact that type II copper in these enzymes binds anions, it is tempting to postulate that its role is to bind peroxide after the first electron transfer step. Peroxide, if formed, is never in fact released from the enzyme. Under steady-state conditions type I Cu^{2+} (associated with the substrate binding site) is reduced, while type III Cu (associated with the oxygen binding site) is oxidized, leading to the situation shown schematically below:

$$\text{SUBSTRATE} \text{---} Cu^+ \text{----}$$
$$[Cu\text{--}Cu\text{--}O_2]^{\sim 4+}$$

i.e. when the enzyme is turning over, a "potential difference" is set up within the electron transfer system which can be likened to an electric circuit!

Let us now reexamine the spectral data given in Table 5-V. We see that those proteins which do not interact with O_2 in any way do not have a band at $\sim$330 nm, but that the other bands are almost identical. It therefore appears that the blue proteins all contain Cu^{2+} in a distorted environment (type I copper), and furthermore, the astonishing similarities between the spectra of all these proteins suggests that the site itself is very similar. There is a possibility that this is a result of a common "ancestor" for every copper-containing blue protein, and that oxidase activity is the result of modifications though the ages. The role of the distorted site is not understood.

It is now time to consider haemocyanin. The protein has a molecular weight of about half a million, and contains varied amounts of copper ranging from ten to twenty Cus/molecule. The metal can be removed from the protein to yield a molecule which does not bind oxygen, but full activity is restored on the addition of Cu^+, suggesting that the oxidation state of Cu in the deoxy form is +1. Further evidence for this assumption is provided by the lack of both esr and visible spectra, and the low magnetic susceptibility exhibited by deoxy-haemocyanin. Upon oxygenation an intense blue color appears, and an esr spectrum corresponding to only 10 percent of the total Cu is observed. The oxygen binds in the ratio of one O_2 to two Cus.

There has been much controversy over the oxidation state of the Cu in oxy-haemocyanin. Lack of an esr spectrum and magnetic susceptibility is not necessarily indicative of a $Cu(I)$ state, since like hemerythrin, the spins of the unpaired electrons on the Cu^{2+} could be coupled together via spin-pairing with O_2, via dipole-dipole interactions, or via a bridging oxy atom within a $Cu^{2+} - Cu^{2+}$ dimeric unit. Another similarity to hemerythrin is the fact that the kinetics of oxygenation again reveal a two-step mechanism pointing to initial O_2 binding followed by a conformational change. A model incorporating these observations is shown in Figure 5-19.

It is difficult to distinguish between the various possibilities shown in Figure 5-19 for the mode of oxygen binding to copper, since unlike the iron case, Mössbauer spectroscopy is not applicable and hence there is no evidence for nonequivalent Cu atoms in the oxy-form. The model would involve a conformational change on oxygenation analogous to that suggested for hemerythrin, which might help to stabilize the product with respect to irreversible oxidation to peroxide and Cu^{2+}. If the surrounding ligands were again strongly covalently bonded, the true oxidation state of the copper might be indeterminate and there might be sufficient electron donation to copper to prevent oxidation. It must be added at this point, however, that inorganic cupric peroxy complexes are extremely unstable.

This leads to our final point, namely the great similarity between the spectra of haemocyanin and the spectra of those oxidases that contain similar types of Cu^{2+} dimers. It is too much of a coincidence for the two classes of proteins to have such similar spectra and chemical characteristics were there not considerable structural similarities also. The band at ~330 nm appears in both kinds of proteins, and seems to be related to the oxygen binding site in each case. The precise explanation of this band is thus of considerable importance in assessing the relationship between haemocyanin and the oxidases. It could be a property of the dimeric nature of the site or a consequence of the $Cu - O_2$ interaction, i.e. $Cu^{2+} \rightarrow O_2$ charge transfer, but the latter explanation, though very attractive for haemocyanin, would suggest that an oxygen molecule was always bound to the type III copper site of the oxidases in a similar manner to that found in oxy-haemocyanin, and only became released upon reduction of substrate; the enzyme would therefore

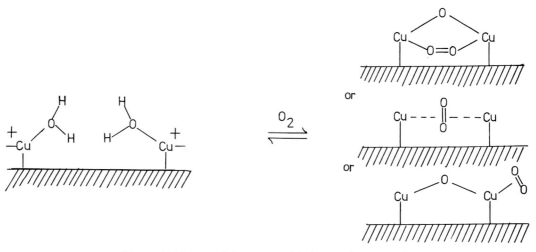

Figure 5-19. Model for oxygen binding to hemocyanin.

not be reoxidized by molecular oxygen, but re-reduced by the substrate, and the diamagnetic character of the $Cu^{2+} - Cu^{2+}$ pair could arise from induced spin-pairing with this bound oxygen molecule. The former proposal would suggest that the O_2 binding site in each case was very similar indeed, since the presence of different ligands should markedly affect the spectrum. The interpretation of the visible and near ultraviolet spectra of haemocyanin, and the blue oxidases clearly presents an interesting and challenging problem for future researchers, as it would clarify many points including the hitherto unexplained distorted site of type I copper found universally in the blue copper proteins. That haemocyanin also shows the 610 nm absorption suggests that it too contains some of this type of copper, possibly the 10 percent detectable by esr.

In the light of the above discussion, it would seem reasonable to propose that haemocyanin is a kind of degenerate oxidase lacking a substrate binding site and hence capable of reversibly binding O_2. Conversely, the oxidases may have evolved from simple O_2 carriers to become the sophisticated and specialized molecules that we know them to be. In order to solve the problem completely we require the detailed structural knowledge that can only be obtained from X-ray diffraction or high resolution nuclear magnetic resonance spectroscopy.

The blue copper-containing oxidases have been chosen as examples because more is known about them than most other electron transfer enzymes, and also because the iron containing enzymes will be dealt with elsewhere. We have attempted to correlate the inorganic chemistry of oxygen with the modifications imposed upon them by the macromolecule trying never to lose sight of basic inorganic ideas. It is indeed a triumph that much of the chemistry and biochemistry of metalloproteins is explicable in terms of these elementary concepts.

N_2 FIXATION

Certain bacteria *(Azobacter, Clostridium pasteurianum)* and blue-green algae are able to fix molecular nitrogen, that is carry out its six electron reduction to ammonia.

$$N_2 + 6H^+ + 6e \rightarrow 2NH_3$$

The MO scheme for molecular nitrogen Figure 5-20 shows that this reaction amounts to the transfer of four electrons into the π^* and two electrons into the $3\sigma^*$ level (thus causing N-N bond cleavage), and we might expect certain similarities to O_2 reduction on the basis of a very similar molecular orbital picture.

A closer examination, however, reveals that in N_2 the π^* levels are unoccupied, and hence arguments concerning the splitting of their degeneracy and the favora-

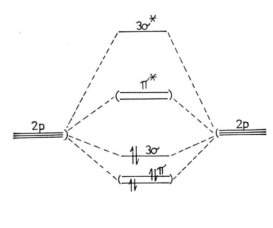

Figure 5-20. Molecular orbital scheme for the nitrogen molecule.

bility of two electron mechanisms no longer apply to the same extent; in fact N_2 bacteria exhibit very different chemical behavior to that of oxidases as will presently become evident. The reaction, although thermodynamically favorable ($\Delta H \sim 418kJ$ mole^{-1}) has a very high activation energy, arising from the considerable difference in energy between the highest filled orbital (3σ) and lowest empty orbital (π^*) so that electron transfer into the latter will be more difficult, and the need for activation even greater than for molecular oxygen. (In fact, the origin of the high activation energies are different in the two cases, that found in O_2 reduction arising from the need to initially promote it to a spin-paired configuration.)

Nitrogenase catalyses the reduction of N_2 to ammonia under the "mild" conditions found in biosystems, while extremes of temperature and pressure are required to effect the reaction in the industrial Haber process. The enzyme is probably composed of two proteins (neither of which are as yet very well characterized) one containing molybdenum and iron and usually termed the Mo-Fe protein, the other containing iron alone. The molybdenum-iron protein contains one (possibly 2) atoms of molybdenum and about sixteen atoms of iron, and is very rich in sulphide, containing twenty cysteine residues and 1 mole of "acid labile" sulphide per Fe. An esr signal resembling that found in typical iron-sulphur proteins ($g = 1.94$) is present in the reduced protein, together with three other resonances at $g = 2.01$, 3.67 and 4.30, which are also present in the oxidized form. To date, no esr signal attributable to molybdenum has been detected. Mössbauer spectroscopy suggests that the iron occurs in the Fe(III) state, but a small shoulder in the spectra has been interpreted as evidence for high spin Fe(II) or Fe(I) species, which may also be indicative of the presence of a unique iron-binding site within the protein. The Fe protein contains two to three irons and two sulphide groups, but shows no esr spectrum. There is no activity unless both proteins are present together.

The enzyme will reduce a number of other substrates, some of which are shown together with their reduction products in Table 5-VII. Some interesting points arise from the nature of these substrates:

1. They all act as ligands for "b" class metals, indicating that a metal in a low oxidation state may be present in the binding site.

2. They are all small rod-like molecules suggesting that the site may well be situated in a narrow cleft, a premise backed up by the inhibition produced by branching in the structure of the substrate.

Although there is still no direct evidence, it is thought that the substrate binding site is either a single Fe or Mo atom, or possibly a Mo-Fe pair, with cysteine, acid labile sulphide, or both, as ligands.

Enzymic activity is also dependent on the presence of ATP and an electron donor thought to be ferrodoxin or flavodoxine. The ATP has been described as the "electron activator," the energy release accompanying its hydrolysis being considered to

TABLE 5-VII

NITROGENASE SUBSTRATES*

Substrate	Reduction Product
Nitrogen N_2	$2NH_3$
Acetylene C_2H_2	C_2H_4
Acetonitrile CH_3CN	CH_4, CH_3NH_2
Methyl Isonitrile CH_3-NC	C_2H_6
Vinyl Cyanide $CH_2 = CHCN$	C_3H_6, C_3H_8, NH_3
Cyanide CN^-	CH_4, NH_3, CH_3NH_2
Hydrogen $\left.\right\}$ Carbon Monoxide	Inhibitors

* Chatt and Leigh: *Chem Revs*, 1:121, 1972.

TABLE 5-VIII

TRANSITION METALS CAPABLE OF
BINDING MOLECULAR NITROGEN

Ti	—	Cr(?)	—	Fe	Co	Ni	—
—	—	Mo	—	Ru	Rh	—	—
—	—	W	Re	Os	Ir	—	—

initiate the process of reduction and setting up a "potential difference" across what appears to be a very extended electron transfer chain within the Enzyme-co-enzyme complex. Such a process is equivalent to an "ATP generated redox potential" which is capable of reducing the N_2 to NH_3, and the analogy to the electric circuit might again be drawn. However, the actual mechanism of N_2 reduction is still completely obscure.

Inorganic chemists became interested in nitrogen fixation when it was discovered quite recently that certain transition metals (shown in Table 5-VIII) can form complexes with N_2. All are intermediate between class "a" and class "b" metals, and there appears to be the same kind of preference found in O_2 complexes for "soft"

TABLE 5-IX

SOME REPRESENTATIVE DINITROGEN
COMPLEXES

Complex	Infrared Stretching Frequency (cm^{-1})
$[Ru^{II}(NH_3)_5 (N_2)] Cl_2$	2,130
$[Os^{II}(NH_3)_5 (N_2)] Cl_2$	2,010
$[Os^{II}O_2 (N_2) (PEt_3)_3]$	2,064
$[Os^{II}Br_2 (N_2)(PMePh_2)_3]$	2,090
$[Os^{II}H_2 (N_2) (PEtPh_2)_3]$	2,085
$[Mo^o (N_2) (PPh_3)_2 (PhMe)]$	2,005
$[Co^o (N_2)(PPh_3)_3]$	2,093
$[Co^I H (N_2) (Ph_3P)_3]$	2,094
$[Fe^o (H_2) (N_2) (PEtPh_2)_3]$	2,057
$[Fe^I (H) (N_2) (Ph_2P-CH_2-CH_2-PPh_2)_2]^+$	2,090

polarizable ligands (Table 5-IX). Every compound shows a strong infra-red absorption due to N-N bond stretching at about 1900 to 2200 cm^{-1}, considerably lower than that found for N_2 in the Raman (2331 cm^{-1}). This lowering of the infrared stretching frequency indicates a decrease in bond order and therefore gives a measure of the charge transfer from metal to the $N_2 \pi^*$ orbitals, showing that, like O_2, N_2 is also activated by coordination to a metal ion, but that the extent of charge transfer is much smaller on account of the high energy of the π^* orbitals. The fact that there *is* an absorption in the infrared region is proof that charge transfer has occurred, and indicates end on bonding for N_2, since an infra red spectrum is only obtained if there is a dipole moment or charge separation within the chromophore (Figure 5-21).

A much greater degree of activation is achieved in bridged dinitrogen complexes (similar to $Co(O_2)Co$ species), provided the two metals between which the N_2 is bonded are sufficiently different. One appears to require "a" class character, the other "b" class character (Table 5-X).

A study of these type of bridged complexes then may lead to a model for N_2 fixation involving two transition metals (*cf.* Fe + Mo in nitrogenase) within a specific site, but as yet few transition metal dinitrogen complex has shown any tendency towards reduction at all.

The mechanism of nitrogen fixation could conceivably be one of two (a) transfer of electrons in pairs followed by protonation to give immine-type intermediates, -NH-

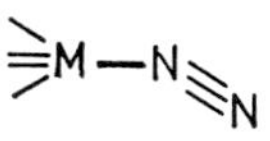

Figure 5-21. Coordination of dinitrogen to a transition metal.

TABLE 5-X

SOME BRIDGED DINITROGEN COMPLEXES*

Complex				Stretching Frequency (cm^{-1})
$Re^ICl(N_2)(PMe_2Ph)_4$	+ TiCl$_3$	$\longrightarrow$	$\rightarrow Re-N_2-Ti\leftarrow$	1,805
	+ CrCl$_3$	$\longrightarrow$	$\rightarrow Re-N_2-Cr\leftarrow$	1,890
	+ MoCl$_3$	$\longrightarrow$	$\rightarrow Re-N_2-Mo\leftarrow$	1,850
	+ MoCl$_4$	$\longrightarrow$	$\rightarrow Re-N_2-Mo\leftarrow$	1,795
	+ excess MoCl$_4$	$\longrightarrow$	$\rightarrow Re-N_2-Mo\leftarrow$	1,680

NH-, a process which could also be described as H-atom transfer, and (b) transfer of six electrons to form nitride N^{3-} and subsequent protonation. Much work has been done on systems containing Ti^{II} . . . N_2 intermediates which will protonate to give good yields of ammonia if very strong reducing agents such as sodium naphthalide

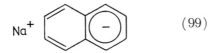

$$ (99) $$

are present, and it is thought that here mechanism (b) is involved. The reducing conditions are by no means mild and biological nitrogen fixation is more likely to proceed by mechanism (a), so such systems will not be considered further. All we can really say about N_2 fixation on the basis of the activation shown by dinuclear metal $-N_2$ complexes is that the

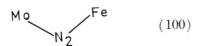

$$ (100) $$

type of site seems possible and even plausible. It is a field which is being very actively pursued, and new developments occur rapidly. We hope that it will not be long before as much is known about N_2 fixation as about the biological reactions of O_2.

We have tried throughout the discussion to show how a knowledge of inorganic chemistry can lead to a greater understanding and appreciation of the complex processes involved in biological systems. While as yet this approach has not proved too successful in the study of nitrogen fixation, considerably more success has been achieved in the elucidation of the mechanism of O_2-macromolecule interactions. One point, however, is worth emphasizing—the time has come when we must adopt a kind of scientific ecumenism and bury any prejudices we may have towards scientists of a "different creed," for it is only by the united efforts of interdisciplinary teams of chemists, physicists, biologists and medical research workers, that really significant advances will be made in these fields of molecular biology and biophysics, together with their extremely important practical applications such as that of combating disease on the molecular level. The potential of such cooperation is vast and must surely present a most exciting challenge to any student of molecular biology today.

REFERENCES

1. Cotton, F. A., and Wilkinson, G.: *Advanced Inorganic Chemistry*, 3rd ed. Wiley, 1972.
2. Phillips, C. S. G., and Williams, R. J. P.: *Inorganic Chemistry*. Oxford U Pr, 1966, vol. II.
3. Taube, H.: Mechanisms of oxidation with oxygen. *J Gen Physiol*, 49:29, 1965.
4. Fallab, S.: Reactions with molecular oxygen. *Angew Chemie*, Int Ed, 6:496, 1967.
5. Wilkins, R.: Uptake of O_2 by Co(II) complexes in solution. Bio-inorganic Chemistry.

Advances in Chem, Series 100, American Chem Soc, 1971.

6. Choy, V. J., and O'Connor, C. J.: Chelating dioxygen complexes of the platinum metals. *Co-ord Chem Revs,* 9:145, 1972/73.

7. Arhland, S.: Factors contributing to (b)-behaviour in acceptors. *Structure and Bonding.* Springer-Verlag, vol. 1, p. 207.

8. Williams, R. J. P., and Hale, J. D.: Classification of acceptors and donors in inorganic reactions. *Structure and Bonding.* Springer-Verlag, vol. 1, p. 249, 1966, New York.

9. Perutz, M. F.: Stereochemistry of co-operative effects in haemoglobin. *Nature, 228:*726, 1970.

10. Bearden, A. J., and Dunham, W. R.: Iron electronic configurations in proteins. *Structure and Bonding.* Springer-Verlag, 1970, vol. 8, p. 1.

11. Malmstrom, B. G.: Copper in biological systems. *Adv Enzymol,* 33:177, 1970.

12. Frieden, E., Saki, S. O., and Kobayashi, H.: Copper proteins and oxygen. *J Gen Physiol, 49:*213, 1965.

13. Chatt, J., and Leigh, G. J.: Nitrogen fixation. *Chem Revs, 1:*121, 1972.

CHAPTER 6

MIXED LIGAND COMPLEXES AND THEIR BIOLOGICAL SIGNIFICANCE

R.-P. Martin and J. P. Scharff

Introduction
Mixed Ligand Complexes
Biological Significance of Mixed
  Complexes
References

INTRODUCTION

Two or more different ligands may compete for the same central metal ion to form mixed ligand complexes. Ligands, excluding solvent molecules, may differ either by their nature, their optical configuration, their ionization state, or their binding sites. Such species are commonly formed in natural and biological fluids where various metal ions interact with various potential ligands. We shall restrict our discussion to the case of ternary complexes since little is known yet about mixed complexes involving more than two different ligands except that some have been shown to be the catalytic entity in several enzymatic processes.[1] Amino acids, peptides, proteins and to a lesser extent nucleic acids and their metabolites, have been studied in this respect.

An illustration of mixed complex formation in biological fluids is given by the experiments of Neumann and Sass-Kortsak[2] who used ultracentrifuge and radioactive tracer techniques to study the competition between albumin and amino acids for the Cu(II) present in human blood serum. Normal human plasma contains close to 15 μM of copper shared between a nonexchangeable fraction (ceruloplasmin), a more labile form called albumin, and some of the twenty-three amino acids present at concentrations less than 1 mM. A rather surprising finding was that when all, or most, of the normally occurring amino acids were present together in serum at physiological concentrations, they apparently bind a much larger proportion of the nonceruloplasmin bound copper in serum than might be expected on the basis of their individual binding capacities. Thus, it was assumed that in addition to the binary complexes of amino acids with copper, so-called mixed complexes are formed, these consisting of one ion of copper and two *different* amino acids. The results suggested that among the various amino acids, histidine, threonine, glutamine and asparagine are most probably involved in this mixed ligand complex formation.

A number of similar mixed species have now been detected by both direct and indirect methods involving physicochemical measurements, i.e. equilibrium analysis of potentiometric data, and even by isolating these mixed complexes in the solid state.[3, 4] Structures have been elucidated both in solution and in the solid state.

We have recently reviewed mixed ligand metal ion complexes involving amino acids and peptides[3] there being little known about the chemistry of mixed coordination involving proteins. In this latter case, in-

vestigations on a molecular basis are difficult (1) since proteins contain a number
of different potential ligand groups (such
as imidazole, amino, carboxyl, phenoxy and
often sulphydryl groups) and (2) because
of the large structural and conformational
effects on the properties of metal binding
sites.

Interest in mixed chelation also arises
from a study of reaction mechanisms in
which the metal ion or bound ligand becomes activated for selective reactions.
Martell has recently reviewed this topic.[5a]
Hydrolysis of α-amino acid esters, stereoselective decarboxylation, Schiff base formation, and amino acid oxidase models provide typical examples. (These aspects are
treated in Chapter 5.)

Metal complexes appear to be a central
problem in biochemistry as, in addition to
their possible role in the transport of trace
metals, they are also found as essential cofactors for many enzymes. Enzymes have
been classified as metal-activated enzymes
and as metalloenzymes, the latter differing
from the former by exhibiting a radical
loss of activity when the metal is removed.
Mechanisms involving metal ions (alkaline
earths, most of the first transition series
ions, zinc, cadmium, and molybdenum) can
be of several types. In some instances, i.e.
Mg^{++} in pyruvic decarboxylase, the metal
ion favors the binding of apoenzyme and coenzyme and in other cases (Ca^{++} for α-amy-
lase), they induce configuration changes
necessary for the enzyme action; some
metals are found as enzyme activators, the
process being accomplished through ternary complex formation between enzyme,
metal ion and substrate, i.e. in galactose
oxidase where copper ions form mixed
complexes with substrates. Finally, metal
ions can act indirectly, for instance when
an inhibitor binds to an enzyme and deactivates it.

This chapter is limited to the *properties*
of mixed ligand complexes and their biological significance.

MIXED LIGAND COMPLEXES

The competition between metal ions
and organic ligands obeys various rules. A
particularly good example system is that of
metal ion—amino acids—peptides or proteins. This system will be taken as an illustration in the following section.

Thermodynamic Parameters

Mixed ligand complexes in general have
been studied using physicochemical methods and thermodynamic and structural view
points.

Quantitative information concerning interactions between a metal ion M, ligand
A and ligand B in a protic solvent may be
conveniently described by the formation
constant, i.e. overall stability constant, of
the species $M_pH_qA_rB_s$ formed according to:

$$pM + qH + rA + sB \rightleftharpoons M_pH_qA_rB_s$$

The stability is measured by the stoichiometric equilibrium constant β_{pqrs} expressed
in terms of concentrations at constant temperature and ionic strength.

$$\beta_{pqrs} = \frac{[M_pH_qA_rB_s]}{M^pH^qA^rB^s}$$

where m, h, a, b represent the concentrations of the free components. Conventionally, parameter q is negative when the number of protons dissociating from ligands A
or B exceeds the maximum number of protons dissociating in the absence of metal
ion or when protons dissociate from the
coordinated water molecules, i.e. when
hydroxy complexes are formed.

These stability constants are usually determined by precise potentiometry rather
than from other physicochemical methods
such as spectrometry, polarography, calorimetry, liquid-liquid partition, or solubility.
However, special mention has been made

TABLE 6-I

THERMODYNAMIC PROPERTIES FOR MIXED COMPLEX FORMATION REACTIONS
Cu + A + B ⇌ CuAB

Ligand A	Ligand B	Log β_{11}	ΔG_{11}	ΔH_{11}	ΔS_{11}
Glycine[*]	α-Alanine	15.05	−85.90	−49.16	123.0
Glycine[*]	Valine	15.06	−85.98	−49.54	122.2
Glycine[*]	Serine	14.66	−83.68	−48.95	116.3
α-Alanine[*]	Valine	15.20	−86.78	−49.29	125.5
α-Alanine[*]	Serine	14.91	−85.10	−48.95	121.3
Valine[*]	Serine	14.84	−84.68	−48.74	120.5
Glycine[†]	Sarcosine	15.59	−88.95	−45.65	145.6
Glycine[†]	Alanine	15.81	−90.21	−51.67	129.3
Alanine[†]	Aminoisobutyric acid	15.77	−90.00	−50.29	133.1
Glycine[†]	Aminoisobutyric acid	15.89	−90.67	−48.07	143.1

ΔG_{11}: kJ mol^{-1}; ΔH_{11}: kJ mol^{-1}; ΔS_{11}: JK^{-1} mol^{-1}.
[*] Ting, P. I. and Nancollas, G. H.: *Inorg Chem*, 11:2414, 1972.
[†] Yingst, A., Izatt, R. M. and Christensen, J. J.: *J Chem Soc*, Dalton trans., 12:1189, 1972.

of polarimetry because amino acids, peptides and their metal complexes often possess molecular optical rotations which are suitable for a thorough analysis.

In the present instance, the stoichiometries of mixed species may be deduced from those of their parent complexes. This holds for copper complexes with monodentate (A = NH$_3$) and bidentate (B = α-amino acids) ligands; for example, CuAB and CuA$_2$B in equilibrium with CuA$_r$ (r ⩽ 4) and CuB$_s$ (s ⩽ 2), or zinc and cadmium complexes (r ⩽ 6, s ⩽ 3) where six theoretical mixed complexes must be expected: MAB, MA$_2$B MA$_3$B, MA$_4$B, MAB$_2$, and MA$_2$B$_2$. Tables 6-I and 6-II provide examples where both A and B are α-amino acids and bidentate ligands. A single mixed copper complex is formed, while nickel is shared between NiAB, NiA$_2$B, and NiAB$_2$. The presence of a third ligand leads to the formation of NiABC, i.e. in the Ni(II)-

TABLE 6-II

MIXED COMPLEXES STABILIZATION $\Delta_{10rs} = \log\beta_{10rs(exp.)} - \log\beta_{10rs(calc.)}$

Systems		Δ_{1011}	Δ_{1012}	Δ_{1021}
Cu-glycine-DL-α-alanine	NaClO$_4$ M-25°C[*]	0.12		
Cu-glycine-L-tyrosine	NaClO$_4$ M-25°C[†]	0.19		
Cu-DL-α-alanine-L-tyrosine	NaClO$_4$ M-25°C[†]	0.50		
Cu-L-histidine-L-threonine	KNO$_3$ 0.1M-25°C[‡]	0.87		
Ni-glycine-DL-α-alanine	NaClO$_4$ M-25°C[§]	0.17	0.27	0.12
Ni-glycine-DL-valine	NaClO$_4$ M-25°C[‖]	0.23	0.38	0.41
Ni-DL-α-alanine-DL-valine	NaClO$_4$ M-25°C[‖]	0.19	0.33	0.07
Ni-glycine-diglycine	NaClO$_4$ M-25°C[§]	0.14	0.22	0.08

[*] Martin, R.-P. and Pâris, R. A.: *CR Acad Sci* Paris, 258:3038, 1964.
[†] Martin, R.-P. and Pâris, R. A.: *CR Acad Sci* Paris, 257:3932, 1963.
[‡] Freeman, H. C. and Martin, R.-P.: *J Biol Chem*, 244:4823, 1969.
[§] Martin, R.-P. and Mosoni, L.: *Bull Soc Chim Fr*, 2917, 1970.
[‖] Scharff, J. P. and Morin, M.: *Bull Soc Chim Fr*, 2198, 1973.

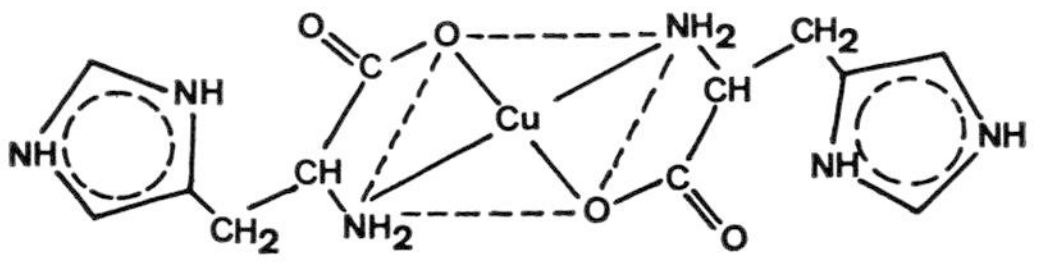

Figure 6-1. Cu(HHis)$_2{}^{2+}$

glycine-DL-a-alanine-DLvaline system studied by Cromer-Morin, Martin and Scharff[23] in 1973.

Mixed complex formation is usually strongly dependent upon the number and the type of donor atoms and the configuration of the ligands. The N(amino), N(imidazole) and O(carboxyl) donor atoms of histidine may either bind to a metal ion simultaneously, or in pairs or even just one at a time. In the physiological pH range the N(imidazole) atom is partly protonated (pK = 6) and equilibria are involved in which metal is found in different complexed forms some of which have N(imidazole) bonded and others not.

In solutions of low pH (pH < 4), it is conceivable that only N(amino) and O (carboxyl) atoms coordinate copper in a glycine-like manner, the protonated imidazole being away from the metal as shown by X-ray structure analysis of solids (Figure 6-1). Imidazole nitrogen hydrogens successively dissociate at higher pHs and then become donors. The first step: Cu (HHis)$_2{}^{2+}$ $\rightleftharpoons$ Cu(HHis)(His)$^+$ + H$^+$ can then lead to the mixed species Cu(HHis) (His)$^+$ which may be described by Figures

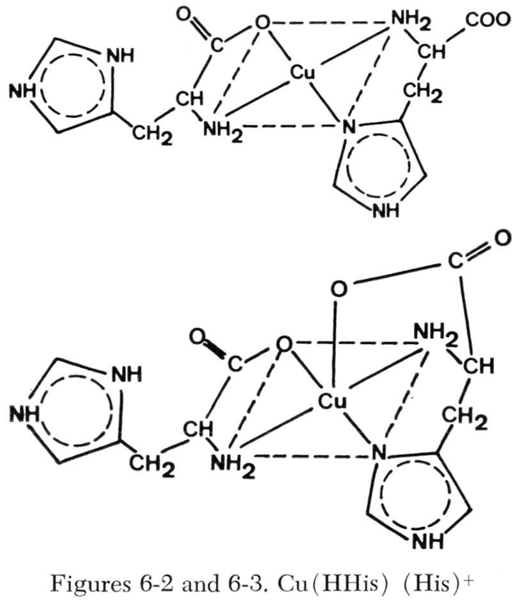

a

$\Updownarrow$

b

$\Updownarrow$

Figures 6-2 and 6-3. Cu(HHis) (His)$^+$

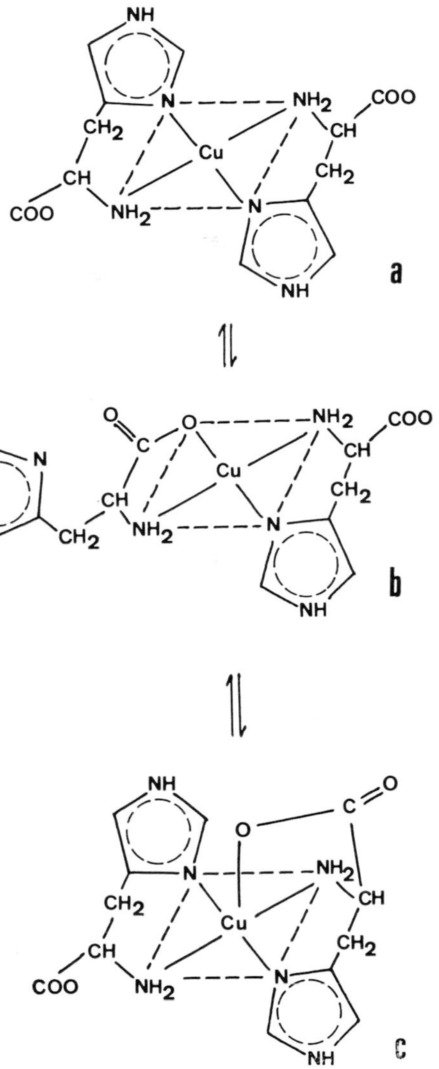

c

Figure 6-4. Cu(His)$_2$

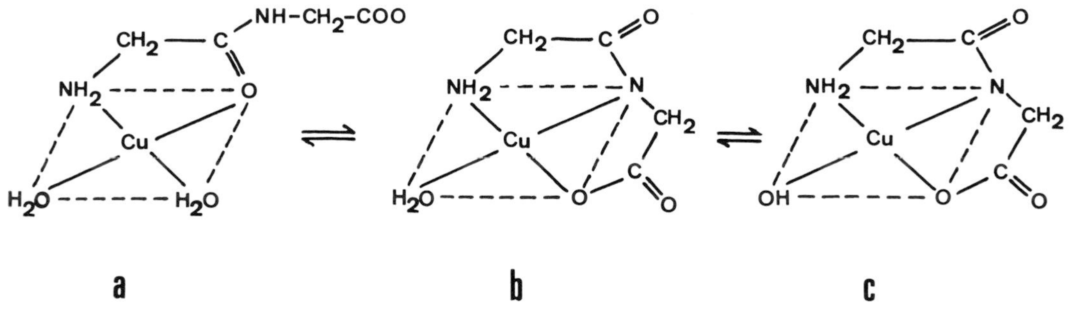

Figure 6-5

Figure 6-5a. CuB⁺; Figure 6-5b. CuH₋₁B; Figure 6-5c. Cu(H₋₁B) OH⁻.

6-2 or 6-3. At neutral pH, histidine bound to copper in Cu(His)₂ can possibly have three different forms according to Figure 6-4. These choices have been resolved using thermodynamic analysis.

The tridentate form of histidine (Figures 6-3 and 6-4) so far has only been observed in solids for the mixed species copper-histidine-threonine (by X-ray diffraction) but it is highly likely that this configuration is maintained in Cu(II)-histidinate complexes in solution. This suggestion is further supported by the recent observation that a stereoselective effect has been detected in Cu(L-His)(D-His).

Spectroscopic determinations also support a similar behavior for L-threonine and copper at pHs > 10 where N(amino) and O(alcoxy) are found in the coordination plane producing a strongly puckered chelate ring that has the O(carboxyl) in a position suitable for binding copper in an irregular axial position.

Peptide complexes have been extensively investigated in both the solution and solid states.[6] As far as transition series metal ions are concerned, N(amino) terminal N, O and N(peptide) donor atom and also O(carboxyl) may all be involved in coordination. The lateral chains also have specific

roles and N(peptide) involvement characterizes Ni(II) and Cu(II) complexes.

Let us consider the 1:1 copper complexes of diglycine (Figure 6-5) and triglycine (Figure 6-6).

Each water molecule in these complexes are exchangeable in solution and in the solid state with imidazole molecules.

Glycine(A) produces mixed complexes of forms 7 and 10 by displacing two water molecules. Both B and B′ are bidentate in *cis* or *trans* position with respect to glycine. Forms 8 and 11 behave quite differently: Mixed chelate formation becomes selective as an oxygen donor atom has to be displaced to give a stable species. The stronger coordinate bond Cu-O(carboxyl) makes the mixed Cu-trigly-gly species more stable.

Such coordination provides a good model for the binding of organic ligands to protein that is mediated by having copper at the terminal amino group. This specificity does not occur when the metal is zinc. Perrin et al.[7] detected two mixed species ZnAB and ZnAB(OH) where A-glycine and B-diglycine or triglycine. The peptide nitrogens do not dissociate and the same donor atoms are involved in both simple and mixed species. Both diglycine and triglycine are bidentate and are bound to zinc

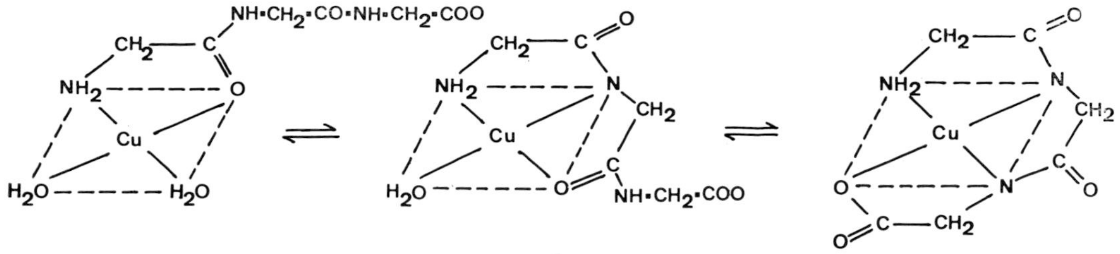

a   b   c

Figure 6-6

Figure 6-6a. CuB'^+ (pH 4-5); Figure 6-6b. $CuH_{-1}B'$ (pH 6-8) Figure 6-6c. $CuH_{-2}B'-$ (pH 8-10). N.B.: Figures 6-1 to 6-6c were not designed to be entirely correct: electric charges are generally omitted for the sake of clarity and for instance in figures 6-5b through 6-6c when a Cu(II) is bound at the N(peptide) atom, resonance occurs between the forms

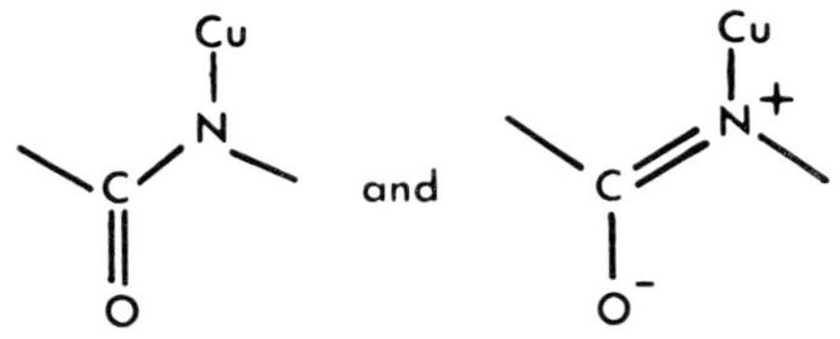

through their N(amino) and O(carboxyl) atoms. Unfortunately, little is known about peptides having functional side chains from their amino acid residues.

Proteins, and especially albumin, are found in considerable amounts in biological fluids (such as human serum, 0.65 mM). The sequences of amino acids provide two main sites for binding copper and zinc to each albumin molecule. It has been demonstrated that these sites bind the metal at least as firmly as the individual parent a-amino acids. Cysteines (in the case of zinc) and histidine compete even more efficiently when these metal ions are bound to albumin. Giroux and Henkin[8] calculated that 98 percent of loosely bound zinc ions are coordinated to albumin and are in equilibrium with amino acids present at physiological concentrations. A fivefold increase in the amount of cysteines present reduces this figure to 74 percent. These au-

thors have not considered the formation of mixed albumin-zinc-amino-acid complexes and their contribution to the fraction of zinc bonded to albumin. Histidine has been shown to be complexed to albumin via copper and investigations of many molecular interactions of this type are in progress in several laboratories. In studies of reaction mechanisms and thermodynamic properties, the main use of formation constants is in calculating the concentration of each complex present at equilibrium. This approach has been made possible by the development of computer programs designed to solve multiple simultaneous equations.

For instance, one of these programs: COMICS (Concentration of Metal Ions and Complexing Species) has been used to calculate the concentrations of species likely to be present in a solution containing Cu(II) and Zn(II) and sixteen of the amino acids found in human blood plasma.[7] The conclusions from Sass-Kortsak's experiments discussed above are confirmed and may be stated precisely: At pH 7.4 (T = 37°C, ionic strength = 0.15M (KNO_3)), most of the Cu(II) present is coordinated to cystine (CySSCy) and histidine (His).

The highest concentrations of complexes are seen to be CuH(His)(CySSCy) 45.4%, Cu(His)(CySSCy)⁻ 39.6%, Cu(His)₂ 10.7%, and the free metal concentrations are $[Cu^{2+}] = 10^{-11}M$ and $[Zn^{2+}] = 10^{-6}M$. It is interesting to note that the first of these mixed complexes is uncharged and so may facilitate its transport across membranes.

These types of calculations have been used by Stumm and Morgan in 1970 in order to obtain a comprehensive picture of the possible chelating influence of naturally occurring organic compounds in natural waters. For their purpose, a hypothetical multimetal-multiligand system was considered. It included organic ligands such as glycine, citrate, nitrilotriacetate, salicylate (which contains functional groups similar to that of humic acids) and nocardamine (a ferrichrome). However, mixed complex formation was not incorporated even though such species are known to be present, e.g. Fe(III)-nitrilotriacetate-sulfosalicylate.

The thermodynamic properties which have to be taken into account are enthalpy ($\Delta H°$), entropy ($\Delta S°$) and free energy ($\Delta G°$) changes.

The enthalpy of complex formation can usually be interpreted as the strengths (energy) of the bonds between the central cation and ligands compared to those between the central cation and solvent molecules, i.e. heat of complex formation—heat of removing water molecules (in aqueous media) from the metal ion and the aquated protonated ligands.

The entropy of complex formation includes the difference between the transitional (and rotational, for polyatomic ligands) entropy lost by the ligands and the corresponding entropy gained by the displaced solvent molecules. There is also a statistical term which will be discussed later.

Strong complex formation expressed by a high value of β_{pqrs} requires a large negative enthalpy and a large positive entropy of complex formation. Naturally, a large negative enthalpy change can counterbalance and override small or negative entropy changes and *vice versa*.

The free energy change, $\Delta G°$, is related to the corresponding thermodynamic equilibrium constant $\beta°_{pqrs}$ by the Gibbs-Helmholtz expression:

$$\Delta G° = -2.303 \ RT \ \log \ \beta°_{pqrs}$$

and equals $-5.707 \log \beta°_{pqrs}$ kJ mol⁻¹ at 25°C.

Several attempts have been made to correlate $\Delta G°$ of mixed complex formation with other parameters characterizing the system. In 1967 Jackobs and Margerum[25] found a relationship between the change in free energy during the formation of mixed ligand Ni complexes and factors such as ion-ion and ion-dipole interactions, chelate ring formation, the nature of coordination bonds formed, and the influences of groups coordinated. Such relations have been verified in several cases (for instance, for the mixed complexes of Ni(glycine)B with B = nitrilotriacetic acid, ammonia or diethylenetriamine). In 1973 Tanaka[26] developed mechanistic considerations of formation constant for mixed ligand complexes of nickel involving amines and amino carboxylates.

Friedman et al.[5b] have used a different approach to quantifying the stabilization of mixed complexes. They determined the free energy change when a central metal ion and two different ligands are in solution of infinite dilution, and also when the ligands are located at distances determined by the coordinate bonds. Such treatments can help to predict the behavior of a given system provided that several analogous systems have already been studied. The fact that this method involves a number of antecedent experimental determinations limits its usefulness.

Much early work was concerned solely with the measurement of formation constants but more recently workers have recorded the dependence of log $\beta^{\circ}{}_{pqrs}$ values upon temperature in order to calculate ΔH° values. It is clearly highly desirable that enthalpy changes should be obtained by direct calorimetric measurement. Then, assuming that the formation constant is known, the entropy change may be calculated. However, much thermodynamic data for mixed ligand complexes is still lacking. This data is very necessary to yield information regarding effects such as the statistical effect, repulsion between unlike ligands, geometric factors, dipole interactions with the solvent, the number and type of bonds present and the effect of chelation. Indirect information concerning these phenomena should be carefully interpreted.

In fact there are but a few publications on the subject of thermodynamics of mixed complexes and we have tabulated known data in Table 6-I. Authors have discussed these results in terms of statistical models and have shown that the enhancement factor term that applies to mixed ligand formation constants does not necessarily arise solely from the entropy term but rather from several sources. These include various steric repulsion terms, specific hydration terms and considerations of the altered charge distribution in the mixed complexes. These detailed investigations provide useful insights into the many activation mechanisms involved in mixed complex formation.

FORMATION OF MIXED LIGANDS COMPLEXES

The process of mixed ligand complex fomation is governed by several general rules drawn from numerous experiments.[3, 4, 5b]

First, it is clear that the stereochemical configurations of the parent complexes, i.e. of the binary species, are of great importance and that steric hindrance can often prevent the formation of mixed complexes. Kida[27] has suggested that in general mixed complexes cannot be formed when their parent complexes have different geometrical configurations, i.e. when one parent binary complex has a planar configuration and the other parent is octahedral. In other instances in which no such obvious physical factors are involved, the following general principles have begun to emerge from the experimental data.

According to Kato,[28] the binding energy diagrams for ligands plotted as a function of their ligand field strength leads to the conclusion that any pair of ligands which form only high-spin, or only low-spin complexes with a given transition metal ion are capable of forming mixed complexes with it. On the other hand, it is difficult to obtain a mixed complex when the metal ion forms a high-spin complex with one ligand and a low-spin complex with another. In this latter case the mixed complex has a tendency to disproportionate into two binary complexes, each having only one type of ligand coordinated to the metal.

Kida[29] has suggested that mixed complex formation depends upon the types of bonds formed between the metal ion and the ligands. This aspect is expected to become more important in the near future as molecular orbital computations for coordination compounds are developed. It has been verified by calculations that the electrostatic effect exerts an influence that promotes the formation of mixed complexes. Similarly, σ-covalency of the coordination bond is expected to affect mixed complex formation. Alternatively, π-bonding ligands can act as acceptors for d electrons from metal ions having full, or almost full, d orbitals, through the ligands using their vacant p orbitals (or π-antibonding orbitals) for this purpose. This phenomenon, the so-called back-coordination or back-donation, lowers the electron density near the central metal ion and hence has an effect opposite to that of σ-bonding ligands. Thus, if ligands are σ or π-bonding, mixed complexes can be formed, while π-bonding ligands tend to exclude non π-bonding ligands and so form unstable complexes.

Thus, it seems that ligands of unlike "types" are liable to form unstable complexes. One such classification of ligand "type" features in the HSAB (hard and soft acids and bases) theory in which Pearson[30] regards metal ions as Lewis-acids and the ligands as Lewis-bases as described in Chapter 3. Jørgensen[31] has extended the principle to the formation of mixed ligand complexes and suggested that "soft bases tend to group together on a given central atom or ion and that hard ligands tend to group together." This mutual stabilizing effect is called *symbiosis*. Many such examples are known to support this, even though several exceptions have also been found. In fact, symbiosis is often counter balanced by other factors. If this were not so, all equilibria of the type $\frac{1}{2}MA_{2r} + \frac{1}{2}MB_{2s} \rightleftharpoons MA_rB_s$, would lie to the left while they can, of course, go in either direction. Fridman[5b] has pointed out that if the "soft with soft, hard with hard" principle is applied without taking note of the structural properties of the complexes, deviations will undoubtedly be observed from the principle. The principle of symbiosis holds true only if the ligands have very different properties, i.e. F^- and CN^- are not compatible. However, as the hardness difference decreases their mixed ligand complex compatibility increases. This increase is not monotonous as with certain optimum differences the mixed ligand complexes are more stable than the compounds containing the same ligands.

An interesting case of mixed ligand complex formation is provided by enantiomeric ligands. Stereoselectivity must occur when the formation constants for the following reactions are found to be different:

$$M(L-A) + (D-A) \rightleftharpoons M(L-A)(D-A)$$

and

$$M(L-A) + (L-A) \rightleftharpoons M(L-A)_2$$

For statistical reasons, 50 percent of the the metal ought to be found as $M(L-A)(D-A)$ in solutions of M and the racemic mixture of ligands A. Potentiometric, or better, enthalpimetric, also circular dichroism, nuclear magnetic resonance and absorption, measurements showed significant deviations from this value of 50 percent. Formation of the racemic species $Co(D-His)(L-His)$ has been shown by NMR and potentiometric titration to be preferred to that of the optically pure complex $Co(L-His)_2$ and $Co(D-His)_2$.

With copper and nickel ions, Barnes and Pettit[32] found significant stereoselectivity to be absent or very small with asparagine. Conversely for histidine complexes stereoselectivity occurred with all metals investigated (Cu^{2+}, Ni^{2+}, Zn^{2+}).

Stereoselectivity amongst mixed complexes of the type MAB has also been detected and measured by comparing equilibria:

$$M(L-A) + (D-B) \rightleftharpoons M(L-A)(D-B)$$

and

$$M(L-A) + (L-B) \rightleftharpoons M(L-A)(L-B)$$

It is apparently absent in the system Cu(II)-(L or D-histidine)-(L-threonine) (histidine is tri or bidentate, L-threonine is bidentate) (Freeman and Martin,[33] 1970). However mixed complex formation between Zn(II), (L-histidine) and (L or

D-aspartic acid) is stereoselective (Riaute and Martin,[34] 1973).

This stereoselective effect is stronger than that observed in the corresponding binary species. However, the few data concerning such systems are sometimes unreliable and many further investigations are clearly necessary. Nevertheless, the adage is often stated—tridentate ligands are necessary for stereoselectivity.

Stability of Mixed Complexes

Mixed ligand complexes can be considered as models for enzyme-metal ion-substrate interactions and investigations into the stability of mixed complexes could result in a better understanding of the formation and of the function of metalloenzymes as biological catalysts. It has been suggested that complex stability may be the principal factor in determining the specificity in metal ion activation of enzymes. However, an examination of data available to date suggests that direct correlations is not usually present. For example, many enzymes (phosphokinases, enolases) have Mg^{2+} as the best coenzyme and Mn^{2+} is the best coenzyme with many peptidases; however Mg^{2+} and Mn^{2+} form relatively weak complexes with both substrates and proteins. So Schubert[35] has suggested that it is not the *absolute* but rather the *relative* stabilities of the enzyme and substrate complexes that must be considered. Specificity usually depends on several factors: the nature of the metal ion, donor atoms, position of lateral chains, and the associated ligand field which dictates both thermodynamic and kinetic behaviors. Most of the research into mixed coordination corroborates the concept that mixed species often have very high stabilities:

The stability of the mixed species with respect to those of the parent binary complexes may be expressed in terms of the quantity $\log \chi_{rs}$. For example, for the reaction $MA_2 + MB_2 \rightleftharpoons 2 MAB$ (formation constant $= \chi_{11}$), the position of this equilibrium, and consequently the tendency for mixed complex formation is determined by the sign and the value of χ_{11}:

$$\log \chi_{11} = 2 \log \beta_{1011} - (\log \beta_{1020} + \log \beta_{1002})$$

Sigel[9] has suggested an alternative expression for mixed complex formation by considering the following equilibria:

$$M + B \rightleftharpoons MB \quad K_{MB}^{M} = [MB]/[M][B]$$
$$MA + B \rightleftharpoons MAB \quad K_{MAB}^{MA} = [MAB]/[MA][B]$$

He uses the difference between the stabilities of the binary and of the ternary complexes to characterize the tendency to form ternary complexes

$$\Delta \log K = \log K_{MAB}^{MA} - \log K_{MB}^{M} = \log K_{MAB}^{MB} - \log K_{MA}^{M}$$

This relationship may be generalized to:

$$\Delta \log K_{rs} = \log \beta_{10rs} - (\log \beta_{10r0} + \log \beta_{100s})$$

We have already mentioned that a statistical factor governs the manner in which mixed ligand species are formed relative to their parent complexes. Sharma and Schubert have generalized this idea for a general complex MA_rB_s and expressed the statistical factor as $S = \frac{(r+s)!}{r!s!}$. Assuming that the stability constants of parent complexes are known, the corresponding β_{10rs} constant can be estimated from the relationship:

$$\log \beta_{10rs(calc.)} = \log S + \frac{r}{r+s}$$
$$\log \beta_{10(r+s)0} + \frac{s}{r+s} \log \beta_{100(r+s)}$$

However, the difference $\Delta_{10rs} = \log \beta_{10rs(experimental)} - \log \beta_{10rs(calculated)}$ is usually found to be positive (Table 6-II). Thus, the stabilization of mixed complexes cannot be explained by purely statistical grounds and this is especially true when the chelate ring formed by the individual ligands are similar in nature. Many other factors such as thermodynamic considerations,[5b] electrostatic factors, the nature of the solvent, the influence of ionic strength, asymmetry of mixed complexes, and ligand-ligand interactions have been discussed in order to explain this extra stabilization. (Detailed studies of all these factors are available in Sigel's "Metal Ions in Biological Systems" and in other publications listed therein.) We do not propose to comment upon all of them, but in connection with biochemical coordination chemistry it is necessary to consider some aspects of the problem.

Acidity constants for the ligands are very strongly dependent upon the polarity of the solvent. It is, therefore, interesting to compare the influence of pH on binding sites of potential biochemical interest. Under physiological conditions it appears that the imidazole group is a much more effective binding site than the amino groups. In 1952, Tanford[36] showed that in the binding of copper ions to serum albumin the principal binding sites were the imidazole groups of the histidine residues of the protein molecule. At pH 7, both the imidazole and the carboxylate groups show approximately the same coordination tendency towards Mn^{2+}, Mg^{2+}, Fe^{2+} and Zn^{2+}; small variations in pH conditions can favor the coordination of either one of these two groups. The phosphate group usually has just one free ionizable proton and at neutral pH this is dissociated to give a negatively charged ligand, the coordination tendency of which is only very slightly influenced by the competition for protons at pH $\geqslant 7$. However, this is not the situation for binding sites containing the SH group; for this group the proton competition is very powerful and only those metal ions having a high coordination tendency towards S^-, i.e. Cu^+, Zn^{2+}, Cd^{2+}, Ni^{2+}, Hg^{2+} can coordinate.[9] Finally, we must

point out that pH gradients in membranes are known to seriously affect the transport of ions and of molecules, a phenomenon which is often dependent upon coordination phenomena.

The stabilization of mixed complexes is considerably influenced by ionic strength variations, a fact of great importance in biological fluids where parameters such as the ionic strength and dielectric constant can be extremely variable. The effects of ionic strength upon the stability of mixed complexes are usually large when the corresponding binary complexes are of different charge.

According to Sigel and McCormick,[9] changes in the polarity of biological fluids can occur through the creation of structured water regions, i.e. in the vicinity of enzymes and of their active sites. Thus, it seems possible that the factors described here are "tools" that are used by nature to favor one side of an equilibrium or the other. Hence, it appears that the variety of roles that metal ions might play in biological reactions can range from weak ionic strength effects to highly specific complex associations.

BIOLOGICAL SIGNIFICANCE OF MIXED COMPLEXES

Mixed Coordination in Metal Ion Catalysis in Biological Systems

Most of the classes of enzymes include several metal-dependent enzymes. According to Vallee and Williams,[10] such enzymes may be classified either as "metal activated enzymes" or as "metalloenzymes." The difference lies in the strength of the binding of metal to enzyme. In metalloenzymes, the metal is slowly exchangeable and usually belongs to the active site. In such cases, geometric factors are of greater importance. The metal ions in metalloenzymes have the advantage of providing an electronic marker at the active site: Electronic, epr, nmr, and to a lesser extent Mössbauer spectroscopy have all been used successfully for such investigations. Substitution of the metalloenzyme metal by other metal ions may provide more suitable characteristics (transition metal ions Mn^{2+} and Co^{2+} favorably replace Mg^{2+} and Zn^{2+} respectively) and such substitutions are usually milder than many organic modifications. A comparative behavioral study of exchanged metalloenzymes sometimes presents some interesting correlations between physicochemical properties and enzyme activity.

As yet, mixed chelate complexes involving amino acids and simple peptides (or denatured proteins) have not provided even approximate models for activation process such as occur in metalloenzymes or metal activated enzymes. However, a future design of such models might investigate the binding capacity of side chains bearing functional or polar groups.

Several reviews have been published concerning the various aspects of metalloenzymes or metal activated enzymes[1, 10–12] and we shall limit our discussion to: (1) the detection of ternary or quaternary species, (2) the geometric and electronic structures of the active site of the enzymes and their complexes with substrates and products, and (3) the affinities and specificities of substrates.

Interactions between Metal Ions and Macromolecules

Interactions between metal ions and macromolecules are of several types:

Metal Ions Participating in the Control of Protein Structures

It has been known for several years that if zinc is removed from yeast alcohol dehydrogenase (by introducing chelating agents) the protein dissociates into subunits. Hydrogen isotope exchange techniques have proved to be very useful for delineating the structural roles of metals in proteins and the problem of stabilization of protein structure. In fact, hydrogen-tritium exchange in metallo- and apo-protein are significantly different. For example, thionein hydrogen exchange is almost instantaneous whereas metallo-thionein, the native protein, retains twenty-six of its ninety-four potentially exchangeable hydrogen atoms (Figure 6-7). Approximately ten of these atoms exchange slowly over the next five hours. Addition of metal ions to the apo-protein partly restores the slow exchange characteristics of an organized structure or of a more compact macromolecule. Thus, metals retard the hydrogen exchange of proteins through structural alterations and it has been shown by other means, e.g. thermal perturbation, that the degree of retardation of exchange is proportional to the influence of metals on structure. Figure 6-8 shows the effect of copper,

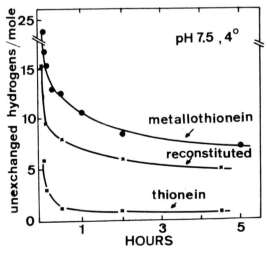

Figure 6-7. Hydrogen-tritium exchange of thionein and metallothionein.[11] (Used with permission. Amer Chem Soc, *Bioinorganic Chemistry.*)

manganese and iron on the optical rotation of conalbumin when this protein is heated. The stabilization of conalbumin against heating has the order iron > manganese > copper and this same sequence holds for the effectiveness of these metals in retarding the hydrogen-tritium exchange of conalbumin.

From studies of the iron-transferrin system, the metal ion appears to affect the exchange of hydrogens that are distributed throughout the molecule and this must influence the total macromolecular configuration. Modifications of optical rotatory dispersion ord spectra of the protein caused by the presence of various metal ions suggests the same conclusion.

These observations suggest that metals, by modulating equilibria between closely related states of protein which have different conformational energies may well influence enzymic activities, protein-protein interactions, or turnover rates for proteins in biological fluids. They may also exert important regulatory constraints upon biochemical reactions. On the other hand, metal ions may split the hydrogen bonds resulting in the formation of two single strands from a double-stranded deoxyribo-

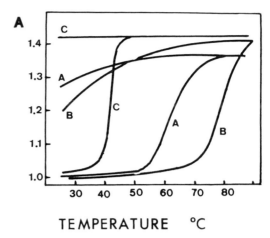

TEMPERATURE   °C

Figure 6-9. Change in absorbance of DNA solution on heating and cooling.[12] (Used with permission. Amer Chem Soc, *Reactions of Coordinated Ligands and Homogeneous Catalysis.*) A. In absence of divalent metal; B. With Mg^{2+}; C. With Cu^{2+}.

nucleic acid (DNA) molecule. Since the single-stranded molecule absorbs much more strongly than the double-stranded molecule, the splitting may be followed by variations of the absorption peak at 260 nm and Figure 6-9 illustrates both this stabilizing effect of Mg^{2+} acting on phosphate groups and the splitting action of Cu^{2+} ions. Metal ion stabilization of quaternary structures of proteins and enzymes merits further research. The structural roles of these cofactors while not as specific as the activation processes appear to be very important to an understanding of the mechanisms of enzymic catalysis.

Metal Ions Acting as Lewis Acid Catalysts

The apparent lack of correlation between enzyme activity and metal complex stability most probably finds its explanation in the specificity of enzymes. Compared to enzymes, metal ions are far less specific and cations having a strong binding capaci-

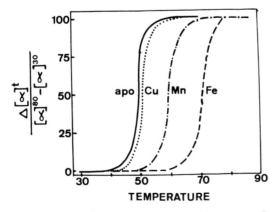

TEMPERATURE

Figure 6-8. Thermal transitions of apo- and metalloconalbumins.[11] (Used with permission. Amer Chem Soc, *Bioinorganic Chemistry.*)

ty when attached to an enzyme are capable of reacting with many ligands as well as with the substrate. Also, denaturation may occur when the metal reacts with groups other than the active site. On the other hand displacement of a metal by another metal ion can sometimes activate the enzyme. Eichhorn[12] reports, however, that metal activation and metal inhibition follow the order of formation constants in enzymic as well as nonenzymic processes and that this order is approximated by the sequence of peaks that are observed in a plot of activity *versus* metal ion concentration. A typical illustration is given by urease in which the order of inhibition correlates with the order of complex stability; only metal ions inhibit this reaction. Enolase catalyzes the addition of a water molecule to phosphoenolpyruvate in the presence of Mg^{2+}, or paramagnetic Mn^{2+}, according to:

$$^-OOC-\underset{\underset{CH_2}{\|}}{\overset{PO_4^{2-}}{\underset{C}{|}}} + H_2O \xrightarrow{\text{enolase}} {}^-OOC-\underset{\underset{CH_2OH}{|}}{\overset{PO_4^{2-}}{\underset{CH}{|}}}$$

(100A)

nmr spectroscopy suggests the formation of a ternary enzyme-metal-substrate bridged complex. A detailed study of such intermediary mixed species produces interesting information concerning the mechanisms of enzyme reactions. In the case of enolase; metal ions activate the reaction but they do so in an order which is the reverse of complex stability ($Mg^{2+} > Zn^{2+} > Mn^{2+} > Fe^{2+} > Co^{2+} > Ni^{2+}$): For enolase, inhibition competes with activation.

In other words, modifications to the stability sequence may occur for two reasons: (1) the best Lewis acids (Cd, Pb, Hg) may preferentially bind, and block, the most active bases, or (2) all the metals may combine with the same center but their different geometries (and their different behaviors in various ligand fields) are likely to generate the activity or inhibition. It is interesting to note that both $Zn(II)$ and $Co(II)$ have the property of entering low symmetry sites quite readily. Such sites have often been found to have catalytic activity.

Mixed Coordination in Metal Activated Enzymes

Physicochemical investigations of metalloenzymes and their interactions with substrates, of their analogues, and of their inhibitors have been extensively developed. They are illustrated by the following examples: Substitution of cobalt for zinc ions native to alkaline phosphatase results in an active enzyme having distinctive optical properties which are generated by the interaction between cobalt and ligands from the protein. It is now possible to distinguish between those metals which are catalytically essential and those which play only a structural role.

Adding four cobalt ions per mole of apophosphatase restores its activity. Actually, activity is only restored after the addition of the final two cobalts. Absorption spectra produced by adding the first two moles of cobalt are simple and indicate that the cobalts are bound to "structural" sites (Figure 6-10; curve "$Co_2P'tase$"). However, following the addition of the second pair of cobalt atoms, multibanded spectra (Figure 6-10; curve "$Co_4P'tase$") occur and correspond to the unusual configuration of the metal; these arise from cobalts bound at the "catalytic" sites. They differ from both octahedral ($Co(H_2O)_6^{2+}$) and from tetrahedral ($CoCl_4^{2-}$) configurations. Such irregular geometry is also observed with carboxypeptidase and carbonic anhydrase.

It is instructive to study the action of inhibitors: Inhibitors, such as phosphate ions,

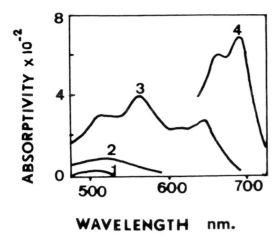

WAVELENGTH nm.

Figure 6-10. Cobalt absorption spectra.[11] (Used with permission. *Bioinorganic Chemistry*.) 1) $Co(H_2O)_6^{2+}$; 2) $Co_2P'tase$; 3) $Co_4-P'tase$; 4) $CoCl_4^{2-}$.

which alter the activity of the enzyme also alter the spectrum of the cobalt in the active site. Two moles of phosphate are bound per mole of protein; then the spectrum is simplified and becomes similar to that of an octahedral cobalt complex. Ulmer and Vallee[11] suggests that vacant coordinating positions are then filled by phosphate. Similar behavior has been observed using circular dichroic spectra.

In the case of carbonic anhydrase, mixed complexes involving anions can be easily detected using spectroscopic measurements of the Co(II) enzyme species. Binding constants are both large and of quite unexpected orders. Thus, halides bind free zinc ions in the order $F^- > Cl^- > Br^- > I^-$, only F^- binding with any appreciable strength, $logK = 2.0$. In the enzyme the order is reversed, $I^- > Br^- > Cl^- > F^-$, and although fluoride binding is normal, iodide is bound more strongly ($logK = 3.0$), three orders of magnitude stronger than it is to free zinc ions. The zinc ions of the enzyme also bind NO_3^- very strongly, $logK = 3.6$ (and also N_3^-, $logK = 4.5$; CNO^-,

$logK = 5.1$, with remarkable strength). The small anion CN^- is bound to the enzyme with approximately the same strength as to zinc ions in free solution (the same holds for the Zn-histidine-CN^- system). Anions CNO^-, N_3^-, NO_3^- are isostructural with the reactants and products of the enzyme reaction (CO_2, CO_3^{2-}, HCO_3^-). The active site which contains positively charged zinc and organic groups (NH^+) presents therefore a remarkable affinity for these similar ligands (even with charge differences) and demonstrates the existence of selective activation.

Alcohol dehydrogenase from horse liver (LADH) in its natural state has four zinc per mole. These play both structural and catalytic roles.[11] This enzyme is inhibited by chelating agents. This may either be irreversible by moving the metal or reversible through binding to zinc to form a mixed complex enzyme-zinc-chelate, e.g. as with 1,10-phenanthroline. Ulmer and Vallee[11] found from absorption and ord spectra that two moles of 1,10-phenanthroline are bound per mole of enzyme. They are only bound to the two active zinc atoms. Titrations (ord) using dipyridyl give the same results.

Cobalt also replaces zinc in LADH and again spectroscopic determinations indicate that the complex has a distorted geometry. According to Ulmer and Vallee it might be looked upon as an "intrinsic characteristic of such systems designed to achieve a condition energetically favorable to catalysis." Thus "such possible distortions and/or the existence of an open coordination position could then signal the existence of an activated energy state, preexisting the entry of substrate, and lowered by the formation of enzyme-substrate or inhibitor complexes. We have termed such a postulated activated energy state, intrinsic to the enzyme, entatic, i.e. in the state of tension or stretch." Entasis implies that the difference in energy between ground state and the transition state for the enzyme reaction is reduced. Mixed complex coordination is particularly important to such activation. The role of functional groups at, or close to, the three dimensional active site is fundamental as far as fixation and activation of the substrates are concerned. This important aspect needs to be developed further.

The study of enzyme mechanisms by nuclear spin relaxation induced by paramagnetic probes led Mildvan and Cohn[1] to suggest four hypothetical coordination schemes for mixed complexes involving enzyme, paramagnetic cation (Mn^{2+}) and the substrates, that is: substrate bridge (A), metal bridge (simple or cyclic) (B), and enzyme bridge (C) complexes. These authors empirically classified a number of enzymes and their coordination schemes according to the experimental enhancement of Mn^{2+} on the proton relaxation rate of water (Table 6-III). Enhancement factors ϵ_b (Mn-Enzyme complex) or ϵ_T (ternary complex with the substrate) are defined from the experimental relaxation rates according to:

$$\epsilon = \frac{1/T_{1(obs)}{}^* - 1/T_{1(o)}{}^*}{1/T_{1(obs)} - 1/T_{1(o)}}$$

where asterisks indicate the presence of a macromolecule; $T_{1(o)}{}^*$ and $T_{1(o)}$ are the relaxation rate in the absence of the paramagnetic ion. Substrate bridge enzymes (E-S-M) are detected when $\epsilon_T > \epsilon_b$: the substrate (nucleotide) is necessary to bring the aquated manganous ion into the macro-

TABLE 6-III

CORRELATION OF THEORETICAL COORDINATION SCHEME WITH EMPIRICAL ENHANCEMENT BEHAVIOR[1]

A. Substrate bridge E-S-M Type I ($\epsilon_b < \epsilon_T$)	B. Metal Bridge E-M-S or E Type II ($\epsilon_b > \epsilon_T$)	C. Enzyme bridge S-E-M Type III ($\epsilon_b \cong \epsilon_T$)
Muscle creatine kinase	Muscle pyruvate kinase	Citrate lyase
Brain creatine kinase	Pyruvate carboxylase	Dopamine hydroxylase
Muscle adenylate kinase	Histidine deaminase	Uridine diphosphate glucose
Arginine kinase	D-Xylose isomerase	pyrophosphorylase
Tetrahydrofolate synthetase	Carboxypeptidase	(S = PPi)
3-Phosphoglycerate kinase	Yeast pyruvate kinase	Tryptophan RNA synthetase
Yeast hexokinase	Phosphoenolpyruvate	(S = PPi)
Uridine diphosphate glucose	carboxykinase	Valine RNA synthetase
pyrophosphorylase	Phosphoenolpyruvate	(S = PPi)
(calf liver)	carboxylase	
(S = UTP)	Phosphoenolpyruvate	
Tryptophan RNA synthetase	synthetase	
(beef pancreas)	Enolase	
(S = ATP)	Phosphoglucomutase	
Valine RNA synthetase	Inorganic pyrophosphatase	
(E. coli)	Ribulose diphosphate	
(S = ATP)	carboxylase (spinach)	
Yeast aldolase	Yeast aldolase	
(S = fructose diphosphate,	(S = glyceraldehyde-3-	
dihydroxyacetone phosphate)	phosphate)	

molecular environment in a metal substrate bridge complex. In the absence of substrates, binding of Mn^{2+} to the enzyme causes significant enhancement of the effect of manganese on water. If E-M-S or cyclic

ternary complexes are formed by adding the substrate, the ϵ factor is reduced ($\epsilon_T < \epsilon_b$).

Addition of substrates to the binary M-Enzyme complex according to E-M + S → S-E-M does not change the enhancement factor ($\epsilon_T \cong \epsilon_b$). Results however, may be modified in the following cases:

1. The relaxation mechanisms for the binary and ternary complexes are different, and therefore the relaxation rate of water protons are significantly modified. Such interference may be detected by working over a wide range of temperature.

2. "Compulsory" rather than random binding of substrate and metal to the enzyme may occur, and produce another limitation. For instance, if metal binding occurs first, followed by the binding of substrate, the system could exhibit proton relaxation rates for water that are higher for binary compared to ternary complexes while no bond is formed between metal and substrate; the system exhibits type II behavior and yet has an enzyme bridge structure.

The assignments of such coordination schemes are not unequivocal and independent measurements are required. Aldolase, in which zinc has been replaced by Mn^{2+}, ought to belong to E-S-M type according to water proton relaxation rates but measurement of the relaxation rate for the C-1 proton of fructose diphosphate indicates that it functions instead as a metal bridge complex.

Additional information may be deduced from nuclear spin relaxation; the distance, r, between a paramagnetic ion and a nucleus can be evaluated from $1/T_{1M}$ or $1/T_{2M}$ (longitudinal and transverse relaxation times). For a proton that is being relaxed by divalent manganese the classical Solomon-Bloembergen[37] equation becomes (at 60 MHz):

$$r(\text{in Å}) = 815 \left[T_{1M} \left(\frac{3\tau_c}{1 + 14.2(10^{16})\tau_c^2} + \frac{7\tau_c}{1 + 6.15(10^{22})\tau_c^2} \right) \right]^{1/6}$$

where $T_{1M} = T_{1p}$ (paramagnetic contribution to the relaxation rate) when the rate of chemical exchange is rapid. If the experimental value of T_{1p} is taken instead of T_{1M} this leads to an upper limit for r. The correlation time, τ_c, is usually set equal to 3×10^{-11} sec. The precision of the r values from nuclear relaxation is ± 10 percent and the agreement with X-ray structural determinations is usually good. Some results are given in the following Table 6-IV.

Water proton relaxation rate measurements also provide some insight into the dynamic aspects of enzymic activity. Aconitase, which transforms citrate to isocitrate, belongs to the class of lyases that are activated by the addition of iron(II) (located at two tight metal-binding sites) and a reducing agent. Kinetic and thermodynamic properties of the metal bridge aconitase-Fe(II)-citrate complex are consistent with its role in catalysis: The rates of formation and dissociation of this complex are at least two orders of magnitude greater than

TABLE 6-IV

DISTANCE BETWEEN PARAMAGNETIC SITE OF PROTEIN AND ATOM OF
BOUND LIGAND AS DETERMINED BY NUCLEAR RELAXATION[1]

Protein	Paramagnetic Center	Ligand		Distance (Å) Relaxitivity (range)	Models
Pyruvate kinase (muscle) . . .	Mn	FPO_3^{2-}	Mn . . . F	3.0-5.8	3.8 ± 0.8
Pyruvate carboxylase	Mn	Pyruvate	Mn . . . CH_3	3.5-6.6	4.0 ± 1.7
		α-Ketobutyrate	Mn . . . CH_3	3.5-6.6	4.2 ± 0.4
			Mn . . . CH_2	3.5-6.6	4.7 ± 1.2
		Oxalacetate	Mn . . . CH_2	3.1-7.9	3.7 ± 0.7
Carboxypeptidase	Mn	Indole-Ac	Mn . . . $^-CH_2^-$	⩽8.3	4.8
		Bu-Ac	Mn . . . $^-CH_2^-$	⩽6.9	4.8
		Br-Ac	Mn . . . $^-CH_2^-$	⩽5.7	4.8
		Methoxy-Ac	Mn . . . $^-CH_2^-$	⩽4.3	4.8
			Mn . . . $^-CH_3$	⩽4.7	4.7
Histidine deaminase	Mn	Urocanate	Mn . . . HC_2	⩽3.5	3.5 ± 0.1
			Mn . . . HC_5	⩽3.6	3.6 ± 0.1
		Imidazole	Mn . . . HC_2	2.9-4.5	3.5 ± 0.1
			Mn . . . HC_5	3.2-4.8	3.6 ± 0.1
Metmyoglobin (seal)	Fe(III)	F . . . H	Fe . . . F	⩽6.1	1.9
			Fe . . . H	2.9 ± 0.1	2.7
		Azide	Fe . . . H	3.1 ± 0.1	2.7

the maximal rate of the enzyme-catalyzed conversion of citrate to isocitrate indicating that the ternary complex forms and dissociates rapidly enough to participate in catalysis. Mn(II) ions compete with Fe(II) for the binding sites and form similar but inactive metal bridge ternary complexes. The study of molecular motions in these complexes do not produce satisfactory explanation.[13]

Role of Metal Complexes in Invasion of Bacteria by Viruses

Organized systems such as cells and viruses may interact with metal ions as mediators. Eichhorn[12] reports that viruses are altered by chemical agents such as metal complexes in which the coordination sphere of the metal is partly filled (Cd (CN)$_3^-$ is active while EDTA complexes are not). The mechanism of alteration is described in Figure 6-11 (a, b, c). It in-

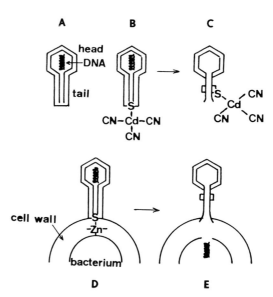

Figure 6-11. Viral invasion of bacterial cell.[12] A. Intact virus; B. Complex formation; C. Alteration of virus and contraction of tail sheath; D. Attachment of virus to bacterial cell wall; E. Injection of DNA into cell.

volves a bond formation between the metal complex and a sulfur bridge of the proteic coating of the virus (b) which then cleaves it (c). This then enables DNA to leave the virus.

Virus T_2 bacteriophage can infect cells of the bacteria *E. coli*, according to a similar scheme (Figure 11 c, d). Studies on *E. coli* have shown that the cell wall contains zinc. When a virus attaches itself to the zinc containing cell wall, the latter is then split by an enzyme and DNA passes from the phage to the bacteria which is an essential step in virus replication. When the zinc has been extracted by an acid, the phage cannot be digested by the cell wall. This property is restored upon reinserting the zinc. This situation has been interpreted as the existence of a zinc complex in the cell wall of the bacterium and this is bound to a sulfur-containing bond of the proteic coating which is then destroyed.

Mixed Ligand Complexes in Storage and Transport Phenomena

The nature of coordination compounds suggests there being an important role for such species in the storage and transport of either metal ions or of donor molecules. Transportation of ions and molecules through membranes from one compartment to another is usually classified as passive, facilitated (in which specific groups are located in membranes and a gradient of affinity is needed), or active (in which the transportation is coupled with an exergonic reaction, e.g. ATP → ADP). Mixed ligand complexes have been suggested to be involved in these mechanisms.

Transport of Metal Ions

Transport and storage metalloproteins control the concentrations of most metal ions found *in vivo*. These proteins not only control the metal concentrations and carry them to the sites that require them, but also they act as catalysts in placing the metal ions into their new molecules. This is the case for transferrin, the serum iron transport protein which mediates the exchange of iron between cellular iron donor and receptor sites. Experiments by Saltman[14] involving both spectral measurements and direct determinations of formation constants have shown the important role of low molecular weight chelates in the transport of iron to transferrin and other iron binding sites. It has been demonstrated that ternary complexes of such chelates (nitrilotriacetate, citrate, EDTate) with iron and transferrin may be formed according to:

$$Ch\text{-}Fe + Tr \rightleftharpoons Ch\text{-}Fe\text{-}Tr \rightleftharpoons Ch + Fe\text{-}Tr$$

and it is expected that the regulation and control of iron metabolism through the intestine as well as across all biological membranes is determined by the ability of iron to be chelated by such ligands. The chemical nature of the ligand is the determining factor for fast iron transfer through the membranes. On the other hand, the ultimate site of deposition of the iron within the various tissues is regulated and controlled both by the availability of the specific binding sites of macromolecules and the ability of the low molecular weight iron chelate to interact with these biopolymers and, ultimately, to transfer the iron to it.

Another example of metal ion transport may be found in the removal of metals from the body. It is well known that ligands such as EDTA, 2-3-dimercaptopropanol (BAL) and D-penicillamine have been used successfully for removing toxic metals. Administering D-penicillamine constitutes a treatment of Wilson's disease (an excessive accumulation of copper in the liver, tissues and brain). In this connection, mixed ligand complexes may play a very important role. For copper, as noted in the introduction, the Cu(II) in ceruloplasmin is not exchangeable *in vivo;* only the albumin and the amino acid bound fraction of serum Cu(II) are exchangeable and considered to be the physiologically important transport form of copper.

The investigations of Sarkar and his group[15] revealed that besides mixed complexes consisting of two different amino acids, copper is also bound in mixed complexes such as human serum albumin-Cu(II)-L-histidine. They postulated that the following equilibria are maintained under physiological conditions:

$$Cu\,(II)\ +\ amino\ acid\ \rightleftharpoons\ Cu\,(II)-amino\ acid$$

$$\Big\updownarrow albumin$$

$$albumin\text{-}Cu\,(II)-amino\ acid$$

$$\Big\updownarrow$$

$$Cu\,(II)-albumin\ +\ amino\ acid$$

$$(100B)$$

Amino acids other than L-histidine may also play a similar role in forming an intermediate ternary complex. The mechanism above might be expected to have an important part in the exchange of Cu(II) between a macromolecule and a low molecular weight substance which, in turn, can readily be transported across biological membranes. To date, this research group has successfully synthesized glycylglycyl-L-histidine methyl ester, the spectral and proton displacement characteristics of which are exactly the same as human albumin. It is anticipated that this tripeptide, as a copper complex, is small enough (compared to the large albumin molecule) to cross biological membranes in the kidneys and can thus be excreted.

Transport of Donor Molecules

Many metalloproteins control substrate, e.g. O_2, CO_2, fixation processes through the formation of mixed ligand complexes:

Transport of Oxygen

Only very few animal species utilize absorbed oxygen without using hemeproteins. Chapter 7 provides a general description of iron containing proteins. For example of a mixed complex, consider the iron(II) in haemoglobin and in myoglobin which is coordinated by four nitrogen atoms of a porphyrin ring, one axial position being filled by an imidazole group from a histidine amino acid residue and the mixed ligand coordination being achieved by the fixation of molecular oxygen (see Figure 6-12). Three possible modes of binding of this dioxygen molecule have been considered[16] according to Figure 6-13.

Recent theoretical discussions favor the bent, monodentate coordination model with an Fe-O-O angle of approximately 120°. However, the mode of oxygen uptake by hemes is still speculative.

We must also note that haemoglobin co-

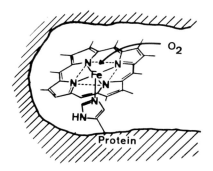

Figure 6-12. Schematic representation of coordination of iron(II) in hemoglobin. (Reproduced with permission of Van Nostrand Reinhold Co., *The Metals of Life*, D. R. Williams, 1971.)

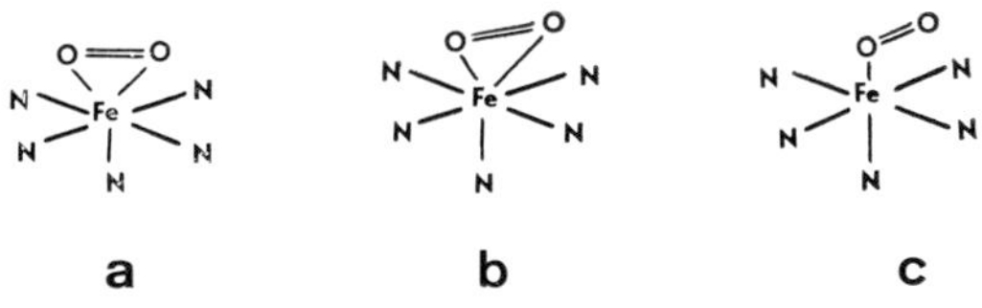

a b c

Figure 6-13. Three favored bonding modes of dioxygen to the iron of the heme subunit: a) symmetrical chelation; b) asymmetrical chelation; c) bent monodentate chelation.[16]

ordinates more strongly to carbonmonoxide than to oxygen and that it is possible to replace the iron in haemoglobin by cobalt (coboglobin) without losing dioxygen absorption properties.

Some copper proteins may play analogous roles in certain molluscs and anthropods which have been known for a long time to have blue blood. This blue color was attributable to a copper containing protein which was one of the largest protein molecules known (MW ranging from 450,000 to 6,680,000)—hemocyanin—this serves as an oxygen carrier in the blood of these animals in place of haemoglobin.

Synthetic Oxygen—Carrying Chelates

Synthetic mixed ligand complexes exhibit similar gas carrier behavior; for example the complex $Fe(dmg)_2(im)_2$; (dmg = dimethylglyoxime, im = imidazole) undergoes the following equilibria in solution

$$Fe(dmg)_2(im)_2 + X \rightleftharpoons$$
$$Fe(dmg)_2(im)X + im$$
$$X = O_2, CO, CN^-$$

A second molecule of X may be absorbed but no oxygen-containing species of either stoichiometry has yet been isolated in the solid state. Earlier studies of O_2 reacting with metal complexes have been made on metal ions of lower oxidation states (Co (II), Cu(I), Fe(II)). Reversible oxygen uptake of such complexes has recently been review by Wilkins.[17] A number of cobalt(II) complexes, including those of

some amino acids (such as histidine) and peptides (such as glycyl-glycine), can interact reversibly with molecular oxygen. It is noteworthy that among the metals of the first transition series only cobalt undergoes such reactions. Iron(II) is irreversibly oxidized and the copper(II) and nickel(II) complexes have very little tendency to react with oxygen. However, Margerum[38] has pointed out that nickel(II) is activated by peptide coordination, e.g. with tetraglycine, so that it reacts with molecular oxygen and catalyzes the oxidation of the ligand to give a number of products including amides or amino acids and of peptides, oxo acids and CO_2.

Biological Nitrogen Fixation

The chemistry of nitrogen fixation by biological systems is still not understood but recent advances in this field are important.[18, 19] Two metalloproteins can be isolated from nitrogenase, one an iron-sulfur protein (molecular weight about 70,000) and the other a molybdenum-iron-sulfur protein (MW about 300,000), both of which are required for the fixation process.

Evidence for mixed complexes involving N_2 as one of the two unlike ligands has been reported by Allen[39] and also by Harrison and Taube[40] for ruthenium ion complexes:

$$[Ru(NH_3)_5(H_2O)]^{2+} \xrightarrow[H_2O, 100 \text{ atmospheres}]{N_2}$$
$$[Ru(NH_3)_5(N_2)]^{2+}$$

Another type of reaction of some importance was discovered by Sacco and Rossi. It is the reversible displacement of dihydrogen or ammonia by dinitrogen:

$$[CoH_3(PPh_3)_3] + N_2 \rightleftharpoons$$
$$[CoH(N_2)(PPh_3)_3] + H_2$$

$$[CoH(NH_3)(PPh_3)_3] + N_2 \rightleftharpoons$$
$$[CoH(N_2)(PPh_3)_3] + NH_3$$

(PPh_3 = triphenylphosphine)

The most common approach to models of these biological processes has been the preparation of stable and isolatable dinitrogen complexes followed by attempts to reduce them. But, in spite of the characterization of numbers of such complexes, to date reproducible reduction has not been accomplished.

No direct biochemical evidence that a metal complex is indeed formed during biological nitrogen fixation has yet been published. Due to the extremely low concentrations of metal in nitrogenase systems, only very accurate methods such as esr, or Mössbauer spectra for Fe, would be able to detect changes in the metal site during nitrogen fixation.

Mixed Chelates in the Storage and Transport of Neurotransmitters

The importance of metal coordination phenomena to an understanding of the binding, storage and transport of catechol amines has been discussed by Colburn and Maas in 1965. They proposed that vesicular binding sites for biogenic amines could involve mixed ligand complexes containing adenosine triphosphate (ATP) and several biogenic amines. Recently Rajan et al.[2] studied the interactions of ATP and various amines with magnesium ions (an important "bio-inorganic" metal ion which is necessary for nerve impulse transmission, for muscle contractions and for the metabolism of carbohydrates). In the pH range 7.0 to 10.5 mixed ligand complexes Mg^{2+}-ATP-amine in the ratio 1:1:1 are formed and their stability constants for the reaction $MgATP + amine \rightleftharpoons Mg(ATP)amine$ are listed in Table 6-V.

The postulated binding sites are the oxygens of catechol amines, ATP is assumed to act as a terdentate ligand through its three phosphate oxygens.

The stimulated uptake of the amines by

TABLE 6-V
COORDINATION OF Mg^{2+}-ATP BY BIOGENIC AMINES[20]

Biogenic Amines	Log of Stability Constants (1M KNO_3, T = 25°C ± 0.1)
Norepinephrine	2.34 ± 0.32
Dopamine	3.05 ± 0.06
Tyramine	2.60 ± 0.04
Phenylethylamine	No interaction
Epinephrine	2.95 ± 0.08
Octopamine	1.93 ± 0.12
6-Hydroxydopamine	2.17 ± 0.14
Amphetamine	2.09 ± 0.05
3-Methoxynorephedrine	2.76 ± 0.09
Norephedrine	No interaction
Ephedrine	No interaction
Phenylephrine	1.90 ± 0.05

the granules can therefore be rationalized as follows: In the presence of Mg^{2+} and ATP, the biogenic amines form ternary chelates (at and above pH7). The ternary chelate-bound amine is transported across the granule membrane as a result of the reaction of the negatively charged mixed chelate with the membrane. The amine transported would be released in the granule and bound to the amine storage site.

Transport of Anions through Membranes

Transport of cations is known to be catalyzed by complex formation with valinomycin type ionophores (cyclic peptides or depsipeptides). Ionophores(I) specific for anions are as yet unknown. However, mitochondria are normally more permeable to anions than to cations. It has been demonstrated that anionic organic ligands may be carried across lipidic membranes by forming a ternary complex at the interface (Figure 6-14a) which diffuses towards the opposite interface where the complexed anion is liberated while the ionophore diffuses back.

In addition, active transport leading to an exchange of anions through mitochon-

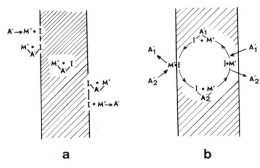

Figure 6-14. a) Scheme for anion transport; b) hypothetical model for mitochondrial ion transport. (Reproduced with permission of Prentice Hall Inc., *The Molecular Basis of Membrane Function*, D. C. Tosteson (Ed.), 1969.)

drial membranes is interpreted again in terms of mixed complex formation according to Pressman and Haynes.[21] They suggested the mechanisms in Figure 6-14b where the energy is provided by the reaction $ATP \rightarrow ADP + PO_4^{3-}$.

Mixed Ligand Complexes of Importance in Soil-Plant Relationships

Metal ions in soil and plants are likely to compete with various natural or synthetic ligands and the concept of ligand-metal-ligand binding may be useful in explaining such effects as the fixation of synthetic chelating agents and of organic matter in soils.[22] Two mechanisms are to be considered; many binary complexes act as dipoles and the positive sites in metal chelates may be bound to the negatively charged clay colloids. The second possibility is the formation of "true" mixed ligand complexes between oxygen groups from the soil particle, metal ion and organic matter (or synthetic ligand) as shown schematically in Figure 6-15.

There is a continuous transition between the purely electrostatic mechanism and that

of coordination through covalent bond formation but, in both instances, metal ions accumulate in the double layer and may adsorb at the solid surface facilitating the approach of the negatively charged organic species to the negatively charged surface. This type of binding may also be expected to occur in the ternary complexes of the high molecular weight compounds (humic acids) which consist of polyanions containing aromatic nuclei with $-OH$ and $-COOH$ functional groups linked through $-O-$ and $-CH-$ bridges. These acids may be fixed on clay by an intermediary of metal ions such as iron, aluminium, or calcium to form hydrophilic mixed complex able to adsorb various cations.

This concept also helps to elucidate the mechanism through which plant roots obtain nutrients from the soil (especially insoluble metals like iron—hypothesis of contact chelation: Figure 6-16).

This is exemplified by the fact that naturally occurring, e.g. ferrichromes, or synthetic, chelating agents, e.g. EDDHA = ethylenediamine-di-(O-hydroxyphenylacetic acid) and present insolubilization of iron even at great dilutions and high pH. The ionic stability of such iron complexes seems to play an important role. For instance, ferrichromes represent a class of naturally occurring heteromeric peptides

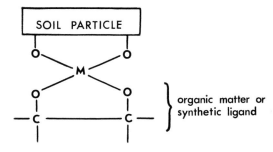

Figure 6-15. Reproduced with permission from.[22]

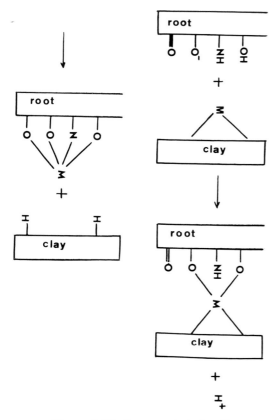

Figure 6-16. Contact chelation.[22]

containing a trihydroxamate as an iron(III) binding center. These natural trihydroxamates bind iron so firmly that iron can be transported to, and into, the cell and donated to the iron enzymes (Neilands,[42] 1964). In the same way, the ferric chelate of EDDHA has a stability constant up to 10^{30} and appears to be the best available treatment of iron chlorosis induced by calcareous soils. As noted previously, mixed ligand complexes are usually extremely stable and they may be expected to give excellent results in such treatments.

Since metal complexes are absorbed by plant roots, there are many examples of synthetic chelating agents being bound to cellular constituents through metals. With plants grown with Fe-(C^{14})EDDHA and Fe^{59}-EDDHA, A. Wallace[22] obtained evidence that an EDDHA-iron-protein mixed complex could be formed.

Hence, the role of mixed ligand complexes appears very important not only to soil and to soil-plant interaction, but also within plants in which they are absorbed and metabolized.

REFERENCES

1. Mildvan, A. S. and Cohn, M.: *Adv Enzymol,* 33:1, 1970.
2. Neumann, P. Z. and Sass-Kortsak, A.: *J Clin Invest,* 46:646, 1967.
3. Martin, R. P., Petit-Ramel, M. M., and Scharff, J. P.: *Metal Ions in Biological Systems.* In Sigel, H. and Dekker, M. (Eds.): New York, 1973.
4. Marcus, Y. and Eliezer, I.: *Coordin Chem Rev,* 4:273, 1969.
5. Beck, M. T. (Ed.): *Proceedings 3rd Symposium on Coordination Chemistry.* Debrecen, 1970. vol. 2, a) Martell, A. E., p. 125; b) Friedman, Y. D., p. 77.
6. Freeman, H. C.: *Advances in Protein Chemistry.* In Anfinsen, C. B., Anson, M. L., Edsall, J. T., Richards, F. M. (Eds.): New York, Acad Pr, 1967, p. 257.
7. Agarwal, R. P. and Perrin, D. D.: Trans. of the Royal Institute of Technology, Stockholm, 1972, vol. 278, p. 385.
8. Giroux, E. L. and Henkin, R. I.: *Biochim Biophys Acta,* 273:64, 1972.
9. Sigel, H. and McCormick, D. B.: *Accounts Chem Res,* 3:201, 1970.
10. Vallee, B. L. and Williams, R. J. P.: *Chem Brit, 4:* 397, 1968.
11. Ulmer, D. D. and Vallee, B. L.: *Adv Chem Ser,* 100:187, 1971.
12. Eichhorn, G. L.: *Adv Chem Ser,* 37:37, 1963.
13. Villafranca, J. J. and Mildvan, A. S.: *J Biol Chem,* 247:3454, 1972.
14. Saltman, P.: *J Chem Edu,* 42:682, 1965.
15. Lau, S. J. and Sarkar, B.: *J Biol Chem, 246:* 5938, 1971.
16. Rayner Canham, G. W. and Lever, A. B. P.: *J Chem Educ,* 49:656, 1972.
17. Wilkins, R. G.: *Adv Chem Ser,* 100:111, 1971.
18. Hardy, R. W. F., Burns, R. C. and Parshall, G. W.: *Adv Chem Ser,* 100:219, 1971.

19. Chatt, J.: *Bull Soc Chim Fr, 2:*431, 1972.
20. Rajan, K. S., Davis, J. M., Colburn, R. W. and Jarke, F. H.: *J Neurochem, 19:*1099, 1972.
21. Pressman, B. and Haynes, D. H.: *The Molecular Basis of Membrane Function.* In Tosteson, D. C. (Ed.).: U.S.A., Prentice Hall, 1969, p. 221.
22. Wallace, A.: *A Decade of Synthetic Chelating Agents in Inorganic Plant Nutrition.* In Wallace, A. (Ed.): Los Angeles, 1962.
23. Cromer-Morin, M., Martin, R. P. and Scharff, J. P.: *C R Acad Sci Paris, 277:*1339, 1973.
24. Stumm, W. and Morgan, J. J.: *Aquatic Chemistry,* New York, Wiley-Interscience, 1970.
25. Jackobs, W. E. and Margerum, D. W.: *Inorg Chem, 6:*2038, 1967.
26. Tanaka, M.: *J Inorg Nucl Chem, 35:*965, 1973.
27. Kida, S.: *Bull Chem Soc, Japan, 33:*587, 1960.
28. Kato, M.: *Z Phys Chem, 23:*391, 1960.
29. Kida, S.: *Bull Chem Soc Japan, 34:*962, 1961.
30. Pearson, R. G.: *J Chem Edu, 45:*581, 643, 1968.
31. Jorgensen, C. K.: *Inorg Chem, 3:*1201, 1964.
32. Barnes, D. S. and Pettit, L. D.: *Proceedings of the XIIIth. I.C.C.C., 1:*234, 1970.
33. Freeman, H. C. and Martin, R. P.: *J Biol Chem, 244:*4823, 1969.
34. Martin, R. P. and Riaute, J.: *Proceedings XVIth. I.C.C.C., Dublin,* 1974.
35. Schuber, J.: *Chimia, 11:*113, 1957.
36. Tanford, C.: *J Amer Chem Soc, 74:*211, 1952.
37. Bloembergen, N.: *Nuclear Magnetic Relaxation,* New York, Benjamin, 1961.
38. Paniago, E. B., Weatherburn, D. C. and Margerum, D. W.: *Chem Comm,* 1427, 1971.
39. Allen, A. C. and Senoff, C. V.: *Cheb Comm,* 621, 1965.
40. Harrison, D. F. and Taube, H.: *J Amer Chem Soc, 89:*5706, 1967.
41. Sacco, A. and Rossi, M.: *Chem Com,* 316, 1967.
42. Neiland, J. B.: *Essays in Coordination Chemistry.* In Schneider, W., Gut, R. and Anderegg, G. (Eds.), Birkhauser, Basel, 1964.

CHAPTER 7

THE CHEMICAL BASIS OF HUMAN IRON METABOLISM

George Winston Bates and Paul Saltman

Introduction
Solution and Coordination Chemistry of Iron
Myoglobin, Hemoglobin, and Cytochrome C: Examples of Functional Iron Complexes
Iron Balance and Iron Deficiency
Intestinal Iron Absorption
Transferrin
Iron Storage
Erythrocyte Iron Metabolism

INTRODUCTION

THE PURPOSE OF this chapter is to provide the student with an introduction to the field of iron metabolism. To set the stage, a description of the solution and coordination chemistry of iron is provided, since it is the chemistry of iron that dictates the nature of biological iron management and utilization. Of special interest here are the chemical difference between ferric and ferrous ions, and the interaction of iron with ligands of biological interest.

In the following section three heme proteins, myoglobin, hemoglobin, and cytochrome C are chosen for a detailed look at the role of iron in a functioning biomolecule.

Iron balance and the pandemic problem

of human iron deficiency are then examined. The remainder of the chapter describes the unique biological mechanisms that have evolved to handle the assimilation, transport, storage, and utilization of iron. The management of iron involves diverse proteins and membrane transport systems, many of which are as yet unknown.

SOLUTION AND COORDINATION CHEMISTRY OF IRON

General topics and background in coordination chemistry are included in other chapters of this text. Our coverage of aqueous iron chemistry will be brief and center about points that will be of importance to our discussion of the function and management of biological iron. As a transition series element, iron is able to lose electrons from partially filled d orbitals to form cations. In aqueous and biological systems, the Fe^{2+} ferrous ion and Fe^{3+} ferric ion are of principal importance. While other oxidation states may exist under specialized conditions, they are not of primary concern from a biological standpoint.

Both ferrous and ferric ions are strong Lewis acids and accept electrons from σ bonding donors. In addition, ferric, and to a greater extent ferrous ions, are able to enter into π bonding arrangements with ligands containing available π orbitals. Both σ and π bonds are important in considering the binding of iron to biomolecules.

The arrangement of ligands about an

* Research from the author's laboratories described above was supported by the United States Public Health Service grant AM 12386-08 and Robert A. Welch Foundation grant A-430.

iron nucleus depends on the nature of the ligands involved. In synthetic systems, complexes with a wide variety of coordination numbers and ligand geometries have been prepared and studied. In biomolecules, 4, 5, and 6-coordinate geometries are found. The most common arangement is the 6-coordinate octahedral geometry and less frequently a 4-coordinate tetrahedral arrangement of ligands.

The spin-states of the ferric and ferrous ions also have an influence on the biological properties of iron complexes. The five d electrons of ferric and six d electrons of ferrous are distributed among the five 3d orbitals. Orbitals will be occupied by a single electron before electrons are paired with opposite spins in an orbital. Energy required to place two electrons in one orbital is referred to as pairing energy. In other words, there is an energy impetus to arrange electrons in the maximally unpaired, or high-spin state. In the case of ferric ion, a maximal spin of five unpaired electrons can be observed as depicted in Figure 7-1. In high-spin ferrous, four unpaired electrons are present since two must necessarily be paired to accommodate the total of six electrons. Low-spin, where there is maximal pairing, results in one and zero unpaired electrons with ferric and ferrous respectively. Intermediate spin states may exist but are not common.

The chemistry of coordination complexes is a function of reciprocal effects between the central metal ion and its ligands. The chemical environment has a pronounced effect on the chemistry of the metal, and on the other hand, the metal affects the types of ligands bound and their reactivity of the bound ligands. This is clearly seen in terms of the effect of ligands on the redox potential and spin-states of the central metal ion.

The five d orbitals are not symmetrically arranged but consist of two orbitals ($d_z{}^2$ and $d_{x^2-y^2}$) directed along the axes of three perpendicular coordinates, and three orbitals directed between the axes. If, as in an octahedral binding geometry, the ligands are pointed towards the metal along the six axes, there will be a repulsion between the ligand electrons and any electrons that are in the metal's $d_z{}^2$ and $d_{x^2-y^2}$ orbitals. Ligands that are effective σ bond donors have "high-ligand fields," resulting in strong electron repulsion effects, which if great enough will overcome the pairing energy barrier and force the metal ion a low-spin configuration. Thus, the spin-state of a metal will depend on the nature of its ligands, that is, its chemical environment.

The chemical environment determines the propensity of the metal ion to lose or gain electrons. The oxidation-reduction potential is directly related to the affinity of the ligands for the two oxidation states in question. The binding affinities, in turn, are a function of the electron configuration of the metal, the types of ligands ("hard" versus "soft" bases), the available spin-states and the geometry imposed by the ligands. The effect of coordination on redox potential is a complicated phenomenon.

An interesting example of the unique function of a biomolecule for a given oxidation state involves the binding of iron to

SPIN STATES OF FERROUS AND FERRIC IONS

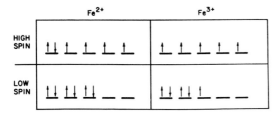

Figure 7-1. Spin-states of ferrous and ferric ions.

transferrin, the serum iron transport protein. Transferrin strongly binds ferric ions, but the affinity for ferrous ion is so weak that it was not recognized until recently. In fact, the difference in the affinity of transferrin for ferric over ferrous is some twenty-seven orders of magnitude! As a result, the redox potential greatly favors the trivalent form and the ferrous-transferrin complex is readily oxidized by a variety of oxidizing agents.

Basolo and Pearson[1] have described and summarized the division of metal ions and ligands into "hard" and "soft" Lewis acids and bases. A base is classified as soft if it has an easily polarizable electron cloud and a low or negligible proton affinity. A hard base has the opposite properties. Hard Lewis acids are those metal ions which have a high net positive charge, small ionic radius, and tightly held outer electron orbitals. The converse is true for the soft Lewis acids. In terms of affinity of ligands for metal ions, the generalization can be made that hard acids have higher affinity for hard bases, and soft acids have higher affinity for soft bases.

The ferric ion, with a positive charge of three, and an ionic radius of 0.67Å is classified as a hard acid, while the ferrous ion with one more electron in its outer d orbitals and an ionic radius of 0.83Å is border line between the hard and soft acids. In terms of ligand association, the generalization can be made that ferric ion displays an especially high affinity for hard oxygen ligands such as water, oxalate, citrate, tyrosyl, aspartyl, and glutamyl residues. Ferrous ion has a high affinity for many soft ligands which can enter into π bonding arrangements including the unsaturated amines bipyridyl and o-phenanthroline which are used as colorimetric iron reagents. It should be emphasized that this

discussion is meant to describe relative affinities. Most chelates and ligands that tightly bind ferric ion will also show an affinity for ferrous ion, and vice versa.

As a consequence of the differences in the nature of the ferrous and ferric ion described above, there is a marked difference in their aqueous chemistry, especially in the neutral pH range. Ferric ion, which has over twice the negative heat of hydration as that of ferrous ion, has a high affinity for the oxygen of the water molecule. As a result of the strong bonding of the oxygen to the ferric ion, the pKa of the water molecule is greatly lowered, and protons are readily released.

$$Fe^{3+} + 6H_2O \rightleftharpoons [Fe(H_2O)_6]^{3+}$$
$$[Fe(H_2O)_6]^{3+} \rightleftharpoons [Fe(H_2O)_5(OH)]^{2+} + H^+$$
$$[Fe(H_2O)_5(OH)]^{2+} \rightleftharpoons [FeH_2O)_4(OH_2)]^+ + H^+$$

The process of proton loss is called hydrolysis and causes the low pH that is observed when simple ferric salts are dissolved in water. The pKa's of the first two acid hydrolysis constants for ferric ion are 2.2 and 3.3, while the pKa of hydrated ferrous ion is estimated to be about 9.5.

The loss of a proton by a coordinated water molecule results in a hydroxyl ligand, which has affinity for a second ferric ion. As a result dimeric ferric aquo complexes can be formed in which the irons are bridged by either one or two hydroxo or oxo groups:

$$2[Fe(H_2O)_5(OH)]^{2+} \rightleftharpoons [(H_2O)_4 Fe \underset{O,H}{\overset{H,O}{<>}} Fe(H_2O)_4]^{4+} + 2H_2O$$

$$(100C)$$

The addition of base causes further release of protons with concomitant linkage of the cations and polymer formation. Spiro and Saltman[2] have reviewed the polymerization phenomenon and its implications for the management of biological iron. These polymeric iron-hydroxyl complexes are of high molecular weight, about 150,000, and are of spherical morphology, and 70Å in diameter. The size range is remarkably narrow. The chemical and physical parameters that make this particle maximally stable are unknown. The iron storage protein ferritin consists of a protein shell surrounding a microcystalline ferric-hydroxide-phosphate sphere which bears a striking physical and chemical resemblance to the synthetic ferric oxo polymers.

Since hydrolysis of ferric iron leads to polymer formation and eventual precipitation, the ferric ion is virtually insoluble at neutral pH in distilled water. In fact, the solubility can be calculated at $10^{-18}M$ at pH 7. However, ferrous ion is soluble to the extent of 0.1 M at neutral pH. Only by displacing coordinated water with the ligands of a strong chelating agent, can ferric ion be maintained soluble in neutral and alkaline solutions. Even with chelates present, an equilibrium exists between monomeric and dimeric and polymeric ferric complexes. This equilibrium depends on the geometry and binding affinity of the chelate. For example, the hexadentate chelate EDTA with an affinity for iron of $10^{25}M$ exists as a mixture of monomeric and dimeric complexes, in which Fe^{3+}-EDTA monomers are probably linked by a linear oxo bridge between the ferric ions. NTA, a tetradentate chelate of lower affinity permits the formation of condensation complexes which remain quite reactive and in equilibrium with monomeric Fe^{3+}-NTA. Citrate, on the other hand, is a tridentate chelate and permits the formation of unreactive polymers of the type described above. It is clear that in the function and management of biological iron that effective and specific complexing agents are required

to maintain the iron in a soluble and functional form. Some of the most important ion complexes are proteins.

Of the twenty amino acids incorporated into proteins, several have side chains that can act as ligands to metal ions. Examples of these are illustrated in Figure 7-2. In addition to these side chains, the terminal carboxyl and amino groups; and the oxygen and nitrogens of the peptide bond are potential ligand donors to the metal ion. It is important to recognize that through the folding and bending of the protein backbone, ligand groups can be brought into a geometry resulting in a multidentate metal binding site of extremely high stability. The environment that the binding site presents to the metal profoundly influences the subsequent reactivity and availability of the bound metal ion.

In discussing metal binding to enzymes, a distinction is made between readily dissociable metal-enzyme complexes, the metal

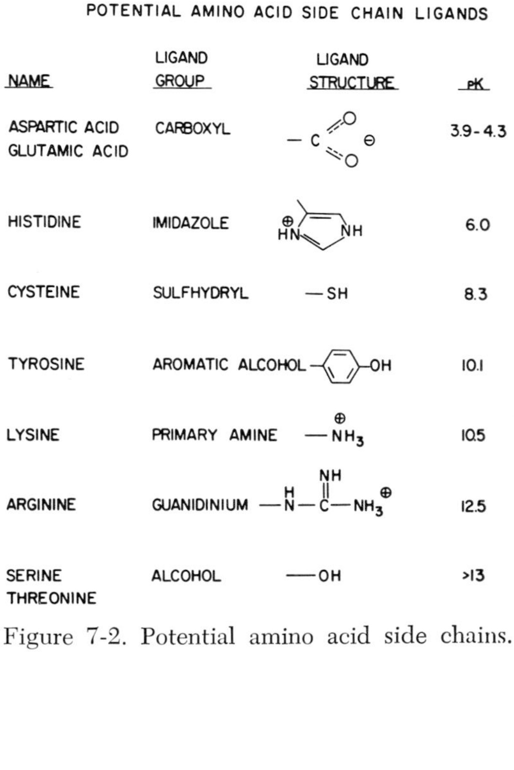

POTENTIAL AMINO ACID SIDE CHAIN LIGANDS

NAME	LIGAND GROUP	LIGAND STRUCTURE	pK
ASPARTIC ACID GLUTAMIC ACID	CARBOXYL	$-C\overset{O}{\underset{O^{\ominus}}{}}$	3.9-4.3
HISTIDINE	IMIDAZOLE	(imidazole structure)	6.0
CYSTEINE	SULFHYDRYL	$-SH$	8.3
TYROSINE	AROMATIC ALCOHOL	(aromatic ring)$-OH$	10.1
LYSINE	PRIMARY AMINE	$-NH_3^{\oplus}$	10.5
ARGININE	GUANIDINIUM	$-N\overset{H}{\underset{}{}}-\overset{NH}{\underset{}{C}}-NH_3^{\oplus}$	12.5
SERINE THREONINE	ALCOHOL	$-OH$	>13

Figure 7-2. Potential amino acid side chains.

activated enzymes; and very tight associations, the metalloenzymes. In the latter, the metal is generally retained during the purification process. The mechanism of action of the two enzyme types may be quite similar.

Mildvan[3] has recently reviewed the field of metalloenzymes and metal activated enzymes, and has emphasized three basic geometries in the binding of metal, enzyme, and substrate. The first is the "substrate bridge" model in which the metal is bound only to the substrate and is, therefore, of the form M-S-E (where M = metal, S = substrate, and E = enzyme). Second is the "enzyme bridge of the form M-E-S. In this case the metal primarily stabilizes a particular conformation of the enzyme. Finally, the "metal bridge" model E-M-S and E-M,

$$\underset{S}{\diagdown}$$

with the metal coordinated to both the enzyme and the substrate. This latter geometry is the most commonly found, especially in the case of iron containing enzymes.

When bound directly to an enzyme, a metal is referred to as a cofactor. Metals may also be a part of coenzymes, complex organic molecules, which in turn are bound to enzymes and participate in catalysis. Nonprotein components, including cofactors, coenzymes, carbohydrates, etc. that are tightly bound to proteins are referred to as prosthetic groups.

MYOGLOBIN, HEMOGLOBIN, AND CYTOCHROME C AS EXAMPLES OF FUNCTIONAL IRON COMPLEXES

The Diversity of Functional Iron Complexes

In the section above we noted several aspects of iron chemistry which are relevant to its biological utility. These included the capability to reversibly bind a variety of ligands; to assume at least two coordination geometries; to exist in two possible spin-states; and to fluctuate between the ferrous and ferric oxidation states. In addition, we noted that the spin-state, coordination geometry, and redox potential are influenced by the chemical environment provided by the ligand field. These properties offer the living state unique opportunities to fashion efficient catalytic and binding centers. Indeed, iron is found as an essential component of a fantastic variety of metabolically active biomolecules. Malmström[4] has pointed to oxidation-reduction reactions and the transport, storage, and utilization of molecular oxygen as central roles of biological iron.

Iron-protein complexes can be placed in three broad catagories according to the bonding of the iron to the proteins. The heme proteins are a class in which the iron is incorporated into a tetrapyrrole ring system, which in turn is bound by the protein. Examples of heme proteins are myoglobin, hemoglobin, cytochrome c and the enzymes cytochrome oxidase, catalase and peroxidase. A second category is the iron-sulfur proteins, so called because iron is bound only to sulfur ligands, including cysteine residues of the protein. This arrangement is involved in a variety of essential oxidation-reduction reactions. A comprehensive review by Orme-Johnson[5] is included in the list of references. The final category of iron-proteins are those in which iron is bound at unique sites provided by various amino acid residues. Transferrin, and aconitase, a citric acid cycle enzyme, fall into this classification.

Rather than survey the broad spectrum of functional iron-proteins, we have chosen to examine three molecules in detail: myoglobin, hemoglobin, and cytochrome c. These proteins are structurally and chemically well characterized and provide unique insight into structural-functional relationships at a molecular level.

Myoglobin

The metalloporphyrins constitute an abundant cofactor class in nature. They are involved in photosynthesis, oxygen transport, the electron transport chain, and as cofactors of hydroperoxidases. In addition, the porphyrin-like corrin ring is found in vitamin B_{12}, cyanocobalamin.

The porphyrins are substituted tetrapyrroles based on the porphyrin XIV structure. Substituents found in positions 1 through 8 are most commonly methyl, ethyl, vinyl and propionic acid groups. The ferrous protoporphyrin IX complex shown in the Figure 7-3 is referred to as heme, and is the cofactor of hemoglobin, myoglobin, catalase peroxidase, and certain cytochromes. Certain variations in the heme structure are found among the cytochromes.

In the evolution of organisms from anaerobic to aerobic environments more effi-

cient energy metabolism became possible since the oxidation of foodstuffs to CO_2 and H_2O can provide far more ATP per fuel molecule than do anaerobic pathways. For larger animals which must depend on vigorous physical activity for survival, mechanisms were required to transport large amounts of oxygen to the tissues, to remove large amounts of metabolic end-products as CO_2, and to store sufficient amounts of oxygen in the active muscles.

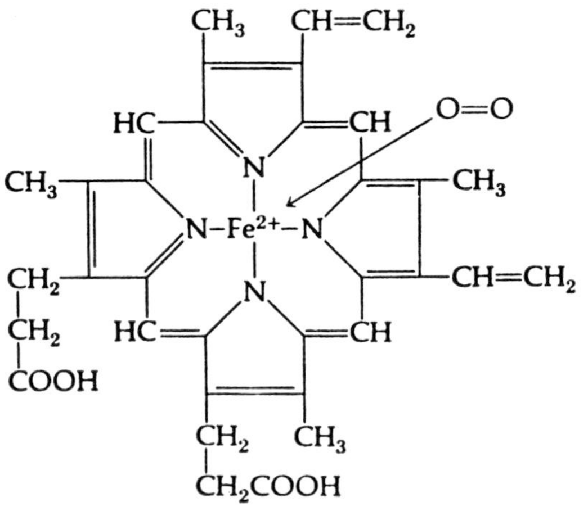

Oxygen has a limited solubility in water. For each 100 ml of air equilibrated water about 0.5 ml of oxygen is dissolved. Hemoglobin, however, when present in the normal physiological concentration of 15 gm/100 ml can bind the O_2 and effectively increase the oxygen saturation level to 20 ml oxygen per 100 ml of whole blood. This represents a forty-fold increase in the capacity of blood to transport oxygen. Myoglobin, in a similar fashion, is able to enhance greatly the level of oxygen held within active muscle tissue so that a reserve of oxygen is always present. In addition, there is good evidence that myoglobin accelerates the diffusion of oxygen through muscle cells and their membranes to increase the gas exchange efficiency.

The mechanisms for reversible oxygen binding are not simple. The ability of a molecule to exchange π electrons with molecular oxygen to form a bond, and to avoid oxidation in the process is limited to a few cases. In nature, hemerythrin (a nonheme ferrous containing protein) hemocyanin, a nonheme copper protein) and various heme proteins exhibit this property. Numerous early attempts at the design and preparation of synthetic oxygen binding agents failed to produce a complex that could undergo sustained reversible oxygenation under physiological conditions. Recent studies, however, using heme groups substituted to mimic the protein environ-

Figure 7-3. Protoporphyrin IX. (Used with permission. Edelstein, S. L.: *Introductory Biochemistry*. Holden-Day, San Francisco, 1973, p. 108).

ment have met with some success and provide insight into the interaction of oxygen, the heme, and the protein.

Heme itself only acts as an oxygen binding agent under special conditions. When dissociated from protein in oxygenated buffer at physiological pH, heme is readily oxidized to the Fe^{3+} form which is referred to as hematin. The globin chain of myoglobin, and the α and β chains of hemoglobin, however, provide hydrophobic pockets for the heme which is essential for reversible binding of the oxygen.

The primary structure of a protein molecule is the sequence of amino acids that forms its polypeptide backbone. Myoglobin is composed of 153 amino-acids, the sequence of which is known in whole or part for several species. Secondary protein structure refers to regularly repeating spatial patterns of the polypeptide chain; examples are the α helix and the β sheet structure formed by hydrogen bonding between antiparallel chains.

The secondary structure of myoglobin is of fundamental importance. While there is no region of β sheet, there are eight regions of α helix. These appear as the tubular segments in the drawing of myoglobin shown on page 47 of Reference 6. The helical segments are referred to by the letters A through H starting at the amino terminal end of the molecule. Regions of "random coil" between the helices are indicated by the two letters of the helices they separate.

The nonregular folding and bending of the molecule responsible for the spatial arrangement of the helices and the overall configuration define the tertiary structure of the protein. The tertiary structure of the molecule is a shallow disc; to be more precise, an oblate spheroid. When viewed from the "front," the heme group is seen to be slipped between the E and F helical re-

gions, and protected at the back by B, G, H and C.

While these "skeleton" drawings are essential for understanding the configuration of the protein, they do not convey the sense of the tight packing of the atoms in the protein molecule. In the actual myoglobin molecule there is room for only four molecules of water. As with other globular proteins, myoglobin is composed of hydrophilic residues primarily located on the exterior where they interact with water, and nonpolar residues, almost exclusively on the interior. The heme group is exposed to the aqueous solution only on the edge containing the propionic acid side chains. The remainder is buried within the hydrophobic pocket provided by the globin. It is this environment that is important in preventing oxidation of the heme iron.

The binding of the heme to the globin is through van der Waals interactions with some contribution by hydrogen bonding. In addition, an imidazole nitrogen from a histidine in the F helix (His F8) is coordinated to the iron of the heme. This histidine linkage is also of importance in preventing oxidation and, as will be discussed below, is a key component in the functioning of the hemoglobin. A second histidine is nearby on the myoglobin E helix and interacts with a bound water of ferri-myoglobin (metmyoglobin).

Hemoglobin

Hemoglobin, as an oxygen transport protein, has a more complex function than does myoglobin. Hemoglobin must efficiently bind oxygen in the alveolar capillaries of the lungs and be able to readily discharge large quantities of the gas in the tissues. The chemical reactivity of hemoglobin is an elegant example of the "form follows function" theory of modern biology. Hemoglobin is composed of four subunits,

2 α's and 2 β's, arranged approximately in a tetrahedral arrangement.[8] The geometry of the association, and interaction of subunits composing a protein molecule is referred to as its quarternary structure.

The primary structure of myoglobin and the two α and β polypeptides, exhibit sequence homology. The β chain most closely resembles myoglobin. It has seven fewer amino acids, most of which are missing in the carboxyl terminal region. The α chain has twelve fewer amino acids than myoglobin. The D helix is missing, and again, as with β, a portion of the carboxyl terminal sequence is missing. The principal elements of the secondary and tertiary structure are identical in the three proteins. The heme is in a hydrophobic pocket bound by hydrophobic linkages and the covalent bond from His F-8 to the iron of the heme.

The quaternary structure of hemoglobin is almost spherical with a hole down the central axis. Each α chain is in extensive contact with one β and in moderate contact with the other β. These contacts are described as $\alpha_1\beta_1$ and $\alpha_1\beta_2$, respectively. There is little contact between like chains, i.e. $\alpha_1\alpha_2$ and $\beta_1\beta_2$. In the important $\alpha_1\beta_1$ and $\alpha_1\beta_2$ contacts the overwhelming bonding force is hydrophobic or van der Waals interactions. Hydrogen bonds and salt linkages are rare and probably contribute little to the stability of the quaternary structure.

The oxygenation curves for hemoglobin and myoglobin expressed as the percentage saturation with oxygen as function of oxygen pressure (or concentration) is shown in Figure 7-4. The most important characteristics are the hyperbolic function of myoglobin and the sigmoidal curve of hemoglobin. The simple binding of oxygen to independent sites results in a hyperbolic

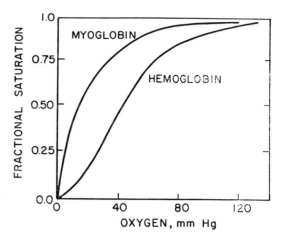

Figure 7-4. Oxygenation of hemoglobin and myoglobin.

curve, while an interaction between the hemes of hemoglobin results in a sigmoidal curve. There is a changing affinity for oxygen as the haemoglobin becomes saturated. The physiological advantage of the sigmoidal binding property comes from a complete loading of the protein with oxygen at a high oxygen pressure in the lungs and a substantially greater release of oxygen in the tissues at lower pressures than would be obtained with independent sites. Let us now turn to the question of the molecular mechanisms providing this site-site interaction, as described by M. F. Perutz.[8]

In a functioning hemoglobin molecule, the heme iron is in the ferrous state. It has six electrons in the 3d orbitals, and a choice of two spin-states: high spin with four unpaired electrons and low-spin with no unpaired electrons. The environment of the porphyrin ring sitting in the hydrophobic pocket is such that the ferrous ion is poised near the crossover point between the two spin-states. A change in the ligand field will induce a change in the spin-state.

In deoxyhemoglobin the iron is in an

unusual five coordinate geometry; bound to the proximal histidine (F8) and the four nitrogens of the tetrapyrrole ring. It is in the high spin-state. Due to the larger ionic radius of this state the ferrous ion cannot be accommodated by the tetrapyrrole ring, but is forced to sit about 0.8Å out of the plane toward His F8.

Molecular oxygen, a good π bond acceptor, increases the ligand field strength and upon binding, forces the iron(II) into the low spin-state. The switch in spin-states decreases the ionic radius and the iron is drawn into the plane of the ring. The immediate effect on the protein matrix is to draw His F8 some 0.75 to 0.95Å closer to the plane of the heme ring.

Steric consideration appear to favor the oxygenation of α subunits first. The movement of the iron and His F8 toward the heme plane results in a movement of the entire F helix toward the center of the molecule and simultaneous breakage of salt bonds between α_1 and α_2. This type of tertiary structure change then is repeated in α_2, and a half-saturated hemoglobin molecule results. The breakage of the salt bonds has released constraints maintaining the hemoglobin in a deoxy conformation, and a change in the relative positions of the subunits can now occur. This quaternary structure alteration results in a movement of the α's away from one another and a closing of a gap between the β subunits. Constraints on the oxygenation of the β subunits are removed and those hemes exhibit a higher affinity for oxygen.

The sequence of events that leads to the expression of the heme-heme interaction of hemoglobin can be summarized as follows: The binding of oxygen to the ferrous heme ring results in a significant change in the ligand field strength experienced by the iron, and results in a pairing of the

electrons in the 3d orbitals. The effect of the change to the low spin-state is to decrease the ionic radius of the iron and to allow it to move into the plane of the heme ring. This movement and a possible shortening of the iron-His F-8 bond changes the tertiary structure of the subunit and alters the subunit-subunit contacts. Subsequent oxygen binding has fewer constraints and there is an increased affinity of the heme for oxygen. This elegant exploitation of the coordination chemistry of iron by hemoglobin was fundamental in allowing the development of large, active aerobic organisms.

Cytochrome c

The cytochromes are a class of heme containing enzymes that are involved in oxidation-reduction reactions. They serve as important components of the mitochondrial electron transport chain. Five cytochromes of varying redox potential in the terminal portion of the chain sequentially accept and donate electrons which eventually are used to reduce molecular oxygen to water. As a result of these favorable redox reactions energy can be stored in the form of adenosine triphosphate (ATP). Cytochromes are also found in the microsomal fraction of the cell and perform a variety of oxidative reactions in which steroids and aromatic compounds are often the substrates.

Of the mitochondrial cytochromes, cytochrome c is the only one that is readily solubilized from the membrane matrix within which the chain functions. As a result cytochrome c is by far the best-characterized protein of the electron transport chain. The enzyme from horses has been determined by R. E. Dickerson and his coworkers at the California Institute of Technology. In addition, the amino acid sequences of cytochrome c from over forty organisms have been determined. A consideration of the structure and function of cytochrome c provides another example of the interaction of iron, protoporphyrin, and protein.

A skeleton drawing of cytochrome c can be seen in Dickerson.[10] The molecule has 104 amino acids and a molecular weight of 12,400 daltons. As with myoglobin, the interior is composed of hydrophobic residues while the hydrophilic residues are generally at or near the surface. Charged groups are only found at the surface.

The heme group sits in a hydrophobic crevice as seen from this view of the "front" of the molecule. The iron of the heme is octahedral, and is coordinated to the four nitrogens of the protoporphyrin ring, an axial histidine,[18] and the sulfur of Met 80. Since cytochrome c does not exchange ligands, but only electrons, it does not require an "open" position in the coordination sphere of the iron.

Cytochrome c has four helical regions and no β sheet structure. The tertiary structure is quite tightly packed in the shape of an egg, with the dimensions 25 $\times$ 25 $\times$ 30Å. On the right and left sides of the molecule are "hydrophobic channels" ringed by positively charged lysine residues. The right side channel is large enough to accept another hydrophobic residue. The left channel is more tightly packed and includes Tyr 74, Trp 74, Trp 59, and Tyr 67.

The function of cytochrome c is to shuttle electrons from a "reductase complex" (associated with cyt c_1) to an "oxidase complex" (cyt a and a_3, known as cytochrome oxidase). In the process of electron exchange the iron of the heme fluctuates between the ferrous and ferric states. The means by which cytochrome c accepts and donates electrons to the two complexes is largely unknown. Dickerson, however, has summarized information from X-ray structure, chemical modification experiments, and amino acid sequences to provide intriguing insight into possible mechanisms of electron transfer through the molecule and associated conformational changes.

Cytochrome c is believed to interact with the reductase via its left hydrophobic channel. The structural importance of this region is emphasized by the fact that the cytochrome c's of most species are very similar in this region. In fact, a stretch of polypeptide (70 to 80) on the left face of the molecule is invariant throughout evolution; virtually all creatures and plants have the same amino acid sequence.

Oxidized cytochrome c is believed to "dock" on the surface of the reductase via the ring of positive charges surrounding the left channel. The reductase then donates an electron to Tyr 74, which in turn passes it on to Trp 59. A negative charge in the region induces Tyr 67 to donate an electron to the heme ring by π electron cloud overlap. At this point there is a positive charge on Tyr 67 and a negative charge on Trp 59, or an ion pair. In the hydrophobic environment of the protein's interior, this is a very high energy situation. Using this energy as a driving force, the molecule undergoes a conformational change (an alteration of the tertiary structure) which brings the two charged aromatic residues into parallelism and allows them to exchange electrons by π cloud overlap.

The conformational change makes the molecule more compact and may cause residues (Phe 82) on the left side to swing in and protect the heme group in its crevice. The oxidation of cytochrome c is believed to occur by interactions close to the heme crevice, and may perhaps involve Phe 82 which is probably close to the heme in the reduced state.

Thus iron combined with the protoporphyrin ring binds and transports molecular oxygen in one protein and binds and transports electrons in another. The protoporphyrin ring modifies and enhances the coordination chemistry of the iron, while the protein binds, protects and leads to the ultimate manifestation of the iron's biological activity.

Iron Balance and Iron Deficiency

Iron is quantitatively the most important of the trace elements. It is present to the extent of 50 mg and 35 mg per kilogram of body weight in men and women respectively, giving a total of some 2 to 5 grams of iron in the normal adult. Approximately two-thirds of this iron is in the hemoglobin of the circulating red cells. Myoglobin accounts for 4 percent, and the important heme enzymes utilize only 0.2 percent of the total body iron. The bulk of the remaining iron, some 30 percent, is storage iron distributed equally between ferritin and hemosiderin. Transferrin bound iron represents only 0.12 percent of the total.

The major pathways of iron flow in humans are depicted in Figure 7-5. Iron enters the lumen of the intestine complexed with a wide variety of dietary foodstuffs. Some of the metal enters the intestinal mucosal cells. The bulk of the iron is excreted in the feces as insoluble complexes. Iron that is transported across the intestine enters the blood where it is sequestered by transferrin

Each transferrin molecule binds two atoms of ferric iron. Two moles of carbonate are also bound. This serves to mediate the transport of iron between the sites of assimilation, storage and utilization. The tissues and cells that are involved in iron storage and in the formation of circulating cells are referred to as the reticulo-endo-

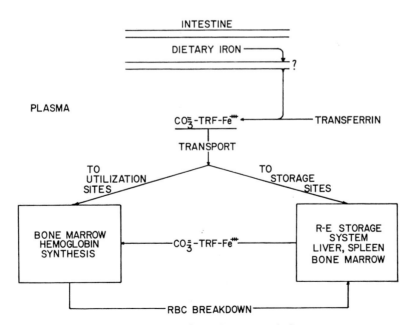

Figure 7-5. Outline of iron metabolism.

thelial (R.E.) system, principally the liver, spleen, and bone marrow. Iron is stored in two molecular forms. Ferritin, which we have mentioned earlier, consists of a protein shell of twenty-four subunits surrounding a polynuclear ion-oxo core. Hemosiderin, a less well defined protein, appears to be an insoluble breakdown product of ferritin although it continues to exchange iron with the rest of the cell. While the greatest amount of storage iron is found in the R.E. system, ferritin is found to some extent in all the cells of the body.

The principal biosynthetic demand for iron is the formation of hemoglobin. Since the red blood cell has a life span of some 120 days and heme is not recycled, there is a constant requirement for newly synthesized heme in the developing erythrocytes of the bone marrow. Iron is donated to these cells by transferrin.

The breakdown of aged red blood cells occurs predominantly in the spleen. The globin and heme are destroyed and the iron becomes incorporated into ferritin to be stored for later use. The body does not normally excrete large amounts of iron but rather is conservative in its use, i.e. humans recycle iron. Iron loss occurs primarily via the intestinal tract from sloughed-off epithelial cells, and by internal and external blood loss. Other factors such as the normal loss of skin tissue, hair, nails, and sweat represent an additional loss of 0.5 mg to 1.0 mg of iron per day. This amount is doubled by women in their childbearing years because of menstrual blood loss, and tripled during pregnancy while iron is being donated to the fetus. Thus, the female daily iron requirement is 2 mg while menstruating, 3 mg while lactating and 7.5 mg while gestating. It is small wonder that the iron supplement advertisements are aimed at the young female segment of the population.

Iron deficiency in various stages is perhaps the most *prevalent* nutritional problem in the world today, affecting the health, vitality, and economic productivity of countless millions. While iron deficiencies have been recognized and treated for centuries, it is only within the past few decades that reasonably large groups have been examined and a true appreciation of the prevalence of iron deficiency has been recognized on a quantitative basis. In the United States, iron deficiency *anemia* is found in approximately 12 percent of nonpregnant and 20 to 58 percent of pregnant women. It is estimated that the iron assimilation from the average American or European diet is generally not sufficient to satisfy the nutritional demands of women in their childbearing years.

Iron deficiency can be considered as a three-stage process:

1. Iron Depletion or Sideropenia:

In this stage the physiological demands for iron have surpassed the dietary intake and a mild depletion of the iron stores has occurred. There is considerable controversy at present as to whether this stage represents a true pathological condition. Symptomology includes an increased efficiency of iron absorption and an absence of hemosiderin in bone marrow samples. Hemoglobin levels are normal. Serum transferrin levels and iron saturation percentage are normal in this stage.

2. Latent Iron Deficiency:

In this stage a more serious depletion of the iron stores has occurred. However, since iron absorption is increased it is possible to maintain sufficient erythropoietic activity and hemoglobin levels are normal. Transferrin iron saturation levels drop from a normal value of 35 percent to below 16 percent.

3. Iron Deficiency Anemia:

In this stage the depletion of iron stores

is complete to the extent that erythropoiesis is impaired and a microcytic hypochromic anemia develops. Transferrin saturation decreases to about 10 percent and total iron binding capacity increases from normal values near 340 μgm% to over 400 μgm%.

While bone marrow biopsy gives the best index of iron stores, it is both painful and too complicated for routine diagnosis of iron deficiency. Hemoglobin levels and hematocrit values and the analysis of serum transferrin values are most commonly used to diagnose the iron status of the individual.

INTESTINAL IRON ABSORPTION AND CONTROL OF IRON BALANCE

Normal individuals on a well-balanced diet have normal hemoglobin and storage iron values. Mechanisms exist to control the overall body iron balance. The primary site of the regulatory control mechanisms that are responsible for iron absorption is the intestinal epithelial cells.

Dietary iron enters the gastrointestinal tract complexed to a variety of ligands and biomolecules including inorganic anions, proteins, amino acids, cofactors, and sugars. In the acid pH of the stomach, many of these complexes are broken down and new complexes formed. The availability of iron for absorption depends upon intralumenal factors and dietary components.

Iron is not absorbed in the stomach but rather in the small intestine, primarily in the duodenum and also the jejunum. The small intestine is an organ that is highly specialized for the absorptive process. The interior surface or mucosa is lined with numerous finger-like projections called *villi* which greatly enlarge the surface area. The mucosal membrane is composed of a layer of absorptive cells, the epithelial cells, which overlay a region of connective tissue laced with an extensive network of blood and lymph capillaries. Several other layers of connective tissue and muscle complete the cross section of the small intestine.

The amount of iron absorbed is in direct proportion to the needs of the individual. A normal individual absorbs about 15 percent of the daily dietary iron. In iron overload this fraction is decreased to 5 to 10 percent and a significant portion of that iron is held by the intestinal epithelial cell rather than being passed on to the circulating iron pool. Since the epithelial cell is sloughed off after two to three days, its iron content is also excreted. In cases of iron deficiency, iron absorption can be in the 25 percent range and very little iron is maintained by the absorptive cells. Clearly, mechanisms exist to adjust iron absorption to physiological needs.

The composition of the diet profoundly influences the level of iron absorption. Components such as phosphates, phytates, carbonate, and oxalate which tend to form inert ferric precipitates greatly decrease the availability of iron. On the other hand, complexing biomolecules, especially sugars, can greatly enhance the level of iron absorption. The role of chelation in iron absorption is critical. If iron is not maintained in a soluble low molecular weight form then the absorptive mechanism intrinsic to the intestinal epithelial cells cannot be operative. An excellent case in point involves the role of carbohydrates in enhancing iron absorption. We have examined the solubilization of iron by sugars and shown that carbohydrates such as fructose can prevent the formation of inert ferric hydroxide polymers. In solution ferric fructose exists as low molecular weight complexes of the form [(Fructose-iron)$_2$O] which are in equilibrium with reactive polymeric species.

As a result of its unique chemistry, ferric fructose is readily assimilated. Studies with humans and laboratory animals have shown iron absorption levels that are double that of the standard iron supplement ferrous sulfate. High levels of sugar iron complexes in the diet can in some cases bypass the normal regulatory mechanisms. Iron storage pathologies can develop in which the liver iron content reaches fatal levels. Bantu tribesmen of Africa show a high incidence of hepatic siderosis caused by the practice of cooking carbohydrate-rich corn paste in cast iron pots. Iron overload is a problem for sweet wine alcoholics who assimilate large fractions of sugar-iron complexes in wine despite the fact that their storage iron depots are already super-saturated. Alcoholics who drink dry wines or whiskys do not exhibit the iron overload problems.

Another important observation relation to iron absorption is that an increase in the level of iron absorption is not observed until two to three days after iron loss by bleeding. Conrad and Crosby[13] have formulated a hypothesis to account for these and other results. It has been suggested that important events occur during the cellular development of the epithelial cells in the crypts of Liberkühn, which dictate the fundamental iron absorption rate of the cells after they mature. In an anemic subject, the cells develop with high capacity to absorb and transport iron while in an iron overloaded subject the converse is true.

The two to three day life span of the epithelial cell then dictates the lagtime that is noted in changing the iron absorption rate.

Several hypotheses have been advanced in order to explain the molecular mechanisms the intestinal epithelial cells employ in regulating iron assimilation. An early concept was that the level of ferritin in the epithelial cells was the principal regulator of iron uptake. This was referred to as the "mucosal block" mechanism. Active transport of iron into the blood has also been proposed. Presumably this "pump" would be controlled by the iron requirements of the organism. Helbock and Saltman[12] have proposed a mucosal-directed active-transport system, which is outlined in Figure 7-6. The important feature of this hypothesis is an active transport system which effectively prevents the iron that has been assimilated by the cell from being passed on to the circulatory system. This pump was found to be under feedback control. Under conditions of iron deficiency in test animals the pump was inoperative resulting in an increased iron flow to the blood. The most important feature of this mechanism is the role of low molecular weight chelates of iron in the passive transport of iron into the blood.

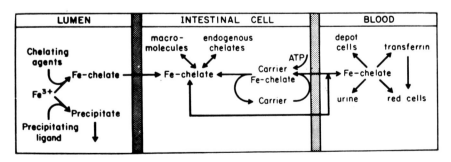

Figure 7-6. Iron transport across the small intestine. (Used with permission. Helbock, H. and Saltman, P.: *Biochim Biophys Acta, 135*:979, 1967.)

TRANSFERRIN

Introduction

The transferrins are a group of related iron-binding proteins that are found in leucocytes and body fluids (serum, cerebrospinal fluid, milk, semen) of mammals and certain other organisms and in the whites of avian eggs. Their primary role is in the mobilization and transport of iron, but they are of fundamental importance for the bacteriostatic action as well. Bacterial growth is limited in the presence of transferrin due to the very low concentration of available iron in the medium.

The nomenclature of the transferrins has become cumbersome. The serum protein was named "siderophilin" meaning iron loving by one group of investigators and "transferrin" since it transports iron by another. The latter name has gained wide use and for the sake of consistency in these days of computer literature it is preferred. The egg white protein was originally named "conalbumin," "ovo-transferrin" is now preferred. The protein found in milk, leucocytes and external secretions is referred to as "lactoferrin," since it was originally found in milk.

Convenient terminology can be adopted from the hemoglobin group. Transferrin (or lactoferrin) is used as the generic term and refers to the protein in any of its several forms. Fe^{3+}-transferrin and apotransferrin refer specifically to the iron filled or metal-free forms. The suffix-CO_3 is used to bring special attention to carbonate binding. In this chapter, we will center our attention on the chemistry and biology of human serum transferrin.

Physical Properties

The transferrins bind two ferric ions per molecule at sites that are composed of amino acid side chains. For each iron bound a carbonate anion is also tightly bound. Serum transferrin has a molecular weight of 76,600 and is an oblate elipsoid with molecular dimensions of 55 by 28Å. The amino acid sequence has not been fully determined although work on this subject is underway for both serum transferrin and ovo-transferrin.

Evidence from ultracentrifugation, dielectric dispersion, viscosity, and chemical reactivity suggests that the transferrin molecule undergoes a significant conformational change upon binding iron. It is suggested that the protein expands slightly and becomes more spherical. The iron-protein is more stable toward chemical denaturation and proteolytic attack. Both apo- and iron-transferrin are relatively stable as a result of some twenty disulfide bonds that cross link the protein chain.

In addition to the two metal and two anion binding sites, transferrin has two heteropolysaccharide chains covalently attached to the polypeptide. These observations and the rather high molecular weight, suggested that transferrin might consist of two identical subunits. Studies using denaturing conditions and disulfide cleaving reagents, however, failed to reveal subunits of the protein. It has also been considered that the single polypeptide chain might be composed of "pseudo subunits," that is similar or identical regions combined in one protein by the process of "gene fusion." Recent studies, however, have shown that there are no extensive identical or even closely similar regions of polypeptide. The best evidence to date is that the transferrin molecule is composed of a single polypeptide chain that is folded in such a way to give two very similar metal ion and anion binding sites.

The Nature of Metal Binding

In recent years a variety of physical techniques have been brought to bear on the question of the nature of metal binding by the transferrins. These techniques have included radioisotopes electron-paramagnetic resonance (epr), equilibrium dialysis, nuclear magnetic resonance, stopped flow spectrophotometry and fluorescence spectroscopy. While a tentative picture of the metal binding site has emerged, there are as yet many questions remaining unanswered.

The two metal binding sites of serum transferrin appear to be identical or nearly so by epr measurements. The sites of ovotransferrin, however, do not give the same epr signal as that of serum transferrin and further the two sites of ovotransferrin show some clear cut differences.

The affinity of the two serum transferrin sites for iron is about equal and is extremely high; about 10^{31} under physiological conditions. The sites do not appear to interact; a random distribution of iron on the two sites has been observed. Evidence obtained using trivalent lanthanide ions as fluorescent probes suggests that the metal binding sites are at the distal ends of the molecule. The amino acid side chains that coordinate the iron appear to be histidyl, tyrosyl and possibly tryptophanyl residues as revealed by various spectroscopic parameters. It is clear that transferrin does not have the binding geometry exhibited by the iron-sulfur proteins.

In addition to iron, transferrin binds copper, chromium, gallium, cobalt, manganese and a variety of other metal ions. In serum, however, only iron is bound to any significant extent. Binding of ferrous ion can be demonstrated by the rapid oxidation of the iron to the ferric state.

In summary, serum transferrin possesses two high affinity iron binding sites composed of amino acid side chains (and possibly carbonate). The sites are well separated and are equal in binding affinity and noninteracting. The binding of iron produces significant changes in the conformation and stability of the proteins. Evidence indicates that the serum transferrin has practically identical sites, while ovotransferrin may have two dissimilar sites on the same molecule.

The Anion Binding Site

While it has been known for two decades that transferrin binds a carbonate anion for each metal ion bound, it was only recently that substantial insight into the nature and function of anion binding has been gained. A controversy that has existed since the discovery of carbonate binding has regarded the binding of the ferric ion in the absence of carbonate, or a carbonate substitute. It now seems clear that the protein has a very low affinity for iron in the absence of a suitable anion. Attempts to form an anion-free complex are complicated by the propensity of iron to polymerize and precipitate at neutral pH in the absence of suitable chelating agents.

There are three possible models that can be considered for metal and anion binding to the transferrin. These are: 1) the anion is coordinated solely to the metal; 2) the anion is bound only to the protein at an "allosteric" anion binding site; and 3) the anion is bound to both the protein and the metal ion, and as such constitutes an intrinsic component of the metal binding site. It is possible to investigate these models by physical techniques and chemical reactivity studies using carbonate substitutes.

There are several lines of evidence which appear to rule out the model 1, and to favor model 3 over model 2. First, the exchange of specifically bound $^{14}CO_3$ with unlabelled bicarbonate in buffer solutions is on the order of days to weeks depending on conditions. If model 1 were functioning an exchange rate of milliseconds or seconds would be anticipated. Second, a rather narrow range of anions are able to act as carbonate substitutes. It has been shown that effective anions must have a carboxyl group and a potential coordinating atom on the α-carbon. Good carbonate substitutes include oxalate, glyoxalate, glycolic acid, thioglycolyic acid, and glycine. Malonate, the three carbon dicarboxylic acid is a fair substitute and is one of the few known exceptions to the α-ligand rule. Inorganic anions such as nitrate, sulfate, and phosphate do not act as substitutes. We have interpreted this information as favoring model three. Perhaps the substitute can interact via ionic or hydrogen bonds with the protein and with the metal ion via the coordinating group on the α carbon.

Finally, the very high degree of interaction between the anion and metal binding functions is significant. The fact that the protein has an affinity of greater than 10^{31} in the presence of carbonate and may not even form a complex in the absence of anions appears to favor model three. Most allosteric interactions enhance ligand binding by a factor of 10^2 to 10^3. The fact that the transferrin exhibits such an enormous binding increment in the presence of carbonate together with the information on exchange rates and the α-ligand rule strongly supports model 3. A diagramatic representation of such a site is shown in Figure 7-7. In this hypothetical scheme, the carbonate is depicted as interacting with positive charges on the protein via two of its oxygens and binding the iron via a coordinate bind through the third oxygen. In the ab-

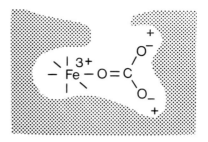

Figure 7-7. Hypothetical Binding Site of Transferrin. (Used with permission. Bates, G. W. and Schlabach, M. R.: *FEBS Letters*, 33:289, 1973.)

sence of a synergistic anion, a ligand of the iron would be absent and charge repulsion between the positive charges of the iron and the protein would occur. The net result would be a far less stable metal binding site.

Iron Exchange Reactions

The exchange of iron between transferrin binding sites and chelating agents provides an insight into possible biological metal exchange mechanisms, the chemical reactivity of transferrin, and also the nature of iron in aqueous solution. We have examined the donation of iron to transferrin by several ferric chelating agents, especially citrate, NTA, and EDTA.

The rates of reaction of human apotransferrin with these iron complexes as monitored spectrophotometrically are shown in Figure 7-8. The reaction with Fe^{3+}-NTA occurs very rapidly. Fe^{3+}-citrate and Fe^{3+}-EDTA are less reactive, requiring twenty hours and one week respectively. The reason for this wide variability in reaction rate can be explained in terms of the chemistry of the ferric chelate complexes. EDTA, a very strong hexadentate chelate effectively prevents polymer formation, but in the process of binding iron, completely surrounds the six coordination sites of the

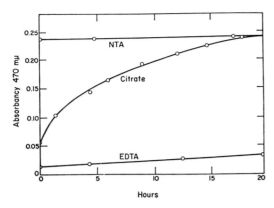

Figure 7-8. Rate of reaction of ferris chelate complexes with apotransferrin. (Used with permission. Bates, G. W., Billups, C., and Saltman, P.: *J Biol Chem, 242*:2810, 1967.)

metal. Since most ligand exchange reactions at transition metal ions follow an Snl mechanism, it is necessary for the EDTA to partially unwrap from the metal before the transferrin can attack coordination sites of the cation.

Fe^{3+}-Citrate is slow for another reason. Citrate is a weaker tridentate chelate and is unable to prevent polymer formation. At a 1:1 ratio of citrate to iron about 75 percent of the iron exists as large polynuclear iron complexes virtually identical to the ferric-hydroxide polymers described above. In fact, the citrate appears to be acting as a coating for the inorganic polymer. The remaining 25 percent of the iron is in a low molecular weight form and is rapidly reactive. Transferrin, however, is unable to attack the 150,000 molecular weight polymer and the rate limiting step of the second phase is relase of iron from the polymer in a low molecular weight complex.

NTA falls between citrate and EDTA in iron binding affinity and the number of coordinating ligands, and as a result provides an optimally reactive complex. This tetradentate chelate effectively prevents inert polymer formation, yet leaves two of the coordination sites of the iron exposed to the solvent. While dimeric and other condensation products do form, they appear to be in rapid equilibrium with monomeric iron.

The reaction of Fe^{3+}-NTA with transferrin has been examined by stopped flow spectrophotometry. The reaction is biphasic, the first phase requiring 0.2 seconds and the second phase ten seconds. Kinetic studies have provided strong evidence that the first phase is an attack by apotransferrin-carbonate on the iron of Fe^{3+}-NTA to form an NTA-Fe^{3+}-transferrin-CO_3 complex. The second phase represents the breakdown of this mixed ligand complex to form the physiological Fe^{3+}-transferrin-CO_3.

The removal of ferric ions from transferrin by chelates is a quite slow reaction, suggesting that the iron is not readily accessible to these agents. Replacement of the carbonate by a relatively weak substitute such as glycine greatly enhances the iron release rate. It is apparent that the anion binding site exerts an important influence on the reactivity of the metal ion.

Biology

Iron Transport

The requirement for a serum iron transport molecule is dictated by the fact that ferrous ion is readily oxidized to ferric ion under physiological conditions and that ferric ion in an unbound state is highly toxic. This toxicity in turn is related to the propensity of ferric ion to polymerize, precipitate, and to form nonspecific complexes with a variety of rather delicate biomolecules. Clearly, there is a need for a high-affinity iron transport agent to protect the organism and to deliver the iron in a specific fashion to cellular iron receptor sites.

The mechanism by which transferrin

picks up the dietary iron that crosses the intestinal wall is not well defined. One hypothesis holds that iron is transported through the mucosal cells by a set of specific membrane and cytoplasmic carriers and is donated to transferrin at the serosal membrane by iron donor sites. While this process may be operating under normal circumstances, there is clear evidence that transferrin is not required for iron assimilation since persons congenitally lacking transferrin readily accumulate the metal. Injection of apotransferrin, *in vivo*, does not alter the rate of iron uptake.

Another school of thought is that iron is transported as soluble ferrous ions and appears in the serum as the divalent cation. It has been suggested that the protein ceruloplasmin (ferroxidase) acts to oxidize the iron to the ferric state, following which it becomes rapidly sequestered by transferrin. Recent chemical studies, however, indicate that transferrin binds ferrous ions, and rapidly oxidizes it to Fe^{3+}-transferrin in the presence of oxygen and other oxidizing agents. Further, if the iron becomes oxidized prior to association with transferrin in serum then a variety of nonspecific binding sites compete for the strong Lewis acid properties of ferric ion. These chemical studies appear to be in conflict with the proposed mechanism of ceruloplasmin action. It should be emphasized, however, that strong evidence exists for the involvement of copper metabolism in iron mobilization. Under conditions of copper deficiency there is increased storage of iron and ineffective mobilization of the metal to the red cells.

A third mechanism is based on the observed high rates of transport of low molecular weight ferric complexes. Iron moves across the mucosal cells and appears in the blood as ferric chelates. The high reactivity of transferrin towards labile iron

compounds would assure the rapid movement of this iron into the specific iron management pathways.

The Donation of Iron to Cells

Under normal physiological conditions transferrin donates iron to all cells of the body. The most intensively studied system is the interaction of transferrin with the developing erythrocyte (red blood cell). As will be discussed below, erythrocytes proceed through a cellular maturation sequence in which hemoglobin is synthesized during an early period and large amounts of iron are required. The final immature stage is the reticulocyte which has no nucleus but continues a low level of hemoglobin synthesis and iron assimilation. Large numbers of these reticulocytes in the circulation (reticulocystosis) can be induced in experimental animals by repeating bleeding or treatment with phenylhydrazine.

Using radioisotopes it is possible to study the reaction of transferrin with reticulocytes. The basic experimental approach is to load transferrin with tracer amounts of ^{59}Fe and to label the protein by introducing ^{125}I or ^{131}I into a limited number of tyrosyl residues. In addition, $^{14}CO_3^{2-}$ can be introduced into the anion binding site and the unbound carbonate removal by gel filtration. This transferrin preparation is then incubated with a reticulocyte population for various times and under various conditions. At the end of the incubation period, the mixture is rapidly cooled and the cells washed by centrifugation and then analyzed for radioactivity.

Using this approach, the following general concepts of iron donation have evolved. Each immature cell possesses about 200,000 specific iron receptor sites. These sites are able to recognize the transferrin molecule. The iron-transferrin binds to the membrane in the form of a protein-iron-site complex. The iron is released to the cell, carbonate is released to the medium and and apo-transferrin is liberated. The exact mechanism of iron donation is currently being investigated by several laboratories. It has been demonstrated that if oxalate is substituted for carbonate at the anion binding site, a greatly diminished iron donation rate is observed. While these results should be interpreted with caution, it seems to lend credence to the long held hypothesis that the anion binding site plays a central role in release of iron to the membrane receptor site. If for example, a proton donor on or in the membrane could cause the release of the carbonate the binding affinity of the transferrin would be greatly lowered and the iron would become readily accessible.

Fletcher and Huehns made a remarkable observation that the two sites of the iron-transferrin molecule behave differently *in vivo*. One site appears to give up its iron to storage depots while the other site is used by developing erythrocytes and other iron utilization cells. This hypothesis is of central importance to theories regarding iron flux in the body and the control of the major iron maintenance pathways. The Fletcher-Huehns hypothesis has received support from several laboratories and is deserving of careful consideration. The study of the overall control and integration of iron metabolism in humans and other higher animals is not well understood at this time. Certainly, transferrin plays a key role in iron metabolism.

IRON STORAGE

Ferritin: Physical Properties

There are several constraints that must be imposed on any biological iron storage molecule. First, the iron must be in a stable chemical form that is not susceptible to attack by the myriad of complexing agents within the cell. Second, the iron must be readily available when needed for biosynthetic reactions. Third, the binding molecule must be both stable and under immediate biological control. Fourth, a maximum amount of iron per volume of storage molecule is necessary.

Transferrin, for example, would be a poor iron storage molecule. For each gram of iron stored, over half a kilogram of protein would be required. The liver could not accommodate that amount of protein simply for iron storage, not to mention the exorbitant expense of biological energy to synthesize all that protein. Clearly, a more compact form of iron is required such as an inorganic ferric micelle. This polymeric ferric hydroxide complex has a spherical shape with a diameter of about 70 Å and a quite narrow size distribution. If cross linking of polymers is prevented by an association of suitable anions on the exterior surface, the polymer remains stable in solution. The chemical environment of the iron itself is quite stable, being linked and cross-linked by a series of oxo bridges. Further, the iron constitutes about 40 percent of the weight of the micelle.

The ferritin molecule consists of a "protein shell" surrounding a core of microcrystalline iron. The protein shell is made up of twenty-four subunits distributed at the vertices of a snub cube. Each subunit has a molecular weight of 18,500, giving the assembled protein shell a molecular weight of 448,000. The iron core is best described as a ferric oxide hydrate and phosphate and has the approximate composition $[(FeOOH)_8 (FeO:OPO_3H_2)]$. The exact atomic structure of the mineral core has not been elucidated to this time. The X-ray diffraction patterns of the ferritin core indicate homogeneous poorly crystalline material. The atomic structure appears to resemble some common ferric oxide hydrate crystalline forms, yet it has unique properties of its own, which may maintain the iron in a form that is readily mobilized. Phosphate is believed to prevent the conversion of the micelle to a more stable, inert mineral structure.

The ferritin molecule can have a variable iron content and have either a single iron crystallite or a small number of separate iron crystallites within the protein shell. Fully loaded, ferritin contains as many as 2,500 iron atoms which make up 25 percent of the molecular weight. Under these conditions, the iron core is about 70Å in diameter and the entire protein 120Å in diameter. Upon removal of the iron core, the iron free apoferritin molecule shrinks somewhat to a diameter of about 100Å.

There is a striking similarity between the ferritin core and the inorganic iron micelle not only in shape and size distribution but also in its chemical and physical properties. In this regard, Mössbauer spectroscopy, spectrophotometry, magnetic susceptibility and X-ray scattering have all pointed to similar features in these iron balls.

The ferritin and apoferritin molecules are extremely stable under a variety of solution conditions. For example, apoferritin does not dissociate at all between pH 2.8 and 10.6, and only slowly at more extreme pH values. The complex of subunits is stable in 8 M urea, and at high temperatures. Some evidence suggests that the ferritin subunits are bound together primarily by hydrophobic interactions. This means a bonding between the hydrophobic residues such as lucine, alanine, and phenylalanine, which are present in abundance. Crichton, however, has indicated that salt linkages and hydrogen bonding may be also important in maintaining the quaternary structure.

Hemosiderin

A second iron storage protein is hemosiderin, so named because it was mistakenly thought for a time to contain iron in the form of heme. It is now known that the iron of hemosiderin, as in ferritin, is in the form of an inorganic iron-hydroxide-phosphate polymer.

Hemosiderin has long been a puzzle to bio-inorganic chemists. It is amorphous, and water insoluble which, of course, greatly hinders its physical characterization. While ferritin subunits and fragments are found to be associated with hemosiderin, the protein content is quite low and up to 45 percent of the hemosiderin by weight is contributed by iron. While the exact relationship of ferritin and hemosiderin has not been clarified at this time, it is generally believed that hemosiderin is formed when there are insufficient subunits to maintain all of the iron as discrete ferritin molecules. Recently, the morphology of the hemosiderin micelle has been examined in some detail and has been found to be identical to ferritin micelles by the techniques of electron microscopy, Mössbauer spectroscopy and X-ray diffraction. The only significant difference noted was that the hemosiderin micelles were found to be smaller by some 10 to 20 percent. These observations bear on the questions of the biological role of hemosiderin, as will be discussed below.

Chemical and Biological Aspects of Iron Storage

The most fundamental unsolved question concerning iron storage involves the mechanism for the deposition and mobilization of ferritin iron. The most popular line of experimentation has been attempts to remove and reconstitute the iron by *in vitro* chemical means. To date, no laboratory has succeeded in completely stripping the iron from ferritin and then reconstituting a stable, crystallizable iron-saturated ferritin molecule, nor has it been possible to achieve a complete dissociation and re-association of the native apoferritin. Let us now focus on the chemical iron exchange experimentation and the biological implications. The addition of ferrous ion to apoferritin under oxidizing conditions leads to an enhanced rate of iron oxidation and the formation of a partially reconstituted ferritin. Michaelis-Menten enzyme kinetics have been observed for the ferritin enhanced oxidation rate. It is suggested that the reaction follows the sequence

$$\text{Apoferritin} + \text{Fe}^{2+} \rightleftharpoons \text{Apoferritin-Fe}^{2+}$$

$$\text{Apoferritin-Fe}^{2+} + \text{O}_2 \rightarrow \text{Ferritin}$$

Ferrous binding sites are proposed to be present on the inner surface of the apoferritin shell. When bound to these sites, the ferrous iron is highly susceptible to oxidation to the ferric state. Initially, there is a lag phase which is believed to be a nucleation step, subsequent iron is deposited and oxidized on the nascent microcrystallites. As the surface area of the iron core grows, a greater rate of iron deposition is possible. This "crystal growth" model would predict a sigmoidal progress curve which is observed experimentally.

In support of the ferrous ion binding site concept of apoferritin is the fact that chemical modification of two cysteine residues and one histidine residue, abolishes the rate enhancement effect. It is suggested that these residues *may* be involved in the ferrous binding. An interesting parallel is noted between the enhanced oxidation rate of ferrous in the presence of apoferritin and apotransferrin as discussed above. Both proteins weakly bind ferrous ion but greatly enhance its oxidation to the ferric form, for which the proteins have a high affinity.

An equally attractive mechanism involving chelation has also received experimental support. A possible route for the biosynthesis of the iron laden ferritin is shown in Figure 7-9. It is well established that the spontaneous polymerization of iron at physiological pH will lead to the formation of iron miceles which are identical by several parameters to the iron cores of ferritin. Perhaps in the cytoplasm of iron storage cells, iron complexed to endogenous chelates undergoes this polymerization to give a "preformed" ferritin core.

It has been possible to show that in the presence of apoferritin and the iron core, subunits become dissociated from the apoprotein and reassociate about the iron micelle. Initially, a partially reconstituted form is obtained which will reach an equilibrium with native ferritin. This mechanism also suggests that hemosiderin need not arise from the breakdown of ferritin but could arise via spontaneous iron polymerization, and in fact be a source of iron cores for ferritin formation.

The fact that iron can be removed from ferritin using a combination of reducing agents and ferrous specific chelates has led to speculation that *in vivo* iron mobilization follows a reduction-chelation route. It has been suggested that flavin mononucleotide may be directly responsible for the reduction of the ferritin iron and an enzyme has been isolated from bovine liver that can catalyze the transfer of electrons from NADH to FMN to augment the iron reduction process. This occurs only under completely anerobic conditions *in vitro*.

It is, of course, not mandatory that iron be mobilized from ferritin via a reduction-chelation mechanism. Investigations have indicated that substantial amounts of iron can be removed from ferritin in the ferric form by certain chelating agents. NTA was

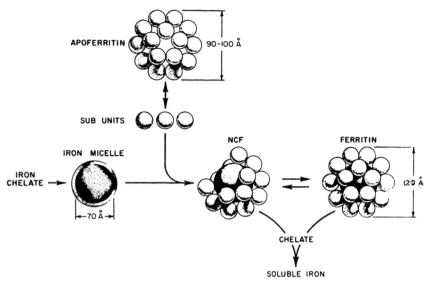

Figure 7-9. Ferritin synthesis and mobilization model. (Used with permission. Pape, L., Multani, J. S., Stitt, C., and Saltman, P.: *Biochem,* 7:606, 1968.)

the most efficient chelate tested, followed by EDTA and citrate. The rate of removal of the iron is not directly related to the binding affinity of the chelate. While endogenous ferric ion chelates have not been identified which are operative in the *in vivo* iron mobilization, diketogulonic acid, the oxidation product of ascorbic acid has been tested and is reasonably effective *in vitro.*

Ferritin occupies a central role in the management of biological iron. Although it exists in largest amounts in liver, spleen, and bone marrow, it can be demonstrated in all tissues of the body. In some tissues, multiple forms of ferritin exist (isoferritins) and differences between normal and malignant cell ferritins have been demonstrated.

The level of ferritin is responsive to the needs of the organism for an iron storage site. Injection or feeding a test animal iron brings on a rapid, transient burst of ferritin synthesis. The relation of ferritin and hemosiderin is not clear. While both are demon-

strable in many tissues, and both are able to donate iron for biosynthesis, it is not clear how iron is exchanged between the two forms. There is increasing evidence that hemosiderin is in fact a breakdown product of ferritin that is called into play when excessive iron stores require an even more compact form of iron storage. In fact, hemosiderin has been observed as a cytoplasmic inclusion surrounded by a limiting membrane. It is anticipated that the future will bring new insights into the mechanisms of iron storage and mobilization.

ERYTHROCYTE IRON METABOLISM

The synthesis of hemoglobin by the developing red blood cells requires a supply of iron and presents a unique opportunity for the study of iron management at the cellular level. The development of the red blood cell occurs primarily in the bone marrow, and is initiated by the action of the hormone erythropoietin on a multipotential stem cell. The earliest recognizable cell of the erythrocyte line is the pronormoblast. The early stages of development are marked by the elaboration of a subcellular structure specialized for the synthesis of protein. The nucleus is large. Mitochondria are in abundance. There is an extensive network of endoplasmic reticulum.

The synthesis of the protein portion of hemoglobin, the α and β globin chains, is directed and controlled via the transcription of the genetic information of DNA in the assembly of a messenger RNA molecule and the subsequent translation of the m-RNA base sequence into the amino acid sequence of the globin chain at the ribosome. The protein synthesis must, to an extent, be coordinated with the synthesis of heme.

The synthesis of protoporphyrin IX, the organic constituent of heme, begins with the activation and condensation of a molecule of glycine and succinate to give the molecule δ-aminolevulinic acid (δ-ALA). The enzyme responsible for this step, δ-ALA synthetase, is the biological control point for heme synthesis. Subsequent steps involve the condensation of two δ-ALA molecules to form a monopyrrole, the linking of four monopyrroles, ring formation, and finally modification of the tetrapyrrole ring's side chains. In the final step, ferrochelatase, a mitochondrial enzyme, inserts a ferrous ion into the center of the ring and the final product heme is formed.

It is apparent that for an efficient system, a coordination between globin production, protoporphyrin biosynthesis and iron assimilation must exist. While there is good evidence that heme plays a central role in regulating both globin synthesis and its own synthesis, it is not clear how iron supply influences these components. Free protoporphyrin levels are high when iron deficiency exists, indicating that this pathway continues despite a lack of sufficient iron to form heme.

As the cell matures, and a full complement of hemoglobin is normally present, a final succession of maturation events occur. There is a gradual disintegration of the complex subcellular structures; the mitochondria become fewer in number, the endoplasmic reticulum diminishes and the nucleus disappears. In this semifinal stage hemoglobin production and iron assimilation are markedly decreased. Finally, the mature red cell loses its subcellular architecture, ceases hemoglobin synthesis, and does not interact with transferrin.

The delivery of iron to the young, active erythrocyte precursor was discussed above. Although there are reports that transferrin actually enters the cell, the consensus of workers in the field is that iron exchange takes place at the membrane via a cellular iron receptor protein. The transmembrane transport of iron may be mediated by a membrane bound carrier protein molecule. Advances in working with these insoluble transport agents are proceeding at a rapid pace and isolation of the membrane transport agents should be forthcoming.

The removal of iron from the transferrin binding site and transport to a region of high iron concentration may be a metabolically active process. ATP is a likely candidate to drive the active transport mechanisms. Reducing agents could provide an energy form to remove iron from the transferrin binding site, possibly in conjunction with a removal of the carbonate.

Once in the cytoplasm of the cell, the iron is either moved directly into the heme fraction or is stored temporarily as ferritin, based upon the availability of iron. When a surplus of iron exists, a larger quantity goes into ferritin. A missing link in erythrocyte iron metabolism has been the intracellular iron transport agent. For the same reasons that dictate the need for a plasma iron transport agent, it would be anticipated that a cytoplasmic "transferrin" would exist. While there have been scattered reports of such a protein, they are marked with conflicting results and no convincing evidence has been presented to date. Perhaps the intracellular iron is transported by low molecular weight chelates, or perhaps the iron is even picked up at the membrane by protoporphyrin IX with the concomitant formation of heme. These and other aspects of intracellular iron management will provide for exciting experimentation.

GLOSSARY

Ceruloplasmin—a serum copper protein proposed to be active in the oxidation of ferrous ion.

Cytochrome c—an electron transport protein of mitochondria.

Erythrocyte—red blood cell.

Ferritin—an iron storage protein in which twenty-four subunits surround a core of iron-oxo polymers.

Hemoglobin—the oxygen transport protein of red blood cells.

Hehosiderin—an insoluble complex of iron and protein believed to be a breakdown product of ferritin.

Hydrolysis—the loss of protons from water molecules in the primary coordination sphere of metal ions.

Metalloporphyrins—a class of cofactors in which a metal ion is held in a substituted tetrapyrrole ring.

Myoglobin—the heme containing oxygen binding protein of muscle.

Reticulocyte—final maturation stage of developing red blood cell.

Transferrin—the serum iron transport protein.

REFERENCES

1. Basolo, F. and Pearson, R. G.: *Mechanisms of Irorganic Reactions,* 2nd ed., New York, John Wiley, 1967.
2. Spiro, Th.G. and Saltman, P.: Polynuclear complexes of iron and their biological implications. *Structure and Bonding.* 1969, vol. 6. pp. 116-156.
3. Mildvan, A. S.: Metals enzyme catalysis. In Boyer, P. D. (Ed.): *The Enzymes.* 1970, vol. 2, pp. 445-536.
4. Malstrom, B. G.: Biochemical functions of iron. In Hallberg, L., Harwerth, H. G., and Vannotti, A. (Eds.): *Iron Deficiency.* London and New York, Acad Pr, 1970, pp. 9-20.
5. Orme-Johnson, W. H.: Iron-sulfur proteins: Structure and function. *Ann Rev Biochem,* 42:159-204, 1973.
6. Dickerson, R. E. and Geis, I.: *The Structure and Action of Proteins.* New York, Harper and Row, 1969.
7. Perutz, M. F.: Stereochemistry of cooperative effects in haemoglobin. *Nature,* 228:726-739, 1970.
8. Perutz, M. F.: The hemoglobin molecule. *Sci Am,* November 1964.
9. Takano, T., Kallai, O. B., Swanson, R., and Dickerson, R. E.: The structure of ferrocytochrome c at 2.45 Å resolution. *J Biol Chem,* 248:5234-5255, 1973.
10. Dickerson, R. E.: The structure and history of an ancient protein. *Sci Am,* April 1972.
11. Harris, J. W. and Kellermeyer, R. W.: *The Red Cell,* rev. ed. Cambridge, Harvard U Pr 1970.
12. Helbock, H. J. and Saltman, P.: The transport of iron by rat intestine. *Biochemica et Biophysica Acta,* 135:979-990, 1967.
13. Conrad, M. E. and Crosby, W. H.: Intestinal mucosal mechanisms controlling iron absorption. *Blood,* 22:406-412, 1963.
14. Bates, G., Hegenauer, J., Renner, J., and Saltman, P.: Complex formation, polymerization, and autoreduction in the ferric fructose system. *Bioinorg Chem,* 2:311-327, 1973.
15. Aasa, R., Malmstron, B. G., Saltman, P., and Vanngard, T.: The specific binding of iron(III) and copper(II) to transferrin and conalbumin. *Biochim Biophys Acta,* 75:203-222, 1963.
16. Bates, G. W. and Schlabach, M. R.: A study of the anion binding site of transferrin. *FEBS Letters,* 33:289-292, 1973.
17. Bates, G. W., Billups, C., and Saltman, P.: The kinetics and mechanism of iron(III) exchange between chelates and transferrin I. The complexes of citrate and nitrilotriacetic acid. *J Biol Chem,* 242:2810-2815, 1967.

18. Roeser, H. P., Lee, G. R., Nacht, S. and Cartwright, G. E.: The role of ceruloplasmin in iron metabolism. *J Crim Invest, 49:*2408-2417, 1970.

19. Frieden, E.: Ceruloplasmin, a link between copper and iron metabolism. *Nutr Revs, 28:*87-91, 1970.

20. Fletcher, J. and Huehns, E. R.: Function of transferrin. *Nature, 218:*1211-1214, 1968.

21. Pape, L., Multani, J. S., Stitt, C. and Saltman, P.: In vitro reconstitution of ferritin. *Biochem, 7:*606-612, 1968.

22. Chrichton, R. R.: Structure and function of ferritin. *Angew Chem Internat Edit, 12:*57-65, 1973.

23. Fischbach, F. A., Gregory, D. W., Harrison, P. M., Hoy, T. G., and Williams, J. M.: On the structure of hemosiderin and its relationship to ferritin. *J Ultrastruct Res, 37:*495-503, 1971.

CHAPTER 8

CELLULAR CALCIUM AND MAGNESIUM METABOLISM

F. L. BYGRAVE

INTRODUCTION

IT WILL BE EVIDENT from even a cursory glance at some of the chapters in this book that magnesium and calcium are the two most abundant bivalent ions in the cells of practically all living systems from microbes to man. Therefore, it is not surprising that much has been written about the bio-inorganic chemistry of these two ions, often with particular reference to the chemical features they exhibit in relation to their respective modes of biological action. What is now clear is that magnesium and calcium play a variety of important structural and catalytic roles in overall cellular metabo-

lism and as such have evolved as intrinsic ingredients in the complex mass of metabolites found in all cell systems. Apart from an ability to antagonize the activating effects of magnesium in many enzymic processes, calcium also plays a vital role in numerous physiological functions such as blood clotting, lactation, nerve conduction, muscle contraction, bone metabolism and the formation of intercellular cement.

This article will consider in some detail an aspect of cellular calcium and magnesium metabolism which is little understood yet is a central part of the chemistry of life. The basis to this consideration lies in the fact that calcium and magnesium are often antagonistic in their effects on biological reactions. The interrelation between the magnesium-calcium concentration ratios inside cell compartments and the rates at which the many and various metabolic processes located therein are able to function will be examined along with the means by which the mammalian cell in particular is able to regulate this ratio. Finally, attention will be drawn to situations in which an imbalance in this ratio could lead to pathological alterations in cellular behavior.

CELL STRUCTURE AND METABOLIC COMPARTMENTATION

It will be necessary in the first instance to briefly recall the substructure of the typical cell and to indicate the subcellular location of some of the major metabolic processes which occur and to which attention will be drawn subsequently in this chapter. Figure 8-1 outlines the major structural features of a typical mammalian cell. The outer *(cell) membrane* provides a barrier which separates the contents of the cell from the extracellular environment. Within the cell are contained at least three membranous structures. The single *nucleus* which houses the large proportion of the cell deoxyribonucleic acid (DNA), the *endoplasmic reticulum,* a membrane network which penetrates much of the cell interior and the *mitochondria.* The latter are cylindrically shaped, double-membraned organelles about 1 to 2μ long and 0.5μ wide. In the liver cell, for example, they number about 1,000. The bulk of the remainder of the cell which is called the *cytoplasm* consists mainly of soluble material of a protein nature, and various lipid droplets.

Like most biological membranes, the cell membrane is freely permeable to only a few select molecules. This membrane permits the entry of, for example, oxygen, glucose, and some ions as well as the exit of lactic acid and carbon dioxide. As we shall see later, there is a need in the membrane for a mechanism which allows for the entry and exist of particular molecules essential for the economy of the cell. Such mechanisms are called "transport" mechanisms and the proteins involved are called "carrier" proteins.

Enzymes involved in the synthesis of proteins and phospholipids are located in the endoplasmic reticulum and in the cytoplasmic soluble fraction. The latter fraction also contains all of the enzymes necessary for the metabolic conversion of glucose into lactic acid. This process, which is called *glycolysis,* is an anaerobic one in that molecular oxygen is not an obligatory component in any of the individual reac-

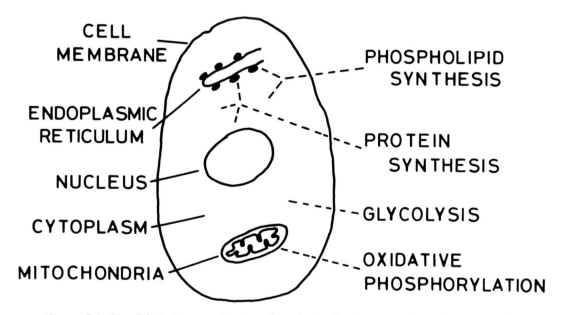

Figure 8-1. Simplified diagram showing the relationship between the substructure of a typical cell and the metabolic processes which occur in the different substructures.

tions that constitute the process. The mitochondria are the "power-houses" of the cell. They contain all of the enzymes necessary for the complete breakdown of pyruvate to carbon dioxide and water. Some seventy enzymic steps comprise the processes known as the *citric acid cycle* and *oxidative phosphorylation*. These processes enable the bulk of the energy, originally present in glucose, to be converted to and conserved in the form of adenosine triphosphate (ATP), the "energy-currency" of the cell.

INTRACELLULAR DISTRIBUTION OF CALCIUM AND MAGNESIUM

Before considering the intracellular distribution of magnesium and calcium, it is worth pointing out that in biological systems such as in higher organisms which have a protected environment known as the extracellular or plasma fluid, the concentration of calcium in this fluid is about $10^{-3}M$ while that inside the cell (cytoplasm) is of the order of 10^{-6} to $10^{-7}M$. However, the concentration of magnesium is about $\times$ $10^{-2}M$ inside the cell and significantly less outside (approximately $10^{-3}M$). Thus the situation exists in which there is a considerable calcium gradient across the cell membrane high on the outside and a less substantial magnesium gradient in the opposite direction across the same membrane. Later it will be shown how the calcium gradient in particular is established and maintained across the biological membrane.

Relatively little reliable information is known about the intracellular distribution of magnesium and of calcium. Data often given considerable attention in this regard are that of Thiers and Vallee[9] for rat liver. These are shown in Table 8-I. Note that each cell fraction shows a characteristic pattern of metal concentration. Although the amounts of both ions are highest in the

TABLE 8-I

METAL CONTENT OF LIVER CELL FRACTIONS[*]

(Mg/g nitrogen)

	Mg^{2+}	Ca^{2+}
Whole liver	8.5	0.99
Nuclei + Residue	8.4	1.3
Mitochondria	10.3	2.7
Microsomes	9.9	1.4
Cytosol	5.3	0.50

[*] R. E. Thiers and B. L. Vallee, "Distribution of Metals in Subcellular Fractions of Rat Livers," *J Biol Chem*, Vol. 226 (1957), pp. 911-920.

mitochondria, it is clear that magnesium is distributed almost evenly between each of the cell fractions. Calcium on the other hand is found in the cytoplasm in only relatively low amounts. It is evident from the data that the ratio of magnesium to calcium is approximately three in the mitochondria while it is approximately ten in the cytoplasm and about seven in the microsomes (endoplasmic reticulum). The significant points being made here are that the *relative abundance* of magnesium and of calcium in the different (liver) cell fractions is not the same, i.e. the magnesium-calcium ratio, for example, is higher in the mitochondria than in the cytoplasm. This information suggests that mechanisms exist within the cell to maintain magnesium and calcium concentration gradients between cell compartments. How the cell is able to carry out this task will be seen presently.

Only a fraction of the total calcium and magnesium which is present inside the cell exists in the "free" or ionized form. Each of these ions form moderately strong complexes with a variety of intracellular proteins, enzymes and metabolites, particularly those containing phosphate or carboxylate groups. The data in Table 8-II illustrate the degree to which calcium and magnesium

TABLE 8-II

LOG STABILITY CONSTANTS OF
METAL COMPLEXES*

$$K = \frac{[ML]}{[M][L]}$$

	EDTA	*ATP⁴⁻*	*ADP³⁻*
Mg^{2+}	8.7	4.58	3.34
Ca^{2+}	10.6	4.45	2.89

* R. M. C. Dawson, Daphne C. Elliott, W. H. Elliott, and K. M. Jones (Eds.), "Metal Nucleotide Complexes," Data for Biochemical Research. (Oxford, Clarendon Pr, 1969), p. 433.

are able to bind to ATP and to ADP, two of the most prevalent cellular metabolites.

A chelating agent commonly used in biological studies is EDTA. This molecule has four carboxylate residues and as is evident from data in Table 8-II, has a very high affinity for each of the two ions, particularly for calcium.

CELLULAR METABOLIC PROCESSES INFLUENCED BY MAGNESIUM AND CALCIUM

Almost all cellular metabolic pathways are subjected in one way or another to the activating and/or inhibitory action of numerous cellular metabolites. These interactions often form the basis to the well-documented "feed-forward" and "feedback" mechanisms of regulation. There is every indication that many of these pathways are also sensitive to fluctuating in the concentration of bivalent metal ions. Thus the glycolytic pathway which comprises some ten enzymic steps (Figure 8-2), contains five enzymes each of which have an absolute requirement for magnesium. These reactions are also potential sites for antagonism by calcium. This point is considered in detail later with regard to the enzyme pyruvate kinase.

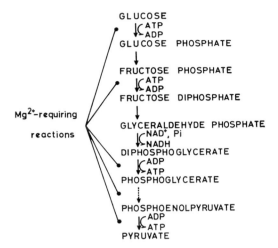

Figure 8-2. The reaction steps in the glycolytic pathway. Those marked with an asterisk have an absolute requirement for magnesium ions and are potentially subject to inhibition by calcium.

The breakdown of pyruvate to carbon dioxide and water and the concomitant formation of ATP from ADP and phosphate which, as we saw previously, takes place within the mitochondria, also occurs via a number of separate enzymic reactions (Figure 8-3). Magnesium is required in many of these processes. (This requirement is difficult to establish in some enzymic steps because the ion is often quite tightly bound to the enzyme protein, e.g. ATP synthetase). Since, calcium is readily accumulated by mitochondria this ion is potentially capable of antagonizing the activating influence of magnesium on many intramitochondrial enzyme reactions.

We have observed that membranes often have multifunctional roles in the cell such as acting as a restricting barrier, permitting the entry and exit of only specific molecules as well as providing a template for numerous enzymic and metabolic reactions. The bulk of the biomembrane is formed from phospholipids and proteins.

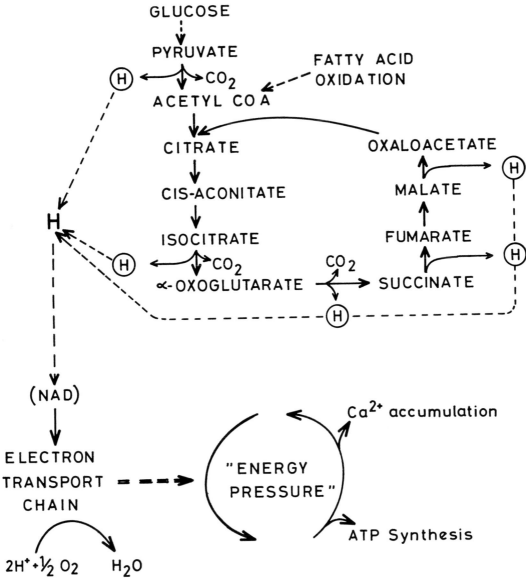

Figure 8-3. An illustration of the functional relationship between glycolysis, the citric acid cycle and respiratory chain activities. The overall (nonstoichiometric) equation is pyruvate + oxygen + ADP + Pi → carbon dioxide + water + ATP.

The way in which these two molecular species interact in the cell so as to form a membrane is largely unknown and is currently the subject of intense biochemical study. Most relevant to the present argument is the knowledge that the biosynthesis of both phospholipids and of proteins involve enzymic steps which have an obligatory requirement for magnesium and are calcium-inhibited. These

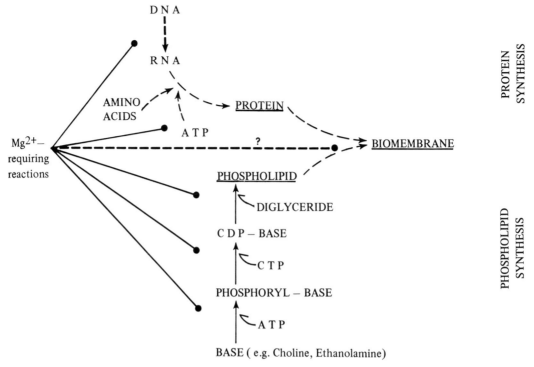

Figure 8-4. Scheme illustrating the essential steps involved in the formation of a biological membrane. Each of the steps marked with an asterisk have an absolute requirement for magnesium and is inhibited by calcium. The abbreviations used are: ATP, adenosine triphosphate; CTP, cytidine triphosphate; CDP, cytidine diphosphate; DNA, deoxyribonucleic acid; RNA, ribonucleic acid.

reactions and the sites at which the ions exert their influence are outlined in Figure 8-4. The relative effectiveness of calcium to inhibit magnesium-stimulated phospholipid synthesis, for example, as shown by the finding that $0.1 \times 10^{-3}M$ calcium completely inhibits the activation brought about by $3 \times 10^{-3}M$ magnesium.

Besides being necessary for enzymes of protein synthesis to function optimally, magnesium ions are also essential for the proper maintenance of DNA, RNA and membrane structure.

It should now be clear that the presence of magnesium is vital for many cellular activities and that the action of magnesium is often counterbalanced by the inhibitory effects of calcium. We can now deduce that it is actually the magnesium-calcium *ratio* which is important in ascertaining the extent to which these reactions proceed.

THE MAGNESIUM-CALCIUM RATIO AND METABOLIC CONTROL

The reason for stressing the magnesium-calcium ratio in the present treatise will become apparent from considering the behavior of a specific enzyme whose activity like many other enzymes is highly sensitive to this ratio.

Pyruvate kinase is an enzyme of the glycolytic pathway which is located in the cytoplasm of the cell. The reaction catalyzed by this enzyme involves the transfer of a phosphoryl group from phosphoenolpyruvate to ADP (Figure 8-5). Numerous

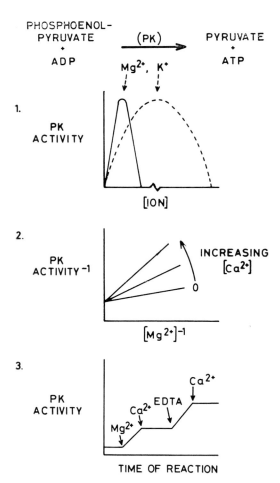

Figure 8-5. Composite scheme illustrating how the activity of a soluble enzyme such as pyruvate kinase can be influenced by the presence or absence of magnesium and calcium. Graph 1. Dependence of pyruvate kinase activity on magnesium and potassium ion concentrations. Graph 2. Variation of pyruvate kinase activity with magnesium concentration in the presence of varying calcium concentrations. Graph 3. Variation of pyruvate kinase activity by the addition of magnesium, calcium and EDTA.

experiments have indicated that *both* magnesium and potassium need to be present in order that the enzyme become fully active.

Like the pyruvate kinase reaction, all phosphoryl groups transfer reactions have an absolute requirement for magnesium and sometimes potassium; they are frequently but not always inhibited by calcium. In some reactions the magnesium combines initially with the adenine nucleotide and this complex then interacts with the enzyme to form the ternary, enzyme-metal-substrate complex. Alternatively, as is the case with the pyruvate kinase reaction, the magnesium combines initially with the enzyme imposing upon it the required "conformation." This enzyme-metal complex then combines with the adenine nucleotide to form the ternary complex.

The first graph in Figure 8-5 relating pyruvate kinase activity to metal ion concentration shows first, that in the absence of either added magnesium or potassium, the enzyme is virtually inactive. As the concentration of each ion is increased independently in the presence of an optimal amount of the other, the activity of the enzyme increases to a maximum value beyond which it begins to diminish. The second point is that considerably lower concentrations of magnesium than of potassium are necessary for producing maximal activity. The concentration-activity profile is more abrupt in the case of magnesium than of potassium. For this reason magnesium is potentially a more suitable candidate for "ionic control" than potassium.

The second graph in Figure 8-5 represents a double-reciprocal (Lineweaver-Burk) plot of information obtained when the magnesium concentration is varied in the presence of different (fixed) concentrations of calcium. The nature of the plot indicates that the calcium antagonizes in a competitive way the activating effect of magnesium. The important inference from this information is that it is the magnesium-calcium *ratio* and not necessarily the *absolute* amount of magnesium and of calcium that govern the extent to which the enzyme reaction will proceed. Furthermore, it is clear that this ratio can be varied by simply changing the concentration of only one of the ionic species, i.e. either that of magnesium or that of calcium.

EDTA is an effective chelator of bivalent metal ions that has a greater affinity for calcium than for magnesium (Table 8-II). The third plot in Figure 8-5 shows how the controlled addition of a suitable calcium chelator such as EDTA is able to "turn on" and "turn off" pyruvate kinase activity solely by changing the concentration of "free" calcium in the incubation medium and thus in essence the magnesium-calcium ratio. Since pyruvate kinase is a component of the glycolytic pathway, it would be expected that added calcium will inhibit the overall rate of glycolysis and that this inhibition would be overcome by adding EDTA to the incubation system. This has proved to be the case and supports the notion that where an individual enzyme is influenced by a metal ion, the pathway of which it is a component, can also be influenced by that ion. In the same way, phospholipid synthesis is inhibited by calcium ions because at least three of the individual enzymes are inhibited by this ion. Moreover, the rates of phospholipid synthesis and of protein synthesis can be altered by simply changing the magnesium-calcium ratio in the incubation medium. I would emphasize that all of the reactions we have been discussing which are influenced by the magnesium-calcium ratio are located in the cell outside the mitochondria. This is a location which is readily accessible to the cell membrane, the endoplasmic reticulum and the mitochondria.

Cell Mechanisms Which Can Modify the Magnesium-Calcium Ratio in the Cytoplasm

The various means by which the cell is able to modify the magnesium-calcium ratio in the cytoplasm shall now be considered. The cell components or membrane systems involved in modifying the calcium ion concentration in the cytoplasm are the cell membrane, the endoplasmic reticulum or microsomal membrane and the mitochondrial membrane. Each of these membranes contain calcium-specific, energy-requiring systems which transport calcium through the membrane against the passive release or "leak" of calcium which occurs in an opposite direction. The high specificity of transport for calcium in each case provides the basis for the ability of the cell to modify the magnesium-calcium ratio in the cytoplasm.

The properties of calcium transport across isolated cell and microsomal membranes are similar. However, they differ from that across the mitochondrial membrane with regard to the nature of the energy source. They differ also in regard to the effect of inhibitors of electron transport and oxidative phosphorylation; the two former systems require ATP breakdown as a source of energy for transport, the latter can be supported either by substrate oxidation or ATP hydrolysis.

In the course of evolution, most cell types have developed at least one of the above three membrane systems to a greater degree than the others. Cells of skeletal muscle for instance possess a highly specialized endoplasmic reticulum which releases calcium upon excitation. The released calcium in turn initiates muscle contraction and the calcium "pump" then reaccumulates calcium thus allowing relaxation to occur. Cells of heart muscle also possess a highly developed endoplasmic reticular system used for the transport of calcium. Extensive reviews on this transport system have been written and may be read elsewhere. Only recently have mitochondria been advocated as a means of varying the magnesium-calcium ratio in the cytoplasm. Yet, it has been recognized for almost twenty years that these organelles are able to accumulate bivalent ions, in particular calcium. The large proportion of studies on mitochondrial ion accumulation have been carried out with mitochondria isolated from rat liver. Data from these studies form the basis to the present description of calcium transport in these organelles, especially in relation to the viewpoint adopted in this chapter.

The information contained in Figure 8-6 illustrates the salient features of the accumulation process. Basically, there is located in the inner of the two mitochondrial membranes, a (lipo) protein or calcium "permease" specifically concerned with the transport of calcium from the cytoplasm to the interior of the mitochondria. It is one of many permeases located in this membrane which are responsible for translocating essential metabolites across the mitochondrial membrane. We saw previously that the accumulation of calcium is a process which requires energy. The necessary energy is derived from a little known but much discussed "high-energy intermediate" of oxidative phosphorylation. Mitochondria require energy not only to accumulate calcium but also to retain it once it has been accumulated. Thus, if an appropriate energy inhibitor is added after the calcium has been accumulated, then all of that calcium is released into the extramitochondrial environment.

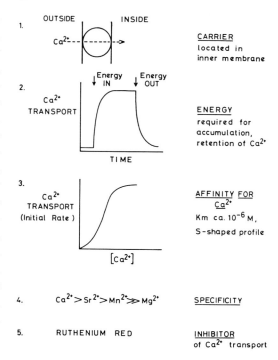

Figure 8-6. The principle features of calcium ion accumulation properties of rat liver mitochondria.

The affinity of mitochondria for "free" calcium is very high. These organelles are therefore capable of reducing the calcium concentration outside mitochondria to a very low level, i.e. approximately 10^{-6}M. Furthermore, when initial rates of calcium accumulation are measured as a function of calcium concentration, a sigmoidal-shaped plot is obtained. So, the accumulation system in mitochondria appears to possess a built-in safety mechanism which can initiate an even faster rate of calcium accumulation once the concentration of calcium in the cytoplasm becomes higher than a "critical" level.

Perhaps the most interesting property of ion accumulation by rat liver mitochondria is the metal ion specificity. Of the two predominant bivalent ions in the cell, only calcium is capable of being transported into the mitochondria at any significant rate. This latter property of rat liver mitochondria, namely their ability to accumulate calcium in preference to magnesium, indicates that these organelles also are able to effectively modify the magnesium-calcium ratio in the cytoplasm.

ROLE OF MITOCHONDRIA IN MODIFYING THE INTRACELLULAR MAGNESIUM-CALCIUM RATIO

Previously we saw how a suitable chelator of calcium such as EDTA could be used to modify the magnesium-calcium ratio in the immediate environment of an enzyme sensitive to this ratio. Also, we have just seen how the ability of many species of mitochondria to accumulate calcium in marked preference to magnesium makes these organelles suitable candidates for varying this ratio inside the cell cytoplasm. How this can be done in an *in vitro* system is depicted in Figure 8-7. The scheme is constructed from data of experiments carried out to test the ability of rat liver mitochondria to vary the magnesium-

Measurement of Pyruvate Kinase Activity–initially in presence of mitochondria but absence of energy-source for Ca^{2+} accumulation; Mg^{2+} and Ca^{2+} also initially absent

MANIPULATION		1	2	3	4
		Add Mg^{2+}	Add Ca^{2+}	Add Energy	Inhibit Energy
Mg^{2+}/Ca^{2+}		High	Low	High	Low
Pyruvate Kinase activity	Low	High	Low	High	Low
Remarks	Reaction slow in absence of Mg^{2+}	Addition of Mg^{2+} initiates reaction	Absence of energy prevents Ca^{2+} accumulation	Ca^{2+} but not Mg^{2+} accumulated by mitochondria	Block in Energy leads to release of Ca^{2+} from mitochondria

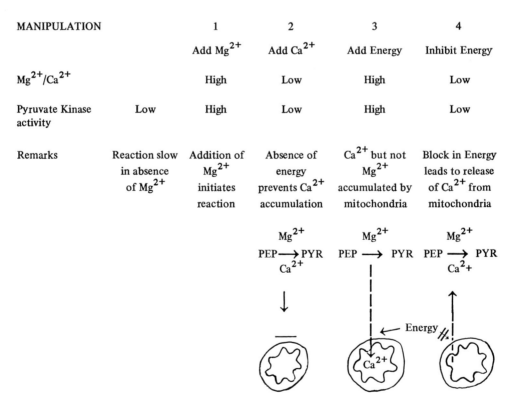

Figure 8-7. Description of how the ion accumulation properties of rat liver mitochondria can be manipulated so as to regulate the rate of ion-sensitive reactions by altering the magnesium-calcium ratio in the environment outside mitochondria.

calcium ratio and thus of any ion-sensitive reactions which might be altered by such a ratio. Again the pyruvate kinase reaction has been chosen as an illustration of a readily measured enzyme which is located outside the mitochondria in the cytoplasm, and whose activity is highly sensitive to the magnesium-calcium ratio.

In the first instance, as the scheme indicates, magnesium and calcium are both absent from the system. Upon the addition of magnesium, the magnesium-calcium ratio becomes infinitely high. Under these circumstances pyruvate kinase activity is maximal. During the third manipulation calcium ions are added to the incubation system. The magnesium-calcium ratio is now relatively low and as a consequence, pyruvate kinase activity is relatively low. Due to an absence of energy for calcium accumulation by mitochondria, the magnesium-calcium ratio is kept low. When an energy source for calcium uptake by mitochondria is added to the incubation mixture, calcium ions move into the mitochondria and the magnesium-calcium ratio increases. As a consequence, pyruvate kinase activity also increases to maximal rates. The final manipulation which may be carried out in this type of experiment, involves adding to the incubation mixture an inhibitor of the source of energy which maintains the calcium inside the mitochondria. When this occurs, all of the calcium which had been previously accumulated by the mitochondria is released into the surrounding medium where the pyruvate kinase reaction is located. The magnesium-calcium ratio again becomes low as does pyruvate kinase activity.

This type of experiment demonstrates how the ion accumulation properties of liver mitochondria, for instance, can be manipulated so as to influence the magnesium-calcium ratio in the cytoplasm and hence the activity of any enzymic reactions which are sensitive to such a ratio.

CALCIUM HOMEOSTASIS

Previous observations may have brought the reader to the conclusion that calcium is a more "mobile" ion than magnesium. This turns out to be the case and provides further support to the argument that the concentration of only one of the two ions needs to be changed in order to vary the magnesium-calcium ratio. We shall now briefly examine the means by which both the extracellular or plasma calcium as well as the intracellular calcium is maintained at dynamic steady-state concentrations. The important fact to appreciate is that the calcium concentrations inside and outside the mammalian cell, while being quite different, are maintained by mechanisms which operate in essentially similar ways.

Plasma calcium does not exist in a homogeneous form, but rather is composed of three major fractions. These are protein-bound or nondiffusible calcium, "free" or ionized calcium, and "complexed" calcium. Only the latter two fractions are diffusible. They constitute approximately 50 percent of the total plasma calcium and determine the concentration of calcium in the plasma outside the cells. The three most important body organs involved in plasma calcium homeostasis are the gut, the bone and the kidney (Figure 8-8). The gut is the site of calcium absorption from the diet, the bone is the major depot of calcium in the body and the kidney is the site at which any excess calcium is excreted. Their role in plasma calcium regulation depends largely on the presence of one or more of the following: parathyroid hormone, calcitonin and vitamin D. Parathyroid hormone and calcitonin are produced by the parathyroid and the thyroid glands, respectively. The production of each is regulated by the concentration of calcium in the blood probably by "feed-back" mechanisms.

Although the precise modes of action of each of these hormones on calcium homeostasis are not known, there is evidence that the parathyroid hormone mobilizes bone calcium, promotes calcium retention in the kidney tubules and, with vitamin D, promotes absorption of dietary calcium from the gut. Calcitonin, on the other hand, antagonizes some of the effects of parathyroid hormone in that it induces movement of calcium from plasma into bone. Again, it appears hardly a coincidence that parathyroid hormone and calcitonin have little effect on magnesium mobilization in plasma.

These same hormones which are important in the regulation of plasma calcium are also important in the regulation of calcium concentrations in the cell. Parathyroid hormone and calcitonin each play significant roles in the transport of calcium across the cell membrane via the calcium-ATPase "carrier." Moreover, there is evidence that a third hormone, cyclic-AMP, is involved in

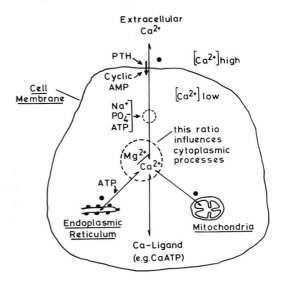

Figure 8-9. Illustration of how the intracellular calcium concentrations can be altered so as to modify the magnesium-calcium ratio in the cytoplasm.

the regulation of intracellular calcium concentrations.

Figure 8-9 represents an attempt to show how the magnesium-calcium ratio inside the cell might be modified by the various mechanisms we have considered. It is evident that there is a most complex network of metabolic reactions at play with each reaction geared to the activity of other reactions many of which are under hormonal control.

CALCIUM AND DISEASE

It has been known for many years that the injection of a dose of magnesium ions into the blood stream of an animal, induces a state of anesthesia; the animal becomes motionless and insensitive to pain. This effect of magnesium, which can be overcome by the injection of an appropriate dose of calcium ions, clearly indicates the response of a complex living system to an

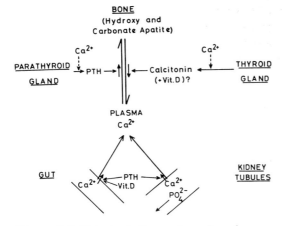

Figure 8-8. Principal sites and modes of regulatory influence of hormones affecting extracellular calcium homeostasis.

imbalance in the magnesium-calcium ratio and also illustrates the importance to the system of maintaining the appropriate ionic balance.

Because calcium homeostasis is so intimately geared to magnesium, phosphate and monovalent ion metabolism and is subjected to the action of several hormones, it is not surprising to find a variety of pathological conditions which in one way or another are potentially attributable to a deranged ionic balance. The following examples illustrate such conditions.

There is a great deal of evidence to suggest that calcification is a pathological condition associated with atherosclerosis. Calcification can be induced by a high dietary calcium or low magnesium intakes and/or the production of hypervitaminosis D. Furthermore, magnesium deficiency increases vitamin D toxicity just as an adequate or excessive dietary consumption of calcium does. Our understanding of the cellular events associated with these circumstances is obscure and obviously needs considerable study.

The toxic effects on living systems induced by many of the heavy metals sometimes reflect an interaction of the metal with a ligand or site on or in the cell to which calcium and/or magnesium would normally bind. Lead, zinc and copper for instance are known to strongly influence energy-linked reactions of isolated mitochondria. The interaction of lead in particular with calcium binding sites probably explains a number of well-established antagonistic effects of lead and calcium on various cellular processes.

An inability of mitochondria from tumor cells to effectively maintain adequate magnesium-calcium ratios in the various cell compartments could explain some of the metabolic lesions seen in cancer cells. Hypercalcemia, for example, is known to be associated with many forms of cancer. Recent work in our own laboratory has shown that although mitochondria from Ehrlich ascites tumor cells can readily accumulate calcium in an energy-dependent process, they are unable to release the calcium in a manner characteristic of "normal" mitochondria. An obvious consequence of this situation could be that the tumor cell maintains a higher than normal magnesium-calcium ratio in the cytoplasm. As a result, all of the metabolic processes located in the cytoplasm, dependent on magnesium and inhibited by calcium, would proceed at rapid, "uncontrolled" rates. Of interest in this regard is that high rates of glycolysis and membrane formation are characteristic metabolic events in many tumor cells.

SUMMARY AND CONCLUSIONS

This chapter has drawn attention to the ways in which magnesium and calcium influence a variety of cellular metabolic activities, in particular those which are located in the cell cytoplasm. Two aspects of the interaction of magnesium and calcium with cellular processes are seen as being particularly relevant to the concept that modification of the magnesium-calcium ratio is a means of regulating intracellular metabolic activities. The first of these is the antagonistic action of the two ions on common metabolic processes; magnesium activation is often counterbalanced by calcium inhibition. The second aspect relates to the ability of the cell to modify the magnesium-calcium ratio in the cytoplasm by removing calcium from that environment via the calcium-specific transport mechanisms present in the membranes of the cell, endoplasmic reticulum and the mitochondria.

The role played by mitochondria in carrying out the task of modifying the cellular ionic environment has been discussed at length. It has also been discussed in relation to current thought on hormonal control of calcium transport systems. Consideration has been given to the need for living systems to maintain an adequate magnesium-calcium ratio. A breakdown in the maintenance of this ratio is seen as accounting for a variety of metabolic diseases.

Although the importance of magnesium and calcium to living systems was recognized almost 100 years ago, we are still a long way from fully understanding the molecular biology of these ions. Of importance in the future will be the need to establish how the magnesium-calcium ratios in the cell are maintained by the numerous hormones which stimulate or depress this ratio. Clearly, the need to research and clarify the relationship between the magnesium-calcium ratio and metabolic activities is vital for our complete comprehension, not only of life itself, but also of a variety of diseases which are associated with life.

REFERENCES

1. Bianchi, C. P.: *Cell Calcium*. London, Butterworths, 1968.

2. Bygrave, F. L.: The ionic environment and metabolic control. *Nature, 214*:667-671, 1967.

3. Heilbrunn, L. V.: *An Outline of General Physiology*. Philadelphia, Saunders, 1952, pp. 528-540.

4. Lehninger, A. L., Carafoli, E. and Rossi, C. S.: Energy-linked ion movements in mitochondrial systems. *Adv in Enzymol, 29*:259-320, 1967.

5. Meli, J. and Bygrave, F. L.: The role of mitochondria in modifying calcium-sensitive cytoplasm metabolic activities: Modification of pyruvate kinase activity. *Biochem J, 128*:415-420, 1972.

6. Moon, J. Y.: Factors affecting arterial calcification associated with atherosclerosis. *Atherosclerosis, 16*:119-126, 1972.

7. Rasmussen, H.: Ionic and hormonal control of calcium homeostasis. *Am J Med, 50*:567-588, 1971.

8. Spencer, T. and Bygrave, F. L.: The role of mitochondria in modifying the cellular ionic environment. *J Bioenergetics, 4*:347-362, 1973.

9. Thiers, R. E. and Vallee, B. L.: Distribution of metals in subcellular fractions of rat liver. *J Biol Chem, 226*:911-920, 1957.

10. Watson, L.: Calcium metabolism and cancer. *Aust Ann Med, 15*:359-367, 1966.

11. Williams, R. J. P.: The biochemistry of sodium, potassium, magnesium and calcium. *Chem Soc Quar Revs, 24*:331-365, 1970.

12. Dawson, R. M. C., Elliott, Daphne C., Elliott, W. H., and Jones K. M. (Eds.): Metal nucleotide complexes. *Data for Biochemical Research*. Oxford, Clarendon Pr, 1969, p. 433.

SECTION II

INSTRUMENTAL APPROACHES TO
BIO-INORGANIC CHEMISTRY

CHAPTER 9

INSTRUMENTAL APPROACHES TO BIO-INORGANIC CHEMISTRY

H. M. N. H. Irving

INTRODUCTION

In approaching the study of bio-inorganic systems, we must first attempt to define our objectives; for quite clearly the techniques applicable to a complex interacting system of component species *in vivo* will not embrace all those that can be applied to a single identifiable species that can be handled *in vitro*. Do we seek, initially at least, quantitative information describing the behavior of a complex system in bulk, i.e. whole organism studies, or are we concerned with extracting the maximum amount of information from a single species and then investigating its reactions with other specified materials—as in metalloenzyme studies? Should we perhaps first study deliberately simplified model systems and investigate simple metal-ligand interactions in aqueous solutions of well defined composition?

Let us consider this second alternative, recognizing that, if the compound to be investigated originally formed part of a complex biological system, quite fundamental problems of preconcentration, separation, and purification will be involved —although this book does not deal specifically with these. For, unless the material we are examining is strictly pure and is available in the same stereoisomeric conformation as that which occurred in the biological environment from which it was derived, it will be risky, and perhaps misleading, to apply information derived from the isolated sample to the behavior of the system as a whole. To cite an obvious case, the X-ray structure of a crystalline solid gives a precise picture of how large numbers of its molecules are arranged in space under the constraints imposed by the particular crystal lattice it has adopted. In solution this, and alternate lattice structures, will be modified to a greater or lesser extent by the intervention of the solvent and so quite different conformations of individual molecules may now present themselves and quite different degrees of inter- or intramolecular association may occur.

Being given a pure and homogeneous specimen of a single molecular species, we shall need to know its ultimate composition, its molecular weight, the detailed stereochemical disposition of the component atoms which determine its size and shape, and the strength of some, at least, of its bonds. For substances available in limited quantity, the classical methods of microchemistry will often give satisfactory empirical compositions for small molecules and then the molecular formula follows if a reasonably satisfactory molecular weight can be secured, by e.g. vapor pressure osmometry, gel permeation, or light-scattering techniques. However, the use of high resolution mass-spectrometry in which the relative intensity of satellite peaks due to the isotopic species ^2H, ^{13}C, ^{15}N and ^{17}O and ^{18}O are measured as well as those due to the various molecular ions usually leads (after due reference to J. H. Beynon and A. E. Williams' "Mass and Abundance Tables for use in Mass Spectrometry," 1963), to an unambiguous molecular formula. The cracking pattern deduced for some mass spectra affords valuable structural information which, supported by classical organic methods of degradation and synthesis and by the wide range of physical technics now available, should lead to a clear picture of the organic moiety.

Naturally, a full X-ray determination of structure will give the most information if an appropriate crystalline form is available and, with the aid of modern linear diffractometers and having access to powerful computers even enormously complex structures yield in time, especially when, as in the case of polypeptides or proteins, the sequence of amino acids and similar detailed information is available from purely chemical studies.

The determination of the nature of the inorganic constituent may present difficulty when only a minute amount of sample is available. Neutron activation analysis is distinguished here by its high sensitivity and the certainty with which particular elements are determined. Radiometric methods are also invaluable in isotopic dilution procedures, while a whole range of spectroscopic techniques are now available for quantitative determinations at the microgram and nanogram level. Finally, the oxidation state of a metallic constituent may yield to magnetic measurements, electron-spin resonance or Mössbauer spectroscopy.

The study of formation constants of proton and metal complexes in solution has developed extensively in the last few decades. In its simplest formulation, the equilibrium constant of the system $M + L \rightleftharpoons ML$ (which is, of course, the formation constant of the metal-ligand species ML) can be determined if the total concentration of metal and ligand is known together with the equilibrium concentration of any one of the three participants M, L, or ML. An enormous range of techniques is now available of which potentiometry is perhaps preeminent (particularly when L is the conjugate base of a weak acid) followed by spectrophotometry over a variety of frequency regions, spectrofluorimetry, Raman spectroscopy, optical rotatory dispersion and circular dichroism, and the use of NMR spectroscopy. The position of equilibrium—and hence the numerical value of the reported formation constant when expressed in concentration units—is normally dependent upon the temperature and influenced by the nature of the medium, especially by the solvent, and by the ionic strength. It is unfortunate that so much of the literature refers to temperatures and ionic strengths unrepresentative of blood plasma conditions ($37°C$ and $0.15M$ NaCl). It is still more unfortunate that many of the numerical data reported by different authors, for the same system, should be so discrepant, that estimates of precision are seldom available and often quite grossly underestimated when they are reported.

As the system based on the components M and L becomes more complex, it becomes necessary to consider polynuclear species of the type M_pL_q and protonated or hydroxylated species such as $M_pL_qH_r$ or $M_pL_qH_{-s}$ derived from them, each specified by an additional formation constant. A wide variety of computer programs have been devised to facilitate the essentially numerical problems in handling experimental data. However, it cannot be too strongly emphasized that the more stability constants needed to represent a system, the more numerous and the more reliable the experimental results must be. Since all experimental results must be subject to experimental error it is necessary for such data to be weighted and, in the statistical treatment involved in all computer programs, it is always a problem to know which chemical species really play no significant role in complex equilibria. This is a problem for which there is, as yet, no satisfactory solution. The need for stability constants of very high precision is most easily recognized when attempts are made to calculate what species are present when two or more metals compete for a large number of ligands, e.g. amino acids, where mixed and protonated complexes may occur, and where the relevant formation constants may not differ much numerically in any case. Here chemical intuition is sometimes allowed to play too decisive a role.

Since many chemical reactions involve heat changes, their measurement by microcalorimetry can afford valuable information on enthalpy changes and changes in heat capacity which, in a well chosen series of experiments, can be correlated with the changes in mean free energy to give valuable indications of reaction mechanism.

Studies of the mechanism of bio-inorganic processes are enormously facilitated by the present availability of a variety of ion-selective electrodes and the recently discovered alkali metal complexes, which will certainly play an increasing role in the study of mechanisms of membrane transport. While some of the most important reactions of bio-inorganic chemistry proceed relatively slowly, others are quite rapid and can only be studied by stopped-flow or relaxation techniques or by sophisticated applications of NMR. There are often a complex sequence of reactions proceeding at different speeds which need to be disentangled; and while oscillating reactions are not of common occurrence in inorganic chemistry, as a whole they seem likely to play an important part in the study of many biological phenomena.

It will be obvious that every possible analytical technique that has been exploited in inorganic, organic, and physical chemistry, considered as separate disciplines, must be and can be adapted to the special problems of biochemical systems, sometimes by merely reducing the scale but more often by drastically rethinking the whole approach. A wide selection of the more important of these techniques are discussed in more detail in Chapters 10 to 14. The breadth of bio-inorganic research is illustrated (1) by the basic disciplines accommodating the authors of the next seven chapters—analytical, inorganic, physical, and electrochemistry, biochemistry and pathology; and (2) by the different chemical species upon which they focus attention. Some chapters discuss total elemental analyses, others free-ion analyses, and yet others details of the bonding and structure. Finally, Chapters 15 and 16 give overall views of the combined uses of all these measurements and disciplines.

CHAPTER 10

FORMATION CONSTANT AND
THERMODYNAMIC ASPECTS

David R. Williams

INTRODUCTION

In 1893 Alfred Werner propounded his theory of metal-ligand bonding and thus founded modern coordination chemistry. Niels Bjerrum then demonstrated (1915) that metal complexes were formed in a stepwise manner and in 1941 his son, Jannik, showed that metal-ligand complexing reactions could be quantified in terms of formation constants.* The invention of the glass electrode in 1941 provided J. Bjerrum and Ido Leden with the opportunity to determine the first formation constants (β) from precise potentiometric measurements.[1] During the ensuing thirty years millions of formation constants have been reported and several compilations published.[2]

Formation constants are related to the

Gibbs free energy of formation by the van't Hoff isotherm,

$$\Delta G^\circ = -RT \ln \beta$$

and this, in turn, is related to the enthalpy (ΔH°) and entropy (ΔS°) of formation by the Gibbs-Helmholtz equation.[3]

$$\Delta G^\circ = \Delta H^\circ - T\Delta S^\circ$$

Enthalpies are most accurately measured by using a calorimeter. The first calorimeter was invented in the eighteenth century and since then the precision of the technique has been increased to embrace the temperature range from 0 to 1,300 K and to be capable of monitoring heat changes as micro as 1 joule dissipated over many minutes. Nevertheless, inorganic chemists have not been quick to adopt this physical technique. Consider the four most basic properties that need to be determined for a newly discovered inorganic compound—its size, its shape, the three dimensional layout of its atoms, and the strengths of the bonds embodied in our new compound. Methods are well established for determining the first three properties—we assume that size approximates to weight and use the ultracentrifuge, we merely look at the molecule to decide its shape (albeit under a different light in the electron microscope) and we bounce X-rays off the atoms in order to determine their three-dimensional configurations. These three approaches have all been invented, developed and ap-

---

* "Formation constant" and "stability constant" are synonymous but the former is used in this chapter to stress that the constants refer to the *formation* of bonds from protons, or metal ions, to ligands.

plied to newly discovered inorganic molecules, and yet, in general, inorganic chemists have avoided the interface between "physical" calorimetry and "inorganic" chemistry.

However, all processes, be they chemical or *biological* are accompanied by heat changes and, as might have been expected, the pure chemist's loss soon became the biochemist's gain. Although the subject is still in its infancy, there have already been great strides toward adapting analytical calorimetry to a level of sophistication such that it could tackle the exceedingly complex energetics of *Homo sapiens*.[4]

Calorimetry has many advantages over existing biochemical techniques (for example, calorimetric measurements do not require the translucent solutions necessary for spectrophotometric methods) and so "biological microcalorimetry" has been es-

that have been made under these three headings.

THEORY

Defining the System

Formation constant and thermodynamic measurements are made on a whole range of systems some of which may not be strictly describable in terms of chemical parameters. For example, if we are dealing with simple ligand-metal ion interactions it is usually quite easy to make correlations between thermodynamic parameters (Gibbs free energy or enthalpy of formation per gram mole, etc.) and the detailed structure of the ligand and its complex. On the other hand, whole body systems cannot be clearly defined from a physico-chemical viewpoint.[5]

We depict the field of bio-inorganic solution chemistry in the following manner:

Discipline	*Produces Expertise In*
Physical chemistry	calorimetry, thermodynamics
Analytical chemistry	enzyme assay, kinetics,
Inorganic chemistry	metal ions,
Biochemistry	ligands,
Physiology	metalloenzymes,
Zoology	whole-body systems.

tablished as a new discipline which, we feel, will have a long and respectable career.

Thermodynamic research into bioinorganic chemistry usually falls into three broad classifications—(1) whole organism studies. These usually examine the "simpler" living species such as nitrogen fixing bacteria, (2) metalloenzyme studies, and (3) simple ligand-metal ion interactions investigated in aqueous solutions and supplemented with computer calculations which simulate the actual *in vivo* systems. We propose to discuss the theory and design of bio-inorganic potentiometry and calorimetry and then to describe the important discoveries

Students of the field must necessarily attain a working knowledge of all the disciplines listed in spite of built-in "barriers" such as the laws of heat-flow and of thermodynamics, the ligand field theory, the structures of enzymes and the Greek and Latin nomenclature. Even though one is working with whole organisms and the system is only loosely definable, it should not excuse one from attempting to attain the highest material purity and instrumental precision for the experimental work, and from using the most exacting thermodynamic relationships upon any data. (We do not, of course, suggest that accurately obtained results from a poorly defined system will have anything other than very large standard deviations in derived enthalpies, entropies, etc). In the following thermodynamics we shall not differentiate between levels of sophistication for different systems since thermodynamic laws apply to all species.

Basic Thermodynamic Parameters

The four basic thermodynamic parameters that are used by solution chemists are the Gibbs free energy ($\Delta G°$), the enthalpy ($\Delta H°$), the entropy ($\Delta S°$) and the heat capacity ($\Delta^{\emptyset} C_p$) of formation (or of reaction):

Considering a generalized reaction of a metal ion, B, displacing a proton, H, from a protonated ligand, A (and remembering that all species in aqueous solution are aquated):

$$pAH(H_2O)_w + qB(H_2O)_x \rightleftharpoons$$
$$A_pB_qH_r(H_2O)_y + (p - r)H(H_2O)_z +$$
$$(pw + qx - y - (p - r)z)H_2O$$

and assuming that the formation constant for this reaction, β, has been previously determined, calorimetrically measuring $\Delta H°$ will provide the $T\Delta S°$ term.

Briefly (detailed examples are given later in the chapter) the uses of these four thermodynamic parameters are:

$\Delta G°$ clearly has uses equivalent to those of β. For example, for a reaction to occur β must be positive, hence $\Delta G°$ must be negative. An additional use of $\Delta G°$ is as the intermediary between $\Delta H°$ measurements and $\Delta S°$ values calculated from the Gibbs-Helmholtz equation.

$\Delta H°$ displays the energetics of several concurrent processes—the heat of forming the metal-ligand complex bond, the heat required to remove $(pw + qx - y - (p - r)z)$ molecules of water from the reactants, the heat required to de-protonate the ligand and the heat evolved when the liberated proton becomes aquated. By judiciously selecting *series* of heat measurements to which most of these factors contribute

a constant amount of energy, trends may be examined, for example, in metal-ligand bond strengths as the ionic radius of the metal ion is systematically varied. $\Delta H°$ values may also be used as a measure of the hardness or softness of a species according to the h.s.a.b. approach.

$\Delta S°$ also has several contributory factors—the overall change in the number of particles (Σ products—Σ reactants), steric strains in $A_pB_qH_r(H_2O)_y$ as product and any stresses or strains relieved by destroying the reactants.

$\Delta^{\emptyset} C_p$ Heat capacities are a relatively recent addition to the solution chemist's armamentarium and so they are described in more detail than the previous parameters. In general terms, the heat capacity of a system is defined as the heat absorbed divided by the change in temperature when a very small change occurs in some prevailing condition or parameter. If the experiments are carried out at atmospheric pressure, equation (101) applies. A con-

$$C_p = (\partial H/\partial T)_p \qquad (101)$$

venient change in the prevailing conditions can be effected by supplying a known quantity of energy from an electrical heater and then accurately measuring the change in the temperature of the system. Equation (101) can then be rewritten in terms of measureable quantities as (102). When

$$C_p = (\Delta H/\Delta T)_p \qquad (102)$$

several species are present, and the change in conditions is applied to the total system, then the relationship becomes (103) when $\emptyset$ denotes the sum of the apparent heat capacities of *all* the species present.

$$^{\emptyset}C_p = (\Delta H/\Delta T)_p \qquad (103)$$

In the case of a pure compound, the apparent heat capacity equals the true heat capacity.

In order to apply equation (103) to a chemical reaction, certain modifications have to be made: (a) the electrical heating is removed and the heat required to produce the temperature change ΔT is now the chemical heat of reaction; (b) if all the quantities of species present are expressed on a molar scale, ΔH is equivalent to the standard heat of reaction $\Delta H°$; and (c) since reactions involve either changes in state, or more usually changes in the species present from "reactants" to "products" we must consider $\Delta^{\varnothing}C_p$, i.e. the sum of the apparent heat capacities of the products minus those of the reactants. Thus, the basic definition becomes (104) and it is

$$\Delta^{\varnothing}C_p = [\Delta(\Delta H^{\circ})]/T \qquad (104)$$

usually this temperature variation of the heats of formation that are used to obtain $\Delta^{\varnothing}C_p$. Alternative approaches involve measuring $^{\varnothing}C_p$ for the calorimetric solutions before and after a reaction has occurred, or studying the temperature variation of the entropies of formation.

Heat capacity data have two main uses: (1) they are invoked when the temperature dependence of ΔG°, ΔH° and ΔS° is calculated, and (2) for many years the configurations of molecules in the solid (especially polymer) states have been investigated through heat capacity experiments. In solution $\Delta^{\varnothing}C_p$ values may now be correlated with configurational changes and with medium effects.

Choice of Background Medium

Whole-body investigations must, of necessity, be performed in a nutrient medium that supports the life of the organism. However, when we study metalloenzymes and simple ligand systems, we are free to choose the medium conditions. For neutral ligands, thermodynamic parameters do not vary greatly as one varies the ionic strength of the medium. However, in biological systems the ligands are rarely neutral and so the two problems that face us are (a) at which standard ionic strength *(I)* to work, and (b) the choice of added electrolyte to maintain this *I* constant.

Facts that are considered under (a) are (1) human blood is isotonic with 0.15 M sodium chloride, and (2) when the sum of the equivalent concentrations of all the positive and negative ions disappearing during a complex forming reaction (A^- + $B^+ \rightleftharpoons AB$) does not exceed 150 mM, Biedermann has shown that all activity coefficients can be held constant by a 3.00 M

ClO_4^- as a background medium. Factors affecting the choice of added electrolyte, (b), are (1) Cl^- has a tendency to form ion pairs, NO_3^- has less tendency and ClO_4^- even less so, (2) K^+ and NO_3^- have similar ionic mobilities whereas Na^+ has a lower value, (3) NH_4NO_3 has been used because it has a theoretical advantage in that its structure least upsets that of water (NH_4^+ is isoelectronic with H_3O^+), (4) the Sillén school in Stockholm has reported ΔG°, ΔH° and ΔS° of hydrolysis of many metal ions in 3M (Na)ClO_4. These are very useful in that they can immediately be incorporated into calculations of the thermodynamic parameters of a metal complexing or metalloenzyme reaction, (5) some background electrolytes are easier to purify than others.

Hence, the literature contains many ΔG°, ΔH° and ΔS° for 0.15 M KNO_3 at 37° (the temperature of human body) and others for 3M $NaClO_4$ at 25° (the temperature preferred by thermodynamicists who usually use $I = 0$ at 25°). In fact, there is no ideal medium for thermodynamic investigations, all media are compromises and although students of the Swedish schools have chosen 3M $NaClO_4$ it must be remembered (1) that ion hydration spheres in the medium are not the same as $I = 0$, (2) ΔH° is not absolutely equal to ΔH^{3M}, and (3) entropy changes are not so evident in high I media.

It is healthy to observe that research is now being reported to clarify these medium effects, that Christensen and Izatt have produced equations for correcting ΔH^I to ΔH^0 [6] and that an appeal has been made for research to reveal the best medium for $\Delta^{\varnothing}C_p$ solution measurements.[11] Thermodynamic standard states and the choice of concentration scales (molal, molar or mole fraction, etc.) have also been discussed recently.[5] Superficially, all this thermodynamics can be a bewildering subject but we must realize that we are motivated not only by a desire to accumulate numbers for a particular system but also to compare parameters with those for other systems reported by other laboratories. There are many revealing patterns amongst series of such parameters and these are best considered through the hard and soft acids and bases approach.

H.S.A.B.

The strengths of metal-ligand bonds are conveniently systematized, using the theory of hard and soft acids and bases (h.s.a.b.)[3, 7] as described in Chapter 3. This approach assumes that all bonds between heteroatoms may be considered as having an acid and a base portion. Essentially, this acidity or basicity is decided by the number of valence electrons associated with a species and the ease with which they can be arranged. Tables classifying acids and bases into hard or soft appear in the literature cited.[3, 7]

The main principle behind the h.s.a.b. theory is that strong bonds are only formed between hard acids and hard bases or between soft acids and soft bases. Hard-soft bonds are either very weak or do not exist. $\Delta H°$ and $\Delta S°$ may be used to define the magnitude of hardness and softness. If we could put all acids and bases in order of

merit, right through from very hard to very soft, we would be in a position to predict the relative strengths of any combination of acids and bases. (It should be noted that it is the relative rather than absolute strengths that really interest us as complex formation in solution involves replacing one ligand, often water, by another.) This hardness-softness order can be established by examining the enthalpies and entropies of bond formation as illustrated in Table 10-I. Both the acid and base are considered to be solvated. Hard-hard reactions are usually endothermic but entropy stabilized, i.e. many solvent-acid and solvent-base bonds need to be broken before the reaction can occur and the energy to do this is greater than any energy liberated in forming the acid-base bond. However, since there are many solvent molecules liberated the $\Delta S°$ term is large. Soft acids or bases are only weakly solvated, if at all and so soft-soft reactions involve very little energy to desolvate the reactants. Hence they are exothermic and have small or negative entropies (because *two* particles—acid and base—are making *one* acid-base).

Hence, in a strongly polar solvent such as water, the enthalpy change upon forming an acid-base bond will be more negative the softer the electron acceptor and donor involved. Thus, measuring the enthalpies of a standard acid with a range of bases gives the softness order for these bases and similar measurements on a range of acids with a standard base gives the order for the acids.

FORMATION CONSTANT MEASUREMENTS

The formation constant for the generalized reaction on page 192 is β_{pqr} and it refers to forming product $A_pB_qH_r$ from individual aquated species:

TABLE 10-I

ENTHALPIES AND ENTROPIES OF
FORMATION IN AQUEOUS SOLUTION
CORRELATED WITH h.s.a.b.
CLASSIFICATION

Compound	$\Delta H°$ kJ mol⁻¹	$\Delta S°$ J K⁻¹ mol⁻¹	h.s.a.b. Classi-fication
H⁺ F⁻	12.2	96.6	Hard-hard
H⁺ CN⁻	− 43.5	29	Hard-soft
Hg²⁺ (CN⁻)₄ .	−248.9	−54	Soft-soft
Hg²⁺ I⁻	− 75.3	− 8	Soft-soft

$$pA + qB + rH \rightleftharpoons A_pB_qH_r$$

Such a generalized constant can describe hydroxy complexes, e.g. Cu $\cdot$ histidinate $\cdot$ OH has log $\beta_{11\text{-}1} = 3.462$, or polynuclear complexes, e.g. the copper cystine polymer shown in Figure 10-1. Any given solution usually contains a whole family of complexes (as we shall see in Figure 10-8), their proportional occurrences being concentration dependent. It should be noted that, although such constants describe the composition of and concentration dependence of the complexes present, formation constants reveal very little definite evidence concerning the actual sites of metal-ligand bonding, i.e. whether through carboxylate or amine. On the other hand, enthalpies and entropies of formation do permit some site identification.

The technique of measuring a series of constants using emf methods is as follows: A thermostatted vessel is set up under an inert atmosphere and contains a solution of known total concentrations (at constant ionic strength). This solution contains a stirrer, a detector electrode, e.g. a glass electrode for [H⁺] or an amalgam electrode for [B²⁺], and a salt-bridge as a standard reference electrode. The total concentration conditions inside the vessel are varied in a stepwise manner (usually through a titration approach) and the electrode emfs *versus* total concentrations are tabulated point by point. A computer program is then applied, e.g. LETAGROP–Swedish for "least squares," or SCOGS–stability con-

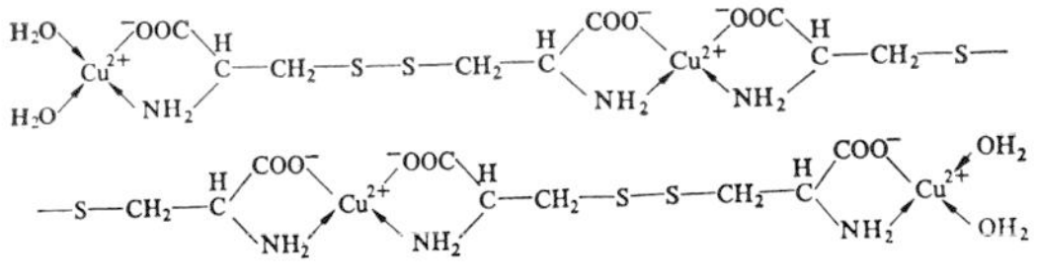

stants of general species to express such data in terms of the "best" constants. Briefly, the computation is as follows:

For any experimental measurement we have

(a) a set of accurately known values,
(b) a measured value or two, and
(c) a set of unknowns.

For example, consider a simple emf titration to determine the concentration pK of an acid ($\beta_{101} = \frac{[HA]}{[H^+][A^-]}$). The emf, E, may be measured on a glass electrode (giving [H⁺]) or a ligand electrode (such as Ag/AgCl for following [Cl⁻] if AH refers to HCl).* The emf for each point in the titration, E^i,

$$E^i = f(T, t, v^i, v_0, T_A, E^O, \beta_{101}, K_w)$$

where T is the temperature, t is the concentration of titre, v^i is the volume added, up to point i, to v_0 of initial solution, T_A is the total ligand concentration, E^O is the standard electrode potential for the electrode arrangement chosen, β_{101} is the required pK, and K_w is the ionic product of the medium. Therefore, we can list our three groupings:

(a) $T, t, v^i, v_0, T_A, E^O, K_w$
(b) E^i
(c) β_{101}

This data is fed into a computer that has been programmed by LETAGROP and SCOGS to make a series of guesses at (c). For each guess, an E_{calc} is computed for each of the n points in the titration and the

---

Figure 10-1. The copper(II)-cystine polymer, A_3B_4.

* Glass or Ag/AgCl electrodes really respond to a_{H^+} and a_{Cl^-} but it is usual when determining *concentration* constants to measure the E^O values by taking electrode readings, not from a_{H^+} buffers, but from solutions of known [H⁺]. Remember also, that E^O is I dependent so should be measured at one's chosen I.

best β_{101} is chosen as being the one that produces U, of the least squares sum,

$$U = \sum_{i=1}^{n} (E^i{}_{calc} - E^i)^2$$

(104A)

to be a minimum, U_{min}.

If (c) consists of several unknown constants, all are varied (guesswise) and for *every* possible combination of guesses a value of U is calculated so that, once again, U_{min} can be chosen.

APPLICATIONS OF FORMATION CONSTANTS

The concept that the usefulness of βs has largely been supplanted by enthalpy and entropy investigations, since *structurally* $\Delta H°$ and $\Delta S°$ arguments are more reliable, is false. Formation constants have the following important uses:

1. as Gibbs-Helmholtz precursors to $\Delta H°$ and $\Delta S°$ determinations;
2. to determine the *compositions* of complexes present in solution, for example Österberg has found (using amalgam and glass electrode potentiometry) that copper(II) glycylhistidylglycinate (A) complexes exist as CuA_2, Cu_3A_4 and $Cu_{15}A_{16}$; and
3. to calculate the *concentrations* of complexes present in a multicomponent system. We shall describe two examples of (3).

The first example occurs when many reactions are taking place simultaneously in the same container, e.g. the blood stream, it is not possible to resolve them into their constituent formation constants *in situ* and then to calculate the effect of altering the system, e.g. changing the concentration of a metal ion or of administering a drug. In-

stead, we must resort to examining each possible metal-ligand reaction *in vitro* and characterizing all these reactions in terms of formation constants. Next, we use computer simulation to combine all these reactions and to calculate the concentration of each complex at equilibrium. These concentrations should then be checked against an analysis of the contents of the reaction vessel. Next, the effect of adding a new reaction can be calculated.

There are two widely known programs for these calculations, COMICS* (Concentration of Metal Ions and Complexing Species) from Dr. Perrin's laboratory and HALTAFALL* (Swedish for "Concentrations and Precipitates") from the late Professor Sillén's laboratory. Both programs are designed to solve multiple simultaneous equations. For example, Perrin et al. have used COMICS to calculate the distribution of copper(II) and zinc(II) complexes of seventeen amino acids present in a blood plasma model at pH = 7.4, 37°C and $I = 0.15$ M.†

The input data consisted of the total concentrations of metals and amino acids (Table 10-II) and their independently determined formation constants (pKs, ligand-metal complex pβs, mixed ligand complex pβs, and hydroxy complexes—159 constants *in toto*). The computed output, which satisfies all the equilibria involved, is shown in Tables 10-II and 10-III.[3, 8]

---

* These programs can handle solid, liquid or gaseous equilibria, e.g. the equilibria involved in the sea and the atmosphere above.[3]

† It must be emphasized that this example refers to a rather simple model (a) in that there are several ternary complexes which are not included. For example, the Cu · histidinate · threoninate of Chapter 18 and (b) complexing to powerful chelating proteins such as serum albumin, is not included. Later models are gradually taking account of these factors.

TABLE 10-II

	(a) (mM)	(b) (μM)
Alanine	0.383	3.0
Arginine	0.086	3.1
Cysteine	0.033	0.0014
Cystine	0.042	0.41
Glutamic acid	0.048	0.48
Glutamine	0.568	20
Glycine	0.205	2.1
Histidine	0.074	1.2
Isoleucine	0.068	0.72
Leucine	0.129	1.4
Methionine	0.026	0.79
Ornithine	0.055	0.0052
Proline	0.205	0.24
Serine	0.107	3.7
Threonine	0.117	5.4
Tryptophan	0.054	1.1
Valine	0.246	2.9
Copper(II)	0.0011	0.34 pM
Zinc(II)	0.046	4.0

(a) Average Total Concentrations of Free Amino Acids and Metal Ions in Human Blood Plasma.

(b) Computed Equilibrium Concentrations of Amino Acid Anions and Free Metal Ions at pH = 7.4.[s]

TABLE 10-III

COMPUTED EQUILIBRIUM CONCENTRATIONS AMONG THE MOST PREVALENT (> 1% OF TOTAL METAL ION) AMINO ACID COMPLEXES AT pH = 7.4[8]

Complex	% of Total Metal Ion
$[Cu \cdot Cystine \cdot His]^-$	48
$[Cu \cdot H \cdot Cystine \cdot His]$	37
$[Cu(His)_2]$	13
$[Cu(Gln)_2]$	0.3
$[Cu \cdot OH \cdot His]$	0.3
$[Cu \cdot H \cdot (His)_2]^+$	0.2
$[Cu \cdot His]^+$	0.2
$[Zn \cdot His \cdot Cys]^-$	24
$[Zn \cdot His]^+$	21
$[Zn(Cys)_2]^{2-}$	16
Zn^{2+}	8.6
$[Zn \cdot H \cdot His \cdot Cystine]$	5.8
$[Zn(His)_2]$	5.7
$[Zn \cdot Gln]^+$	3.3
$[Zn \cdot H \cdot Cys \cdot His]$	2.2
$[Zn \cdot His \cdot Cystine]$	1.7
$[Zn \cdot H_2 \cdot His \cdot Cystine]$	1.5
$[Zn \cdot Gly]^+$	1.4
$[Zn \cdot H \cdot (Cys)_2]^-$	1.3
$[Zn \cdot Thr]^+$	1.3
$[Zn \cdot Ser]^+$	1.0
$[Zn \cdot Ala]^+$	1.0

It may be seen that for copper, out of a possible fifty-two complexes, only three account for 98 percent of the bound copper. For zinc, however, the metal is more evenly distributed over half the possible sixty-eight complexes. The success of histidine, cystine and cysteine in complexing the majority of the copper and zinc ions arises from their tridentate characters (all the other amino acid anions are bidentate).

The second example involves ligand pollution studies. In the United States there is a current dilemma concerning the substitution of nta (trisodium nitrilotriacetate) for phosphates (usually pentasodium tripolyphosphate) in detergents (*ca* 8% nta). Ten to 90 percent of the nta ingested (from drinking water, etc.) can be absorbed into the system, the ligand eventually becoming deposited in bones. Preliminary tests using rats have failed to rule out the possibility of nta being a carcinogen.

There are two aspects to the nta pollution problem: (1) What is the effect upon the normal equilibria present in rivers, lakes and reservoirs and the endemic plant and fishlife therein? (2) In which of its many metal complexes does nta enter *Homo sapiens* from detergenated washing water or from polluted drinking water?[*]

To illuminate these problems, Childs[9] has computed the equilibrium concentrations for a model lake water system (Figure 10-2). The salient results were (1) the nta outcomplexes most of the other ligands, (2)

[*] Ligands and metal complexes involved in carcinogenesis are discussed in Chapter 19.

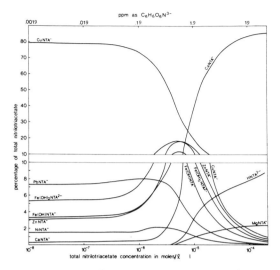

Figure 10-2. The percentage distribution of nitrilotriacetate metal complexes in a computer model of lake water at pH $= 8.$[9]

the distribution of nta complexes is both pH and total nta concentration dependent. These model system studies are invaluable when one considers the safe limits for nta both in detergents and in the environment as a whole, the rate of nta degradation in the presence of metal ions, and the particular complexes of nta present in blood plas-ma. Meanwhile, the nta carcinogeneity tests continue!

FUTURE USES OF FORMATION CONSTANTS

These two examples of the uses of formation constants are just the tip of the iceberg. Many other studies are currently being performed on model systems designed to deepen our understanding of the metal-ion activation of enzymes, of the masking and removal of toxic cations, of kinetic studies of ion transport, of metal ion deficiency or excess diseases, e.g. S. A. K. Wilson's disease, and of the effects of adding complexing drugs to systems. All these studies are feasible with existing computer simulation programs. Meanwhile, programs are being expanded to handle multiphase systems where the relative volumes of phases and their distribution coefficients are to be handled and this can eventually lead to multienzyme computations.

CALORIMETRIC MEASUREMENTS

In general, each calorimetric investigation involves the matching of an instrument

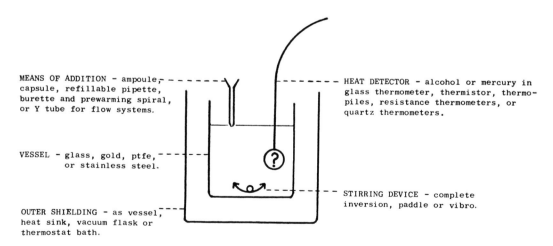

Figure 10-3. The basic essential components of a calorimeter.

to a chemical system. Consequently, there are almost as many calorimeters reported in the literature as there are tables of enthalpies of formation. We do not propose to describe the design of these instruments as details are readily available from other sources[3, 4, 6] (Figure 10-3). Any well-appointed laboratory should be capable of constructing a calorimeter, or alternatively, several are available commercially, their choice covering wide ranges of phase, temperature or concentration.

The minimum number of experimental measurements that are necessary to solve for n unknown heats of formation is n and the most efficient way of accumulating this data is to employ a titration technique. The raw data from such an investigation usually appear as tables or graphs of resistance or temperature *versus* amount of titrant added. The measured data are converted into "measured joules" by appropriate calibration experiments which measure the calorimeter response per unit of known energy (usually electrical heating) and then into "corrected joules recorded" by computing corrections for such side reactions as the heat of formation of water from its ions, the heats of deprotonating the ligand, heats of hydrolysis of metal ions, and heats of dilution. Computer programs have been compiled to take the effort out of these corrections. The resulting "corrected joules recorded" versus "ml added" thermogram usually has the form shown in Figure 10-4 and it is instructive to lay this smooth curve alongside a "species distribution" versus "ml added" graph showing the extensive complex interchange activity caused by adding just a few ml of titrant. These latter curves are calculated from known formation constants and the HALTAFALL or COMICS programs.[3] Finally, it is a rela-

tively easy step to convert these experimental values into the $\Delta H°$ for forming the

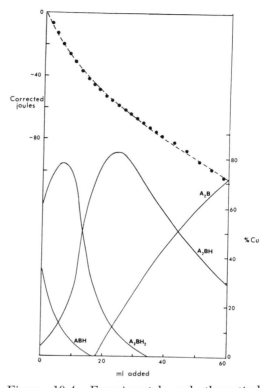

Figure 10-4. Experimental and theoretical thermograms and species distribution curves for the copper(II)histidinate system. The diagram shows the calorimetric titration results of the heat liberated (in joules) when 50.00 ml of histidinate⁻ (A) = 60.00 mM, copper²⁺ (B) = 19.14 mM, and protons (H) = 97.78 mM was titrated with a solution of A = 60.00 mM, B = 19.14 mM and H = −18.00 mM. The experimental points are shown as full circles and are plotted above species distribution plots for the four major copper complexes present during this titration. The curves represent the percentage of total copper present in each complex. The broken line is the theoretical thermogram calculated from these species distribution plots and $\Delta H°$ for forming these complxes. $\Delta H°$ values were obtained by subjecting the full circles, and similar titrations, to the SOLV program.

complexes present; convenient programs for doing this are the LETAGROP or SOLV least square approaches.[10]

The literature contains compilations of enthalpies of formation, but here we might pause and recall that we are dealing with *biological* solutions and so do not always have formation constants for, and can positively identify, the ligand donor sites of the system under investigation. However, these drawbacks can be circumnavigated: (1) Should $\Delta G°$'s (and βs) be unknown and not conveniently measured by the more usual potentiometric method they may be calculated from calorimetrically determined $\Delta H°$ and $\Delta S°$. This is achieved by applying the "entropy titration approach" to one's experiments. This is a procedure that produces both the enthalpies and entropies of formation from the same series of calorimetric titrations,[3,6] and (2) an alternative approach to this kind of problem is to construct enthalpic curves from one's calorimetry. An enthalpic curve is a plot of the "corrected joules recorded" *versus* $\bar{Z}$, the average number of ligands attached to a metal ion or substrates bonded to a metalloenzyme (Figure 10-5). Many interesting deductions can be made from the shapes of the curves alone even without a knowledge of the molecular weight of the ligand or metalloenzyme. Reference 12 shows how conclusions concerning symbiosis effects (a h.s.a.b. term for the mutual stabilizing influence of two or more ligands attached to a metal ion), ionic strength effects, and differentiation between inner and outer sphere complexing (ion pairing) can all be deduced from the shapes of experimentally determined enthalpic curves.

APPLICATIONS OF CALORIMETRIC INVESTIGATIONS[10]

Whole Organism Studies

Bacterial Calorimetry

One of the distinctions between life and death is that living organisms must of necessity be constantly undergoing energy transformations. If some of this energy appears as heat liberated or absorbed it can be monitored calorimetrically. As stated in the introduction, bacteria are some of the

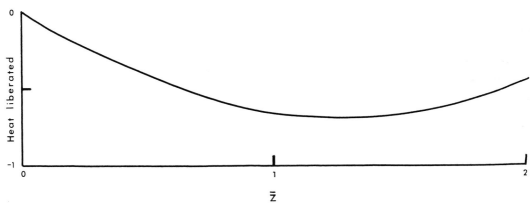

Figure 10-5. An enthalpic curve for a symbiotic system—cadmium fluoride. The acid, Cd_{aq}^{2+} is classified according to the h.s.a.b. approach as "borderline" hard/soft. When the first water molecule is displaced by the harder fluoride base the product CdF_{aq}^+ symbiotically becomes hard and then able to accept the second fluoride ligand with a more favorable enthalpy, i.e. the second bond appears stronger than the first. Units = J (g ion)$^{-1}$.

most important and "simplest" species that might be studied as whole organisms.

Three general points should be noted: (1) Microbial degradation always involves the evolution of heat (sometimes leading to spontaneous combustion); (2) when bacterial growth is curtailed by, for example, excluding an essential nutrient or by adding an antibiotic, this is paralleled by a quantitative decrease in the amount of heat monitored; (3) in spite of their complexity, Prat has demonstrated that the heat characteristics of any particular type of bacteria are essentially reproducible.

Higher Organisms

Prat has also studied the thermochemistry of higher organisms and reported that their heat production per unit weight is smaller than that for unicellular organisms.[4] Nevertheless, (1) the thermograms of multicellular organisms are still specific, i.e. characteristic of the particular plant or animal, (2) these thermograms are susceptible to external agents, both physical and chemical, and (3) their shape is also influenced by the age of the organism. Very little work has been reported for metal dependent studies, but clearly the field is a wide and promising one. With the present emphasis upon the environment and pollution, investigations into the thermochemistry of metal dependence and of metal poisoning, for example, by cadmium, lead, and mercury, and of the relationships between calcium and senescence would all be very desirable. Recently, there have been reports of some interesting changes in the energy metabolism of small animals following injury. This work could have important implications concerning chemical injury caused by administering metal complexes as drugs.

The literature, to date, does not contain much writing about the subject, but never-

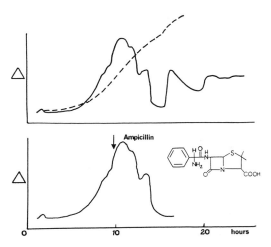

Figure 10-6. *Escherichia coli* growing in a complex medium that contains traces of several metal ions. The growth is monitored by calorimetry (———) and turbidity (- - - -).

theless writing of much importance, and we have extracted an example system that demonstrates (1) that microcalorimetry is not just an alternative instrument for the biologist but one that can improve upon an existing technique (experiment 1 in Figure 10-6), and (2) the high resolving power and reproducibility of bacterial calorimetry (compare experiments 1 and 2).

Calorimetry Versus Turbidity

In experiment 1, the growth of *E. coli* has been monitored by the usual method (turbidity) and by the flow calorimetry approach. Clearly, the calorimetry is far more specific and categorizes the growth into several distinct regions. In the second experiment, the calorimetric response is an exact reproduction of experiment 1 until an antibiotic is added after ten hours and then each step of the biocidal action is crisply recorded by the calorimeter. (Note that *E. coli* is a zinc dependent bacteria and that ampicillin is a good ligand for sequestering zinc ions).

Metalloenzyme Systems

General Description

Metalloenzymes are composed of peptide chains (one or several) and transition series metal ions (often in even numbers).[3] The molecule has a definite geometry and is held together by intra- or inter-peptide bonds (frequently these are hydrogen bonds). The metal ion occurs in a cleft or depression (called the "active site") and the amino acid residues lining the approach to this active site are specifically "designed" to assist the approach of the substrate and the exit of the products. The metal ion in the active site is in contact with but a small proportion of the total number of amino acids. (The remainder are concerned with holding the active site in the correct geometry, with creating the correct electrostatic environment for the enzyme action to occur and with providing the outer surface of the enzyme with a layer of charges, or hydrophobic groups, such that the enzyme as a whole is water, or lipid, soluble). These active site amino acids hold the metal ion's bonds "entatic," i.e. somewhat stretched so that a vacant bond is readily available to capture the incoming substrate. Finally, the three dimensional arrangement of polypeptide chains may be different with and without the substrate attached.*

Clearly, calorimetry has a prosperous future in examining all these aspects of enzyme activity—the heats of solvation of the enzyme as a whole, the solvation characteristics of the active site, the entropies of the entatic state of the active site, the energetics of enzyme-substrate interactions, and the

heat capacity changes associated with allosteric effects. We shall select examples of this type of study from three general areas.

Enzyme Assay

Calorimetry has been used as a means of assaying enzymes, enzyme substrates and even the amount and type of disease present. Figure 10-7 shows the calorimetric assay of horse serum cholinesterase: Spectral methods gave the activities of different amounts of pure enzyme and then a calibration curve of calorimetric response, Δ, *versus* activity was constructed. Unknown quantities of enzymes were then assayed by measuring their Δ's and referring to the calibration curve. This work used a flow calorimeter and an additional technique

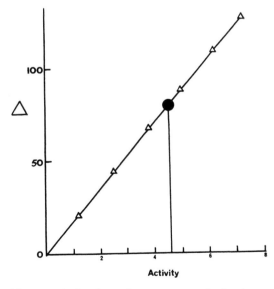

Figure 10-7. The calorimetric method of assaying enzyme activity. Calorimetric response, Δ, is plotted against units of enzyme activity (amplification = 10^4) and for known amounts of enzyme and then Δ from an unknown quantity of enzyme ($\bullet$) is plotted on the line and the amount of activity is read off the horizontal axis intercept.

---

* Excellent examples of all these principles may be seen in pictures of haemoglobin and oxyhaemoglobin printed in most textbooks of biochemistry and described here in Chapter 7.

called the "amplification effect" (the cholin-esterase assay measurements were performed in a tris(hydroxymethyl)amino-methane buffer; the enzymic process releases protons which instantaneously react with the buffer to give a large heat effect). Conversely, when an excess of enzyme was used, the assay method has been employed as a means of assaying the substrate.

Calorimetry as an Indicator of Health and Antibody Activity

Levin has monitored the heat produced by leucocytes and thrombocytes for normal man and for patients exhibiting thyroid dysfunctions. He has also studied the metabolic enthalpy patterns in human erythrocytes. These patterns can be used to differentiate between anaemias—haemolytic and pernicious anaemias increase the heat outputs compared to normal patients, while sideropenic anaemia cells exhibit a normal production of heat.

Metal dependent antigen-antibody reactions have been researched using calorimetry; (for example, haemocyanin-antihaemocyanin produce 167.4 kJ (mol of antibody)$^{-1}$ and $\Delta S° \doteq -418.4$ J mol^{-1}K^{-1}). However, in general, the $\Delta S°$ of antigen-antibody interactions are positive.

Small Molecules-Enzyme Reactions

A fruitful field for future research is that of calorimetic investigation into small ligand-metalloenzyme and into metal ion-protein (or apoenzyme) interactions (we must remember that not all metalloenzyme reactions involve metal ions being tightly bonded in the active site—some ions commute). Such studies would be of value to enhancing our understanding of enzyme-substrate interactions, of metal toxicity studies and membrane transport phenomena. A good example of such studies exists in the metal ion-apocarbonic anhydrase work of Henkens, Watt and Sturtevant. Carbonic anhydrase is a metalloenzyme that contains a zinc ion in its active site. (The enzyme occurs in erythrocytes and is responsible for catalyzing the reaction between carbon dioxide and water and thus assisting venous carbon dioxide transport.) The Yale University researchers removed the zinc ions from bovine carbonic anhydrase, i.e. they produced the apoenzymes, and then measured the enthalpies of reinserting not only the zinc but also divalent cobalt, copper, cadmium and nickel ions. In all instances, $\Delta G°$ was large and negative and $\Delta H°$ was positive i.e. the complexes were entropy stabilized. For *in vivo* reasons, the zinc system was of particular interest especially since the thermodynamic parameters, when compared to those for amino acids and other amine ligands, suggested that it was carboxylate groups that bind the zinc ion most firmly to the apoenzyme, while it had been traditional to assume that the zinc in carbonic anhydrase is held mainly by bonds to imidazole and primary amino nitrogens.

The Sturtevant school has also investigated the energetics of hemerythin oxygenation: A deoxygenated solution of protein was mixed with a buffered oxygenated solution in a flow calorimeter.

$$HrFe(II)_2 + O_2 \rightleftharpoons HrFe(II)_2O_2$$
$$\Delta H°_{25°, pH=7} = -38.5 \text{ kJ(mol monomer)}^{-1}$$

The oxidation, as distinct to oxygenation above, of

$$HrFe(II)_2 + mH_2O \overset{-2e}{\rightleftharpoons}$$
$$[HrFe(III)_2(H_2O)_m]$$

was also studied and it is interesting to note that the enthalpies of oxidation of hemerythrin and cytochrome C, each having ferrous ion active centers, are almost equal ($\Delta H°_{25°, pH=7} = 51.0$ and 59.0 kJ (mol Fe)$^{-1}$ respectively). Cytochrome C is not considered to undergo large conformational changes on redoxing and, thus $\Delta^{\varnothing}C_p$ is found to be small (-251.0 J mol^{-1}). We should note that these enthalpies of formation are pH dependent. For example, Fe(III)cytochrome C possesses a proton that ionizes with a pK of 9.3 and the overall effect is a $\Delta H°$ versus pH curve that is sigmoidal.

$\Delta^{\varnothing}C_p$, and to a certain extent $\Delta S°$, measurements are frequently more useful than $\Delta G°$ or $\Delta H°$ when discussing the processes through which bio-polymers unfold. In a similar vein, heat capacity data have been used to measure the amount of water trapped between the polypeptide chains in hydrated haemoglobin (32.4 g H$_2$O per 100 g dry weight).

The Predilection for Iron Protein Research

By far the majority of reported studies from an admittedly modest literature have been concerned with iron proteins (for example, the two most recent reports are those for the enthalpy of oxidation of sperm whale myoglobin and of haemoglobin-haptoglobin interactions). Their history reaches back fifty years to the first studies of the reactions of blood gases and haemoglobin by Hartridge and Roughton, and there is a very good reason for this predilection for haemoglobin: Until recently, only very small amounts of metalloenzymes were available and so feasible studies were strictly limited to readily available enzymes. However, during the last five years, the design of modern microcalorimeters and the development of new methods for isolating enzymes has opened the doors to a wide range of metalloenzyme calorimetry studies.

Simple *In Vivo* Ligand-Metal Ion Interactions

Metal Complexes in General

Understandably, there appears to be a reciprocal relationship between the molecular complexity of a system (compare the complexities of whole organisms to that of metalloenzymes and to simple ligands) and the quantity of thermodynamic data that have been reported for the system. Thus, thermodynamic parameters for whole body-metal ion reactions are only sparsely distributed throughout the literature, while the reports of simple ligand-metal ion data are voluminous. These include the energetics involved when metal ions react with amino acids, peptides, hormones, nucleic acids, nucleotides, carboxylic acids, carbohydrates, lipids, simple anions, administered drugs and even the solvent water. There are also figures for the heats of hydrolysis and of dissolution of both ligand and metal ions—these are necessary for applying corrections to raw data in order to produce enthalpic curves and enthalpies of reaction. Usually, one has a selection of values reported for any given reaction and so one chooses enthalpies determined at the same ionic strength and temperature as is currently being studied. In general, calorimetrically determined $\Delta H°$ and $\Delta^{\varnothing}C_p$ are more reliable than values derived from the temperature dependence of formation constants.

The heat properties of pharmaceuticals is a subject of current interest. It has been shown that they can be assayed, regardless of dosage form (gelatine capsules or tablets, etc.), by aqueous solution calorimetric titration against metal ions. This might be compared to the more usual methods which first extract the actual constituent from the tablet, etc. and then use rather time-consuming, nonaqueous titrations. For example, the theophylline content of aminophylline (a suppository used in treating asthma) can now be thermometrically titrated against metal ions. With the increased use of low temperature surgical interventions, uses of curarizing agents, and sleep-inducing drugs that are strongly hypothermizing, $\Delta H°$ and $\Delta°C_p$ values are becoming very useful for medical chemistry computations.

Histidine Complexes in Particular

However, we ought to be conversant with the thermodynamics of the normal *in vivo* processes occurring in the healthy organism before we attempt to investigate the reactions introduced by disease or by administered drugs. The power of thermodynamic arguments in unravelling the intricacies of *in vivo* reactions is best illustrated by studying one ligand in some depth. We have chosen histidinate the anion of the amino acid histidine:[13]

$$(105)$$

Gibbs free energies, enthalpies and entropies of forming histidinate complexes with divalent metal ions from the first transition series have been reported. The zinc(II)histidinate system has anomalous properties and the origin of these anomalies has been traced through the thermodynamic parameters reported. In brief, the zinc(II) histidinate complexes are less stable than the corresponding cobalt(II)histidinate, while for all other amino acid systems the converse situation arises. It materializes that the zinc(II)(histidinate)$_2$ complex is tetrahedral (both amine nitrogens being bonding) and the carboxylate group is just loosely associated with the metal ion

(proved by comparison with the thermodynamics for histamine). This conclusion was reached because, relative to other comparable metal ions, the zinc bonds are enthalpy weaker (ligand is bidentate instead of tridentate) and entropy stronger (six membered chelate rings are less strained when tetrahedral rather than octahedral).

For many years, biochemists have quoted the complex formation between copper(II) and histidine as an example of chelation *in vivo*. Nevertheless, the structures of the complexes formed *in aqueous solution* have never been clearly established. We now propose to demonstrate the usefulness of calorimetry in elucidating such structures.

The histidinate ligand has three readily available coordinating groups and the cupric ion prefers distorted octahedral (Jahn-Teller) coordination. However, models show that three normal histidinate-copper bonds cannot occur. As may be expected, potentiometric investigations have shown (1) that a range of copper-histidine complexes exist in aqueous solution, and (2) that the distribution of metal and ligand between these complexes is pH dependent. Such a situation suggests the possibility of tautomerism. Herein lies the danger of examining the system by techniques that do not monitor aqueous solutions: Crystallization or freezing might well create conditions in which the alternative to the *in vivo*, the less stable, isomer occurs. Hence nonaqueous/ambient temperature data ought to be extrapolated only with extreme care to *in vivo* discussions.

Figure 10-8 shows the structures of copper-histidine complexes as suggested by thermodynamic analysis. We shall describe the logic behind the first two complexes as an example of the method:

AB's suggested structure involves hista-

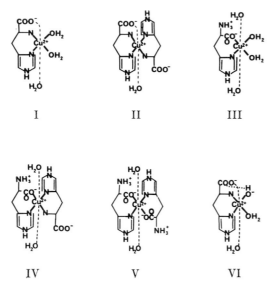

I   II   III

IV   V   VI

Figure 10-8. Suggested structures for the copper(II)-histidinate complexes present in aqueous solution. Full lines represent planar bonds to copper, broken lines are axial bonds with the exception of O . . . . . H in VI which is a hydrogen bond.

mine-type planar nitrogen donors and a more weakly bonded apical carboxylate. The original suggestion for imidazole-copper bonding came from formation constant comparisons, e.g. β_{AB}(histidine) $\doteq \beta_{AB}$(histamine) $>> \beta_{AB}$ (for another amino acid such as phenylalanine). Our enthalpies conform to this pattern (-43.89, -43.05 and -21.47 kJ mol^{-1} respectively). Further, the existence and approximate magnitude of the carboxylate-copper interactions may be seen from a comparison of the heats of protonating the two donor groups common to histidine and histamine and the heats of copper complexing these groups $-\Delta H^\circ_{AH_2}$ (histamine) $> -\Delta H^\circ_{AH_2}$(histidine) and yet $-\Delta H_{AB}$(histamine) $< -\Delta H^\circ_{AB}$ (histidine) ($81.71 > 77.03$; $43.05 < 43.89$ kJ mol^{-1} respectively). β arguments based upon these lines would have been rather dubi-

ous indeed since they include both enthalpy and entropy contributions, the latter having an appreciable difference between histamine and histidine because of the presence of a carboxylate group.

A$_2$B has the suggested structure as shown and the histamine-type bonding is based upon enthalpy arguments parallel to those of AB. Although the one apical, loosely bonded, carboxylate is expected to persist through to A$_2$B, the fact that $-\Delta H^\circ_{AB+A} \rightarrow$ A$_2$B is lower than for adding a histamine molecule to a copper(II) ion (40.0 and 43.05 kJ mol^{-1} respectively) suggests that the second carboxylate is not involved in bonding to the metal ion. It is noteworthy that such structures have been reported from crystallographic studies and nmr has suggested that A$_2$B is a tautomeric equilibrium between Cu(4N donors) $\rightleftharpoons$ Cu(3N and 1O donors). Similar arguments, often using entropy as well as enthalpy figures, have been invoked to suggest the other structures shown in Figure 10-8.

Lanthanide ions have been used as a probe for following calcium in biological systems and the $\Delta^\varnothing C_p$ of forming the Ln^{3+}-histidine complexes may be quoted as an excellent example of the uses of heat capacity measurements in revealing ionic aquation conditions. Figure 10-9 shows a sharp "flick" in the middle of the series in keeping with either a change in the states of hydration of the metal ion, or of the complexes, or a change in the coordination number of the metal ions. Grenthe and Ots have suggested "hydration equilibria" as

$$(BA_j(H_2O)_x \rightleftharpoons BA_j(H_2O)_y + (x - y)(H_2O)$$

being the main source of such inflections in $\Delta^\varnothing C_p$ for the lanthanide complexes. Conductivities, enthalpies and entropies show similar trends but much less succinctly thus

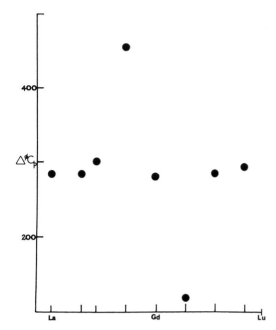

Figure 10-9. $\Delta^{\varnothing}C_p$ for the formation of lanthanide(III)-histidine complexes at 31°. Units = J mol^{-1}K^{-1}.

supporting Wadsö's comment that "$\Delta^{\varnothing}C_p$ data are among the most elucidative properties that can be studied."[4]

The energetics of optically active metal complexes of histidine have also been investigated: Amino acids *in vivo* have the L configuration, the reason for this being unknown. However, D-amino acids have been incorporated into some drugs and Barnes and Pettit have discovered some interesting properties among the heats of formation of bis complexes. As may be expected, metal^{2+}(L-His)$_2$ $\equiv$ metal^{2+}(D-his)$_2$, but rather surprisingly) (Ni^{2+} or Zn^{2+}) (L-his) (D-his) are more stable than their optically pure D or L analogues. However, Cu^{2+} (L-his)(D-his) is less stable. It has been suggested that this situation arises because of either copper's square planar coordination tendency or of the dynamic tautomer-

ism of copper complexes discussed earlier.

Thus, just one amino acid has given food for thermochemical thought and reasons for more calorimetric investigations.

FUTURE USES OF CALORIMETRY

We foresee future developments in bioinorganic calorimetry in topics for diverse disciplines: (1) The theories necessary for kinetic investigations using calorimetry are already available and there could well be an almost exponential increase in kinetic and related metalloenzyme studies; (2) drug screening by calorimetry—the preliminary screening of transition metal anticancer drugs is through the induction of lysis in lysogenic bacteria, the process being monitored by turbidity measurements (Figure 10-6 shows how calorimetry could improve the precision of such monitoring); (3) drug membrane studies—enthalpy measurements can be used to determine the distribution coefficients of species between nonaqueous solvents and water—equations have been reported relating these coefficients with cancer drug action; (4) a study of the thermodynamics of the regulation processes whereby humans live at 37°, fish can live at 0° and lizards can live at any temperature from 5 to 35°. The long-range outcome of such studies is a deeper understanding of the relationships between longevity, metabolism and ambient temperatures; (5) in the instrumental field, there is a tendency to incorporate calorimeters into other pieces of equipment. For example, Berger has coupled calorimetry with stopped-flow spectral measurements and Brown has produced multiple calorimeters to speed up the output of results and to perform many experiments on unstable systems before they decompose.[4, 10]

REFERENCES

1. Beck, M. T.: *Chemistry of Complex Equilibria.* London, Van Nostrand, 1970.
2. Sillén, L. G. and Martell, A. E. (Eds.): Stability constants. Chem Soc London publications number 17 and 25, 1964 and 1971.
3. Williams, D. R.: *The Metals of Life—An Introduction to the Solution Chemistry of Metal Ions in Biological Systems.* London, Van Nostrand, 1971.
4. Brown, H. D. (Ed.): *Biochemical Microcalorimetry.* New York, Acad Pr, 1969.
5. Wadsö, I.: Biochemical thermochemistry; Jameson, R. F.: Thermodynamics of metal-complex formation; *Thermochemistry and Thermodynamics.* Vol. 10 of M.T.P. International Review of Science, Chemistry Series 1. London, Butterworths.
   Sillén, L. G.: *Co-ordination Chemistry,* Martell, A. E. ed. London, Van Nostrand, 1971, p. 491.
6. Hill, H. A. O. and Day, P. (Eds.): Physical methods in advanced inorganic chemistry. In Christensen, J. J. and Izatt, R. M.: *Thermochemistry in Inorganic Solution Chemistry.* New York, Interscience, 1968, Chap. 11.
7. Pearson, R. G.: Hard and soft acids and bases. *J Chem Educ, 45:*581, 643, 1968.
8. Hallman, P. S., Perrin, D. D. and Watt, A. E.: The computed distribution of copper(II) and zinc(II) ions among seventeen amino acids present in human blood plasma: A model for biological systems. *Biochem J, 121:*549, 1971.
   Perrin, D. D. and Agarwal, R. P.: Multimetal-multiligand equilibria. In Sigel, H. (Ed.): *Metal Ions in Biological Systems.* New York, Marcel Dekker, 1974, Chap. 4, p. 167.
9. Childs, C. W.: Chemical Equilibrium Models for Lake Water Which Contains Nitrilotriacetate and for "Normal" Lake Water, Proc. 14th Conf. Great Lakes Res., 1971, p. 198.
10. Sigel, H. (Ed.): Metal ions in biological systems. In Williams, D. R.: *The Thermochemistry of Bio-inorganic Systems.* New York, Marcel Dekker, 1974, Vol. 4, Chap. IV.
11. Ashcroft, S. J. and Mortimer, C. T.: *Thermochemistry of Transition Metal Complexes.* New York, Acad Pr, 1970.
    Christensen, J. J. and Izatt, R. M.: *Handbook of Metal-Ligand Heats and Related Thermodynamic Quantities.* New York, Marcel Dekker, 1970.
12. Davidson, F. C., Sloan, J. P. and Williams, D. R.: The shapes and applications of enthalpic curves. *J Appl Chem Biotechnol, 21:* 300, 1971.
13. Jones, A. D. and Williams, D. R.: *J Chem Soc* (A), 1550, 3138, 1970; 3159, 1971; (Dalton), 790, 1972.

CHAPTER 11

APPLICATIONS OF KINETIC STUDIES

Alan D. B. Malcolm

Introduction
Flow Methods
Relaxation Methods
Nuclear Magnetic Resonance
Examples of Applications of Kinetics to Bio-Inorganic Systems
Conclusion

INTRODUCTION

THE STUDY OF the rates of chemical reactions has probably provided more information about the mechanisms of these reactions than has any other technique: The order of a reaction can often give a guide to its molecularity, the variation of reaction rate with temperature leads to its activation enthalpy and hence to information concerning the transition state, and the variation of rate with pH gives information about the role of the proton or hydroxyl ion in reaching this transition state.

It will be assumed that the reader is familiar with the mathematical treatments used for 1st and 2nd order reactions and for reversible reactions (two general texts which cover this background are given at the end of this chapter).

There are two features which distinguish biochemical reactions from conventional organic reactions—their high velocity and their high specificity. In inorganic chemistry some metal ligand reactions have rate constants of the same order of magnitude as those of biochemical reactions but it is the exception rather than the rule to find a ligand with as great a specificity for a metal as has an enzyme for its substrate.

There is considerable evidence from studies on enzymes that these two features are very closely linked and that the same factors give rise to both the high velocity and the high selectivity. It therefore seems likely that a study of kinetics will provide a better understanding of both these remarkable features of biological reactions.

FLOW METHODS

A consequence of these high rate constants (10^5 sec^{-1} and 10^8 M^{-1} sec^{-1} for first and second order reactions have often been observed) is that the use of concentrations high enough to be measured by conventional spectroscopic techniques results in reactions with very short half-times (less than 1 second). Conventional mixing in a spectrophotometer cuvette is obviously unsuitable for kinetic studies on such a reaction. The earliest attempt to overcome this problem was made by Hartridge and Roughton in their attempts to measure the rates of binding of ligands to haemoglobin. By flowing the two reactant solutions together rapidly and then passing the mixture along an observation tube, they observed the solution (and estimated, colorimetrically, its composition) at varying distances from the point of the mixing.

As may be imagined, the above method consumes large quantities of material and

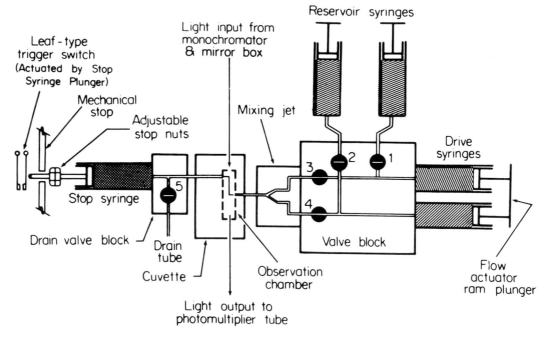

Figure 11-1. Line diagram of a typical stopped flow apparatus. (Reproduced with permission from Bulletin 131 of the Durrum Instrument Corporation.)

this inhibited its widespread use in biochemistry, a subject where even small quantities of material are often difficult to obtain.

This led to the development of the stopped flow apparatus, a diagram of which is shown in Figure 11-1.

By driving the two reservoir syringes pneumatically it is possible to observe the mixture within about 1 millisecond of mixing. The solution hitting the stop syringe activates the trigger switch which starts the recorder (usually a storage oscilloscope although increasingly the data is fed immediately into a computer for analysis). It was with the aid of such an apparatus that the production of a complex between enzyme and substrate (in this case catalase and hydrogen peroxide) was observed directly for the first time.

To illustrate how the data are treated in order to obtain the individual rate constants, let us consider the simple case of an enzyme with only one substrate and one product.

$$E + S \underset{k_{-1}}{\overset{k_{+1}}{\rightleftharpoons}} ES \overset{k_{+2}}{\rightarrow} E + P$$

As usual we can write an equation for the net rate of change of the ES complex in terms of its rate of formation and rate of destruction

$$\frac{d}{dt}[ES] = k_{+1}[S]([E]_0 - [ES]) - [ES](k_{-1} + k_{+2})$$

where $[E]_0$ is the concentration of enzyme at t = 0.

$$\text{or} \quad \int_0^{[ES]} \frac{d[ES]}{k_{+1}[S][E]_o - [ES](k_{+1}[S] + k_{-1} + k_{+2})} = \int_0^t dt$$

which gives

$$-\frac{1}{k_{+1}[S] + k_{-1} + k_{+2}} \, \ln_e \left(1 - [ES] \frac{k_{+1}[S] + k_{-1} + k_{+2}}{k_{+1}[S][E]_o}\right) = C + t$$

This may be rewritten as

$$[ES] = \frac{k_{+1}[E]_o[S]}{k_{+1}[S] + k_{-1} + k_{+2}} \left(1 - e^{-(k_{-1} + k_{+2} + k_{+1}[S])t}\right)$$

or, since $K_m = (k_{-1} + k_{+2})k_{+1}$

$$[ES] = \frac{[E]_o[S]}{[S] + K_m} \left(1 - e^{-k_{+1}(K_m + [S])t}\right)$$

Now the rate of product formation is given by

$$\frac{d[P]}{dt} = k_{+2}[ES]$$

and integration now gives

$$[P] = \frac{k_{+2}[E]_o[S]t}{K_m + [S]} + \frac{k_{+2}[E]_o[S]}{k_{+1}(K_m + [S])^2} \left\{e^{-k_{+1}(K_m + [S])t} - 1\right\}$$

We can use this relationship in two ways: (1) When t is small (such that $k_{+1}(K_m + [S]) \ll 1$) then the exponential may be expanded to give

$$e^{-k_{+1}(K_m + [S])t} = 1 - k_{+1}(K_m + [S])t + \tfrac{1}{2}k_{+1}^2 (K_m + [S])^2 t^2$$

and hence we have

$$[P] = \tfrac{1}{2}k_{+1}k_{+2}[E]_o[S]t^2$$

Thus a plot of log [P] against time will give a straight line of slope $k_{+1}k_{+2}[E]_o[S]$. $k_{+2}[E]_o$ is the maximum velocity for the reaction (V_{max}) and hence if [S] is known it is easy to calculate k_{+1}. (2) When t is very large the above relationship reduces to

$$[P] = \frac{k_{+2}[E]_o[S]t}{K_m + [S]} - \frac{k_{+2}[E]_o[S]}{k_{+1}(K_m + [S])^2}$$

and a plot of [P] against t is now a straight line. (This is, of course, the so-called "steady state" of the reaction.) This line cuts the abscissa at $t = 1 \left/ k_{+1}(K_m + [S]) \right.$

Assuming that K_m (the Michaelis constant for the reaction) is known, k_{+1} may be calculated. During the first few milliseconds after mixing the amount of product formed follows the graph shown in Figure 11-2.

Once k_{+1} is known, if the dissociation equilibrium constant, K_d, for the binding of substrate to enzyme is also known, then k_{-1} may be calculated from

$$K_d = k_{-1} \left/ k_{+1} \right.$$

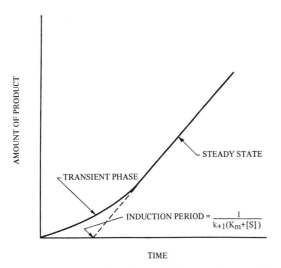

Figure 11-2. Product formed during the initial period after mixing enzyme and substrate.

This technique of presteady state kinetics using stopped flow apparatus has been widely used. For example, the rates of ligand binding to haemoglobin have been measured in this way. Unfortunately the method has several limitations: (1) It is not possible to mix solutions in times shorter than a millisecond and this means that very fast reactions ($k_{+1} > 10^7$ M^{-1} s^{-1}) cannot be measured in stopped flow equipment (There is a limit to the extent to which a reaction may be slowed down by lowering the concentration of reactants—the concentration must be high enough to be detectable by some means such as spectrophotometry.); (2) first-order reactions with $k > 10^3$ s^{-1} cannot be studied and since a great deal of biological control is thought to occur by means of conformational changes (which may well give first-order kinetics) this is obviously a considerable disadvantage; (3) it is also a pity that dissociation rate constants cannot be determined directly since equilibrium binding constants for biological systems cannot always be measured precisely. (4) Finally,

even the simplest enzyme-catalyzed reaction will proceed through more than one intermediate and presteady state kinetics is only capable of providing information about steps prior to, and including, the rate limiting step.

RELAXATION METHODS

Many of these shortcomings may be overcome by applying relaxation kinetics: The reactants are taken mixed, at equilibrium, and external parameters, for example, temperature or pressure, are rapidly changed such that the system is no longer at equilibrium. Then the shift to the new equilibrium position is now followed.

Before deriving the equations used in an analysis of such data it may be useful to consider some of the ways in which the external parameters of the system may be changed:

(1) The easiest and most commonly used method is by so-called "temperature jump." Any reaction which has a nonzero enthalpy will have an equilibrium constant which varies with temperature. The temperature of a solution may be raised by discharging a condenser through it. The energy dissipated is given by $\frac{1}{2} CV^2$ (where C is the capacitance of the condenser and V is the voltage to which it is charged). For example, a 50 nanofarad condenser charged to 30 kilovolts can provide 22 joules—sufficient to heat 1 ml of an aqueous solution by 5°C. More important, the half-time taken for the discharge of a condenser is given by $\frac{1}{2} RC$ (where R is the resistance of the solution through which this discharge is taking place). For a 50 nF condenser and a 1 cm cube of 0.5M sodium chloride solution (which has a resistance of rather less than 100 ohms) this heating time is 2.5 μsec. Hence, provided that the shift to the new equilibrium position has a time con-

stant slower than about 10 μsec this method will be able to monitor it. Reactions faster than this (and there are many in biology) may be studied by a more elaborate apparatus where the heating time is as short as 20 nanoseconds.

(2) A reaction with a nonzero volume change may be perturbed by a pressure change. This may be achieved by starting with the system at high pressure, bursting a diaphragm to release the pressure and then following the chemical relaxation to the new low pressure. If the reaction is accompanied by a change in the electric dipole moment of the molecule than a change in the electric field strength applied to the system can be used to perturb the equilibrium. This method has been used for studying reactions involving charge separation such as metal-ligand reactions of which binding to haemoglobin is a good example.

Let us now consider how the data are treated to provide the rate constants using a simple enzyme/substrate binding

$$E + S \underset{k_{-1}}{\overset{k_{+1}}{\rightleftharpoons}} ES$$

where the dissociation constant is given by

$$K_d = k_{-1} / k_{+1},$$

at any time we may write

$$\frac{dS}{dt} = k_{-1}ES - k_{+1} E \cdot S$$

(square brackets to indicate concentrations have been omitted for convenience in writing these equations)

Let the equilibrium concentrations after the temperature jump be $\bar{E}$, $\bar{S}$ and $\overline{ES}$. Provided that the jump is a small one, the change in the equilibrium constant will not be large and thus at any time after the jump the various concentrations will only be slightly

different from equilibrium. Let this difference be designated x. We may therefore rewrite the above equation as

$$\frac{d(\bar{S} + x)}{dt} = k_{-1}(\overline{ES} - x) - k_{+1} (\bar{E} + x)(\bar{S} + x)$$

This may be multiplied out and remembering that

$$\frac{dS}{dt} = O \text{ and that } k_{-1}ES = k_{+1}E_oS$$

the equation reduces to

$$\frac{dx}{dt} = -k_{-1}x - k_{+1}(\bar{E} + \bar{S})x - k_{+1}x^2$$

If the displacement from equilibrium is small, then terms in x^2 may be ignored

$$\frac{dx}{dt} = -k_{-1}x - k_{+1}(\bar{E} + \bar{S})$$

or $x = x_o \exp - (k_{-1} + k_{+1} [\bar{E} + \bar{S}])t$

Thus, the return to equilibrium follows a simple exponential law characterized by a so-called relaxation time given by

$$\frac{1}{\tau} = k_{+1}(\bar{E} + \bar{S}) + k_{-1}$$

Thus, by measuring τ as a function of $\bar{E}$ and $\bar{S}$, the rate constants may be obtained. The ratio k_{-1} / k_{+1} may be compared with the value of K_d used to calculate $\bar{E}$ and $\bar{S}$.

There are several important points to be made about relaxation kinetics:

(1) Every chemical transformation will be associated with a relaxation time. Thus, a system in which several equilibria coexist will have as many relaxation times as there are equilibria. If these times are similar in magnitude, then there may be analytical problems in resolving them but it does mean that steps beyond the rate limiting step may be studied.

(2) A further consequence of this is that

although many reactions of interest are not associated with a readily measurable spectral change, many of these may be coupled to a reaction which may be observed. Consider, as an example, a reaction in which a proton is released.

$$A + B \rightleftharpoons C + H^+$$

If an acid base indicator is added to this system, an extra equilibrium exists:

$$Ind + H^+ \rightleftharpoons Ind\ H$$

There will now be two relaxation times—the faster from the indicator reaction and then as the slower reaction adjusts to its new equilibrium position, the indicator reaction will adjust again to allow for the protons being produced in this reaction. Hence, we will observe the spectral change of the indicator, but the relaxation time will be that of the reaction we wish to study.

(3) It has been seen how k_{+1} and k_{-1} may both be determined provided that K_d is already known. Since the ratio of the rate constants provides a value for K_d it should be possible to determine k_{-1} and k_{-1} without knowing K_d previously. Simple algebra shows that for the reaction $E + S \rightleftharpoons ES$

$$1/\tau^2 = k_{+1}^2(E_o - S_o)^2 + 2k_{+1}k_{-1}(E_o + S_o) + k_{-1}^2$$

This may be further simplified if it is arranged experimentally that $E_o = S_o$ when

$$1/\tau^2 = 2k_{+1}k_{-1}(E_o + S_o) + k_{-1}^2$$

Thus k_{+1} and k_{-1} may be obtained from the plot of $1/\tau^2$ against $(E_o + S_o)$ and so K_d (and hence ΔG°) for the reaction may be obtained directly from relaxation kinetics.

(4) The temperature jump method may only be applied to reactions having nonzero enthalpies. (This follows from the van't Hoff equation $\frac{d}{dT}\ln K_d = \frac{\Delta H^\circ}{RT^2}$). From the magnitude (amplitude) of the concentration change on raising the temperature, it is possible to calculate the enthalpy of the reaction. It may be shown that

$$M = \Gamma\ \left(\frac{\Delta H^\circ}{RT^2}\right) \delta T \bullet \delta E \quad (105A)$$

(where M is the amplitude of the relaxation), and

$$\frac{1}{\Gamma} = \frac{1}{E} + \frac{1}{S} + \frac{1}{ES} \quad (105B)$$

(δT is the increase in temperature, and $-\delta E$ is the molar extinction change for the reaction). From a series of temperature jump experiments it is now possible to obtain k_{+1}, k_{-1}, K_d, ΔG° and ΔH°. The latter two parameters will give ΔS° and repeating the experiment at different temperatures will give the activation enthalpies for k_{+1} and k_{-1} and also of course provide a check on the value of ΔH°.

The three methods mentioned so far are all examples of so-called "step perturbations." The equilibrium is disturbed only once during each experiment. It is also possible to apply a periodic perturbation. A sound wave is a periodic pressure fluctuation and will thus affect the equilibrium of a chemical reaction involving a volume change. When the frequency of the sound wave is low compared with $1/\tau$ for the chemical relaxation, then the reaction will keep in phase with the oscillatory pressure. As the frequency of the sound wave is increased, the chemical reaction will "lag" behind the pressure wave. Qualitatively, it can be understood that a measurement of the energy absorbed by a sample from the sound wave as a function of frequency can lead to a determination of the relaxation time for that system.

For the three methods of step perturbation, the most popular method of detection is light absorption (visible or ultraviolet). However, fluorescence has also been used and, in the case of the electric field jump, the change in electric conductivity is an obvious means of detection.

NUCLEAR MAGNETIC RESONANCE

Finally, there is one method capable of measuring the very fast rates found in biological systems—that based on nuclear magnetic resonance. If the chemical shift of the ligand protons is different in the complex and in the free ligand, then the broadening of the nmr spectrum on addition of the metal ion may be used to provide values for the rate constants of the metal-ligand reaction. For example, when the proportion (P) of total ligand present as the complexes is small the equation is

$$\Delta \nu = \Delta \nu^o + \frac{P4\pi (\Delta \delta)^2 \tau}{1 + 4\pi^2 (\Delta \delta)^2 \tau^2}$$

(where $\Delta \nu$ and $\Delta \nu^o$ are the half widths of the nmr signal in the presence and absence, respectively, of the metal; $\Delta \delta$ is the difference in the chemical shifts for the free and complexed ligand and τ is the relaxation time for the system). These P values may then be used to calculate the rate constants.

EXAMPLES OF APPLICATIONS OF KINETICS TO BIO-INORGANIC SYSTEMS

Much of the data obtained by the above methods is described in Chapters 4 and 5 of this book. However, it is worthwhile to mention here some examples of problems which have been studied by these techniques.

The *in vivo* essential metals may be crudely classified into four categories:

(1) The alkali metals, sodium and potassium, tend to form weak complexes having high velocity constants both for their formation and for their dissociation. Consequently, one of their important biological

functions is as charge carriers—the voltage change accompanying a nervous impulse is accompanied by a passage of K^+ out of the cell and of Na^+ into the cell. Before the nerve cell may be triggered again, the high internal K^+ concentration and low internal Na^+ concentration must be restored. It is still not known how the cell achieves this specificity of ion transport, but the discovery of cyclic ligands such as valinomycin and the macrolide actins (Figures 11-3 and 11-4) has provided a possible model.[*] These ligands are far more specific in their affinity for metal ions than are most non-cyclic ligands—for example, valinomycin binds potassium ions approximately 10^3 times more strongly than it binds sodium ions. Their specificity seems to arise from the ionic radius of the metal ion and the size of the hole produced by the carbonyl oxygens. Even more important is the fact that the exteriors of these molecules are hydrophobic and thus provide a means of "dissolving" normally hydrophilic ions in environments such as the lipids of biological membranes. These are just the properties which a biological ionophore must have and therefore a study of the kinetics of these interactions could well provide a test of some theories of ion transport.

A temperature jump experiment, using methanol as solvent and murexide as a metal ion indicator, showed that the relaxation time for potassium-valinomycin binding is less than 5 microseconds and thus was too short to be determined precisely by temperature jump. However, by observing the reaction at 224 nm (the amide absorption band) without the indicator, two relaxation times were then observed. One was again too fast to measure, but there was also a slower one. This has been in-

---

[*] These systems are discussed further in Chapters 15 and 19.

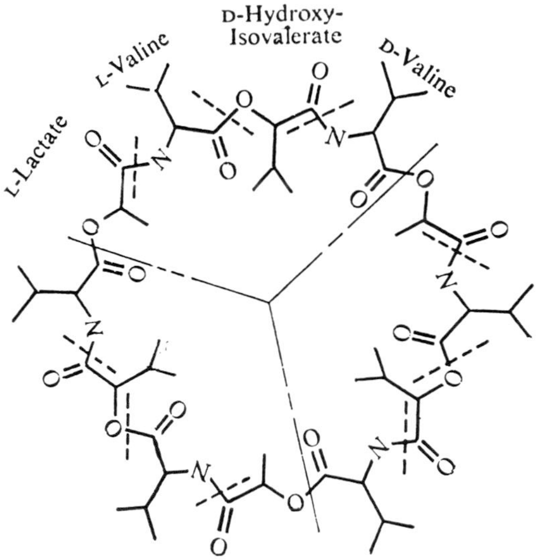

Valinomycin

Figure 11-3. Valinomycin.

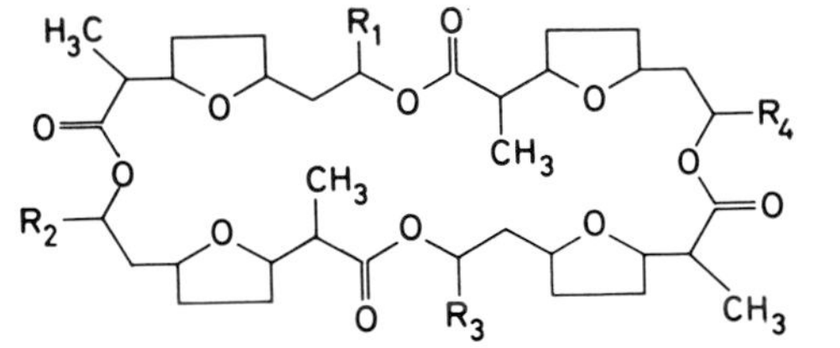

$R_1 = R_2 = R_3 = R_4 = CH_3$ Nonactin

$R_1 = R_2 = R_3 = CH_3$ $R_4 = C_2H_5$ Monactin

$R_1 = R_3 = CH_3$ $R_2 = R_4 = C_2H_5$ Dinactin

$R_1 = CH_3$ $R_2 = R_3 = R_4 = C_2H_5$ Trinactin

Figure 11-4. The macrotetrolides.

terpreted to mean that initial complex formation is very fast ($> 3.10^8$ M^{-1} sec^{-1}) but that there is a subsequent rearrangement of the complex to a more stable conformation. Additionally, sound absorption studies for the sodium-valinomycin system do indeed suggest the existence of two equilibria

$$\text{Val} + \text{Na}^+ \underset{k_{-1}}{\overset{k_{+1}}{\rightleftharpoons}} [\text{Val Na}^+] \underset{k_{-2}}{\overset{k_{+2}}{\rightleftharpoons}} [\text{Val Na}^+]^1$$

Where

$$k_{+1} \cong 10^8 \text{ M}^{-1} \text{ sec}^{-1}$$

$$k_{-1} \cong 2.5.10^7 \text{ sec}^{-1}$$

$$k_{+2} = 5.2.10^6 \text{ sec}^{-1}$$

$$k_{-2} = 2.3.10^6 \text{ sec}^{-1}$$

It has also been shown by this method that the alkali metal ion with highest affinity for valinomycin—rubidium—not only has the highest association velocity constant but also has the lowest dissociation rate constant.

A comparative study of the rates of dissociation of potassium from various cyclic ionophores has been made using the line broadening of the protons in the ligand. The order of the rates, valinomycin < monactin < nonactin is exactly the opposite of the order of effectiveness for transport in synthetic membranes. It seems therefore, that the dissociation of the metal from the ionophore cannot be the rate limiting step in membrane transport.

(2) The second category of ions may be termed "structure formers" and includes Mg^{2+} and Ca^{2+}. These ions form thermodynamically stronger and kinetically less labile complexes than do the alkali metals.

The kinases (enzymes transferring a phosphate group form adenosine triphosphate to another molecule) invariably require magnesium as a cofactor. One of the functions of the Mg^{2+} ion is partially to neutralize the tetranegative charge on the ATP by binding to the triphosphate. Ca^{2+} is also capable of binding to the phosphate groups and, in fact, the equilibrium constants for the two metals are very similar. It is somewhat surprising, therefore, to find that while Mg^{2+} is a cofactor, Ca^{2+} is often an inhibitor.

The rate constants for reaction of these metals with ADP and ATP have been determined by the temperature jump method (using 8-hydroxyquinoline as an indicator for the metal). The results are shown below

	$k_{+1}(M^{-1}sec^{-1})$	$k_{-1}(sec^{-1})$
$Mg^{2+} + ADP^{3-}$	3.10^6	$2.5.10^3$
$Mg^{2+} + ATP^{4-}$	$1.2.10^7$	$1.2.10^3$
$Ca^{2+} + ADP^{3-}$	$>2.5.10^8$	$>3.5.10^5$
$Ca^{2+} + ATP^{4-}$	$>10^9$	$>2.10^5$

The higher rate constants for the Ca^{2+} complexes mean that they have a shorter existence than the corresponding Mg^{2+} complexes. It may be, therefore, that a long lifetime of the metal-nucleotide complex is required for catalysis (perhaps to enable a slow conformation change to occur in the protein) and this would explain the antagonistic effects of the two metals.

Mg^{2+} not only binds to the phosphate groups in mononucleotides, but also binds to the same groups in nucleic acid, thus causing an electrostatic stabilization. The integrity of the ribosome and the conformation of transfer RNA which is susceptible to aminoacylation by the appropriate ligase are both dependent on a high concentration of magnesium ions. The rate of the conformational change induced in phenylalanine transfer ribonucleic acid by magnesium ions has been studied by the stopped flow technique. EDTA is added to the complex to remove the Mg^{2+} and by following the increase in optical density (at 260 nm) that arises from the unfolding of the tRNA, it has been found that the unfolding is a single first-order process having a rate constant of 40 sec^{-1} (at 42°).

(3) The role of zinc in biological systems is that of a Lewis acid. In enzymes such as carboxypeptidase or alkaline phosphatase the Zn^{2+} withdraws electron density from the bond to be broken thus effecting catalysis. Unfortunately, Zn^{2+} has no readily measurable spectral property of its own, as it does not change its valence state during the course of the catalyzed reaction and enzyme bound zinc exchanges very slowly with zinc ions in solution.

Although the zinc ion is usually tightly bound to the enzyme, it is possible to remove it by exhaustive dialysis of the enzyme against a chelating agent such as EDTA. This has been done for carbonic anhydrase which catalyses the hydration of carbon dioxide and the rate of recombination of the apoenzyme with the zinc has been measured by stopped flow. This can be observed either from the change in the ultraviolet absorption spectrum or from the liberation of protons or from the return of the enzyme's ability to hydrolyze esters such as p-nitrophenylacetate on addition of the zinc. All three methods of detection give the same second-order rate constant of $10^4 M^{-1}s^{-1}$. Several points of interest may be mentioned. It would be reasonable to expect that the binding step itself producing the UV difference spectrum and the identicality of the three rate constants shows that there is no step (such as a conformation change) between the initial binding and the return of enzyme activity. The numerical value of the rate constant is some two or three orders of magnitude lower than those for zinc and small polydentate ligands (measured by relaxation methods). This suggests a possible difference in coordination between the zinc in these complexes (as is also postulated in the entatic state theory) and this suggestion is confirmed by the UV spectrum of the cobaltous substituted enzyme which is that of the tetrahedrally coordinated metal—most small polydentate ligands form a hexa coordinated complex with zinc. Also, the association rate constant is very similar for apo-carboxypeptidases binding to zinc, where the metal is also known to be tetracoordinate.

(4) The metals iron, copper, cobalt and molybdenum are also tightly bound to the enzymes in which they function. Their variable valence results in their use as redox catalysts and this variable valence results in readily measurable spectral changes during catalysis. For example, iron is found in haemoglobin and the cytochromes, copper in the phenol oxidizing enzyme laccase, molybdenum in xanthine oxidase, while cobalt is part of the coenzyme Vitamin B_{12}.

As already mentioned the ferrous protein, haemoglobin, was the first to be studied by fast reaction techniques and the first direct demonstration of a complex between enzyme and substrate was achieved by studying the rate of binding of hydrogen peroxide to catalase. This reaction has an association velocity constant of between 5×10^6 and $10^7 M^{-1} sec^{-1}$ (depending upon the source of the catalases).

Of all the allosteric proteins, haemoglobin has been the one most intensively studied. Since both the protein conformation and the electronic spectrum of the ferrous ion change on oxygenation, it has been possible to use relaxation kinetics to study the conformational change. The equilibrium between oxygen and isolated α or β subunits of haemoglobin (and also myoglobin) shows only one relaxation time on carrying out a temperature jump experiment. Since there is only one relaxation time, it may be assumed that there is only one chemical equilibrium, and hence that the binding of oxygen to single subunits is a simple one-step process. However, a similar experiment on intact haemoglobin reveals three relaxation times. The dependence on oxygen concentration of these three times suggests that the two fastest are associated with oxygen binding. The slowest represents some change in the structure of the protein. The association velocity constants for binding of oxygen are about $5.10^7 M^{-1} sec^{-1}$ and the rate for the structural change is about 11 sec.$^{-1}$. Attempts have been made to fit these data to the two different models for transitions within an allosteric protein. It is a reflection on the sensitivity of the experimental technique required that the data can be made to fit either the concerted model of Monod et al. or the sequential model of Koshland et al. equally well.

It has long been known that the binding of oxygen to haemoglobin is pH dependent —as O_2 binds, a proton is released—the so-called Bohr effect. An electric field jump on the haemoglobin-oxygen system shows a relaxation time considerably shorter than the three mentioned above. This is probably (but by no means certainly) the result of ionization of the so-called Bohr proton (thought to be on the C terminal histidine) induced by the electric field.

CONCLUSION

These examples of bio-inorganic kinetics illustrate two principles: First, a variety of different techniques are available for obtaining the required data, but, secondly, the conclusions which may be drawn from these kinetics experiments are, so far, rather tentative. It is true that kinetics will never prove any reaction mechanism but it is also true that no mechanism can be considered proven until it satisfies all of the kinetic parameters obtained by the methods described in this chapter.

REFERENCES

1. Moore, W. J.: *Physical Chemistry.* Engelwood Cliffs, Prentice-Hall, 1963.
   Frost, A. A. and Pearson, R. G.: *Kinetics and Mechanism,* 2nd ed. New York, Wiley 1961.
   Hague, D. N.: *Fast Reactions.* London, Wiley,

1971. (An introduction to chemical kinetics is provided in the above publications.)

2. Havsteen, B.: Rapid Reactions: Flow and Relaxation Methods. In Goodwin, T. W. (Ed.): *Instrumentation in Biochemistry.* Biochemical Symposium No. 26, 1966.

Kustin, K. (Ed.): *Methods in Enzymology.* Acad Pr, 1969, vol. 16.

Eigen, M.: New looks and outlooks on physical enzymology. *Quart Rev Biophysics, 1:3,* 1968.

Caldin, E. F.: *Fast Reactions in Solution.* Oxford, Blackwell, 1964. (A more detailed description of the apparatus mentioned here will be found in the above publications.)

3. Eigen, M. and Hammes G. G.: Elementary steps in enzyme reactions. *Adv Enzymol, 25:1,* 1963.

4. Diebler, H. et al.: Kinetics and mechanism of reactions of main group metal ions with biological carriers. *Pure and Applied Chemistry, 20:93,* 1969.

5. Schechter, A. N.: Measurement of fast biochemical reactions. *Science, 170:273,* 1970.

6. Chock, P. B.: Relaxation methods in enzymology. *Biochimie, 53:161,* 1971.

7. Gutfreund, H. F.: Transients and relaxation kinetics of enzyme reactions. *Ann Rev Biochim, 40:315,* 1971.

CHAPTER 12

SELECTIVE ION-SENSITIVE ELECTRODES

G. J. Moody and J. D. R. Thomas

Introduction
Basic Principles
Bio-Inorganic Applications
Conclusion

INTRODUCTION

Glass electrodes have been standard items of laboratory equipment for about forty years, but it was only about fifteen years ago that glasses of modified composition were developed for selective response to cations, other than hydrogen. The present widespread interest in selective ion-sensitive electrodes is of even more recent origin and stems from the commercial availability in 1966 of the fluoride solid-state lanthanum fluoride crystal membrane electrode to coincide with the demand for a convenient method of determining trace fluoride in water fluoridation and dental health programs.

Constructional principles of the new-generation ion selective electrodes are similar to the classical glass electrode, namely: Internal reference element—internal reference solution—sensor membrane. However, certain solid-state membrane and coated wire electrodes lack a conventional internal reference system and depend instead on direct wire contact to the back of the membrane.

A convenient practical classification of ion selective electrodes may be based on sensor membrane material types to include glass, homogeneous and heterogeneous solid-state materials, liquid ion-exchangers, carrier complexes, semiconducting organic charge-transfer complexes and ion-radical salts, and electroactive hydrophobized graphite. Coated wire, micro, combination, and enzyme electrodes are subsidiaries to this classification, as are the selective membrane permeator electrode systems for gases.

Detection limits and the main interferences of most commercially available electrodes are summarized in Table 12-I.

Application of the new-generation solution ion-sensitive electrodes in the biological field has been stimulated by the electrolyte composition of biological fluids (Table 12-II). Thus, in extracellular fluids, sodium is the principal cation with chloride as the major anion. In intracellular fluids, potassium is the major cation and phosphate the principal anion—except in erythrocytes where chloride predominates. Of special interest is ionized calcium, because of its importance in various biochemical and physiological processes such as bone formation, nerve conduction, membrane phenomena, muscle contraction and relaxation, cardiac conduction and contraction, blood coagulation, and enzyme activation. The calcium electrode has facilitated the study of calcium ion activity on membrane alkalinization in mitochondria, on mitosis and on the rate of cell division.

On the wider biological scale, interest in

TABLE 12-I

COMMERCIAL SELECTIVE ION-SENSITIVE ELECTRODES

Electrode	Lower Molar Detection Limit[e]	Principal Interferents
H^+ (G)*	10^{-14}	OH^- at pH $\sim$ 13
Na^+ (S)	10^{-6}	H^+
Na^+ (G)	10^{-6}	H^+; Ag^+
K^+ (G)	10^{-6}	H^+; Ag^+; NH_4^+; Li^+; Na^+
K^{+a} (L)[b]	10^{-6}	Cs^+; Rb^+
NH_4^+ (G)[c]	10^{-6}	K^+; Na^+; Rb^+; Li^+
NH_3 (Diffusion)	10^{-6}	Volatile amines
Ca^{2+} (L)	10^{-5}	Zn^{2+}; Fe^{2+}; Pb^{2+}; Cu^{2+}
$Ca^{2+} - Mg^{2+}$ (L)	10^{-5}	Zn^{2+}; Fe^{2+}; Cu^{2+}; Ni^{2+}; Ba^{2+}; Sr^{2+}
Cd^{2+} (S)	10^{-7}	Ag^+; Hg^{2+}; Cu^{2+}
Pb^{2+} (S)[d]	10^{-7}	Ag^+; Hg^{2+}; Cu^{2+}
Ag^+ (S,P)	10^{-7}	Hg^{2+}
Cu^{2+} (S)[d]	10^{-8}	Ag^+; Hg^{2+}
F^- (S)	10^{-6}	None except OH^-
Cl^- (S,L,P)	10^{-5}	S^{2-}; Br^-; I^-; CN^-
Br^- (S,P)	10^{-5}	S^{2-}; I^-; CN^-
I^- (S,P)	10^{-6}	S^{2-}; CN^-
CN^- (S,P)	10^{-6}	S^{2-}; I^-
SCN^- (S)	10^{-5}	$S_2O_3^{2-}$; S^{2-}; I^-; Cl^-; Br^-
NO_3^- (L)	10^{-5}	ClO_4^-; I^-; ClO_3^-; Br^-
BF_4^- (L)	10^{-5}	I^-; ClO_4^-
ClO_4^- (L)	10^{-5}	I^-
S^{2-} (S,P)	10^{-17}	None
SO_2 (Diffusion)	10^{-6}	Cl_2; NO_2

* G = Glass; S = Homogeneous Solid State; L = Liquid Membrane; P = Pungor Heterogeneous Type. Diffusion = a diffusion electrode, which senses gases indirectly, comprises a gas permeable, but ion-impermeable, membrane to separate the analytical solution from the internal solution. The gas species, for example, sulphur dioxide, diffuses through the membrane to effect a change in the activity level of an ion, H^+ in this case, comprising the internal solution, $SO_2 + H_2O \rightleftharpoons H^+ + HSO_3^-$
This activity change is then detected by an internal selective-ion sensitive electrode, in this case the classical glass pH electrode.

[a] Normally based on valinomycin.

[b] The Corning electrode is based on potassium *tetra*(p-chlorophenyl) borate exchanger.

[c] The Beckman electrode comprising a solid organic sensor is 1,000 times more sensitive to NH_4^+ than Na^+.

[d] Orion lead and copper liquid membrane types now withdrawn because of inferior selectivity performances compared with Orion solid-state counterparts.

[e] These figures will vary depending on definition of limit.

applying ion selective electrodes takes in mineralized tissues, dental materials, plant materials and enzyme reactions. The more chemical provinces of foods, fertilizers, soil, water supplies, air and stack gasses will not be covered in this account.

Investigations of biological systems with glass electrodes are normally based on well-established principles and have been critically discussed elsewhere. For example, a careful survey by Khuri[1] stresses the importance of avoiding carbon dioxide loss and describes electrode modifications for use *in situ, in vivo* and *in vitro*. Only the

TABLE 12-II

INORGANIC ELECTROLYTE COMPOSITION ($10^{-3}M$) OF CERTAIN BODY FLUIDS[*]

	Plasma	Cell Fluid (Muscle)	Erythrocyte Fluid
Na^+	153	10	15
K^+	5	148	150
Ca^{2+}	2.5	—	—
Mg^{2+}	1	20	1.5
Cl^-	110	—	74
HCO_3^-	28	8	27
HPO_4^{2-}	1.5	67	—
SO_4^{2-}	0.5	—	—

[*] Used with permission. E. W. Moore, *Glass Electrodes for Hydrogen and Other Cations: Principles and Practice*, G. Eisenman, ed. New York, Marcel Dekker, 1966.

more recent nonglass electrode developments will be discussed and may be facilitated by a preamble on the principles of ion selective electrodes.

BASIC PRINCIPLES[2, 3]

General

A selective ion-sensitive electrode as part of a potentiometric cell assembly contributes ($E_M + E_{M'}$) to the cell emf, E_{cell}, which is made up of various junction potentials:

$$E_{cell} = E_M + E_{M'} + E_R + E_{R'} + E_j \quad (106)$$

Except for E_M which corresponds to the potential arising at the junction of the ion selective electrode membrane with the test or standardizing solution, the remaining contributory potentials are assumed to remain constant. Hence, E_{cell} is related to changes in E_M although some of the junction potentials can be troublesome, especially those of the reference electrode.

Without entering into the mathematical derivations, the ion selective electrode potential, E, with respect to the reference electrode is given by

$$E = \text{Constant} \pm S \log [a_i + K_{ij}(a_j)^{z/y}] \quad (107)$$

This means that the electrode responds selectively to an ion, i, of charge z and activity a_i in the presence of an interfering ion, j, of charge y and activity a_j. K_{ij} is the selectivity coefficient. The constant term incorporates various junction potentials (except for E_M) as well as the standard potential characteristic of the ion selective electrode. The second term on the right hand side of equation (107) takes the positive sign for cations and the negative for anions. When there is Nernstian behavior, S corresponds to $\frac{2.303\ RT}{z\ F}$.

The near-Nernstian calibration slope of selective ion-sensitive electrodes normally extends over a wide p(Ion) range, and the fact that electrodes can reach down to $10^{-5}M$ and beyond is scientifically useful, for even the $10^{-5}M$ level corresponds to 0.19 parts per million fluoride and 0.40 parts per million calcium. As seen in Table 12-I, many electrodes operate beyond $10^{-5}M$, while in some cases metal buffers may be used to exploit a more extended linear calibration range. An instance of such an extension to $10^{-13}M$, or less under appropriate pH conditions, is that for a precipitate-based copper(II) Selectrode. The pCu buffers used here for the pCu 7 to 18 range depend on NTA (nitrilotriacetate) or EDTA (ethylenediaminetetraacetate) as ligands and are based on pH 4.75 acetate and pH *ca* 9 borate buffers. Leaching of membrane sensor material and other factors restrict the universal application of this facility for extending the linear calibration range.

Specificity is rather exceptional for selective ion-sensitive electrodes and a common interference effect is the falloff in linear calibration. This is illustrated in Figure 12-1 (curve A) which also demonstrates a

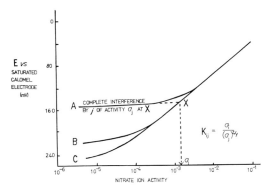

Figure 12-1. Influence of chloride ions on the response of a nitrate-sensitive (Orion 97-07-02 nitrate liquid ion-exchanger) poly(vinyl chloride) electrode showing generalized steps in the calculation of K_{ij} (i = NO_3^-, j = Cl^-). Curve A, calibration in the presence of 5×10^{-1}M chloride ions; Curve B, calibration in the presence of 5×10^{-3} Chloride ions; Curve C, normal calibration.

common method of obtaining K_{ij} whereby the emf response is measured for solutions containing a fixed amount of interferent with varying activities of the primary ion for which the electrode is designed. K_{ij} is calculated from

$$K_{ij} = \frac{a_i}{(a_j)^{z/y}} \quad (108)$$

where a_i and a_j are values appertaining to the intercept of that part of the calibration curve (of near-zero slope) corresponding to complete interference by the interferent, *j*, with that of slope corresponding to more or less unfettered response by the primary ion, *i*. K_{ij} values of less than unity calculated in this way mean that the electrode preferentially responds to the primary ion, *i*.

The selectivity coefficient, K_{ij}, frequently varies with the level of interferent and K_{ij} values are more meaningful if the activity of the interfering ion for which they were determined is also quoted. Then it is a simple matter to calculate by equation

(108) the useful limit of the electrode for ion *i*.

An alternative mixed-solution method of expressing selectivity involves varying interferent activity, a_j, at constant primary ion activity, a_i. This method is normally used for expressing the pH range over which selective ion-sensitive electrodes are useful.

Activity Coefficients in Mixed Electrolyte Systems

Apart from true interference, all ions to various degrees affect the ionic strength of solutions and hence the activity of the primary ion. This can influence the extent to which the ion is sensed by the electrode due to the dependence of the activity coefficient, *f*, on ionic strength, μ, as shown by

$$\log f = -\frac{A z^2 \sqrt{\mu}}{(1 + \sqrt{\mu})} \quad (109)$$

where *z* is the valence and A is a constant. Equation (109) is, of course, only one of many modifications of the Debye-Hückel equation but it is a form that is commonly employed in the selective ion-sensitive electrode field. A more complicated version for higher ionic strengths proposed by Bates, Staples and Robinson[4] is based on the Robinson-Stokes hydration theory by which it is assumed that water bound to ions is no longer part of the bulk solvent and that the Debye-Hückel expression [first term of equation (110)] gives the true activity coefficient (mole fraction scale) of the solvated ions:

$$-\ln f_{\pm} = \frac{A z_+ z_- \sqrt{\mu}}{1 + Bd \sqrt{\mu}} + \frac{h}{\nu} \ln a_w + \ln[1 + 0.018(\nu - h)m] \quad (110)$$

Here A and B are constants of the Debye-Hückel theory, *d* an ionic size parameter, μ

the ionic strength, a_w the water activity, v the number of ions produced by a mole of the electrolyte and h the hydration number, that is, the number of moles of water bound to one mole of electrolyte irrespective of distribution among cations and anions.

Conventions for calculating activity coefficients lead to problems of internal consistency in that multiple pathways exist to the activity coefficients of single ions. However, this is not very significant at an ionic strength of 0.1M. Despite the recommendation that cation-responsive electrodes be standardized in solutions of the corresponding completely dissociated chloride salts and anion-responsive electrodes in solutions of completely dissociated sodium (potassium for fluoride) salts of the anions, completely dissociated salts other than those of chloride and sodium, respectively, do not usually affect calibration. But this is a feature that merits further study.

In actual measurement situations, mixed electrolyte systems are frequently the rule, especially with biological systems. The variation of activity of one electrolyte in the presence of others is thus of prime importance. The interaction between ions, mainly between those of opposite charge can lead to a variation of activity coefficients with composition of mixed electrolyte solutions and thus contradict the Debye-Hückel formula prediction that log $f_\pm$ is proportional to $\sqrt{\mu}$. In dilute solutions the conflict is minimal, but at higher concentrations, $> 0.1M$, the influence of ion interactions can, among other things, alter the significance of the ionic size parameter, d, in equation (110).

Studies with an Orion 92-20 calcium ion electrode on mixed calcium chloride-calcium nitrate solutions and with nitrates and chlorides of sodium and potassium have

shown that within experimental error, the mean activity coefficients of calcium chloride in mixed solutions up to an ionic strength of 0.3M are equal to those which would hold for pure calcium chloride of the same ionic strength.

Mean sodium chloride activity coefficients have been determined with a selective sodium-sensitive electrode/silver-silver chloride electrode system in mixed sodium chloride-calcium chloride solutions within the range of sodium chloride and calcium chloride levels (0.05M to 0.5M) encountered in extracellular fluids. These show that at constant ionic strength, log f_{NaCl} varies linearly with the ionic strength of calcium chloride in the mixture in accordance with Harned's rule:

$$\log f_{NaCl\ (mixture)} = \log f_{NaCl\ (pure)} - a_{12}\mu_{CaCl_2} \quad (111)$$

The activity coefficients of sodium chloride in solutions containing $5 \times 10^{-3}M$ calcium chloride (0.5M total ionic strength) were shown to be close to those of corresponding pure sodium chloride solutions, that is, the effect of calcium chloride on sodium chloride activity in extracellular fluids would be negligible. The reverse is not the case, for calculations involving the alternative version of Harned's rule:

$$\log f_{CaCl_2\ (mixture)} = \log f_{CaCl_2\ (pure)} - a_{21}\mu_{NaCl} \quad (112)$$

show that the activity coefficients of calcium chloride in mixtures typical of extracellular fluids were considerably less than those of corresponding pure calcium chloride solutions.

Comparatively low levels of other ionic components in serum, like potassium and magnesium which comprise less than 6 percent of the total ionic strength of serum, have a negligible effect on the activity coefficient of calcium ion. Of course, the activity coefficients of these ions can be expected to be affected by the high sodium level as with calcium.

Interference Elimination and Ionic Strength Adjustment

The application of selective ion-sensitive electrodes, especially with biological systems can become complicated owing to the presence of colloidal electrolytes, the loss of carbon dioxide and the complexation equilibria by buffers and other adjusters.

Interference can frequently be overcome. For example, a rather involved buffer system for determining soil nitrates consists of 0.01M aluminium sulphate, 0.01M silver sulphate, 0.02M boric acid, 0.02M sulphamic acid and sufficient 0.10M sulphuric acid to give a pH of 3.0. The low pH maintained the equilibrium bicarbonate low and the water-extractable organic acids undissociated; silver precipitated chloride; aluminium complexed anions of organic acids, and sulphamic acid destroyed nitrite quantitatively.

An assessment of ionic strength is obviously desirable, but because of the powered term in $\mu = \frac{1}{2} \sum cz^2$, conductance monitoring may not always be satisfactory, unless, of course, the individual ionic contributions (concentration, c, and valence, z) are known from previous analyses. Such information is useful for preparing calibration standards of the same ionic strength range as the sample under examination and is the basis of synthetic sea water samples as calibrants. Use of calcium chloride standards in 0.150M sodium chloride is a similar ploy in calibrating calcium ions electrodes for serum/plasma calcium ion determinations within the $0.5 \times 10^{-3}M$ to $2 \times 10^{-3}M$ range.

However, situations where the ionic strength can be mimicked are not the rule; frequently there is considerable overall variation in composition between samples. In these cases, it is helpful to add an "ionic strength adjuster" to both sample and calibrating solutions in order to achieve a common ionic strength level to release the ion of interest from complexes and to buffer samples to regions of minimal pH interference.

Although virtually any ionic substance may serve as an ionic strength adjuster, it must not complex with the primary ion, and the selectivity coefficient, K_{ij}, must be negligible; j in this case being an ion strength adjuster constituent. Several systems have been described and an important one is TISAB (Total Ionic Strength Adjustment Buffer) used in fluoride determinations. It has a pH of about 5 and is composed of 1M sodium chloride, 0.25M acetic acid, 0.75M sodium acetate and 0.001M sodium citrate. Adjusters have been periodically criticized, for example, TISAB for its carboxylate ion content (acetate and citrate) in the potentiometric titration of fluoride with lanthanum nitrate. Modification has been made and the citrate (acetate is retained) has been replaced by DCTA (1,2-diaminocyclohexane,N,N,N′,N′-tetraacetate), although it has been shown that the lanthanum fluoride electrode is only minimally affected by these carboxylate ions.

Known Addition and Gran's Plot Methods

The above methods for using selective ion-sensitive electrodes are based on simply measuring the electrode potential, E, with respect to that of a reference electrode and relating to activity/concentration by a calibration graph. Use in potentiometric titrations depends on the change in E with the usual added titrant. However, there are other techniques by which the electrodes can be used; these include differential and null-point potentiometry, but of greater interest in the present context are known addition/subtraction and Gran's plot methods.

In the *standard addition to sample method*, the emf, E_o, between the selective ion-sensitive electrode and suitable reference electrode is measured for the sample solution of volume, V_o, and total molar concentration, C_o, of the sought species:

$$E_o = Constant \pm S \log x_o f_o C_o \quad (113)$$

where S is the calibration slope, f_o the activity coefficient and x_o the fraction of uncomplexed ions. A new emf, E_1, is measured after adding a small volume, V_1, of a standard solution (concentration C_s) of ions of the species sought, where $C_s \sim 100 C_o$:

$$E_1 = Constant \pm S \log x_1 f_1 \frac{(V_o C_o + V_1 C_s)}{(V_o + V_1)}$$
$$(114)$$

f_1 and x_1 correspond to the new activity coefficient and fraction of free ions, respectively. An essential assumption is that $x_o \sim x_1$ and $f_o \sim f_1$. Hence, the difference between equations (113) and (114) simplifies to

$$E_1 - E_o = \Delta E_1 = \pm S \log \frac{(V_o C_o + V_1 C_s)}{C_o (V_o + V_1)}$$
$$(115)$$

and C_o can be resolved

$$C_o = \frac{C_s}{10^{\pm \Delta E_1 / S} (1 + \frac{V_o}{V_1}) - \frac{V_o}{V_1}}$$
$$(116)$$

or for no allowance for volume change when V_1 is small in relation to V_o

$$C_o = \frac{C_s V_1}{(10^{\pm \Delta E_1 / S} - 1) V_o} \quad (117)$$

For *analate addition* purposes, that is,

sample addition to standard, the corresponding version of equation (116) is

$$C_o = C_s [10^{\pm \Delta E_1 / S} \cdot (1 + \frac{V_s}{V_{o1}} - \frac{V_s}{V_{o1}}]$$
$$(118)$$

where V_{o1} is the volume of sample solution added.

The *known-subtraction method* depends on lowering the level of uncomplexed ions by adding a complexing agent when the structure of the equations depends on the nature of complexation (or precipitation), that is, 1:1 or 1:n.

An interesting variation of the standard addition to sample method is a modification of Gran's method[5] presenting potentiometric titration data in linear form by using a semiantilog plot. The principle is apparent in equation (114) (taking $x_1 = 1$) and assuming constant activity coefficient, now taken into the Constant term

$$(V_o + V_1) \, 10^{\pm \Delta E_1 / S} =$$
$$(V_o C_o + V_1 C_s) \, 10^{\pm Constant / S}$$
$$(119)$$

where V_1 is the volume of standard solution added. A plot of the left hand side of equation (119) *versus* V_1 gives a straight line which intercepts the abscissa for a V_e value and where $C_o V_o = -C_s V_e$ so that the concentration of the ion under test, C_o, can be easily calculated.

Computation of the left hand side of equation (119) is avoided in the semiantilog Gran's Plot Paper supplied by Orion Research, Inc. and which has been corrected for limited volume changes, that is, for $(V_o + V_1)$. Hence, all that is required is to plot the emf response on the ordinate against the volume of standard addition on the abscissa when the intercept, V_e, is easily discernible (Figure 12-2). Apart from restriction on volume changes, the Orion

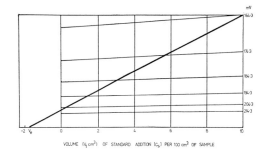

Figure 12-2. Illustration of volume-corrected (10 cm³ addition of standard to 100 cm³ sample) semiantilog Gran's plot for cell with selective ion-sensitive electrode exhibiting univalent Nernstian response pattern.

Gran's Plot Paper is dependent on true Nernstian response by the electrode.

BIO-INORGANIC APPLICATIONS

Serum, Plasma and Whole Blood Calcium Ion Studies

Serum calcium exists in three forms, namely, nondiffusible protein-bound (30 to 55% of total), complexed but diffusible (5 to 15% of total) and ionized. The ionized form seems to play an important part in physiological processes and its measurement can be a very useful diagnostic aid as in hyperparathyroidism where, in most cases, the calcium ion activity is raised even when the total calcium may be normal. Multiple myeloma and renal failure may also be diagnosed. It is not surprising, therefore, that much interest centers on the calcium ion selective electrode. In all its applications it is second only to the fluoride electrode, and its acceptability in serum/plasma calcium ion studies is supported by the agreement between "normal" ionic calcium obtained with the electrode (1.00 to 1.26 mM) and the Maclean and Hastings frog heart method (1.06 to 1.31 mM). The chemical method based on the color reac-

tion of calcium ions with the dye murexide is too pH-dependent with the added complication of albumen-murexide complexation.

When using calcium ion selective electrodes, technique is of prime importance, often accounting for the dismay experienced by some investigators on its utility. There have also been complaints concerning expensive and cumbersome membrane replacement. Such criticisms should not belittle the well-established successes of the electrode; in any case, some criticisms can be minimized by trapping the liquid ion exchanger sensors in PVC or other inert matrices.

The need for careful sampling and subsequent processing has been highlighted by Moore's studies on technique, selectivity calibration, freezing, pH and other factors.[6]

Calcium ion levels are usually quoted in terms of concentration (Table 12-III) with calibration of the electrode being made in standard solutions of calcium chloride in 0.15M sodium chloride. This largely avoids the indeterminate nature of activity, and providing the pH is above 5.5, there is no interference from hydrogen ions. In this respect, the pH of venous blood is of the order of 7.3 to 7.4, but the buffering effect of carbon dioxide can cause difficulties—a loss apparently causing a decrease and again an increase in ionic calcium respectively. This also emphasizes caution in the use of buffers unless their effects on calcium ionic activity is known.

The pH effects on heparinized whole blood resemble those in corresponding sera with a fall in pH leading to an increase in ionic calcium. However, the ionic calcium level is less in whole blood than in serum, and even at the normal venous pH of 7.3 to 7.4, the difference is significant, being 0.045 mM at pH 7.32. This lowering is account-

TABLE 12-III

TOTAL AND IONIC CALCIUM IN HUMAN SERUM, PLASMA AND BLOOD

(Ionic levels determined with calcium ion selective electrode)

Total Calcium(mM) Mean	Range (± 2SD)	Ionized Ca(mM) Mean	Range (± 2SD)	Ionized/Total × 100	No. of Subjects for Ionized Ca	Reference
Serum						
2.41	0.56	1.18	0.31	49.2	17	a
2.45	0.24	0.99	0.08	40.5	23	b
2.39	0.15	1.24	0.09	50.8	21	c
2.41	0.10	1.10	0.06	45.6	22	d
2.48	0.29	1.14	0.13	45.8	52	e
3.04	—	1.39	—	45.7	2	f
2.64	0.42	1.21	0.25	45.6	16	g
2.29	0.24	1.22	0.09	53.4	{231 for total / 397 for ionized}	h
2.44	0.08	1.25	0.03	51.2	20	i
Plasma						
2.43	0.11	1.00	0.06	41.2	22	d
2.65	0.41	1.16	0.22	43.8	50	g
2.71	0.06	1.31	0.02	48.2	8	j
Whole Blood						
—	—	1.08 (Estimate)		—	6	e
2.72	0.03	1.34	0.02	49.2	8	j
2.42	0.05	1.11	0.02	45.7	13	j adults
2.45	0.04	1.09	0.02	44.4	7	j children
2.60	0.04	1.24	0.02	47.7	23	j newborn

[a] Raman, A.: *Biochem Med*, 3:369, 1970.

[b] Hattner, R. S., Johnson, J. W., Bernstein, D. S., Wachman, A. and Brackman, J.: *Clin Chim Acta*, 28:67, 1970.

[c] Robertson, W. G. and Peacock, M.: *Clin Chim Acta*, 20:315, 1968.

[d] Sachs, Ch., Bourdeau, A.-M., and Balsan, S.: *Ann Biol Clin*, 27:487, 1969.

[e] Moore, E. W.: *J Clin Invest*, 49:318, 1970.

[f] Moody, G. J., Oke, R. B. and Thomas, J. D. R.: *Unpublished results*, 1970.

[g] Oreskes, I., and Hirsch, C., Douglas, K. S. and Kupfer, S.: *Clin Chim Acta*, 21:303, 1968.

[h] Li, T-K. and Piechocki, J. T.: *Clin Chem*, 17:411, 1971.

[i] Fuchs, C., and Paschen, K.: *Deut Med Wochenschraft*, 97:23, 1972.

[j] Radde, I. C., Höffken, B., Parkinson, D. K., Sheepers, J. and Luckham, A.: *Clin Chem*, 71:1002, 1971.

able by a calcium-heparin complex. Although small amounts of heparin (up to 30 units per cm³ of blood) are claimed to have no significant effect on calcium ion activity, indiscriminate use of heparin should be avoided when collecting whole blood samples.

Temperature-induced variations in protein binding affecting ionic calcium levels reinforces criticisms over the lack of thermostatted flow-through calcium ion selective electrodes. However, there is an abundance of electrode and electrolyte data at room temperature to support the utility of determining ionic calcium at 25°C rather than at 37°C. In this respect, ionic calcium can be reliably determined in previously frozen serum.

Apart from being an empirical diagnostic aid in hyperparathyroidism, cirrhosis and hypercalcemia of cancer, calcium ion selective electrodes represent a major advance in studying the physical chemistry and physiology of calcium metabolism. They have been an important aid in resolving the components of total serum calcium; thus total calcium may be determined in serum ($\sim$ 2.7 mM) and ultrafiltrates by the EDTA method and ionic calcium with a calcium ion selective electrode (a flow through electrode has been used). This allows resolution of protein-bound (nondiffusible) calcium ($\sim$ 1.2 mM) diffusible complexed calcium ($\sim$ 0.35 mM) and ionic calcium ($\sim$ 1.15 mM). Such studies have shown an apparent lack of correlation between ionic and total calcium, with the ionized calcium for different subjects lying within the narrow range of 0.94 to 1.33 mM, and only a slight variation for any individual over several months. Ultrafiltration studies coupled with the calcium ion selective electrode thus show that variation in total serum calcium is almost entirely accountable by variation in protein-bound calcium and have led Moore[6] to suggest hormonal regulation of ionic serum calcium levels.

Other Ion Studies in Serum, Plasma and Blood

The possibility of ion-binding has been explored for sodium and potassium. For sodium, glass electrode studies on serum indicate a correlation with flame photometer sodium levels amounting to no sodium ion binding. This is based on the assumption of 96 percent by volume for serum water content giving a mean f_{Na^+} of 0.747 compared with 0.750 for f_{NaCl} in pure solution at a sodium concentration of 1.40 $\times$ 10^{-1}M.

Earlier glass electrode studies for potassium have had conflicting results but serum is an unfavorable fluid for accurate potassium glass electrode measurements since the sodium : potassium ratio is of the order 35 to 40/1. To counter this, there is, at best, only a selectivity of about 10 for potassium ions over sodium ions in glass electrodes, so that a separate sodium ion concentration measurement is needed to correct the reading of the potassium glass ion electrode. This constraint is not inherent in the newer valinomycin-based potassium electrode with its potassium over sodium ion selectivity of the order of 5,000; furthermore, the selectivity of potassium ions over protons of about 18,000 renders buffering unnecessary. Liquid ion-exchanger potassium ion-sensitive electrodes are also superior to glass in terms of sodium and hydrogen ion selectivity. Independent studies with these and the valinomycin electrodes agree with potassium levels ($\sim$ 4.5 mM) determined by flame photometry and atomic absorption spectroscopy and suggest minimal ion-binding.

Chloride is the most important natural anion in the blood and chloride-responsive electrodes may be used for determinations in serum without removal of proteins.

With the recent interest in fluoridation of water supplies, serum fluoride has become rather important. One difficulty, however, is access to low fluoride serum for preparing standards. In these circumstances, human serum from a normal young adult who has not drunk fluoridated water for twenty-four hours is acceptable when the inorganic fluoride will be 0.5 μM or less; fluoride-free water may also be used with little loss of accuracy.

That some fluoride is bound in biological fluids is indicated by the higher levels in human serum ashed diffusates compared with unashed diffusates. Also, nonionic fluorine in plasma has been established, while administration of fluorine-based anesthetics also leads to nonionic fluorine. In general however, the broad agreement of fluorine levels determined by the fluoride electrode and those obtained by fluorescence of a Morin thorium complex confirms the convenience of the fluoride electrode—provided the fluoride level is not below the 10^{-6}M limit of the electrode. Gran's plot and known addition procedures may be used for isolated samples, but further careful studies are required before such procedures can be fully accepted.

Similar to fluoride monitoring of toxicity after methoxyflurane anaesthesia, a knowledge of bromide ion levels is sometimes required in toxicology and in treatment-control with bromine-based sedatives. Towards this end the use of a bromide ion-selective electrode depends on two blood samples, one before and one after bromide ion administration. A quadratic expression describing the relation between millivolt readings (E) and bromide ion concentrations is used:

$$Br_i^- = a_0 + a_1E_i + a_2E_i^2 \quad (120)$$

where a_0, a_1 and a_2 are constants determined by additions (i) of bromide ions to the first sample.

Nonblood Biological Fluids

Cerebrospinal fluid, gastric juice, urine, saliva and sweat are the principal nonblood biological fluids suitable for examination by selective ion-sensitive electrodes. Urine presents problems since conditions of pH, sodium levels and ionic strength vary considerably. Consequently, each urine sample needs to be first analyzed for sodium, potassium and ammonium and standards prepared to mimic the urine in ionic strength and sodium concentration. About 50 per-

cent of the calcium in urine is ionized, the remainder being soluble citrate, phosphate, sulphate and oxalate calcium complexes. Unlike serum calcium, urinary ionized calcium ($\sim 50 \pm 8$ to 9%) seems to show a marked dependence on the total calcium.

Problems of pH in gastric juice may be overcome by bringing the samples to pH 6.2 to 7.2 with sodium hydroxide prior to ionized calcium analyses. In this way, the calcium selective ion-sensitive electrode used by Moore and Makhlouf[7] has shown that the appearance of calcium in gastric juice involves both "secretory" and "non-parietal" components. Furthermore, there is a close correlation between ionized and total calcium with the calculated regression line for gastrin being

$$[Ca^{2+}] = 0.02 + 0.80 \, [Ca] \quad (121)$$

This indicates that calcium is ionized mostly in pure secretory fluid and suggests that if calcium enters the juice physically bound to pepsin, (P), it is displaced from the protein molecule by hydrogen in very acid juice:

$$CaP + H^+ \rightleftharpoons Ca^{2+} + HP \quad (122)$$

Cerebrospinal fluid and saliva present no problems in terms of pH, and ion determinations on saliva could benefit from a wider application of combination (sensor and reference) electrodes. Such a device can be developed even for calcium with the availability of sensor membranes wherein exchanger is trapped in a PVC matrix. To date, however, it is the solid-state silver chloride combination electrode that has won its spurs in cystic fibrosis studies. The disease can be readily diagnosed[8] from the elevated chloride ion levels in patients' sweat (~ 100 mM) compared with a "normal" ~ 25 mM. Reasonably accurate monitoring of newborns, infants and children may be made with a sodium glass electrode, but the more recent direct-reading skin chloride electrode/reference electrode combination assembly is a more convenient way of detecting the one case per 2,000 live births. The *in situ* measurements require less sweat than conventional methods and the assembly is available complete with ion meter and sweat-stimulating attachment. The 80 percent detection rate is highly competitive and, as with alternative methods, the borderline cases can only be properly identified by careful clinical and repeated laboratory observations. It is the individuals who fail to sweat sufficiently that are difficult, and here sweat stimulation by pilocarpine iontophoresis may be more convenient than the thermal method as no preparatory steps are required and the equipment is ready for immediate use. The danger from skin burns is eliminated by a current-limited (< 1.5 mA) battery power source.

Fluoride level in biological fluids is normally low, but industrial exposure to fluorine materials, medical treatment such as methoxyflurane anaesthesia, fluoride treatment of bone diseases, haemodialysis in areas with fluoridated water and dental health and water fluoridation programs have provided the impetus for using the lanthanum fluoride membrane electrode for screening purposes, especially through urinary fluorine in the 10^{-5} to 10^{-3}M range. The usual method is based on direct fluoride determination following pH and ionic strength adjustment. This is justified by recovery values and by comparison with microdiffusion methods. There are discrepancies however, in certain studies between fluoride levels measured directly by fluoride electrode and those by microdiffusion from perchloric acid and the matter merits further study.

A further interesting application of the fluoride electrode is the determination of 5-trifluoromethyl-2'-deoxyuridine and its metabolite, 5-trifluoromethyl-uracil in urine rendered possible by the differential hydrolysis rates of the carbon-fluorine bonds in the two compounds.

Intracellular Fluids

Intracellular and related fluids are difficult to study because of their relatively minute volumes and it is here that microcapillary electrodes will play a special role. A microcapillary version of the potassium liquid ion-exchanger electrode has been used *in situ* for following potassium ion gradients along the proximal convoluted tubule of a rat kidney. The mean tubular fluid to plasma potassium ion concentration ratio falls significantly from 0.89 for the first convolution to 0.81 for the last convolution of the proximal tubule.

Ionic activities inside living cells attract the fascination of biologists and ion selective microelectrodes can cope with such difficult situations. The magnitude of the problem can be appreciated from the $\sim 10^{-5}$ μl volume of a frog ventricle cell (0.01 mm diameter). Glass micro capillaries with 0.5 μm diameter may be pulled from Pyrex® glass capillary, and after hydrophobizing the terminal 200 μm with organic silicone compound, sensing liquid ion exchanger is introduced into the tip of the capillary pipette and the remainder filled with aqueous electrolyte and inner reference system. Such electrodes, successful for measuring potassium and chloride (but not calcium), have been used with *Aplysia* neurons to prove their long-term stability (> 8 hours inside cell often with more than one cell penetration). However, the small-sized tip and low conductivity of the ion exchangers lead to high resistance (10^9 to 10^{10} ohms) thus demanding a high input impedance voltage measuring device ($> 10^{13}$ ohms).

Milk

Selective ion-sensitive electrodes have a possible role in monitoring milk both for its calcium content and as a possible vehicle for health fluoridation programs. The electrode method appears to give higher ionic calcium levels ($\sim$ 2.7 mM) than the more conventional ion-exchange resin method ($\sim$ 2.5 mM) possibly because of the smaller pH changes in the former case. However, in passing from raw milk to sterilized milk ($\sim$ 2.25 mM) and pasteurized milk ($\sim$ 2.05 mM), there is a fall in ionic calcium level.

Protein-bound fluorine may be released by forming insoluble Amido black-milk protein complexes at pH 2.0 and the fluoride then determined following filtrate adjustment of pH with a citrate buffer. Of course, milk as a vehicle for health fluoridation means that the fluoride levels are only of the order of 1 ppm and where the fluoride levels are considerably lower ($\sim$ 0.01 ppm) some measurements may be necessary in the nonlinear range of the electrode. Concentration and microdiffusion methods can, however, be used to bring fluoride levels into the linear range of the electrode.

Mineralized Tissues and Dental Materials

The tedious diffusion methods of fluoride determinations can now be avoided with the lanthanum fluoride electrode, thereby simplifying analytical methods associated with dental health interests. Thus, 1-20 mg sample of enamel, dentine, bone or calcium phosphate may be placed in a disposable plastic tube, dissolved in 1.00 cm³ of 0.50M perchloric acid, the pH adjusted to about 5.6, calcium complexed with 4.00 cm³ of 0.50M sodium citrate, and the fluoride determined with a calibrated fluoride elec-

trode. This kind of procedure with smaller samples may be used in biopsy procedures for fluoride levels in the outer 1 to 2 μm layers of enamels where mean concentrations of fluoride vary between 400 and 2,500 ppm in the anterior teeth of different persons. The tooth layer distribution of fluoride may also be studied, for example, in a study of thirty-three children the first 0.5 μm layer of teeth had 2,000 to 5,800 ppm fluoride with 1,500 to 4,800 ppm in the second 0.5 μm layer.

As long as care is taken over interference patterns, sequential procedures may be used for bromide, followed by chloride (at pH near to 2.5) and fluoride at pH 5.4 with 2M perchloric acid/2M citric acid for solid dissolution and 2M trisodium citrate for subsequent pH adjustment. Specially designed dissolution and diffusion chambers are easily made to cope with submicrogram and even nanogram quantities of fluoride in small samples such as dental plaque.

Fluoride figures obtained directly with the fluoride ion electrode on toothpastes containing either sodium fluoride or tin(II) fluoride match those obtained by a diffusion/thorium nitrate titration procedure. Because of the existence of tin(II) fluoride complexes, ($\beta_3 \simeq 10^9$) separate calibration curves are necessary for sodium and tin(II) fluorides; also, possible interference by base materials should not be ruled out.

Plant and General Biological Materials

Except for nitrate levels in plant nutrition studies, most interest concerning ion-selective electrodes in the plant and general biological field centers on fluoride, mainly on environmental grounds. There is nothing unusual here in procedures with normal fusion and diffusion/dissolution prior to application of the fluoride electrode and simply a measurement on an aqueous extract for nitrate.

There is the matter of possible interference of nitrate determination by chloride, bicarbonate and nitrite. Chloride may be removed with a silver form ion-exchange resin, but a more comprehensive interference eliminator is a pH 3 buffer[5] already described above and containing silver sulphate along with aluminium sulphate, boric acid and sulphamic acid. A limitation occurs in the detection limit for present-day nitrate electrodes which lies between 10^{-4} and 10^{-5}M.

Complexes of Biological Significance

An interesting application of selective ion-sensitive electrodes is seen in the determination of formation constants of copper(II) complexes with glycine, glutamic acid and tris(hydroxymethyl)aminomethane. Another is that saturated solutions of calcium carbonate under atmospheric carbon dioxide have been shown to contain 80 percent Ca^{2+} with the other 20 percent dissolved calcium being made up of $CaCO_3$ and $CaHCO_3^+$ species.

Of the more obvious significance in the dental and mineralized tissue field is the complexation of calcium ions with various inorganic phosphates with the results for tetrapolyphosphate, trimetaphosphate and tetrametaphosphate obtained by calcium electrode studies matching those of pH titrations.

The difference, ΔE, in emf between cells of the type

as given by

$$\Delta E = \frac{3RT}{2F} \ln \frac{f_\pm^o}{f_\pm}$$

where $f_\pm^o$ and $f_\pm$ are the mean ionic activity coefficients of calcium chloride in the absence and presence of amino acid, respectively, may be used to calculate salting-in parameters for amino acids by calcium chloride. Such parameters calculated by Briggs, Lilley, Rutherford and Woodhead[10] for glycine, β-alanine, γ-amino butyric acid, ϵ-amino caproic acid and glycyclglycine range from 0.759 to 1.663 mol^{-1} kg.

Among the most exciting studies of biologically-based complexes must be those of the association constants for adenosine triphosphate complexes,[11] for example, $KATP^{3-}$ and $NaATP^{3-}$. The constants may be obtained from measurements with valinomycin potassium-sensitive and glass sodium-sensitive electrodes, respectively. Values obtained (2.2×10^2 liter mol^{-1} for $KATP^{3-}$ and 2.3×10^2 liter mol^{-1} for $NaATP^{3-}$) are substantially greater than those previously obtained (*ca* 10 to 15 liter mol^{-1} for $MATP^{3-}$) from indirect measurements. Calcium electrode studies of the calcium ATP system have shown the existence of Ca_2ATP ($K_2 = 1.10 \times 10^3$ liter mol^{-1}) in addition to the previously established $CaATP^{2-}$ complex, with the "electrode" value of K_1 being 2.34×10^6 liter mol^{-1} compared with 2.0×10^3 by pH filtration and 3.2×10^4 by spectrophotometry.

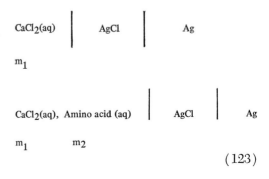

and

(123)

Enzymes and Their Substrates

Selective ion-sensitive electrodes will have an important role in enzyme systems. The principle is the "sandwiching" of an enzyme- or substrate-soaked matrix between the sample solution and a sensor electrode which detects one of the enzymatic degradation products. For example, an ammonium ion-sensitive glass electrode coated with an enzyme-gel (urease-acrylamide) held in place with a nylon net, or urea protected by cellophane, gives electrodes for measuring urea and urease activity respectively. In these cases, the enzymatic reaction

$$CO(NH_2)_2 + 4H_2O \xrightarrow{\text{urease}}$$
$$2NH_4OH + H_2CO_3$$

(124)

produces ammonium ions for detection by the glass electrode. In determining urea in blood and urine, interferences from sodium potassium and other species have to be minimized by adding Dowex 50W-X2 ion exchanger to the sample. Urea levels so obtained are within 2 to 3 percent of those by standard spectrophotometric methods.

Similarly, other nitrogeneous compounds, such as glutamine, asparagine, and D- and L-amino acids, in the presence of their corresponding enzymes—glutaminase, asparaginase and amino acid oxidase—give ammonium ions which may be measured by the glass electrode.

Enzyme electrodes have also been based on nonglass membranes, for example, the polycrystalline sensing element of an Orion 94-06 cyanide electrode covered with a thin polyacrylamide gel senses amygladin since cyanide ions are produced in stoichiometric proportion to the amygladin in the sample solution:

The substrate selective Corning Model No. 476200 acetylcholine ion-sensitive electrode may be used to determine acetylcholinesterase activity. The selectivity towards choline, acetyl-, propionyl-, acetyl β-methyl-, butyryl-, valeryl-, and benzoylcholine may also be used in a kinetic method for characterizing cholinesterases. This is possible because both cholinesterase and acetylcholinesterase hydrolyse acetylcholine, while only cholinesterase will hydrolyse butyrylcholine and only acetylcholinesterase will hydrolyse acetyl β-methylcholine. The different substrates thus permit characterization, but it is necessary to reequilibrate the acetylcholine electrode when the substrate is changed, or to use a separate equilibrated electrode for each substrate.

Selective ion-sensitive electrodes have a potentially wider role in the enzyme field, for example, the fluoride electrode may be used to monitor fluoride ion released by the inactivation of chymotrypsin by diphenylcarbamyl fluoride. The method may be adapted for assaying chymotrypsin with possible $\pm$ 3 percent precision in the 2×10^{-5} to 5×10^{-6}M concentration range.

CONCLUSION

Selective ion-sensitive electrodes have given a fresh impetus to the study of the physico-chemical characteristics of biologically-based systems. However, considerable research effort is still required for a complete understanding of the underlying mechanisms of these electrodes and of the full implications of activities. It is only with continued experimentation that these objectives can be achieved so that this new dimension can be fully utilized in routine

$$C_6H_5\overset{\displaystyle |}{\underset{\displaystyle OC_{12}H_{21}O_{10}}{C}}HCN \xrightarrow[\text{enzyme}]{H_2O} 2C_6H_{12}O_6 + C_6H_5CHO + HCN$$

(125)

monitoring and research in biological systems.

REFERENCES

1. Khuri, R. N.: In Durst, R. A. (Ed.): *Ion-Selective Electrodes*. Special Publication 314, National Bureau of Standards, Washington, D.C., 1969. (This book includes excellent contributions by eminent authors.)

2. Moody, G. J. and Thomas, J. D. R.: *Selective Ion-Sensitive Electrodes*. Watford, England, Merrow Publishing, 1971. (Covers principles and practical aspects.)

3. Moody, G. J. and Thomas, J. D. R.: Selective Ion-Sensitive Electrodes. In Bark, L. S. (Ed.): *Selected Annual Reviews of the Analytical Sciences*. Saville Row, London, The Society for Analytical Chemistry, 1973, vol. 3, p.59. (A comprehensive review with detailed references.)

4. Bates, R. G., Staples, B. R. and Robinson, R. A.: *Analyt Chem, 42:*867, 1970.

5. Gran, G.: *Analyst, 77:*661, 1952.

6. Moore, E. W.: *J Clin Investig, 49:*318, 1970.

7. Moore, E. W. and Makhlouf, G. M.: *Gastroenterol, 55:*465, 1968.

8. Kopito, L. and Shwachman, H.: *Pediatrics, 43:*794, 1969.

9. Larsen, M. J., Kold, M. and von der Fehr, F. R.: *Caries Res, 6:*193, 1972.

10. Briggs, C. C., Lilley, T. H., Rutherford, J. and Woodhead, S.: Paper No. 36 presented at the IUPAC Sponsored International Symposium on Selective Ion-Sensitive Electrodes, UWIST Cardiff, April, 1973.

11. Mohan, M. S. and Rechnitz, G. A.: *J Am Chem Soc, 92:*5839, 1970; *94:*1714, 1972.

CHAPTER 13

CRYSTALLOGRAPHIC TECHNIQUES

C. Glidewell

INTRODUCTION

THIS CHAPTER will attempt to show how crystallographic techniques may be of value in the investigation of bio-inorganic problems, pointing out some of the uses and limitations. No description is given of the successive stages in a crystallographic structure determination, namely data collection, phase determination, and structure solution and refinement, as these are all fully reviewed elsewhere.[1, 2]

In general, the X-ray investigation will disclose the atomic arrangement within the crystal employed. The arrangement found probably is, but may not always be, the same arrangement as in all other crystals of the substance; somewhat different arrangements may occur in different crystalline modifications. It may also be similar to the arrangement in solution. Even if the structures in solution and in the solid are thought to be different, the solid structure may of course provide quite a useful guide to the possible situation in solution.

If the material studied is not crystalline then it is probably impossible to derive an atomic arrangement. For amorphous solids or fluids, the best that can be deduced is a set of vectors up to about 7Å which allows identification of nearest and next-nearest neighbors, but reveals nothing about the long-range structure.

While these may appear to be rather severe restrictions, it should be recalled that the majority of chemically stable molecular species undergo, at most, only conformational changes on change of phase from solid to solution, and that no technique exists at present for the determination of the structure of a complex molecule tumbling randomly in solution. With careful spectroscopic comparisons between phases, useful deductions can often be made about structures in solution from structures in solids.

The topics discussed have been divided into a number of groups, which are arranged in increasing order of molecular complexity. An attempt has been made to keep the number of examples in each group fairly small, lest the trees obscure the forest.

ABSOLUTE CONFIGURATION AND CONFORMATION OF METAL COMPLEXES

A most striking phenomenon in biology is the occurrence of one enantiomer of an optically active compound to the almost complete exclusion of the other. Thus, almost all naturally occurring a-amino acids belong to the L-series, having S configurations at the α-carbon, while most naturally occurring monosaccharides belong to the D-series, having R configurations at the carbon α to the primary alcohol group.

As a consequence of this natural asymmetry, the enantiomers of those chiral metal chelates which are biologically active may exhibit different activities. It is important, therefore, to be able to determine absolute configurations, that is, actual atomic arrangements in space, both of organic molecules such as natural amino acids and of metal chelates such as tris-bidentate complexes.

When an atom scatters X-rays whose frequency is not close to a characteristic atomic absorption frequency, the atomic electrons, which are the principal scattering agent, behave more or less as free electrons, and the scattering amplitude from the atom is wholly real. If the freqency of the incident X-rays is close to an absorption frequency (usually of the electrons having principal quantum number, n = 1), then the electrons cannot be regarded as free and it is found that the scattering amplitude from the atom is no longer wholly real. It is the imaginary component of the so-called anomalous scattering, which occurs near atomic absorption frequencies which provides the key to the determination of absolute configurations.

Consider the contributions to the structure factor, F_h, of two atoms A and B, positioned at r_A and r_B respectively, and having scattering factors f_A and f_B which are real:

$$F_{-h} = f_A \exp 2\pi i \; h \cdot r_A + f_B \exp 2\pi i \; h \cdot r_B$$
$$= (f_A \cos 2\pi \; h \cdot r_A + f_B \cos 2\pi \; h \cdot r_B)$$
$$+ i(f_A \sin 2\pi \; h \cdot r_A + f_B \sin 2\pi \; h \cdot r_B)$$

Similarly

$$F_h = (f_A \cos 2\pi \; h \cdot r_A + f_B \cos 2\pi \; h \cdot r_B)$$
$$- i(f_A \sin 2\pi \; h \cdot r_A + f_B \sin 2\pi \; h \cdot r_B)$$

so that

$$\left| F_h \right| = \left| F_{-h} \right|$$

This relationship is known as Friedel's Law, and is usually assumed to be valid.

However, if the atom A is scattering anomalously, we must write $f_A = f_A{}^o + f_A' + if_A''$, where f_A' and f_A'' are the real and imaginary components of the anomalous contribution to the scattering factor. So that now

$$F_{-h} = (f_A{}^o + f_A' + if_A'') \exp 2\pi i \; h \cdot r +$$
$$f_B \exp 2\pi i \; h \cdot r_B$$
$$= [(f_A{}^o + f_A')\cos 2\pi \; h \cdot r_A - f_A'' \sin$$
$$2\pi \; h \cdot r_A + f_B \cos 2\pi \; h \cdot r_B]$$
$$+ i[(f_A{}^o + f_A')\sin 2\pi \; h \cdot r_A + f_A'' \cos$$
$$2\pi \; h \cdot r_A + f_B \sin 2\pi \; h \cdot r_B]$$

whereas

$$F_h = [(f_A{}^o + f_A')\cos 2\pi \; h \cdot r_A + f_A'' \sin$$
$$2\pi \; h \cdot r_A + f_B \cos 2\pi \; h \cdot r_B]$$
$$- i[(f_A{}^o + f_A')\sin 2\pi \; h \cdot r_A - f_A'' \cos$$
$$2\pi \; h \cdot r_A + f_B \sin 2\pi \; h \cdot r_B]$$

so that

$$\left| F_h \right| \neq \left| F_{-h} \right|$$

In a noncentrosymmetric crystal having at least one anomalous scatterer, we can in principle distinguish a set of atoms at positions r_i from a set at positions $-r_i$: if for the positions r_i

$$\left| F_h \right| > \left| F_{-h} \right|$$

then for the positions $-r_i$

$$\left| F_h \right| < \left| F_{-h} \right|$$

Comparison of the observed values of

$$\left| F_h \right| \quad \text{and} \quad \left| F_{-h} \right|$$

with those calculated for both enantiomers should indicate which one is correct.

If the structure is centrosymmetric, then for an atom A at r_A there will be another atom A at $-r_A$, and the contribution to the structure factor from this pair will be

$$F_h = 2(f_A^\circ + f_A' + if_A'') \cos 2\pi\, h \cdot r_A$$

so that

$$|F_{\underline{h}}| = |F_{\underline{-h}}|$$

in *all* centrosymmetric structures. Since a chiral compound can only crystallize in a centrosymmetric space group when it is a racemate, this will present no difficulty.

Several approaches to the determination of the enantiomer are available in practice. If data of high quality, for example from a diffractometer, are available, then one can simply take a small number of pairs

$$\left|F_{\underline{h}}\right| \text{ and } \left|F_{\underline{-h}}\right|$$

and compare the observed values with those calculated. When only visual data of rather low accuracy are available, one can compute a value of the generalized reliability index R_G for each enantiomer using all the observed data, and then apply a statistical test to determine whether one configuration fits the data significantly better than the other.

If the crystal contains a fragment whose absolute configuration is already known, this defines the absolute configuration of the whole. When this latter method is extended by chemical and spectroscopic correlations, a small number of crystallographic investigations can serve to determine the absolute configurations of very many molecules.

The earliest determinations of configuration using anomalous scattering were of

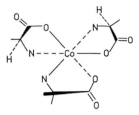

Figure 13-1. The absolute configuration of α-(+)-tris-L-alaninatocobalt(III).

(+)tartaric acid, from which were deduced, by chemical correlations, the absolute configurations of D-glyceraldehyde and the D-series monosaccharides, and of (−)isoleucine, which gave the absolute configurations of the L-amino acids. From these, many more have been derived, such as that of Vitamin B_{12} from the configuration of the D-ribose component. From the configuration of L-alanine, the configuration of (+)tris(L-alaninato)cobalt(III) was found to be that in Figure 13-1. This complex is thus of the same handedness,[4] Λ (Figure 13-2) as complexes such as (+)tris-ethylenediaminecobalt(III), determined by comparison of pairs of

$$\left|F_{\underline{h}}\right| \text{ and } \left|F_{\underline{-h}}\right|'$$

and (+)L-glutamato-bisethylenediamine-cobalt(III), determined from the known configuration of L-glutamic acid.

Inextricably entwined with the question

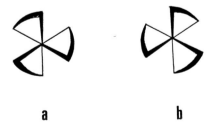

a   b

Figure 13-2. Absolute configurations of tris-bidentate octahedral species, (a) Λ and (b) Δ.

of configuration is that of the conformation of the chelate rings. Corey and Bailar[3] applied the principles of conformational analysis to coordinated ethylenediamine, and deduced a strain-free conformation (Figure 13-3), which was in remarkable agreement with that found experimentally in (+)trisethylenediaminecobalt(III). In isolation, neither this conformation, designated δ, nor its enantiomer, designated λ, is preferred to the other, but in square planar bisethylenediamine complexes, two distinct conformations are now possible, δδ(≡λλ) and δλ(≡λδ), and consideration of the eclipsing of NH_2 protons in δλ suggests that δδ is the more stable. In octahedral trisethylenediamine species, four distinct conformations are possible, δδδ, δδλ, δλλ, and λλλ, in all of which the H . . . H repulsions are different, leading to a different energy for each. For molecules of configuration Λ, the stabilities decrease δδδ > δδλ > δλλ > λλλ, while for the enantiomeric con-

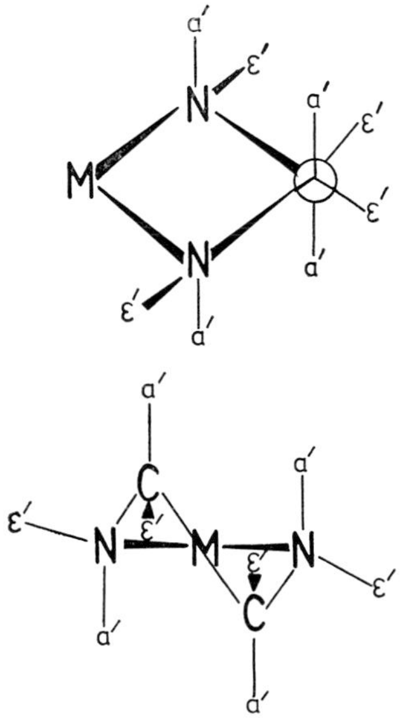

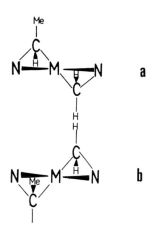

Figure 13-4. Coordination of R-propylenediamine: (a) δ conformation with the methyl group axial and (b) λ conformation with the methyl group equatorial.

figuration Δ, the reverse order holds. A calculation based on a simple H . . . H repulsion potential gave estimated energy differences, for square planar complexes of $E_{δλ} - E_{δδ} \simeq 1$ kcal/mole and for Λ octahedral complexes, $E_{λλλ} - E_{δδδ} \simeq 1.8$ kcal/mole.

In each conformation, δ and λ, the hydrogen atoms can be divided into those which are approximately equatorial (e′) and those which are approximately axial (a′): as in cyclohexanes, the overall potential energy is least if substituents can achieve equatorial positions. For example, R-propylenediamine when chelated has an equatorial methyl group only in the λ conformation, while in δ rings the methyl group is axial (Figure 13-4). Since the λ conformation is more stable in the configuration Δ than in configuration Λ, this accounts for the observed Δ(λλλ) structure of the tris-R-propylenediaminecobalt(III) ion.

Although for the absolute configuration Λ of trisbidentate species, the conformation δδδ is the most stable, each of the other three possibilities has been observed in

Figure 13-3. The δ conformation of coordinated ethylenediamine, viewed (a) along the C-C bond and (b) normal to the C-C bond.

crystals: Λ ($\delta\delta\lambda$) and Λ ($\delta\lambda\lambda$) in the two types of cation found in $[Cr^{III}en_3][Ni^{II}(CN)_5] \cdot 1.5H_2O$, and Λ ($\lambda\lambda\lambda$) in the cation of $[Cr^{III}en_3][Co^{III}(CN)_6] \cdot 6H_2O$ (en = ethylenediamine). The stabilization of these conformations has been ascribed to hydrogen bonding between NH_2 protons in the cation and either water molecules or anions. It seems clear, therefore, that in this instance no reliable deduction can be made from crystallographic results concerning conformational stability, as this may be severely perturbed by intermolecular or interionic forces, and that the conformation observed in a particular solid may reflect neither thermodynamic stability nor the most common conformation in solution.

The value of crystallographic studies here is, first of all, in demonstrating that the possible conformations which occur in metal chelates are only those predicted by conformational analysis. Secondly, in revealing the connection of conformation with configuration when the ligand contains bulky substituents, so that conformational analysis may be applied to these problems with confidence. Thirdly, in determining unambiguous absolute configurations on which to base chemical and spectroscopic correlations.

Speculations have been made about the origin of the natural occurrence of sugars and amino acids in only one enantiomer: It is possible that the initial asymmetric synthesis occurred on the surface of a crystal. This type of surface reaction has been suggested for the dehydration of amino acids to form the earliest polypeptides, and of nucleotides to form primitive polynucleotides, in each case on the surface of an apatite (a rather common group of minerals). Such a mineral will not do for an asymmetric synthesis, as the apatite minerals crystallize in space group $P6_3/m$, and hence

are centrosymmetric. However, a very common mineral whch is noncentrosymmetric, and therefore capable of initiating an asymmetric synthesis, is quartz.

COMPLEXES OF GROUP IA AND IIA METALS

The extensive use of X-ray studies in establishing the basis for selective complexing of alkali metals by antibiotics and synthetic macrocylic ethers is fully described in Chapter 15 on membrane transport. Here we need only record the conclusion concerning macrocycles and cryptates, namely that if a metal ion is of about the same size as the cavity in such a cyclic ligand, then it is more strongly complexed than a metal ion which is either too big or too small to fit the cavity. Plots of formation constant against the ratio (radius of metal ion)/(radius of cavity) for the complexes of dicyclohexyl-18-crown-6 with many uni- and dipositive cations show maximum values of formation constant when the radius ratio is between 0.8 and 0.9: outside this range the complexes are much less stable.

MODELS FOR N_2 AND O_2 COMPLEXATION

For neither the fixation of molecular nitrogen nor the transport of molecular oxygen in biological systems is a detailed description at the molecular level yet available. Structural studies on small inorganic complexes can show how the nitrogen and oxygen molecules may be bound to transition metals, and can thus provide a starting point for discussion of the biological systems.

Complexes of Molecular Nitrogen

In all mononuclear complexes of molecular nitrogen so far characterized structurally, the ligand is bound in the end-on configuration, similar to the isoelectronic carbon monoxide, rather than the side-on configuration, similar to acetylenes. The $M-N_A-N_B$ angle is always very close to 180°, with the N_A-N_B distance in the range 1.03Å to 1.16Å, little different from the value in the free ligand, 1.098Å.

When molecular nitrogen bridges two transition metals, one metal atom is bonded to each nitrogen atom, in marked contrast to the bridging character of CO and CNR where both transition metals are bonded to the carbon of the ligand. While structural data for complexes containing bridging dinitrogen molecules are rather scarce, in the cationic species $[(NH_3)_5RuN_2Ru(NH_3)_5]^{+4}$ the Ru-N-N angles are 178.3° and the N-N distance is 1.12Å, suggesting that the RuN_2Ru fragment should be regarded as a true dinitrogen complex, $M \leftarrow N \equiv N \rightarrow M$, rather than as a diazo compound

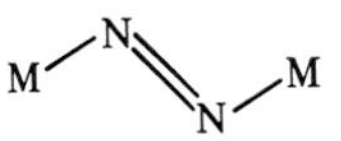

$$M^{\diagup N}\diagdown_{N}{}^{\diagup M} \qquad (125A)$$

Even when coordinated to two transition metal ions, the dinitrogen molecule appears to be little perturbed.

Although no complex is known containing dinitrogen bridging two metal ions with one ligand atom, a complex has been described in which dehydrodiimide (or protonated dinitrogen?) bridges two platinum atoms (Figure 13-5). The N-N distance is 1.18Å, σ 0.09Å. In view of the rather large standard deviation, which may reflect crystal disorder, it is not possible to decide whether the N-N bond length is closer to that in N_2 or that in RN = NR; it is not significantly different from either.

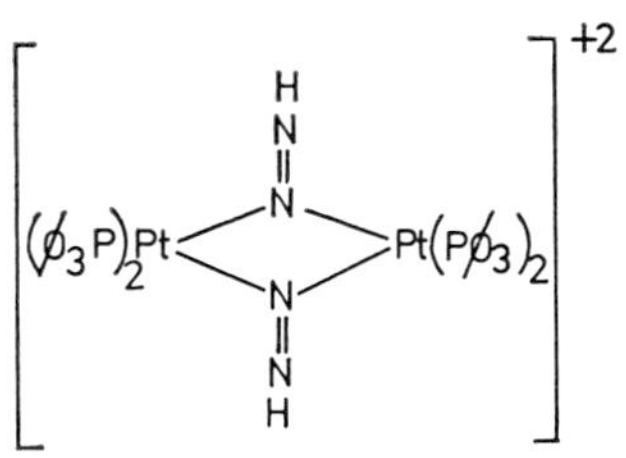

The natural nitrogen-fixing enzyme nitrogenase contains both iron and molybdenum, but its structure is completely unknown. It will reduce not only nitrogen but molecules such as acetylene, methylacetylene and methylisocyanide, but not dimethylacetylene; in fact, only rod-like molecules are reduced, suggesting that the active site may possibly be a long narrow cavity in the enzyme. The possible presence of two metal atoms in the active site is of interest, for while all structurally characterized dinitrogen complexes have the ligand bound end-on, in all complexes of acetylene or monosubstituted acetylenes, the organic ligand is bound side-on. One metal atom could be at the bottom of the cavity (binding dinitrogen) and one could be at the side (binding acetylenes).

Complexes of Molecular Oxygen

Two principal types of molecular oxygen complex have been characterized: In one class the oxygen molecule is bound side-on to the metal atom, and in the other it forms a bridge between two metal atoms as in (126)

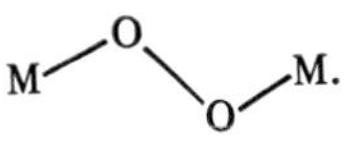

$$M^{\diagup O}\diagdown_{O}{}^{\diagup M}. \qquad (126)$$

A number of the platinum metals have been found to form complexes of the first type. The square planar d^8 complex bis(triphenylphosphine) iridium(I) carbonyl chloride, $(\phi_3P)_2IrCOCl$ reversibly takes up one mole of oxygen to yield an approximately trigonal bipyramidal adduct (Figure 13-6): The analogous iodide also takes up one mole of oxygen, but irreversibly. In the adduct of the chloride, the O-O distance is 1.30Å, and in that of the iodide

Figure 13-5. The bis(μ-dehydrodiimide)bis[bis(triphenylphosphine)platinum(II)](2+) ion.

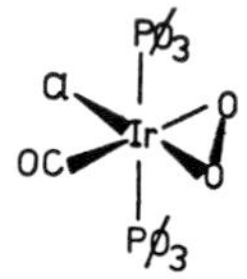

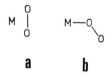

Figure 13-6. 1:1 oxygen adduct with bis(triphenylphosphine)iridium(I) carbonylchloride.

1.47Å (compare O_2 1.21Å, O_2^- 1.28Å and O_2^{-2} 1.49Å). Similarly the cation Rh $(\phi_2PCH_2CH_2P\phi_2)^+$ reversibly forms trigonal bipyramidal RhO_2 $(\phi_2PCH_2CH_2P\phi_2)^+$ having an O-O distance of 1.42Å, while its iridium analogue irreversibly forms IrO_2 $(\phi_2PCH_2CH_2P\phi_2)^+$ in which the O-O bond length is 1.63Å, much longer even than in the peroxide ion.

The suggestion has been made that the irreversibility of oxygen uptake in $(\phi_3P)_2$ IrCOI and $Ir(\phi_2PCH_2CH_2P\phi_2)^+$ is due to a higher electron density of the metal, and hence a more generous $M \rightarrow O_2$ donation, than obtains in $(\phi_3P)_2IrCOCl$ or $Rh(\phi_2 PCH_2CH_2P\phi_2)^+$. This suggestion is consistent with the higher atomic number of iridium, and requires an order of ligand donor capacity P,I > Cl.

While the gross geometry of this type of oxygen complex may be readily understood in terms of electron donation to the metal from a $2p\pi_u$ orbital of the dioxygen, accompanied by donation from the metal to a ligand $2p\pi_g^*$ orbital, the O-O bond lengths do not admit of an easy interpretation. While one might draw direct comparisons with O_2, O_2^- and O_2^{-2} and describe these complexes as $(\phi_3P)_2Ir^{II}(O_2^-)COCl$, $(\phi_3P)_2$

$Ir^{III}(O_2^{-2})COI$, and $Ir^{IV}(O_2^{-3})$ $(\phi_2PCH_2 CH_2P\phi_2)^+$, this seems inappropriate as they are all found to be diamagnetic.

Oxyhaemoglobin, in which the structure of the Fe-O_2 fragment is at present unresolved, is diamagnetic. A number of geometries have been proposed (Figure 13-7). Figure 13-7a illustrates all of the simple mononuclear inorganic oxygen adducts so far examined, and could account for the diamagnetism of oxyhaemoglobin if it is accepted that the degeneracy of the $2p\pi_g^*$ orbitals is split sufficiently on coordination to cause spin-pairing. The $2p\pi_g^*$ orbital in the M-O_2 plane is of higher energy than the one normal to the M-O_2 plane; the out of plane orbital can become doubly occupied, and the in plane orbital can become available for $M \rightarrow O_2$ π-donation (Figure 13-8).

While the simple models may prepare us for an unusual M-O_2 geometry in oxyhaemoglobin, it seems clear that no simple treatment of the structural and magnetic properties of these oxygen adducts is adequate.

The best-characterized molecular oxygen complexes of the binuclear oxygen-bridged type which act as reversible carriers are probably the dicobalt decammine ions $[(NH_3)_5CoO_2Co(NH_3)_5]^{+n}$, where n = 4 or 5. The cation having n = 5 has a *trans* planar structure with an O-O distance of 1.31Å, and may be regarded as a superoxo complex of Co(III); the n = 4 cation is then regarded as a peroxo complex as its O-O distance is 1.47Å, and its dihedral angle between Co-O-O planes is about

$$M \overset{O}{\underset{O}{|}} \qquad M\text{---}O_{\diagdown O}$$

a   **b**

Figure 13-7. Two possible metal oxygen geometries proposed for oxyhaemoglobin.

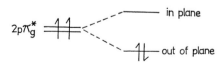

Figure 13-8. Spin-pairing in an oxygen molecule symmetrically bound to a metal atom.

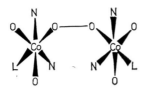

Figure 13-9. Coordination about the metal atoms in $[Co(salcn)]_2O_2L_2$, where L = dimethylformamide.

146°. Notice that $[(NH_3)_5CoO_2Co(NH_3)_5]^{+4}$ cannot be regarded as a complex of unperturbed dioxygen, as $[(NH_3)_5RuN_2Ru(NH_3)_5]^{+4}$ was regarded as a complex of unperturbed dinitrogen.

A number of other cobalt species such as Co(histidine)$_2$, Co(dipeptide)$_2$ and Co (tripeptide)$_2$ take up oxygen reversibly giving products of stoichiometry Co:Ligand:$O_2 = 2:4:1$. None of these has yet been crystallized in a form suitable for X-ray examination, but it is thought that they may have cobalt-oxygen interactions similar to those in the dicobalt decammines.

Co(salen) [N,N'-ethylenebis (salicylideneiminato)cobalt(II)] takes up molecular oxygen in nonaqueous solvents such as pyridine or dimethylformamide (DMF): The structure of the complex $[Co(salen)_2]_2O_2 \cdot$ 2DMF has been determined and is shown in Figure 13-9. The dihedral angle between Co-O-O planes is 110°, which together with the observed diamagnetism suggests that $Co^{III}(O_2^{-2})Co^{III}$ is a reasonable description of the oxygen bridge; however, the O-O distance (1.34Å) is closer to that found in superoxides than that in peroxides.

As with the mononuclear complexes of iridium, the binding of oxygen in binuclear complexes is poorly understood in detail, but the available structures plainly establish the overall geometry of the second type of carrier. This binuclear MO_2M geometry may be of importance in the nonmammalian oxygen carriers haemerythrin

and haemocyanin, in each of which one molecule of oxygen is carried per two atoms, of iron in haemerythrin and of copper in haemocyanin.

PORPHYRINS AND CORRINS

Porphyrins

The oxygen carriers haemoglobin and myoglobin, and the cytochromes, contain haem groups, that is iron(II)-protoporphyrin-IX (Figure 13-10), with the iron in a square planar environment.

Haemoglobin consists of four polypeptide chains and four haem groups, and the overall configuration is known in some detail from crystallographic studies. The iron (II) atom of the haem is coordinated axially by an imidazole group from a histidine in one of the peptide chains, and by a water molecule, which on oxygenation is displaced by molecular oxygen.

There are several features of the oxygenation of haemoglobin, such as the cooperative interaction of the four haem groups, and the change of spin state from high-spin S = 2 in haemoglobin to low-spin S = O in oxyhaemoglobin, whose understanding requires structural data of reasonably high accuracy, preferably for a number of iron porphyrins in order to dis-

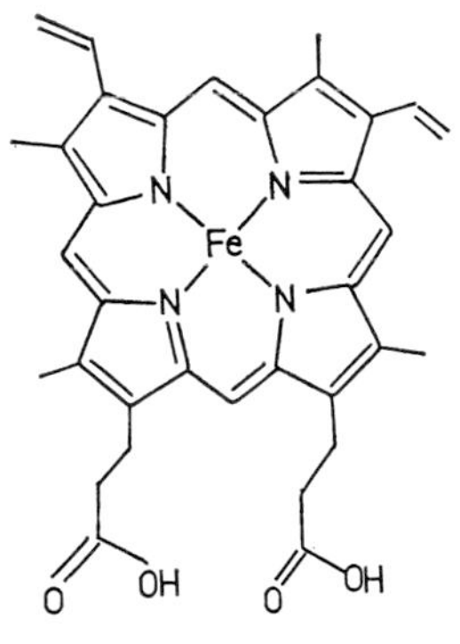

Figure 13-10. Haem. Ferrous protoporphyrin-IX.

cover the general structural characteristics of this class.

Crystallographic studies of metalloporphyrins have provided a detailed explanation of some aspects of haemoglobin function.[6] These studies have been on synthetic metallo-porphyrins with no peptide present, whose advantages are: (1) they possess smaller asymmetric units and hence, (2) greater accuracy is possible for less labor, and (3) they offer the possibility of studying metal ions and spin states not found in natural systems. From a number of determinations, it is clear that while the porphyrin macrocycle is readily deformed normal to its plane (normally with S_4 symmetry), it is very resistant to deformation in the plane, and the radius of the central cavity can thus vary rather little. If the metal ion has a radius less than that of the cavity, it will be situated at the center of the N_4 group; if the metal ion is larger than the cavity, it will be displaced from the N_4 plane.

Thus, in chloroferrictetraphenylporphine, $ClFe^{III}TPP$ (Figure 13-11), which contains high-spin iron(III), the iron is displaced by 0.38Å from the N_4 plane and has square pyramidal coordination, while in the low-spin bis(imidazole)ferrictetraphenylporphine cation, $Im_2Fe^{III}TPP^+$, the iron is approximately coplanar with the nitrogen atoms of the ligand and is octahedral overall. The reason for the difference in geometry is that the radius of low-spin Fe(III) is less than that of the corresponding high-spin ion, since in the low-spin ion the d_σ-orbital $d_{x^2-y^2}$ is unoccupied.

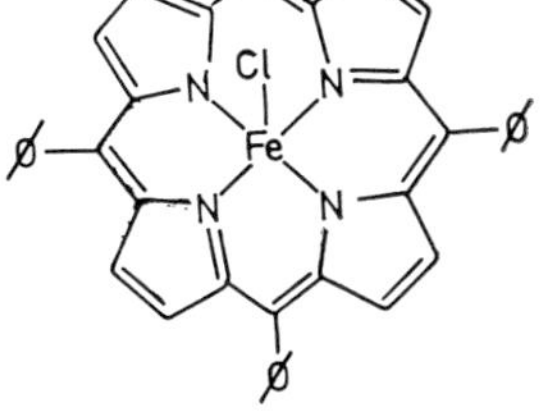

In derivatives of Fe(II), which is larger than Fe(III) in the same spin configuration, we may expect high-spin Fe(II) to be displaced by more than the 0.38Å found for $ClFe^{III}TPP$: in high-spin deoxyhaemoglobin, the iron atom is found to be displaced by about 0.75Å from the N_4 plane, whereas in the low-spin iron(II) species bis(piperidine)ferroustetraphenylporphine, the iron atom is coplanar with the porphine ring.

From these structures of iron porphyrins, and from several others, such as chlorohaemin (high-spin Fe(III), nonplanar FeN_4), methoxyferrricmesoporphyrin-IX dimethyl ester (high-spin Fe(III), nonplanar), oxyhaemoglobin (low-spin Fe(II), planar), carboxyhaemoglobin (low-spin Fe(II), planar) and cyanomethaemoglobin (low-spin Fe(III), planar), the generalization can be made that irrespective of whether the iron is in oxidation state II or III, then the iron is coplanar with the ligand if it is low-spin, and if it is high-spin it is markedly displaced from planarity. The displacement in iron(II) species is greater than 0.50Å, in iron(III) species usually less than 0.50Å.

When haemoglobin is oxygenated, an oxygen molecule displaces the axial water molecule. Although the geometry of the FeO_2 fragment is unknown, it is not unreasonable that the dioxygen, by virtue of its π-acceptor properties, should effect a large enough increase in the ligand field strength to force the iron to spin-pair. The rather smaller spin-paired iron can subsequently move some 0.75Å into the plane of the porphyrin, still bonded to a histidine imidazole, through which the shift of the iron is transmitted to the peptide chain.

Figure 13-11. Chloroferrictetraphenylporphine.

The iron shift is probably the origin of the conformation changes attending oxygenation, and these changes are themselves probably responsible for the cooperative interaction between the haems. Hence, deductions made from the geometries of some rather simple iron porphyrins have led to an interpretation of an aspect of haemoglobin function.[7]

Another metallo-porphyrin derivative of great importance is chlorophyll,[8] the energy-fixing agent in photosynthesis: the macrocyclic ligand differs from the porphyrins proper in that ring D is partially reduced, and a fifth ring E is introduced (Figure 13-12). The crystal structure of chlorophyll itself is undetermined for lack of suitable crystals, but from analogous compounds, some useful and plausible deductions can be made.

The structure of the closely related but metal-free compound, methyl pheophorbide (Figure 13-13), shows that the whole of the macrocyclic ring is planar, apart from the two saturated carbon atoms in ring D. The N_4 group is not square, having $N_1 \ldots N_3$ of 4.06Å and $N_2 \ldots N_4$ of 4.23Å, which could arise either from the protonation of N_2 and N_4, or from the presence of ring E: The effective radius of the cavity is thus 2.03Å.

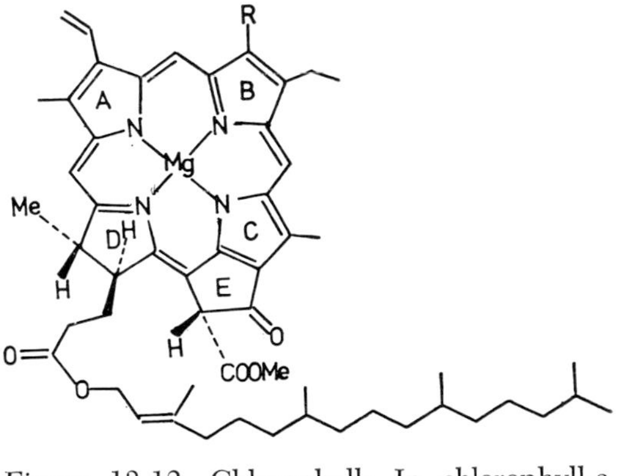

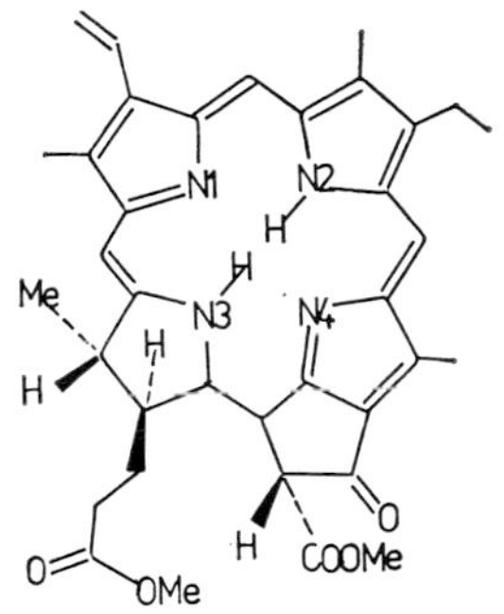

Figure 13-13. Methyl pheophorbide.

In vanadyl deoxophylloerythroaetioporphyrin (Figure 13-14), where ring D is unsaturated but ring E is now partially reduced, the whole of the ligand macrocycle is planar, but the vanadium atom is displaced by 0.48Å from the N_4 plane. The N_4 group is once again no longer square, almost certainly this time a consequence of the presence of ring E; $N_1 \ldots N_3$ is 3.98Å,

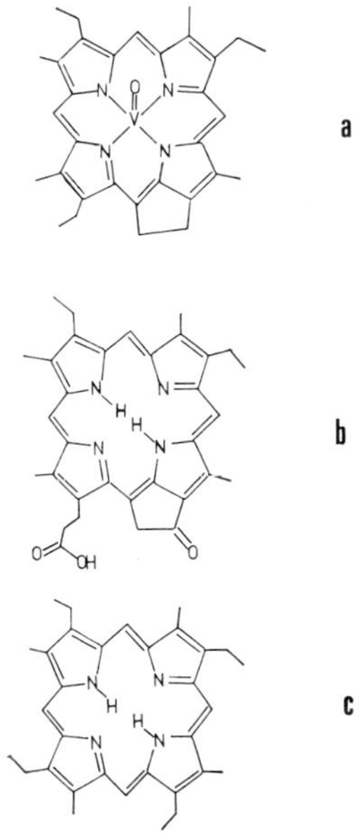

Figure 13-14. (a) Vanadyl deoxophylloerythroaetioporphyrin. Phylloerythrin (b) and aetioporphyrin (c) indicate how the vanadyl complex is named.

Figure 13-12. Chlorophyll: In chlorophyll-a, $R = CH_3$ and in chlorophyll-b, $R = CHO$.

while $N_2 \ldots N_4$ is 4.11Å, giving an effective radius for the cavity of 1.99Å, apparently too small to accommodate a vanadium atom. It may be noted in passing that both this compound and methyl pheophorbide have been found in sedimentary deposits, and are probably both degradation products of chlorophyll (chemical fossils?).

The only magnesium compound similar to chlorophyll whose structure is known in aquomagnesium tetraphenylporphine, where the magnesium is displaced by 0.27Å from the N_4 plane: since the cavity radius in this case is 2.05Å, we may anticipate a displacement of the magnesium in chlorophyll itself of about 0.39Å. The magnesium ion in the tetraphenylporphine complex is square pyramidal, and so will it be in chlorophyll. The magnitudes of the displacement of the magnesium from the N_4 plane, and of the Mg-OH$_2$ distance (2.10Å in the tetraphenylporphine), together with the van der Waals' contact distances of the water molecule, 1.5Å, and of the ligand normal to its plane, 1.7Å, clearly place limits on the stacking frequency of chlorophyll molecules in photosynthetic lamellae; a closest approach of 5.7Å normal to the plane of the ligand is indicated.

Corrins

Of the naturally occurring corrins, by far the most important is vitamin B_{12}, cyanocobalamin.[9] While hydrolytic degradations provided evidence for the presence in the molecule of cyanide, 1-amino-2-propanol, 5,6-dimethylbenzimidazole, 1-D-ribose-3-phosphate, and a number of primary amide groups, an apparently intractable problem was posed by the hexacarboxylic acid, still containing forty-six carbon atoms, which was also isolated from the hydrolyses. X-ray analysis revealed the structures of this acid, and subsequently of the entire

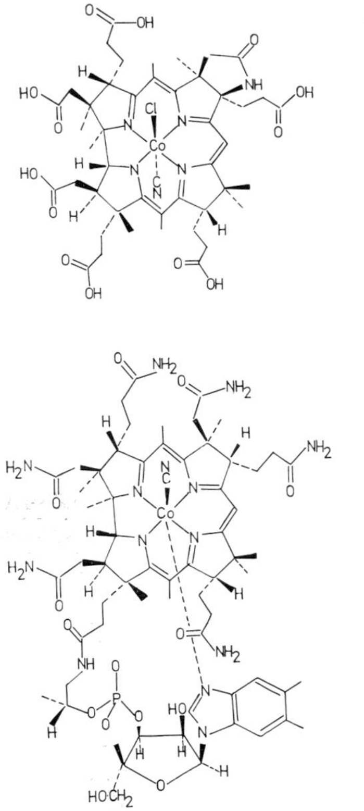

Figure 13-15. (a) The hexacarboxylic acid obtained from the hydrolytic degradation of Vitamin B_{12}. (b) Vitamin B_{12}, cyanocobalamin.

vitamin (Figure 13-15) including the absolute configuration deduced from that of the D-ribose fragment. These structure determinations provide good examples of the power of the X-ray method to take over a structural investigation from the point at which chemical and spectroscopic methods are fully stretched.

A third compound in the B_{12} series to be investigated by X-ray analysis is 5'-deoxyadenosylcobalamin (Figure 13-16), a coenzyme for a number of hydrogen transfer enzymes, which was revealed as a cobalt σ-alkyl derivative, unusual both as a naturally occurring organometallic and as a stable σ-alkyl. It may be doubted whether

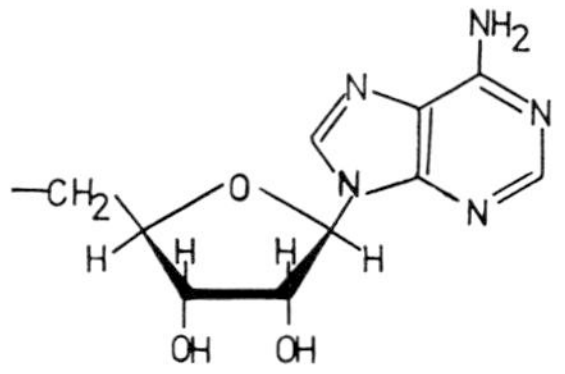

Figure 13-16. The 5'-deoxyadenosyl group which replaces the cyanide ligand in Vitamin B_{12} to yield 5'-deoxyadenosylcobalamin.

this feature of the structure would have been fully accepted in the absence of an X-ray structure determination.

Many properties of vitamin B_{12} are mimicked by rather simple compounds such as cobaloximes, axially substituted derivatives of bisdimethylglyoximatocobalt, and some of these have been used as models in mechanistic studies of vitamin B_{12} chemistry. Of importance, therefore, is the structure of one such, methyl-carboxymethyl-(bisdimethylglyoximato)-pyridine-c o b a l t (Figure 13-17). The relevant structural data for this molecule and for some B_{12} species are collected in Table 13-I, and show that, at least using a structural criterion, the cobaloxime is a plausible model.

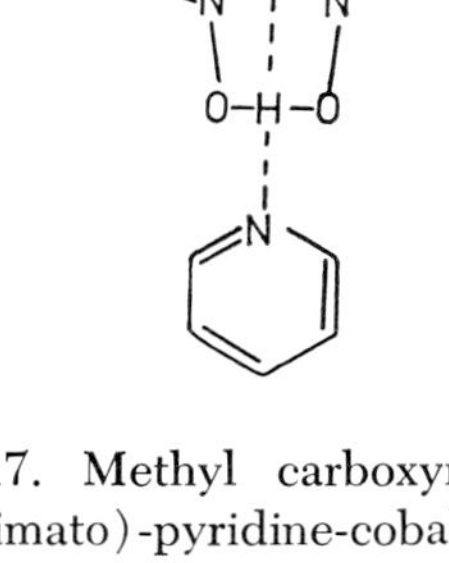

Figure 13-17. Methyl carboxymethyl-bis(dimethylglyoximato)-pyridine-cobalt.

TABLE 13-I

SELECTED BOND DISTANCES (Å) FOR COBALAMINS AND A COBALOXIME ANALOGUE

	Vitamin B_{12}		5-deoxyadenosyl-cobalamin	Cobaloxime
	Dry	*Wet*		
Co-N_{eq} (mean) ...	1.91	1.86	1.95	1.88
Co-N_{ax}	2.07[a]	1.97[a]	2.23[a]	2.04[b]
Co-C_{ax}	2.02[c]	1.92[c]	2.05[d]	2.05[e]

[a] Co-benzimidazole.
[b] Co-pyridine.
[c] Co-CN.
[d] Co-5'-deoxyadenosine.
[e] Co-CH_2COOCH_3.

In addition to these fairly straightforward applications, mention must be made of the use of X-ray studies in the problem of corrin, and in particular vitamin B_{12}, synthesis. One approach constructs a seco-corrinoid in which rings A and D are not joined, complexes it via the four pyrrole nitrogens to a metal ion as a template and then attempts a photochemical $(1, 15)$ $\pi \rightarrow \sigma$ isomerization (Figure 13-18). The structures of the Ni (II) complex, which does not cyclize, and of the Pd(II) and Pt(II) complexes which

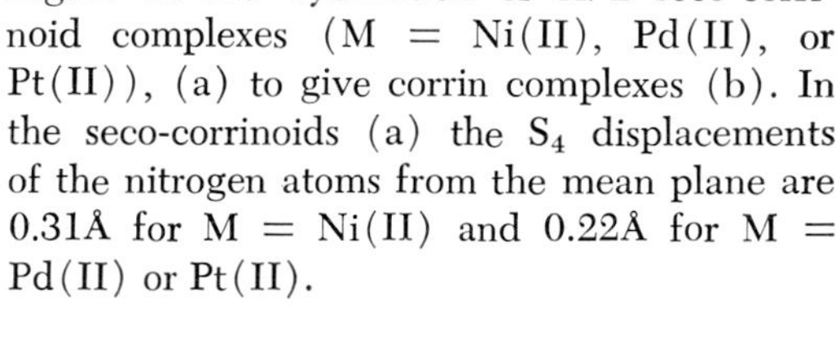

Figure 13-18. Cyclization of A/D-seco-corrinoid complexes (M = Ni(II), Pd(II), or Pt(II)), (a) to give corrin complexes (b). In the seco-corrinoids (a) the S_4 displacements of the nitrogen atoms from the mean plane are 0.31Å for M = Ni(II) and 0.22Å for M = Pd(II) or Pt(II).

cyclize readily, suggest that while the steric factor, namely a greater tetrahedral distortion of the nominally square planar complex in the case of nickel, is important, it is not uniquely so. Although the migrating hydrogen atom is identically placed in the palladium and platinum seco-complexes, but is further from its destination in the nickel derivatives, the platinum complex cyclizes less readily than its palladium analogue.

METALLOPROTEINS

Upon moving into the more complex metalloproteins, we encounter new difficulties, caused principally by the scale of the determination. Even a molecule so complex as cyanocobalamin, vitamin B_{12}, contains fewer than 100 nonhydrogen atoms, even a fairly modest protein may contain a thousand or two.

Protein crystals may contain up to 30 percent of the mother liquor of crystallization; while this suggests that a conformation derived from an X-ray study of such a crystal may be a good representation of the solution conformation, it does render the crystals very soft, so that the atoms can execute large amplitude thermal vibrations leading, through the Debye-Waller temperature factor $\exp(-2B\sin^2\theta/\lambda^2)$, to a rapid falloff in diffracted intensity with increase in $\sin\theta$. One consequence of this is that the number of data may be severely restricted, usually to such an extent that resolution of individual atoms is impossible.

A further problem to be overcome in handling many thousands of data is that of phase determination, where techniques useful for "small" molecules are inappropriate: A Patterson synthesis will be essentially featureless, and there is an insufficient number of intense diffraction maxima for the employment of direct methods. The method of multiple isomorphus replacements must

be used and if, in addition to those for the native protein, intensity data are available for at least two isomorphous crystals containing heavy atoms at different sites, whose relative positions are obtainable from modified Patterson syntheses, then the phase of any reflection can, in principle, be determined unambiguously. Typical heavy atoms used are silver, iodine, platinum, mercury and lead, either in organic or in inorganic form, which can often be introduced into the structure simply by diffusion through the occluded mother liquor. The large polycations $Nb_6Cl_{12}^{+2}$ and $Ta_6Cl_{12}^{+2}$ have also been suggested for this purpose.

As previously mentioned, it is unlikely that a structure determination on a protein will enable the recognition of specific atoms in the peptide chain. The Fourier synthesis will reveal a continuous ribbon of electron density, with side-arms of varying sizes. Provided that the amino acid sequence has been determined by chemical means, then this can be fitted to the electron density since the side-chains of those amino acids containing sulphur atoms or aromatic groups should be fairly easily recognized. While conventional least-squares refinement of protein structures is not possible because of underdetermination, use of typical values of bonded distances, interbond angles and minimum nonbonded contacts, in addition to the sequence information enables plausible structures to be produced at the atomic level.

The most extensively studied of the non-haem metalloproteins is probably the enzyme carboxypeptidase-A.[10] This enzyme, which cleaves the peptide bond at the C-terminal end of a peptide chain, consists of a single chain of 307 amino acids, and a zinc ion. In the resting enzyme, the zinc ion is coordinated approximately tetrahedrally by imidazole nitrogens from two histidines (amino acids numbers 69 and 196, counting from the N-terminal end), by a carboxyl oxygen from glutamic acid—72, and by a water molecule. In the complex of the enzyme and the substrate glycyltyrosine, the water molecule bound to zinc is displaced by the carbonyl oxygen of the peptide bond being cleaved. The structure of the enzyme-glycyltyrosine complex, shows that as the substrate is bound, the enzyme undergoes substantial conformational changes in the region of the active site: The guanidinium group of arginine-145 moves about 2Å towards the active site and appears to bind the C-terminal carboxylate group of the peptide, the OH of tyrosine-248 moves the surprisingly long distance of 12Å towards the substrate and is probably the proton donor in the peptide hydrolysis, and the carboxylate group of glutamic acid-270 moves about 2Å closer to the peptide bond, toward which it probably acts as the nucleophile.

The mechanism inferred from these structural changes is shown in Figures 13-19a and 13-19b. It rationalizes other chemical

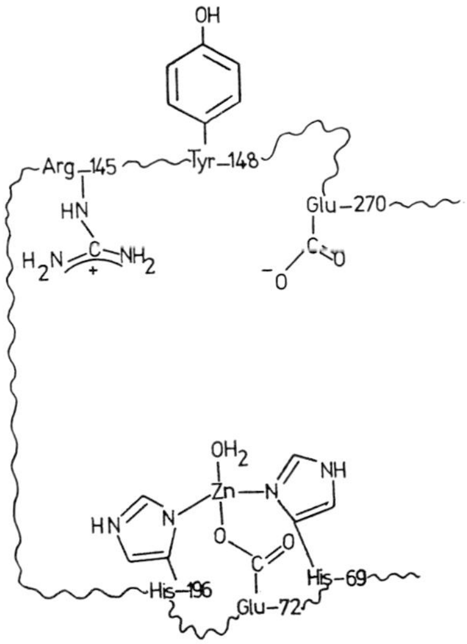

Figure 13-19a and b. Schematic representation of the active site in carboxypeptidase: (a) the resting enzyme, (b) the hydrolysis of glycyltyrosine.

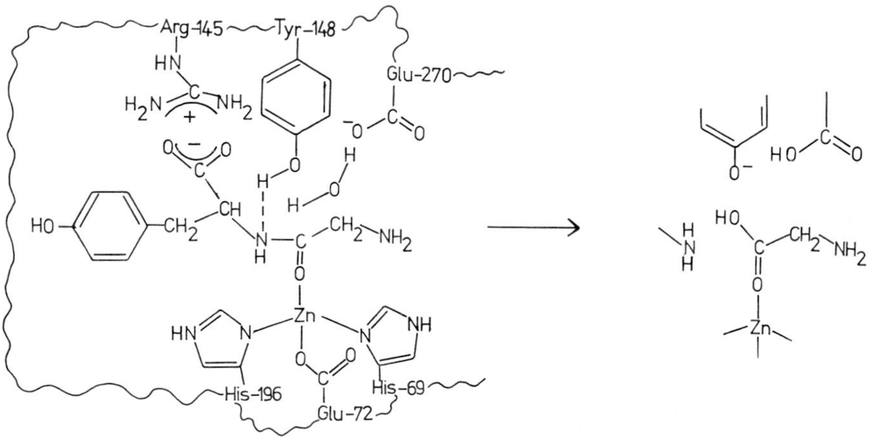

and structural evidence: If the OH groups of tyrosines or the guanidine groups of arginines are protected, then peptidase activity is reduced. If the zinc is replaced by mercury, peptidase activity ceases. The metal ion in the mercury enzyme is displaced by about 1Å from its position in the zinc enzyme, since it requires longer metal-ligand distances, and when the substrate is bound to the mercury ion, the peptide bond to be cleaved is too far away from glutamic acid-270 for this to act as nucleophile, and hence no hydrolysis occurs.

While some details of the mechanism of this hydrolytic cleavage are uncertain, for example whether glutamic acid-270 acts directly as the nucleophile forming a labile anhydride, or via a water molecule, the basic mode of action of this enzyme is beautifully revealed by the two structures, of the resting enzyme and of the enzyme-substrate complex.

Another enzyme containing zinc, carbonic anhydrase, which catalyzes primarily the reaction

$$CO_2 + H_2O \rightleftharpoons H^+ + HCO_3^-$$

but also catalyzes ester hydrolyses, has been studied by X-ray methods, but the study is much less advanced than that of carboxypeptidase-A. The single zinc ion in the molecule is situated at the bottom of a deep crevice in the protein, in a site rather distorted from regular tetrahedral and is coordinated by three nitrogen atoms from the peptide, at least two of which are from histidines, and probably by a water molecule also. No high resolution structure is yet available, either of the resting enzyme or of its complexes with substrates or inhibitors: When such data are known, it should be possible to deduce the enzyme's mode of action, to determine the action of

inhibitors like cyanide and azide, and to rationalize the variation of activity with metal ion; $Zn^{+2} > Co^{+2} > Mn^{+2}$, with most other metal ions inactive.

LARGER INORGANIC ARRAYS

Clathrates

A number of anaesthetic gases form clathrate hydrates, of two distince stoichiometries $8M \cdot 46H_2O$ (for M = cyclopropane, xenon, nitrous oxide) and $M \cdot 17H_2O$ (for M = chloroform, ethyl chloride). Crystallographic examination shows that each class of hydrate is based on groups of twenty water molecules joined by hydrogen bonds into pentagonal dodecahedra. In the 8:46 group, these dodecahedra are stacked into a three dimensional array via hydrogen bonds with tetrakaidecahedral (14-hedral) holes between them; the dodecahedra can accommodate guest molecules whose diameter is less than 5.1Å, while the tetrakaidecahedran can hold molecules up to 5.8Å in diameter. The stacking of the dodecahedra in the 1:17 group is slightly different, forming hexakaidecahedral (16-hedral) holes capable of taking molecules up to 6.7Å in diameter, so that gases with small molecules such as nitrous oxide form 8:46 hydrates, while larger molecules which cannot fit the 5.7Å holes, form 1:17 hydrates.

Water in either of these two molecular arrangements is, in the absence of guest molecules, of higher potential energy than it is in the ice structure, but in the presence of a clathrated species this difference is more than offset by the dispersion energy of the guest-water interaction. If the dodecahedral set of holes is also occupied, by a so-called "help-gas," the stability of the hydrate increases further: For example, the melting point of $CHCl_3 \cdot 17H_2O$ rises from 1.4°C to 14.7°C when exposed to an atmosphere of xenon, which enters the set of 5.1Å holes, giving a limited composition of $CHCl_3 \cdot 2Xe \cdot 17H_2O$.

It is thought that in solution the polar side chains of proteins may cause a considerable ordering of the water molecules into one of the two kinds of structure found in crystalline hydrates. The presence of small nonpolar molecules of an intruding species could, by entering the dodecahedral cavities and acting as a help-gas, increase the stability of the water structure and make it rather more static. If such a static water structure is formed around a nerve fiber, then it may interfere with the conduction of the nervous impulse by preventing the movement of sodium and potassium ions through the nerve cell membrane. This is the basis of the Pauling-Miller theory of anaesthesia, a breakdown of nervous conduction caused by hydrate formation, which is consistent with several lines of evidence:

(1) Most anaesthetics have rather small molecules.

(2) The potency of an anaesthetic gas increases with a decrease in temperature (and so with an increase in the stability of the hydrate structure).

(3) Even xenon acts as an anaesthetic (it forms double hydrates, but enters into few chemical interactions under biological conditions).

(4) The partial pressure of gases required for anaesthesia decreases linearly with the equilibrium partial pressure of their crystalline hydrates (and so with increasing hydrate stability).

Structural investigations of clathrate hydrates have provided the basis for a molecular theory of anaesthesia based only on the steric properties of the anaesthetic agent. Recall that the steric properties of

metal ions were used to account for the selective complexing of the alkali metals by macrocyclic ligands and for the conformational changes on oxygenation of haemoglobin.

Ferritin

Ferritin is the principal iron storage protein found in liver, spleen and bone marrow. The iron is present as $Fe(III)$, in a spherical core some 75Å in diameter, of approximate composition $Fe_9^{III}O_9(OH)_8H_2PO_4$, and containing about 2,000 iron atoms. Preliminary X-ray data have been interpreted in terms of both octahedral and tetrahedral sites for iron.

On the basis of magnetic susceptibility and Mössbauer data, a reasonable model for the iron core of ferritin is provided by hydrolyzed ferric nitrate; mild alkaline hydrolysis of ferric nitrate yields a polymer of approximate composition $[Fe^{III}(OH)_{2.5}(NO_3)_{0.5}]_n$, where $n \simeq 1,200$. Electron microscopy reveals that this polymer consists of discrete spherical aggregates of diameter about 70Å. When a radial distribution function, analogous to a one-dimensional Patterson function, was calculated from the X-ray diffraction of a solution of the hydrolyzed ferric nitrate, probability maxima were found at 2.1Å and 3.5Å, identified with Fe-O and Fe . . . Fe distances respectively: No evidence was found for direct Fe-Fe bonds. The coordination number of the iron was deduced as four.

While it seems clear that only a detailed X-ray study can reveal the structure of ferritin's iron core, this is by no means easy with some 2,000 iron atoms to be placed. To determine phases for the reflections, several heavy atom derivatives will be required. The electron densities of the heavy atoms must be substantially higher than those of the other atoms, if they are to be useful for phase determination: The ratio $\sum f^2_{\text{light atoms}}/\sum f^2_{\text{heavy atoms}}$ should probably not exceed twenty-five (f is the atomic scattering factor which at $\sin \theta = 0$ is equal to the number of electrons in the atom). For the ferritin case $\sum f^2 \simeq 1.2 \times 10^6$, while for the $Ta_6Cl_{12}{}^{+2}$ ion $\sum f^2 \simeq 3.5 \times 10^4$, and for $Bi_9{}^{+5} \sum f^2 \simeq 6.2 \times 10^4$: These ions might just be adequate. Although heavier ions such as polytungstates are known, they are almost certainly too large to form isomorphous crystals, and there is real doubt whether this structure could be solved either by multiple isomorphous replacement or by methods involving anomalous scattering at several wavelengths.

Bone

Almost certainly the most common inorganic materials in animals are found in bones and shells. Shells are made largely of calcium carbonate and the inorganic component of bone has very approximately the composition of calcium phosphate: X-ray analysis showed that it had the structure of the mineral hydroxyapatite $Ca_5(PO_4)_3OH$. The composition, like that of any mineral, may be slightly variable. For example, carbonate can replace phosphate to some extent, if accompanied by substitution of sodium for calcium. In addition, hydroxide can be replaced by fluoride to yield apatite itself, $Ca_5(PO_4)_3F$, which is less acid-soluble than hydroxyapatite, and is found in the enamel of teeth.

Of the salts of common biological metals, the carbonates and phosphates of magnesium and calcium are less soluble than those of sodium and potassium; this is simply a consequence of the double charge on the group II metals. In salts of large anions, those of larger cations are generally less soluble than those of smaller cations, and

the calcium salts are less soluble than those of magnesium. Calcium salts are therefore the most likely candidates for structure formation, a process assisted by the rejection of calcium ions from the cells.[5]

PROSPECTS

In some areas of bio-inorganic structural research, likely lines of advance are reasonably discernable. We may expect to see more structural studies of metalloenzymes, both in the resting state and in combination with substrates, inhibitors and perhaps products, leading for each enzyme to a view of its action such as we now have for carboxypeptidase-A. Such studies should help rationalize also enzyme inhibition and the behavior observed on changing the metal. In particular, we may anticipate work aimed at connecting structure with function on some of the copper proteins, such as uricase or haemocyanin, and on some molybdenum species such as xanthine oxidase or nitrogenase, and a continued effort on nonhaem iron proteins such as the ferredoxims. These structural investigations should take as an additional criterion of completeness the interpretation of esr data, and for the iron species of Mössbauer data also.

Data collection at low temperature may become more popular and this should give an increase in resolution through the reduction of thermal vibrations. If at 30°C the observed data go out to $\sin \theta/\lambda = 0.20$ Å^{-1}, then at −80°C, they will go out to $\sin \theta/\lambda = 0.25$ Å^{-1}.

An area of considerable potential lies between the present domains of X-ray crystallography on the one hand, and electron microscopy on the other. At present the best practical resolution attainable by electron microscopy is about 20Å; this could conceivably be improved to 5Å, which is about the lower limit of resolution for useful X-ray studies. It should therefore be possible to define at an atomic level some of the structures visible using the electron microscope, provided that they are sufficiently ordered to give interpretable diffraction patterns.

There are areas where X-ray techniques can make a substantial contribution. One problem currently attracting wide interest, to which X-ray methods probably have rather little to offer at present, is that of membrane function. Electron density syntheses depend on the presence of long-range order in a moderately rigid structure, and radial distribution functions from non-ordered matter can only give information about the short-range structure. A possible opening is the investigation of diffraction patterns from liquid crystals in which each molecule is labelled with a heavy atom. A Patterson map based on the heavy atoms might give information on the molecular arrangement in the liquid crystal and so aid a possible interpretation, assisted by model building, of the diffraction pattern from the unlabelled crystal. This in turn might aid the structural elucidation of some models for membranes, in particular the lipid bilayer.

REFERENCES

1. Bunn, C. W.: *Chemical Crystallography*. Clarendon Press, Oxford, 1961.
2. Wolfson, M. M.: *X-ray Crystallography*. Cambridge U Pr, New York, 1970.
3. Corey, E. J. and Bailar, J. C., Jr.: Stereospecific effects in complex ions. *J Am Chem Soc,* 81:2620, 1959.
4. IUPAC: Nomenclature of absolute configurations based on the octahedron. *Inorg Chem,* 9:1, 1970.
5. Williams, R. J. P.: The biochemistry of sodium, potassium, magnesium and calcium. *Quart Rev,* 24:331, 1970.
6. Hoard, J. L.: Some aspects of heme stereochemistry. In Rich, A. and Davidson, N. (Eds.): *Structural Chemistry and Molecular Biology.* Freeman, San Francisco, 1968.
7. Perutz, M. F.: Haemoglobin: The molecular lung. *New Scientist and Science Journal,* 50: 676, 1971.
8. Marks, G. S.: *Heme and Chlorophyll.* Van Nostrand, New York, 1969.
9. Pratt, J. M.: *Inorganic Chemistry of Vitamin B_{12}.* Acad Pr, Oshkosh, 1972.
10. Lipscomb, W. N.: Three-dimensional structures and chemical mechanisms of enzymes. *Chem Soc Rev,* 1:293, 1972.

CHAPTER 14

GENERAL ANALYTICAL METHODS

Gordon S. Fell and Hamilton Smith

INTRODUCTION

ALMOST EVERY POSSIBLE TYPE of analytical technique has been applied to the determination of metals in biological systems. Bowen (1966),[1] has listed seventy-eight elements in vertebrate blood determined by more than ten different procedures, and since then various new procedures have been developed. It is clearly difficult to deal with all of these methods in this chapter and only two method types will be discussed in some detail and a brief mention made of some others.

First of all, neutron activation analysis will be described since this very powerful method is not generally dealt with in most texts and yet it is capable of extreme sensitivity and high specificity. Secondly, the basis of the various branches of atomic spectroscopy will be outlined, since these techniques are by far the most commonly used in practice.

NEUTRON ACTIVATION

Introduction

Activation analysis is basically different from other analytical techniques and may be described properly as a new dimension in analytical practice. Normal methods of analysis depend on the properties of the electron shells of atoms but activation analysis makes use of the properties of the nucleus. The result of this is that elements which have closely related chemical properties appear quite different when the nuclear properties are examined and an easy method of distinguishing one from the other is available.

The stable nucleus is immune to the usual processes of chemistry but its properties may be revealed, to some extent, by the process of induced radioactivity. In this process the material under investigation is placed in a flux of nuclear particles or photons. The most used process is that of thermal neutron activation in which very low energy (thermal) neutrons such as are available inside an atomic reactor, are employed. The irradiated atoms capture the low energy neutrons and emit γ-rays. The resulting atoms are unstable and decay to some stable species with the emission of β^- and/or γ-rays of characteristic energy and half-life. The half-life is the time taken for half of the activity to disappear.

Thermal neutron activation analysis requires a number of points to be considered before application to any given problem. These are outlined under the relevant headings below.

Selection of the Technique

Neutron activation is the method of choice for the analysis of about two thirds of the chemical elements when high sensitivity is required. It is, therefore, ideal for trace element analyses, especially in biological materials where the building elements (C,H,O,N) of the matrix do not become appreciably radioactive when subjected to thermal neutron irradiation. This does not exclude the technique when major constituents are being determined, as simple modifications to the activation process can reduce the sensitivity by any required degree. Table 14-I gives a list of sensitivities easily available using standard equipment and the irradiation facilities offered at most research reactors.

If higher sensitivities are required reactors with thermal neutron fluxes of the order of 10^{14} n/cm^2/sec, separation techniques and detectors are available which can increase many of the above sensitivities by a factor of 10^4.

A disadvantage of the technique is that it requires the handling of radioactive isotopes. However, when tissues are being investigated the total amount produced is small and only slight modifications to the normal laboratory are required. The detection of stray radiation and the detection and determination of the sought for species requires the use of special apparatus. This equipment can be of moderate price when chemical separation techniques

are used but can become expensive if the more sophisticated but less sensitive techniques of γ-spectrometry are used.

In general, activation analysis is a straightforward procedure using either simple chemical separation processes backed up by recovery correction techniques and detection of the radiation from a single element, or γ-spectrometry, i.e. placing the radioactive tissue in a detector. This means that relatively nonskilled personnel can carry out the analysis of large numbers of samples without any loss of accuracy.

Sample Handling

Activation analysis is most often used for the analysis of small samples for trace elements and because of the very small amounts of trace elements present contamination before irradiation is a serious problem. After irradiation, contamination is not a problem. Strict attention to clean handling or preferably no handling at all, will keep this problem in bounds. Complete nonhandling can only be achieved in special circumstances, e.g. the irradiation of seed in their natural containers.

The materials from human subjects (or indeed other biological systems) which are analyzed can be divided roughly into

TABLE 14-I
TRACE ELEMENT SENSITIVITY*

Sensitivity Range (g)	Element
1 to 9 × 10^{-12}	Dy, Eu
1 to 9 × 10^{-11}	In, Lu, Mn
1 to 9 × 10^{-10}	Al, As, Cu, Ho, Ir, La, Pr, Re, Sm, V.
1 to 9 × 10^{-9}	Au, Br, Co, Er, Ga, Ge, Hg, I, K, Na, Nb, Pd, Rh, Rb, Sb, Sc, Sr, Ta, Tb, U, W, Y, Yb.
1 to 9 × 10^{-8}	Ag, Ba, Cd, Cs, Ce, Cl, Gd, Hf, Mo, Nd, Ni, Os, Ru, Se, Si, Te, Tm, Ti, Zn.
1 to 9 × 10^{-7}	Bi, Cr, Mg, P, Pt, Sn, Tl, Zr.
1 to 9 × 10^{-6}	Ca, Fe, Pb, S.

* The values quoted here are for irradiation in a thermal neutron flux of 10^{-12}n/cm^2/sec for 1 week or to saturation, whichever is the shorter.

two sections. The main bulk of the subject which may be described as wet and the covering materials, such as skin and nail, which may be described as dry.

Wet tissue for analysis must be removed from some body, usually not by the analyst, and this is the stage at which there is the greatest contamination risk. The contamination comes from a variety of sources:

(1) the surface of the body (dirt, dust, industrial contamination, cosmetics),
(2) the gloves of the pathologist (disinfectants, talc),
(3) the instruments used (metallic dust from the instrument, metal being dissolved by body fluids, contamination from surrounding organs),
(4) the dissecting table (dust, disinfectant). The dissected out material should be placed in clean polyethylene containers without the addition of preservatives which are rich sources of contamination.

If a large enough sample reaches the analyst he can dissect out a piece of uncontaminated tissue from the center of the mass using plastic or preferably silica instruments. In certain cases when only a few elements are to be looked for, usually following chemical separations, then a knife, which does not contain the sought for elements, can be used.

Dry tissue is useful for irradiation purposes, and therefore the material under investigation is usually dried at a pressure of about 0.1 mm Hg. This may be a freeze-drying process or the moisture may be removed by a suitable dessicant which does not contain the sought for elements. The resulting dry tissue is reasonably stable and can be stored in polyethylene containers for some years. Hair, nail and skin scales do not require drying, but they have some problems of their own in so far as they contain not only trace elements deposited in them from the bloodstream but also those from the surrounding environments. For example, zinc and aluminium may be present from cosmetics, copper from hair sprays, arsenic from detergents, gold from jewelry in the form of rings and antimony and mercury from some industrial environments. These widely varying and often unexpected sources of contamination make the interpretation of trace elements in hair difficult.

Once the samples have been prepared and weighed, preferably in a dust free atmosphere, they require packing for irradiation. The samples of tissues which usually weigh under 50 mg are placed in containers or foil made of polyethylene or alumnium or in silica ampoules. If there is any doubt about contamination from these sources they must be tested under the proposed working conditions. Aluminium may be cleaned with benzene and then very dilute nitric acid followed by pure water. As aluminium containers are not usually sealed they are not suitable for the irradiation of substances which may give off volatile materials. Silica is usually cleaned with a mixture of sulphuric and nitric acids (3:5) followed by rinsing with pure water. When the greatest care is required a silica tube can be heated to near melting in an oxygen/gas flame and the resulting zone gradually moved along the tube expelling any impurities before it. Polyethylene is simply washed with pure water or pretreated with acid as for silica. Samples may be sealed in polyethylene tubes or simply wrapped in polyethylene sheet.

Irradiation is normally carried out in containers, supplied by the reactor operators, which can hold reasonable numbers of samples, e.g. 50 of 50 mg + standards.

When high neutron fluxes ($10^{14}\text{n}/\text{cm}^2/\text{sec}$) are required the available volume is usually much smaller.

The Activation Process

Neutron activation is carried out by irradiating the material under investigation with neutrons, generally in an atomic reactor. The neutrons most often used are those of low energy (thermal neutrons). Normally there is a small known percentage of epithermal and fast neutrons present which may interact to produce different isotopes.

Sodium, in the form of the salt of an organic acid, placed in the reactor undergoes the following reaction with thermal neutrons.

$$_{11}^{23}\text{Na} + {_0^1}\text{n} \rightarrow {_{11}^{24}}\text{Na} + \gamma$$

This equation is usually shortened to the following form:

$$_{11}^{23}\text{Na}(\text{n},\gamma)_{11}^{24}\text{Na}$$

The carbon hydrogen and oxygen of the organic acid are not affected significantly but some of the sodium atoms (perhaps less than one in 10^6) capture a neutron to form an unstable isotope. The decay of this isotope is the basis of the detection and measurement.

Unfortunately, the activation process cannot be chosen. All possible reactions take place to some extent depending on the irradiation conditions of the reactor used. For example, a sample of manganese placed in a reactor will be exposed to thermal and fast neutrons at least. The required reaction for activation analysis is

$$_{25}^{55}\text{Mn}\ (\text{n},\gamma)\ _{25}^{56}\text{Mn}$$

If any iron is present in the sample then a small proportion may undergo the following reaction:

$$_{26}^{56}\text{Fe}\ (\text{n},\text{p})\ _{25}^{56}\text{Mn}$$

In this example the capture of a neutron is followed by the emission of a proton, i.e. the mass number of the nucleus remains the same but the charge decreases by one unit. Similarly, any cobalt in the sample is activated to a small extent as follows:

$$_{27}^{59}\text{Co}\ (\text{n},a)_{25}^{56}\text{Mn}$$

In this example a neutron is captured and an a-particle (mass 4, charge $+2$) is ejected.

These two processes are termed first-order reactions and both produce ^{56}Mn.

A second-order reaction which may also occur is as follows:

$$_{24}^{54}\text{Cr}(\text{n},\gamma)_{24}^{55}\text{Cr} \xrightarrow{\beta^-} {_{25}^{55}}\text{Mn}(\text{n},\gamma)_{25}^{56}\text{Mn}$$

In this reaction $^{54}\text{Cr}(2.4\%$ abundance) is converted to ^{55}Cr which being unstable decays by β^- emission ($t_{1/2} = 3.5$ min.) to ^{55}Mn a stable isotope. This stable manganese is then available for further neutron capture to yield ^{56}Mn.

First- and second-order interferences of the above type seldom present any difficulty in the estimation of trace elements in biological material but can give false results if one of the interfering elements is a major component of the matrix. Two reactions which may cause problems in some biological materials are:

$$^{32}\text{S}(\text{n},\text{p})^{32}\text{P} \text{ and } ^{35}\text{Cl}(\text{n},\text{p})^{35}\text{S}$$

Tissues which concentrate an element should also be examined carefully. For example, if thyroid tissue is irradiated in a neutron flux rich in fast neutrons the iodine may react to produce antimony and result in falsely high antimony levels.

Selection of Activation Condition

In theory, but not usually in practice, quantitative estimations of the sought for elements can be made by calculations based on conditions of the activation process and measurement of the induced radioactivity. The usual practice is to irradiate a standard along with the samples. Quantitative estimations are then made by direct comparison of the resulting radiation. The reason for the use of the irradiated standard is the variation possible in a number of the values required in the calculation. For example, the neutron flux varies from point to point throughout the reactor.

The main considerations in deciding on the activation conditions are half-life of the expected isotope, neutron capture cross-section, natural abundance and ease of detection of the resulting radiation.

The half-life is the time taken for the activity of a radioactive material to decay to half of its original value. A half-life may be any length of time but the usable ones vary from minutes to weeks. When an element is irradiated in a uniform flux some of the atoms become radioactive, i.e. form unstable isotopes which immediately begin to decay. The result is that the activity does not increase indefinitely during irradiation but reaches a steady state (saturation) where the rate of decay equals the rate of formation. An element that is irradiated for one half-life of the expected isotope reaches one-half of the saturation value, another half-life adds half as much

eral isotopes) present and the longer the irradiation goes on for the more activity is produced. In general, an irradiation of one half-life of the sought for isotope is enough. Tables of half-life are readily available.[5, 6]

Neutron capture cross-section is a measure of the apparent size of a nucleus with respect to neutron capture. The larger the cross-section (the unit is called a barn) is, the greater the chance of capture becomes. The cross-section is different for each element and for each isotope of it. Lists of cross-sections are also readily available.[5, 6]

Most natural elements consist of a mixture of isotopes and the natural abundance is the measure (%) of any one of the isotopes in the mixture. As each isotope undergoes a different nuclear reaction in any given set of conditions the ultimate expected radioactivity depends on the original amount of a particular isotope.

The isotopes produced by neutron activation usually emit β^- particles and γ or X-rays. The energy of this radiation varies from very low to very high. Normally, detectors handle only a limited range of energies and it is necessary to choose isotopes with radiation characteristics which fit the available equipment.

The following two examples may help illustrate these points.

1. *Manganese*

Mn isotopes produced on irradiation of Mn	— ^{56}Mn
Half-life of isotope produced	— 2.58 hour
Thermal neutron capture cross section of parent isotope	— 13.3 barns
Natural abundance of parent isotope	— 100%
Principal energies of isotope produced	— β^- — 2.86 Mev
	— γ — 0.845 Mev

activity again and so on till at ten half-lives about 99.9 percent of saturation is reached. In biological materials there are a large number of trace elements (most with sev-

In this example there is only one choice of isotope and irradiation for one half-life is feasible. The thermal neutron cross section is large indicating a good isotope yield and good sensitivity for the analysis. This is not reduced by a low natural abundance and as the energy of radiation is within the range of most standard equipment detection and measurement is simple.

2. *Calcium*

	^{45}Ca	^{47}Ca	^{49}Ca
Ca isotopes produced on irradiation of natural Ca	—		
Half-lives of isotopes	— 165 days	4.7 days	8.8 mins.
Thermal neutron capture cross section	— 0.014 barns	0.25 barns	1.1 barns
Natural abundance of parent isotope	— 2.06%	0.0033%	0.185%
Principal energies of isotopes produced — β^-	— 0.25 Mev	0.66 Mev	1.95 Mev
— γ	— ——	1.31	3.10

In this example there is a choice of three isotopes. Examining ^{45}Ca first notice that irradiation for a half-life is not feasible. A week is usually the largest time considered. Assuming this irradiation time, the expected activity would then be 7/165 of the activity for one half ($t_{1/2}$), i.e. a loss in sensitivity by a factor of about 24. The cross section is low indicating low yield and low sensitivity. Coupled to this is a low natural abundance, i.e. only one fiftieth of the available calcium is taking part in the reaction. The energy of radiation is also low, requiring the use of low-energy detectors. Turning now to examine ^{47}Ca, the irradiation for a half-life is feasible so there is no loss of sensitivity here. The cross-section is still low but much better than for ^{45}Ca. However, the natural abundance is very low indicating a large loss in sensitivity as only 0.0033 percent of the available calcium is taking part in the reaction. Combining these factors, it is found that the overall sensitivity for both isotopes is of the same order, that for ^{47}Ca being marginally lower. However, ^{47}Ca gives off radiation which is easily detected by standard equipment and as a result may be a slightly better choice. The third isotope ^{49}Ca can be irradiated for a half-life and the cross-section is even better than for ^{47}Ca but the natural abun-

dance is still very low. Combining the factors shows that the sensitivity will be of the order of 200 times better than either of the other two isotopes and as the energy of radiation is suitable for detection and mea-surement by standard equipment this is the isotope to choose. One other factor—the half-life of the isotope—has a bearing on the choice and is discussed in the next topic.

Selection of Analytical Technique

Two techniques are available in neutron activation analysis. The first involves a chemical separation so that only the sought for element remains after the final process. The activity in the separated element can be detected and measured by either β^- or γ detectors depending on the radiation given off. The second makes use of the γ-energy spectrum to give a separation and detection and estimation depends on selection of the correct energy band. Combinations of the two methods are available where part chemical separation and part spectrometry is used.

Radiochemical separation techniques require the destruction of the sample, followed by several chemical separation steps to give a radiochemically pure product. As there is the possibility of loss at each stage of the process it is usual to add a fixed amount of the element under investigation (say about 10mg) to the samples and standard before processing. When the final pure material is separated the relative recoveries are calculated by adjusting the results to a common recovery value. This allows any of the losses incurred at the various stages to be corrected and is one of the big advantages of activation analysis techniques.

The radiochemical technique is basically a combination of four steps as follows:

1. *Matrix destruction*—In biological systems this consists of a wet digestion technique. Dry ashing techniques in general are not satisfactory due to loss of volatile compounds and contamination problems. In the wet process a mixture of concentrated sulphuric and nitric acids is most useful. It is well worth while spending some time selecting the best mixture for the elements being investigated.

2. *Initial separation*—This step separates the sought for element or elements from the digestion mixture but usually contains various other carried over materials. Hold-back carriers are added so that as many as possible of the very low concentration materials which may have significant activity will not be carried through the separation.

3. *Scavenging separation*—This stage is inserted without hold-back carriers and is designed to encourage as many low concentration materials as possible to follow a side reaction and hence be removed from the system leaving the sought for material behind.

4. *Final separation*—This stage with suitable hold-back carriers leaves the sought for material essentially free of contamination. If the final product is not in a form suitable for recovery estimations then further processing is required.

A survey of the literature shows that the technique usually uses a series of simple techniques, e.g. four or five precipitation steps, or fewer but more complex processes, e.g. distillation. Simple precipitation techniques are useful because only centrifuge tubes and simple containers are contaminated with radioactivity. More complex equipment may be kept out of action for a few days until radioactive contamination is removed. This is expensive and requires a large amount of storage space for radioactive materials collected during decontamination.

Radiochemical separations can be designed for either single or multielement analyses. The maximum number for one person to handle in a day is about fifty.

Techniques in γ-spectrometry require the use of statistical evaluations to allow for background radiation and, more important, the mutual interference of the energy spectra from different isotopes. This interference and the degree of selection required can be modified by a partial separation using radiochemical techniques but results in destruction of the sample. The nondestructive nature of γ-spectrometry is one of its attractions in many biological and other applications.

The method consists of short (minutes) and long (days) irradiations of the one sample followed by short and long delays between various scans on the spectrometer. This gives changing pictures of the elements depending on their half-lives. The energy peaks on the spectra are identified by means of a previously made calibration and measured against known standards or a single standard. However, this single standard technique does require a large amount of calculation and precise timing of the irradiation and counting procedures.

In conclusion, radiochemical separations

give the highest sensitivity and accuracy, but multielement analysis is cumbersome. Lower levels of sensitivity and accuracy are realized with γ-spectrometry, but the technique is nondestructive and simple to apply to multielement analysis.

The selection of the technique is also governed by the half-life of the isotope under investigation. If the half-life is short, then only γ-spectrometry can be used as it is possible to tranfer samples direct from the reactor to the counting device within a fraction of a second. Radiochemical processes take time and though methods are available for fast (5 to 10 mins.) multistage separations many usable isotopes have half-lives significantly below this period. In general, this type of rapid chemistry limits the number of samples that can be handled in one day and processes extending beyond three half-lives are not suitable. Processing should ideally take less than one half-life.

Detection of Radiation

The activity of the isotopes produced by the n,γ reaction consists mainly of β^- and γ radiation and these only will be considered.

Beta-counting may be carried out with a plastic scintillator in a light-tight container but it is more usual to use a Geiger counter. Particles (β^-) can pass only thin sections of material and the counter must be a model so constructed that the radiation can reach the active volume. Aluminium and window counters or glass-liquid counters are suitable for a large number of applications. Most Geiger counting involves the assay of the total activity present in the sample. Generally, only β^- particles are detected in this manner as they give virtually 100 percent efficiency for particles entering the tube. Occasionally γ-rays may be detected by Geiger counting but the efficiency is very low (less than 1%).

Gamma-counting may be carried out using either scintillation detectors (sodium iodide crystals activated with thallium) or solid state detectors (lithium drifted germanium). The scintillation counter gives reasonable sensitivity, but a relatively low degree of separation of γ-energies. The solid-state counter gives low sensitivity, but a relatively high degree of separation. The choice of instrument depends on the energy selection required, the time available for counting and the expected radiation level. These detectors may be used for counting all the γ-radiation (usually a scintillation crystal is used) or for gamma spectrometry in conjunction with a "pulse height analyzer" which sorts out the different energies into a spectrum.

A β^-- or γ-counter left on for some time without a sample being present still records a number of disintegrations. This is due to the normal background (about 0.2 counts/sec for a normal Geiger counter, variable for γ-counters depending on discriminator and analyzer settings) and must be subtracted from any measurements. Normally a background check is made between samples, but may be less frequent when a series of samples is being measured.

Radioactive nuclei disintegrate randomly in time and independently of each other. The result of this is that, other errors being eliminated, the observed count rates of any sample will be distributed at random about the average value. A radioactive assay giving a value n therefore has an error of $\pm \sqrt{n}$. For example, a measure of 10,000 counts has an error of $\pm$ 100 counts, i.e. 1 percent. In practice, 10,000 counts are often impossible to achieve in a reasonable time, so poor statistics must be accepted. However, it is necessary to ensure that any count is significant with respect to the background.

Half-life plays an important part in radioactive assay. If a series of measurements are made on a batch of samples the count rate of each sample is falling with the known half-life. Accurate comparison with each other or with the standard cannot be made without correction for this decay especially if the overall counting time is significant with respect to the half-life. A simple way of correcting the values is to draw a graph of time against activity on semilogarithmic graph paper as follows. Place a mark at 1.0 on the semilogarithmic scale and call this time −0 on the linear scale. Place a second mark at 0.5 on the semilogarithmic scale at time $t_{1/2}$ (1 half-life) on the linear scale and draw a straight line between both points. The decay factor for any time can then be read from the graph. The time to correct to is a matter of choice but is usually the time of the first count, i.e. time 0.

The performance of the whole measuring equipment is also a factor for consideration. After each detection there is a period of time when no more disintegrations are registered. If this "dead time" is low then high count rates can be handled with only a small error but if the "dead time" is high as in Geiger counters then corrections must be made at relatively low count rates. For example, a Geiger counter with a dead time of say $300\mu s$ requires corrections for any count rate above 5 c/s. Some of the more sophisticated γ-counters have built in correction devices.

Calculations

Calculations dealing with the production and decay of an isotope are widely available and outside the scope of this article. The important point to remember about

production is that half of the possible activity is produced after irradiation for one half-life. Up to that point there is a linear relation with time. Beyond it the activity increases as described earlier. Decay problems are simplified when it is remembered that a plot of activity against time on semilogarithmic graph paper is a straight line.

The calculation of the concentration of an element in a sample is made by counting the radiation from the processed material or the selected energy band from the spectrum. The usual procedure consists of a number of steps as follows:

1. Counts/time of samples, standard and background, carefully noting the time at which each count takes place.

2. Change to counts per unit time.

3. Correct counts by subtracting the background counts.

4. Correct for "dead time" if necessary using the following equation.

$$N_c = \frac{No}{1 - No\ T}$$

Where N_c is the correct count rate (c/s)
No is the observed count rate (c/s)
T is the dead time (s)

5. Select a reference time in order to make decay corrections. Usually the time of counting the first sample is selected as time 0 and the others related to it. The correction can either be calculated or obtained from a decay graph as already described. The calculation is based on the following equation.

$$N = No\ e^{-\lambda t}$$

Where No is the number of radioactive atoms present at time = 0
λ is the disintegration constant of the isotope
t is the time lapsed since time = 0

6. If radiochemical separations were used and a constant weight of carrier was added

to each sample and the standard, then a recovery correction should be made. The recovery is calculated for each sample and standard and a representative weight, color intensity, etc., is chosen. Each of the count rates is adjusted to the count rate which would have been observed had the chosen representative recovery been obtained.

The concentration of the element in the sample is then calculated by relating the count rate and weight of the sample to the count rate and weight of the element standard as follows:

$$\text{Elements in sample } (\mu g/g) =$$

$$\frac{\text{Sample count rate (c/s)}}{\text{Sample wt. (mg)}} \times$$

$$\frac{\text{Standard wt. (mg)}}{\text{Standard count rate (c/s)}} \times 10^6$$

Interpretation

The great advantage of activation analysis is the high sensitivity available. Over half of the elements can be usefully detected and measured better by this technique than by any other. Many can be detected below 10^{-9}g and a few even below 10^{-12}g using the normal irradiation facilities available in many reactors. There are a few reactors available which offer much higher neutron fluxes $(n/cm^2/s)$ and, therefore, higher sensitivity of detection.

It should be noted that this high sensitivity does not mean high accuracy. Though there is certainty of identity and detection in very small amounts indeed the accuracy is at best 1 percent when chemical separation is used and then only when a great deal of care is taken. With a normal batch of 30 to 50 samples analyzed for an element, an accuracy of about 3 percent is expected. When gamma-spectrometry is used then the accuracy falls to about 6 percent at best, i.e. when a dominant gamma-photopeak is being evaluated.

The unavoidable error in activation analysis is in the estimation of the quantity of radiation. This is usually expressed in disintegrations per unit time. There is a variation in the count rate which is statistically predictable and the error of any counting is usually accepted as $\pm \sqrt{n}$ so that for 1 percent counting accuracy a total count of 10,000 disintegrations is required. The importance of this in the interpretation of gamma-spectra is plain. A small gamma-peak sitting on a high background of radiation from other sources may be found to have a counting error significant with respect to its own size and, therefore, there may be little value in its determination.

Two methods of evaluating γ-spectra are available in the literature.[2, 3]

A further advantage of activation analysis is that the identity of the measured element is certain. This is established by examining the energy characteristics and half-life of the radiation.

When a result is obtained from such a technique its importance must be evaluated and to do this a base reference of known values is required. This should consist of a random selection of the same material analyzed in the same manner. The results are used to calculate mean values and standard deviations. Two types of value distribution can be applied, roughly, to most situations. These are the normal distribution where the values are held by some controlling mechanism at much the same value, e.g. essential trace elements in tissue, and the log-normal distribution where the values have a skew distribution, i.e. a concentration around some value with a distinct tail of higher values, e.g. nonessential trace elements in tissue. The application of the two types of distribution to trace elements in tissue is discussed by Liebscher and Smith.[4]

Examples of Applications

Neutron activation analysis has been applied extensively to the estimation of the elements in all types of organic and inorganic matrices. This work is described extensively in the literature covering most of the fields where analysis is used. The best source of reference which lists almost all the work reported under matrix, element and author headings is the National Bureau of Standards.[5] Other abstract services are available,[6] and some textbooks list complete methods.[7]

The following examples have been selected to illustrate the application of the technique in various ways to biological materials. In all the examples less than 5 mg of dry tissue is used.

A. *Arsenic in tissue*—a chemical separation without recovery correction and therefore requiring accurate working. This is unusual in neutron activation analysis techniques.
  1. Irradiate samples and standard for one day at $10^{12}n/cm^2/sec$.
  2. Place irradiated samples in digestion tubes and digest with 8 ml of a mixture of concentrated nitric and sulphuric acid (5:3).
  3. Digest till all the nitric acid is removed.
  4. Wash the digested samples into a 200 ml flask.
  5. Add 4 ml concentrated hydrochloric acid and 2 ml concentrated sulphuric acid.
  6. Add 10 μg arsenic carrier.
  7. Add 5 ml of 15% sodium iodide and 0.4 ml of 40% stannous chloride in 50% hydrochloric acid.
  8. Dilute to about 150 ml and place

the flask in a boiling water bath for five minutes.
  9. Add 10 g of zinc pellets, seal the distillation apparatus and allow the reaction to continue for fifteen minutes.
  10. Arsenic in the form of arsine is removed by passing the evolved gas through 1 ml of 1.6% mercuric chloride in a trap. Any hydrogen sulphide is extracted earler by passing the gas through a cotton wool filter impregnated with lead acetate.
  11. Add 5 ml of 0.001N iodine in 40% sodium iodide to the solution in the trap.
  12. Make the solution up to 20 ml and count 10 ml.

B. *Mercury in tissue*—a chemical separation consisting of precipitation steps with recovery. This is the usual neutron activation analysis technique.
  1. Irradiate samples and standard for one week at $10^{12}n/cm^2/sec$.
  2. Place irradiated samples in the digestion tubes and add 10 mg mercury carrier.
  3. Add 2 ml of a mixture of concentrated sulphuric and nitric acids (1:1).
  4. Digest until the brown fumes disappear. If there is any charring remove with a drop or two of nitric acid.
  5. Allow to cool slightly and transfer the digested samples to 50 ml centrifuge tubes by rinsing with water.
  6. Neutralize with 40% sodium hydroxide using congo red indicator.
  7. Place in a hot water bath and add 2 ml of 1% ascorbic acid to precipitate mercury.
  8. Spin down the mercury and wash with water and acetone spinning down each time.

9. Pour off the surplus acetone. Place the tubes in the hot water bath for a short time to dry the mercury precipitate and then dissolve it in 0.5 ml of concentrated nitric acid. Make sure all brown fumes are removed.
10. Dilute and add 0.5 ml of 10% silver nitrate.
11. Add 2 ml of 10% sodium iodide to precipitate silver iodide.
12. Spin down and filter into fresh centrifuge tubes rejecting the silver iodide.
13. Neutralize the solution with 10% ammonia using universal indicator paper.
14. Add 3 ml of copper ethylenediamine complex. This is a mixture of 1 part 10% copper sulphate with 10 parts 10% 1:2 ethylenediamine.
15. Spin down precipitate, wash twice with cold water and once with isopropanol. Transfer the precipitate as a slurry with isopropanol onto previously weighed stainless steel planchettes, dry with infrared lamp and reweigh.
16. Count over the photopeak.

C. *Selenium in tissue*—a partial chemical, partial physical separation technique. This type of separation depends on chemical reduction of interfering activities so that a γ-spectrum of the sought for element can be obtained reasonably free of other photopeaks. Interference from short-lived isotopes and sodium can be almost removed by a delay of one week before beginning the separation.
1. Irradiate samples and standard for one week at 10^{12}n/cm^2/sec.
2. Place irradiated samples in digestion beakers.
3. Add 2 ml of concentrated sulphuric acid and 20 mg of selenium carrier.

4. Heat until white fumes of SO_3 appear and remove any charring with a few drops of concentrated nitric acid.
5. Transfer to centrifuge tubes and add 10 ml of 1% ascorbic acid to precipitate selenium.
6. Heat in a water bath for ten minutes and then wash twice with water and once with ethanol spinning down each time.
7. Transfer the precipitate as a slurry with ethanol to weighted aluminium planchettes.
8. Reweigh and count over the 0.14 Mev photopeak of selenium—75. There are three major photopeaks but this one has the least interference from other γ-activity.

D. *Zinc in tissue*—a purely physical separation. The separation depends on a long delay to allow much of the activity to decay and then estimation by counting over a suitable photopeak.
1. Irradiate samples and standard for one week at 10^{12}n/cm^2/sec.
2. Leave the irradiated material to decay for three months.
3. Count samples over the 1.11 Mev photopeak.

E. *Copper, zinc, cadmium and mercury in tissue*—a chemical separation technique for several elements using ion-exchange separations. This type of analysis offers the advantage of simultaneous analysis combined with the high sensitivity of chemical separation techniques.
1. Irradiate samples and standard for week at 10^{12}n/cm^2/sec.
2. Place the irradiated samples in digestion flasks with 10 mg each of copper, zinc, cadmium and mercury carriers.
3. Digest samples with 2 ml of a mixture of concentrated sulphuric and

nitric acids (1:1) till complete and no nitric acid remains.

4. Dilute remaining solution to 36 ml with 0.5M hydrochloric acid.
5. Pass this solution through the ion-exchange column (7 cm by 1 cm diameter of IRA-400 (Cl form)).
6. Wash the column with 25 ml of 0.12M hydrochloric acid in 10% sodium chloride.
7. Elute the copper with 20 ml of 0.5M hydrochloric acid followed by 20 ml of 0.12M hydrochloric acid in 10% sodium chloride.
8. Elute the zinc (3 ml/min) with 40 ml of 2M sodium hydroxide in 2% sodium chloride.
9. Elute the cadmium with 50 ml of 1M nitric acid (3 ml/min).
10. Strip the mercury from the column with 5% ethylenediamine solution.

The process thus far gives four fractions each containing one of the sought for elements slightly contaminated with each other and similar chemical species. These fractions are then subjected to further separations as follows. That for copper is extensive because many of the major contaminants from biological material appear in this fraction.

Copper

1. Add one drop each of 10% cobalt nitrate, 10% manganese nitrate and 10% ammonium dihydrogen phosphate as carriers.
2. Transfer to centrifuge tubes and add 3 ml of 10% sodium sulphite.
3. Add 1 ml of 10% potassium thiocyanate.
4. Heat on a boiling water bath, spin down precipitate and wash twice with hot water, spinning down each time.
5. Dissolve the precipitate in 0.5 ml of concentrated nitric acid.

6. Add two drops of phosphate carrier and ten drops of 10% ferric chloride, dilute and add three drops of 10% calcium chloride.
7. Neutralize with concentrated ammonia, adding slowly at first and when the precipitation of the iron is complete add a further excess.
8. Add 3 ml of 10% sodium carbonate.
9. Heat on a boiling water bath, filter and reject the precipitate.
10. Neutralize to pale blue color with concentrated acetic acid.
11. Add 0.5 ml of concentrated nitric acid and repeat the thiocyanate precipitation as above.
12. Wash the resultant precipitate and dissolve in 0.5 ml of concentrated nitric acid.
13. Neutralize to a faint precipitate with 5% sodium hydroxide.
14. Add four drops concentrated nitric acid and heat.
15. Add 4 ml of 2% quinaldic acid solution.
16. Wash the resulting precipitate with water and acetone and transfer as a slurry in acetone to preweighed aluminium planchettes.
17. Dry under infrared lamps, weigh and count the activity using an end-window Geiger counter.

Zinc

1. Make the eluate just acid using hydrochloric acid.
2. Add 0.5 ml of 10% ferric chloride carrier.
3. Add 3 ml of 5M sodium hydroxide and heat on a water bath.
4. Centrifuge and filter supernatant solution. Reject the precipitate.
5. Neutralize the filtrate with concentrated acetic acid (phenolpthalein end point).
6. Add 1 ml excess of concentrated acetic **acid.**

7. Heat on a water bath and add excess 2% quinaldic acid.
8. Wash the precipitate with water and acetone, transfer and count as for the copper precipitate.

Cadmium

1. Make the effluent alkaline with 5M sodium hydroxide and centrifuge.
2. Dissolve the precipitate in 1 ml of concentrated hydrochloric acid, dilute to 20 ml.
3. Add 5 ml of 5% thiourea and 8 ml of a solution of ammonium reineckate (saturated at 18°C) in 1% thiourea.
4. Wash the precipitate with 1% thiourea and ethanol.
5. Transfer to preweighed aluminium planchettes as a slurry with ethanol, dry, weigh and count with a Geiger counter.

Mercury

1. Add 1 ml of 10% sodium iodide and 1 ml of 10% copper sulphate to the mercury eluate. This precipitates the mercury iodide-copper ethylenediamine complex.
2. Wash the precipitate with water and isopropanol.
3. Transfer as a slurry in isopropanol to preweighed stainless steel planchettes.
4. Dry, weigh and count over a suitable mercury photopeak.

ATOMIC SPECTROSCOPY

Electronic Origins of Atomic Spectra

Energy changes in the outer shell electrons of atoms are detected by the accompanying emission or absorption of light usually in the UV/visible range, 200 to 800 nm. In Figure 14-1, a schematic diagram illustrates how thermal or electromagnetic energy can be used to cause electrons to be promoted to higher energy levels, and how this can give rise to the various branches of atomic spectroscopy. Because of the simpler nature of the electronic structure of atoms compared to more complex molecules, the emission and absorption of light by atoms is seen as discrete lines in the UV/visible spectrum, rather than the broad bands associated with molecular spectra.

The emission intensity of any given atomic spectral line is determined by the ratio $N_J : No$, where N_J and No are the number of atoms in the higher energy state S_J and the lowest or ground state energy So. The numerical value of this ratio $N_J : No$ is determined by the temperature T of the energy source causing the excitation, and by the energy of the transition $So \rightarrow S_J$. This energy corresponds to a particular frequency or wavelength of light for each element, known as the *resonance wavelength*. This energy matches that of the most probable electronic transition, and

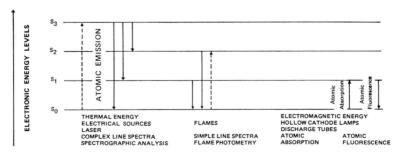

Figure 14-1

TABLE 14-II

Resonance Line, nm	N_J/N_o	
	2,000°K	5,000°K
Na 589.1	4.44×10^{-4}	6.82×10^{-2}
Ca 422.7	9.86×10^{-6}	1.51×10^{-2}
Zn 213.9	7.29×10^{-15}	4.32×10^{-6}

therefore gives rise to the most intense emission line in the atomic spectrum. A list of values for the ratio $N_J : N_o$, at the resonance wavelengths of a few biologically important metals, at two different temperatures is shown in Table 14-II.

As the temperature T increases a greater proportion of atoms reach the higher energy state, and the electrons will emit the resonance wavelength upon return to the ground state. It should be noted that even in the most favorable case, e.g. Na only a small proportion of atoms do reach the higher energy levels and as the wavelength of the resonance line decreases towards the ultraviolet, that very few atoms ever reach the state N_J at all, and therefore the emission of radiation will be weak in such cases, e.g. for Zn.

Atomic Emission Spectroscopy

A basic requirement of any method of studying atomic spectra is the rapid breakdown of any naturally occurring molecular or ionic compounds of the metal to be analyzed. This allows the production of a population of "free atoms" sometimes known as an "atom cell," which is then able to undergo the electronic excitation process. When very high temperature sources, such as electrical arcs or sparks are employed, a great deal of energy is imparted to the outer shell electrons which are promoted to a considerable number of possible higher energy levels. Upon return to the ground state, energy is emitted as a whole series of closely spaced bright lines

in the UV/visible spectrum. Study of such spectra is called *emission spectrography.* This technique requires a very high degree of resolving power in the optical system used for analysis, and therefore very sensitive light detection systems. Components of an emission spectrograph are outlined in Figure 14-2. There is a complex interaction of the various other components in the sample to be analyzed with the metal atoms present, resulting in alterations in the line intensities observed. These interference effects are not fully understood and are difficult to measure and in practice this means that standardization is best done by comparison of line intensities with material similar in general composition to the test samples, but of known and agreed metal composition. However, as a very rapid and sensitive method of multielement qualitative analysis, emission spectrography is most useful. The technique is not often applied in biological work except by specialist laboratories due to the very high capital cost of the equipment.

Applications

Large scale survey of the metal content of soils and plant material is done by this method. By linking the readout system to a laboratory computer it is possible to obtain a printout for about twenty-two elements by the spectrographic analysis of a small sample of dried soil or plant material. In this way deficiencies or excess of some

COMPONENTS OF AN EMISSION SPECTROGRAPH

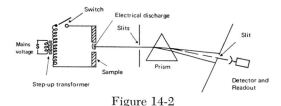

Figure 14-2

metals which may affect the growth of plants or the well being of domestic livestock, can be detected. Thus in various areas of the world deficiencies of B, Co, Cu, Fe, Mn, Mo and Zn have been found in pasture lands, and the correction of such trace metal deficiencies has rendered millions of previously barren acres of land in Australia suitable for agriculture.

Direct analysis of the elemental content of human tissue taken from different body sites has also been carried out by spectrographic analysis. Lists of the metal content of the various organs of the body are published[8] and are useful for reference when an abnormality is being investigated.

A recent development, involving spectrographic analysis has been the use of a *laser microprobe.* This is able to vaporize selectively a very small (50 microns) portion within a single cell. The spectrographic analysis of the vaporized material then gives information about the intracellular content of various metals. This method can achieve the simultaneous assay of eight metals with detection limits of 10^{-13}g for Pb and 10^{-15}g for Mg from the volatilization of very small (10^{-8}g) amounts of intracellular material.

When the source of thermal energy used is in the form of a flame, the technique of analysis is known as *flame emission spectrophotometry.* A flame is a readily available source of energy, which is fairly reproducible in its effects. A considerable practical advantage of flames is that they can readily accept liquid samples so that metals in solution may be analyzed. Also of importance is the fact that flames are less energetic (2,000 to 3,000° K) sources of energy than electrical discharges (5,000 to 8,000° K), and therefore the energy supplied to the electrons of the atoms being analyzed is less, causing fewer electronic

COMPONENTS OF A FLAME PHOTOMETER

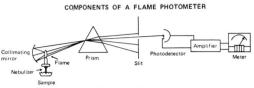

Figure 14-3

transitions to take place. This greatly simplifies the atomic spectra which are observed in flames which consist mainly of the most intense resonance line. As can be seen for the values of N_J : No at 2,000° K given above, even for the more easily excited electrons of the alkali metal series, few atoms in fact reach the higher energy state and are therefore able to radiate light. Flame photometry is therefore more limited in sensitivity of detection than spectrographic analysis. However, the instrumentation required for flame photometry can be of relatively low cost since less elaborate optical systems and light detectors are required to analyze the simpler atomic spectra produced in flames. The components of a flame photometer are shown in Figure 14.3.

The sample, in solution, is introduced to the flame by a device called a *nebulizer.* The most common type employed is one which premixes the sample with the fuel and oxidant and carries this mixture as a fine mist to the flame for combustion:

Various mixtures of fuel and oxidant are available and some commonly used flames are listed in Table 14-III.

PREMIX CONSUMPTION BURNER AND NEBULIZER

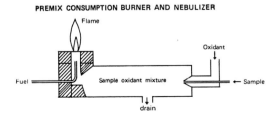

Figure 14-4

TABLE 14-III

TEMPERATURES OF PREMIX GAS
COMBUSTION FLAMES

Fuel	Oxidant	Temperature °K
Propane	Air	2,200
Acetylene	Air	2,450
Acetylene	Nitrous oxide	3,200

Flame Characteristics

Each flame used has its own spectrum of light emission. In Figure 14-5, the spectrum of an oxy-acetylene flame is shown with the resonance lines for some metals superimposed. In some cases the background flame emission could cause a variable interference in the assay of a metal whose principal line was close to the emission wavelength of a flame component. Also shown in Figure 14-5, is the existence of chemical species within the flame such as $CH\cdot$ or $C_2\cdot$ radicals. This fact is of use since chemical reactions may take place in the flame which increase the number of free atoms in the flame. Elements which tend to form stable oxides will be reduced when fuel rich flames with excess hydrocarbon radicals are used. For example, the determination of calcium is more sensitive in a fuel rich or luminous flame since reactions are possible such as:

$$CaO + C_2 \rightarrow Ca + CO + C$$

The choice of a suitable type of flame and the correct mixture of gases in the flame is therefore of great practical importance. The functions of the flame are complex and may be summarized as follows:

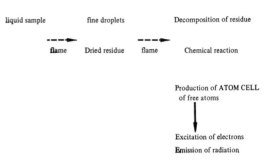

Types of Flame Emission Spectrophotometer

An enormous range of instruments are sold commercially. Summaries of the currently available models are given in the *Annual Reports of Atomic Spectroscopy.*

The very simplest systems, called flame photometers, use glass or interference filters to isolate the desired emission lines, such as Na 589 nm or K 769 nm. A simple direct current photo cell is employed as a light detector and a moving coil galvanometer as read out.

Much more elaborate machines are also sold in which very high quality prisms or gratings are used and sensitive photomultiplier tubes used as light detectors. This type of equipment is required to separate closely spaced lines such as Mg 285.21 and Na 285.28. In addition, a wavelength scanning facility is available making sequential metal analysis feasible.

Applications

A very important use of flame-emission spectrophotometry is in the assay of the

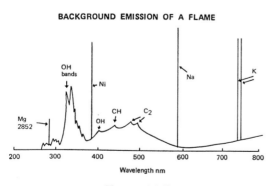

BACKGROUND EMISSION OF A FLAME

Figure 14-5

alkali metals such as Na and K. As has been discussed above these metals are among the most suitable for such determination and are of very considerable biological importance. For example, in routine hospital laboratories many hundreds of measurements are made each day of the concentration of Na and K in blood serum or in urine. The normal concentrations in serum are Na 135 to 145 mmol/1 and K 3.5 to 5.0 mmol/1, and values falling outside these ranges may require investigation and urgent treatment of the patient. A high degree of precision is needed in this analysis and is achieved by use of an "internal standard" technique. The serum sample to be analyzed is diluted, about 1/500 with a dilute solution containing enough lithium ions to give a constant emission line at 670 nm. The Na line at 589 nm and K at 769 nm are isolated with separate interference filters and their light intensities measured as a ratio to that of the reference lithium signal at 670 nm. In this way short term fluctuations in flame conditions or nebulizer efficiency will be common to the required test signals and the internal lithium standard. A precision of better than 1.5 percent can be attained, and using fast automatic analyzers a sample rate of 300 per hour is possible.

The use of the wavelength scanning mode of more elaborate flame spectrophotometers makes it possible to determine several elements in a single sample. Milk samples can be diluted, the protein removed by precipitation, and the supernatant fluid analyzed sequentially for Cu, Mg, Mn and Sr.

Atomic Absorption Spectrophotometry

This has been described as one of the success stories of modern analytical chemistry. Applications of this comparatively low cost method to metal analysis in biological materials are so numerous that the technique must be one of the most rapidly developing, being suitable for metal analysis in soil or plant extracts, food analysis, environmental studies and in many aspects of hospital clinical chemistry.

Basic Principles of AAS

The technique is based upon the absorption of light by free atoms present in a suitable vapor phase or atom cell. As was shown in Figure 14-I, light of a suitable frequency can provide energy to promote electrons from the ground state S_0 to the first excited state S_1, and in this way is itself absorbed. This electronic transition is the resonance transition and the most intense absorption of light will occur when its frequency or wavelength is equivalent to this resonance transition energy. The degree of absorption of light is directly proportional to the number of atoms in the unexcited ground state. As was shown above, Table 14-II, the ratio N_J/N_O is very small even for the most readily excited electrons of the alkali metals. Also, as the energy of the resonance transition increases, that is the corresponding spectral line decreases in wavelength towards the ultraviolet, then by far the greatest proportion of atoms in an atomic vapor are in the ground state N_O. For Zn even at 5,000°K only a few atoms in every million reach the higher energy states N_J. However, all of the atoms in the ground state N_O, are capable of absorbing the correct resonance light energy and therefore AAS is particularly suited to the measurement of such "heavy metals" as Zn, 213.9 nm, Cu 324.8 nm, Pb 217 nm, Cd 228.8 nm, where the resonance line falls below 400 nm, and by far the greatest number of atoms in the

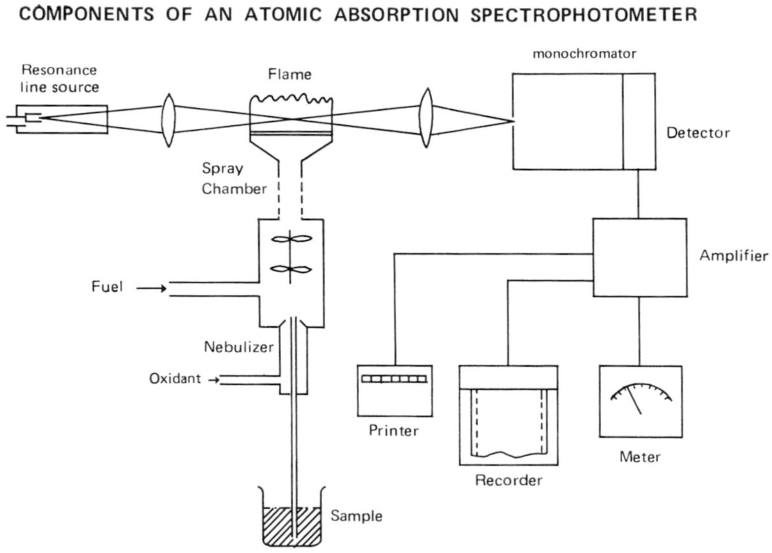

COMPONENTS OF AN ATOMIC ABSORPTION SPECTROPHOTOMETER

Figure 14-6

"atom cell" remain in the ground state. This factor is not the sole reason why AAS is more suitable than flame emission for some elements. Minor changes in flame temperature will considerably affect the small population of excited atoms, and therefore light emission, while the percentage change in the very large population of ground state atoms is very small. This is reflected practically in the good signal to noise ratios generally found in AAS methods, allowing increased sensitivity of detection. The components of an atomic absorption spectrophotometer are shown in Figure 14-6.

Light Source

The successful practical application of the principle of atomic absorption was not possible until a suitably monochromatic source of resonance wavelength light become available. The "target" for light absorption of an atom in the ground state is known as the line width, and in practice, in a flame, this line width is "broadened" by various physical influences to 0.01 nm. Low pressure gas discharge tubes are devices which can emit radiation of a narrow enough range of wavelengths to permit significant absorption by the "line width" of atoms in a flame. A special hollow cathode, with some of the desired metal deposited on it is held in a vacuum tube filled with an inert gas such as neon or argon at low pressure. When an electrical discharge is then passed from an anode to the hollow cathode, atoms of the cathode coating gain energy and emit the resonance line for that element. Other spectral lines are also emitted but these are generally well spaced out and easily separated from the desired resonance line. Usually one such hollow cathode lamp is needed for each metal to be estimated. Although dual element and multielement lamps are sometimes available.

Production of the Atom Cell

The most usual method of producing the desired atomic vapor is by use of a suitable flame. Similar fuel/oxidant mixtures are used to those described in Table 14-III,

TABLE 14-IV

APPROXIMATE LOWER USEFUL WORKING
LIMIT (10% ABSORPTION),
AIR/ACETYLENE FLAME*

	Wavelength nm	$\mu g/ml$
Ca	422.7	0.8
Mg	285.2	0.07
Li	670.8	0.35
Cu	324.7	0.9
Zn	213.9	0.18
Pb	283.3	5.0
Cd	228.8	0.25

* Used with permission. Perkin-Elmer, *Analytical Methods for Atomic Spectroscopy*, 1971.

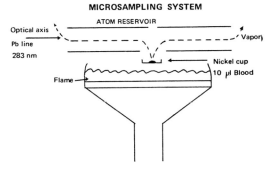

Figure 14-7

but specially constructed long path length burners (5 to 10 cm) are used in order to increase the path length for light absorption. Metals in solution are introduced to the flame via a nebulizer, and the functions of the flame are identical to those described for flame emission analysis, except of course that the final process of excitation is not required. Once more the choice of fuel/oxidant mixture and the nature of the flame to be used will depend upon the metal to be analyzed. Reducing flames should be used with metals tending to form stable oxides, and in certain cases the nitrous oxide/acetylene flame is needed where refractory metal oxides occur, as in AlO or BaO and some others. In addition, the higher temperature of the N_2O/acetylene flame can be used to overcome "chemical interferences" such as stable phosphate formation which would decrease the number of free atoms in the atom cell. Thus calcium analysis, in the presence of phosphate anions can be improved by the use of this type of flame. A list of the lower working limits for some important metals is given in Table 14-IV.

Microsampling System

Although the conventional nebulizer/flame system is suitable for many analyses

there are occasions where severe limitations in the amount of material available for analysis require the use of a microsampling device. More than 90 percent of a liquid sample sprayed into a flame goes to waste, and never reaches the flame. A system is now in use whereby very small samples (10 to 50 μl) of dried material are introduced directly into the flame, placed on cups, boats or ribbons of a suitably inert metal such as tantalum or nickel. In the "Delves cup" version of this system, shown in Figure 14-7, small (10 μl) samples of previously dried blood are prepared in suitable batches in nickel crucibles. These are introduced by a mechanical device to a precise positon in an air acetylene flame, a few mm below a hole in a hollow quartz tube which functions as an atom reservoir, increasing the lifetime of the atom cell in the flame. The absorbance of the Pb line at 283 nm is followed by using a fast-response pen recorder. The first signals obtained comes from nonspecific absorption by the smoke produced as organic material is destroyed and is followed immediately by the lead absorption peak. This type of method is in general use for the assay of Pb in biological samples.

Nonflame Sources of Atom Cells

Devices based on electrically heated *graphite furnaces* or *carbon rods* are now alternative systems for production of atom vapor. The graphite furnace consists of a replaceable graphite tube aligned in the optical axis of the spectrophotometer. This can be heated by electrical inductance to very high temperatures. The tube is held in a water cooled block and can be filled with an inert gas mixture, usually argon or nitrogen. A controlled cycle of heating steps successively dries, chars, then volatilizes the sample placed within the furnace tube. Although expensive, such furnaces allow the detection of very small amounts of metal. For example, 2×10^{-12}g Zn or Cd, 200 to 300×10^{-12}g Ni, Ti and V may be measured in 20 μl samples. A similar type of device based on volatilization from a heated graphite rod is also manufactured. This is less costly to install, but is perhaps not quite so sensitive and precise as the furnace.

Optical System

The output of most hollow cathode lamps is essentially monochromatic and for some applications very simple prisms or gratings can be used. However, it is probably an advantage to obtain a spectrophotometer with a good quality monochromator since in the case of certain metals such as the transition elements, Fe, Co, Ni, the output of the hollow cathode lamps is quite complex, consisting of a number of lines near to the desired resonance wavelength, requiring good resolution in the monochromator. In addition, of course, the flame emission facility available on most modern instruments requires good quality in the optical system.

Both "double beam" or "single beam" optical pathways can be employed. Since fluctuations in lamp intensity can take place and affect the precision of measurement, the more elaborate and expensive double beam systems are preferred where the highest possible accuracy and long-term sta-

bility are required. In addition "dual beam" optics are also used, and in this system two light sources are used and the absorption wanted expressed as a ratio to that of the nonabsorbed line of the reference lamp. Thus, during measurement of Ca, a Sr lamp can be used to give a reference line. This system makes allowance for changes in flame conditions, such as increased scatter or background absorption, and can allow very precise analysis. Unfortunately, only a limited number of pairs of elements are known where suitable line comparisons can be made and whose behavior in the flame is not too dissimilar.

Light Detection and Readout

An obvious complication in atomic absorption spectrophotometry is the necessity to remove the unwanted *emission* from the flame which will also occur at the resonance wavelength, where we wish to determine absorption. The light output from the hollow cathode lamp is modulated, mechanically or electrically, and the amplifier of the light detection system tuned only to detect signals "chopped" at the correct frequency. Thus, the steady emission from the flame, say at 589 nm for Na, is not measured, while the absorption of 589 nm Na light from the pulsed light source can be detected.

Readout systems are enormously varied, including simple meters, pen recorders, strip printers, teletypes or digital voltmeters. It is an advantage to have the signals converted into absorbance rather than transmittance units, and most instruments have a scale expansion device which will allow direct readout in concentration units.

Interference Effects

Very few spectral interference effects are found in atomic absorption and the resolution of Na 285.28 nm and Mg 285.21 nm lines is no longer a problem. However, chemical effects either in the flame or in the sample solution which influences the proportion of atoms in the flame able to undergo absorption do exist and can cause difficulty. *Stable compound formation* is the best-known type of chemical effect and arises when the molecular or ionic compound of the metal to be analyzed is not fully broken down by the flame. Examples are the effects of oxyanions such as phosphate, silicate or aluminate upon the measurement of alkaline earths, i.e. Ca, Mg, Sr. Another example is the formation of stable oxides in the flame which occurs in Al, V or B determination. In addition, proteins may bind metals sufficiently firmly to affect the absorption process, and this can be noticed as a "depression" when calcium is measured in biological material. These effects are overcome in a number of different ways. "Releasing agents" such as La^{3+} salts can be used which will form more stable complexes with interfering $PO_4^=$ anions than, for example, the Ca atoms being analyzed. Other chemicals used are Sr^{2+} and EDTA. Highly reducing flames may inhibit the formation of stable oxides, or high temperature flames such as N_2O/acetylene used to disrupt the stable compounds. High temperature flames however, may themselves introduce an interference effect by causing *ionization* of the metal under study. [Thus, $Ba^0 \xrightarrow{heat} Ba^{2+} + 2e$, this process being 88 percent complete in the N_2O/acetylene flame.] This in turn is dealt with by adding a second, more easily ionized metal to the diluent before analysis. Potassium ion salts are often used and the reaction $K^\oplus \longrightarrow K^+ + e$ appears to limit the ionization effect.

In practice these and many other interference effects can be a problem, but a large and detailed literature now exists which gives a range of methods for overcoming them.[11]

Applications of Atomic Absorption

Some examples in the field of clinical chemistry will be described, the procedures being similar to those used in other areas of research.

Calcium

Interferences

Phosphate is almost always present in biological material at high enough concentrations to cause a depression of Ca absorption. Dilution of serum, urine or solutions of digested food, feces or tissue, with a solution containing 0.1 to 1% La^{3+} is generally satisfactory. Use of a "reducing flame" increases sensitivity by avoiding oxide formation. The N_2O/acetylene flame is sometimes employed, where very high concentrations of phosphate are expected, or extreme sensitivity is required. Dilution of samples in 1% K^+ solutions is then required to prevent severe ionization effects.

Instrumental Conditions

Ca line at 422.7 nm selected. Flame, air/acetylene, faintly luminous.

Sample Preparation

Serum diluted 1/50 with 0.1% La^{3+}. This extent of dilution avoids serious effects from protein. Urine samples are diluted 1/50 to 1/200 using 1% La^{3+} because of the higher phosphate content of urine. Food, feces or tissue samples are dry ashed at 550°C, the residue dissolved in dilute acid, and dilutions prepared as required using 1% La^{3+}.

Standards

Stock solution of 1 mg Ca/ml. Diluted in water to give working solutions within the range of 0 → 30 mg Ca/100 ml. These are then diluted 1/50 with La^{3+} of appropriate strength. Absorbance is linear up to concentrations equivalent to 40 mg Ca/100 ml using this procedure.

Clinical Applications

Normal serum Ca 8.5 to 10.3 mg Ca/100 ml. Normal urine Ca depends on dietary intake but usually $> 50 < 300$ mg Ca/1. Changes in these values occur in various bone diseases. Diagnosis can require "metabolic balance" studies, i.e. measurement of dietary intake and total body losses (urine + feces), over a period of many weeks.

Magnesium

Interferences

No serious interferences are noted and simple dilution of serum, urine or digested food, feces and tissue is all that is needed.

Instrumental Conditions

Mg line at 285 nm. Flame, air/acetylene.

Sample Preparation

Serum diluted 1/100 with 0.1 M HCl. Urine diluted 1/200. Solid material, dry ash at 550°C, dissolved in acid, diluted as required.

Standards

Stock standard 1 mg/ml. Dilutions in water within range $0 \rightarrow 4.0$ mg/100 ml. Final dilution 1/100 with dilute acid. Linear absorbance up to 4.0 mg/100 ml by this procedure.

Clinical Applications

Serum Mg normally within range of 1.5 $\rightarrow$ 2.2 mg/100 ml. Urine magnesium reflects dietary intake, normally $> 50 < 200$ mg/1. Deficiency of magnesium can occur in hospital patients. This is usually due to increased losses from the gastrointestinal tract following surgery and requires treatment by infusion of magnesium salts.

Lithium

Interferences

Slight ionization ($\sim 5\%$) occurs in the air acetylene flame. Therefore, apparent enhancement of Li absorption occurs where the test sample contains variable amounts of potassium. Dilution of serum or urine with solutions containing excess potassium removes this effect.

Instrumental Conditions

Li line at 670 nm. Flame, air/acetylene.

Sample Preparation

Dilution (1/50 for serum, 1/200 urine) in aqueous solutions containing 200 ppm K^+.

Standards

Stock standard 10 mmol Li/1. Prepared in water from Li_2SO_4. Working standards prepared in water at concentrations of 0, 0.5, 1.5, 2.0, mmol/1. Then diluted 1/50 with 200 ppm K^+ solution. Method linear in absorbance up to 4.0 mmol Li/1.

Clinical Application

Lithium carbonate is used as a drug to treat manic-depression. Serum Li should be kept above 0.6 mmol/1 and below 1.5 mmol/1. Serious toxic side effects occur above 2.0 mmol/1, regular checks of blood Li concentration are needed.

Lead

Interferences

No serious chemical interferences. Optical scattering in the flame can be a problem. This may be decreased by selecting 283 nm Pb-line one rather than the more sensitive one at 217 nm. Or concentration methods involving chelation, solvent extraction may be used to semipurify Pb from other substances.

Blood Pb, using Delves Microsampling Device Instrumental Conditions

Pb line 283 nm. Triple slot burner, air acetylene flame, microsampling accessory fitted.

Sample Preparation

Whole blood samples (10 μl) placed in duplicate in nickel crucibles, then dried for two min. at 140°C on a hot plate. The cup is then placed on the holder and inserted into the flame. The absorbance at 283 nm is recorded using a fast response pen recorder.

Standards

Pb is added to whole blood and standards containing 0, 20, 40, 80, 120 μg Pb/ 100 ml are prepared in duplicate. A calibration graph of the peak height of the Pb signal plotted against Pb concentration is then linear up to about 120 μg Pb/100 ml.

Clinical Applications

Blood Pb analysis is used to detect patients suspected of suffering industrial or environmental exposure to lead. Normal values were found to be 25 ± 7 μg Pb/100 ml in a series of 158 people not known to be exposed to a Pb hazard. A group of industrial workers at risk had values of 55 ± 20 μg Pb/100 ml. Confirmation of Pb poisoning, however, requires additional biochemical tests and clinical signs such as anemia or severe abdominal pain.

Atomic Fluorescence

When an atomic vapor is irradiated by an intense light source of the resonance line wavelength, light of an identical wavelength will be reemitted by the electronic process shown in Figure 14-8. Therefore, when a light detector is placed at right angles to an atomic vapor in a flame during

COMPONENTS OF AN ATOMIC FLUORESCENCE SPECTROPHOTOMETER

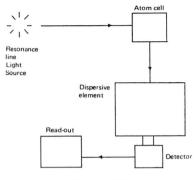

Figure 14-8

its irradiation by a suitably intense line source, emitted light will be detected and this is called atomic fluorescence.

Production of the Atom Cell

Conventional flame/nebulizer systems are in use, but high background signals from the flame can limit the sensitivities achieved. Nonflame sources such as furnaces or carbon rods are very suitable for fluorescence studies.

Applications

Although still in the early stages of commercial development, atomic fluorescence offers considerable gains in sensitivity for certain metals. Using a graphite furnace, detection limits of 10^{-12}g Ag and Mg, 10^{-11}g Pb and Bi, and 2×10^{-14}g Zn can be achieved, giving perhaps the best sensitivities of any spectroscopic technique. Serious interference effects in biological samples can however cause difficulty.

OTHER METHODS OF METAL ANALYSIS APPLIED TO BIOLOGICAL MATERIAL

Colorimetric Analysis

Prior to the instrumental techniques described above, considerable effort was made to develop reagents which would form colored complexes with metals in solution and allow their assay by molecular absorption spectrophotometry. Although limited in sensitivity (10^{-4}g of metal) and of dubious specificity in some instances, many useful reagents were found which are still of value. This is especially true where the method can be used in one of the modern high-speed automatic analysis systems. Thus, the colored dyestuff formed by calcium with the reagent cresolphthaleine complexone allows the routine assay of Ca in serum and urine at a rate of 300 samples/hour, the accuracy of the method being comparable to atomic absorption analysis. The determination of Fe in blood serum is very conveniently done by a color reaction with tripyridyl triazine, an autoanalyzer can be used and a sample rate of 40/h is possible, with a precision of around 10 percent within the normal range for serum Fe, 50 to 150 μg Fe/100 ml serum.[12]

Fluorimetric Analysis

By reaction of metal ions with other organic reagents, complexes may be formed which will fluoresce in solution, and provide a very sensitive method of analysis. Although limited in application by severe quenching or interference effects, some methods may be useful. The screening of workers using Be metal is necessary, since this toxic substance can be very dangerous. A multistep method for Be in urine has been developed which ends with the condensation of Be and the reagent morin, to form a fluorescent complex.[13]

Electroanalytical Methods

Polarographic methods, based upon the current-voltage changes as metals ions are discharged from solution on to a suitable electrode have long been useful for multi-element analysis. Conventional polarogra-

phy, using the dropping mercury electrode is rather limited in sensitivity and to a degree is being replaced by a technique called anodic stripping voltammetry. A stationary electrode (usually a wax impregnated Hg coated graphite rod) is held in a liquid cell holding the test solution, another reference electrode and a platinum electrode. A variable negative potential is then applied to the graphite electrode and the metals in solution "plate out" on to the electrode. By then reversing the applied potential the metals are "stripped" off the electrode in an order related to their electrochemical potential. The order of release gives a qualitative analysis of a mixture and the peak height of the current flowing as each metal is given off is proportional to its concentration.

The equipment is now available commercially and its considerable sensitivity and range makes it a very useful additional technique. From $0.1 \rightarrow 1 \times 10^{-9}$g of Bi, Cd, Cu, Au, In, Pb, Te and Zn can be measured. New types of electrode will eventually extend this range. A limitation, when applied to biological material, is the requirement that all organic substances must be previously removed, usually by wet ashing with strong acid mixtures. Highly purified chemicals must be used or serious contamination of the test material results.[9]

Electron Microprobe Analysis

Electron probe, X-ray microanalysis is a means of identifying chemical elements and measuring their concentrations in very small samples (1 micron) by exposure to a beam of high energy electrons. The X-rays generated by the electrons are then identified by X-ray spectroscopy. The method is based on a modified electron microscope and is of high capital cost but offers extreme sensitivity for all elements of atomic number greater than 8. Detection limits of 10^{-16}g are possible and much vital information about the distribution of elements within small biological samples can be obtained. For example, electron microprobe scans of sections of the human aorta can demonstrate in detail the layers of increased mineralization occurring in atherosclerotic disease.

Similarly the elemental profile of a section of a human tooth can be measured and such studies are of value in dental research, for example on the effects of fluoride. It is clearly of great value to be able to correlate the chemical distribution of elements with the structural detail obtained by light or electron microscopy on the sample.[14]

LIMITATIONS OF ANALYTICAL TECHNIQUE

Although the variety of modern methods is great and seemingly capable of tackling most analytical problems, there are certain limitations to their effective use, and at times considerable doubt as to the meaning of the answers obtained.

Contamination of the Sample

The very great sensitivity of some methods poses a problem in itself. Hitherto, unsuspected sources of inadvertent contamination of the material being analyzed must be carefully avoided if a useful result is wanted. Sample collection and storage may be suspected of adding unwanted metal. For example, blood samples collected by venipuncture using steel needles can add unwanted amounts of Cr and Ni, falsifying the result. Storage of dry soil samples in brown paper sacks was found inadvisable since significant amounts of boron were transferred from the paper to the soil. Many common laboratory materials such as glass, rubber and polyethylene are rich sources of metal contaminants. In ideal circumstances all sample handling and instrumental analysis should be done under "clean room" conditions to avoid laboratory dust, a potent source of trouble. Some techniques such as neutron activation or electron probe analysis where the "reagents" used are neutrons and electrons, avoid the real problem of contamination by the chemical reagents used during other methods of analysis. Much work on the subcellular distribution of metals using conventional methods of biochemical isolation of cellular components, such as density gradient centrifugation ran into severe difficulty because of the unavoidable contamination by metal centrifuge rotor heads or the sucrose used for density gradient formation.

Biological Role of Metals

Analytical data alone can give little direct information as to the biological importance of a metal. Most methods only measure total concentration of metal present in a sample and give little information as to the naturally occurring metal complexes present which are biologically important. Concentration measurement alone does not help greatly when deciding if a metal has an essential biochemical function. The list prepared by Bowen[1] of the elemental content of vertebrate blood shows that some seventy-eight elements can be detected in the red cell. When this data is recalculated in terms of numbers of *atoms per cell*, it is found that very few metals are present at a level of less than one atom per cell. Only a few elements such as Ra, Ac, Pr, and Po can be excluded on this basis as essential elements.

The evidence that a metal may have a vital biochemical function can at present only be derived by difficult bioassay procedures.

Laboratory animals are reared in ultra-clean conditions, plastic isolator cages are used and the air supply carefully purified. Specially formulated and highly refined diets are given lacking only in the metal under study. Animals are maintained on this diet for many weeks and their growth compared to a control group fed on identical diet, but with the addition of the metal being investigated. The rate of growth and general well-being of the two groups of animals is compared, and if the metal is one of the essential ones, dramatic failures of growth and maturation can be seen. In this way quite unsuspected metals, such as tin, can be shown to be an essential metal for the growth of the rat. Similar studies are being conducted for numerous metals especially where some doubt exists as to the dietary requirement of the metal.[10]

In conclusion, it must be noted that while present analytical methods can provide vital initial information about metals in biological material, the full understanding of bio-inorganic chemistry requires analytical data to be supplemented by information about the biochemical functions of metals. Also, as much information as possible needs to be collected on the dietary form of metals, their absorption, tissue distribution and excretion in the intact animal species being investigated.

REFERENCES

1. Bowen, H. J. M.: *Trace Elements in Biochemistry.* Academic Press, 1966.
2. Covell, D. F.: *Analyt Chem.* 31:1785, 1959.
3. Liebscher, K. and Smith, H.: *Analyt Chem.* 40:1999, 1968.
4. Liebscher, K. and Smith, H.: *Arch Environ Health,* 17:881, 1968.
5. National Bureau of Standard note 467 Activation Analysis, A Bibliography published by US Department of Commerce.
6. *Neutron Activation Analysis Abstracts.* London, Science and Technology Agency.
7. Bowen, H. J. M. and Gibbons, D.: *Radioactivation Analysis.* Oxford U. Pr.
8. Tipton, I. H. and Cook, M. J.: *Health Phys,* 9:103, 1963.
9. IUPAC: In Kemulka, W. (Ed.): *Analytical Chemistry.* London, Butterworths, 1969, p. 449.
10. Frieden, E.: *Sci Am,* 227:52, 1972.
11. Price, W. J.: *Analytical Atomic Absorption Spectrometry.* Heyden and Son, 1972.
12. Sandell, E. B.: *Colorimetric Determination of Traces of Metal,* 3rd ed. New York, Interscience, 1959.
13. Undefriend, S.: *Fluorescence Assay in Medicine and Biology.* New York, Acad Pr, 1962, p. 388.
14. Hall, T. A.: The microprobe assay of chemical elements. In Osler, G. (Ed.): *Physical Techniques in Biological Research,* vol. 1, Part A, Optical Techniques, 2nd. ed. New York, Acad Pr, 1971, pp. 158-266.

ADDITIONAL READINGS

1. Heath, R. L.: *Scintillation Spectrometry and Gamma-ray Spectrum Catalogue.* A.E.C. Research and Development Report, Physics, T1D-4500, Phllips Petroleum Co., Atomic Energy Division.
2. Crouthamel: *Applied Gamma-ray Spectrometry,* 2nd ed. Reviewed by Adams, F. and Dams, R. New York, Pergamon.
3. Soete, D. De, Gijbels, R. and Hoste, J.: *Neutron Activation Analysis.* Wiley-Interscience, New York.
4. Bowen, H. J. M. and Gibbons, D.: *Radioactivation Analysis.* Fair Lawn, Oxford U Pr.
5. Lederer, C. M., Hollander, J. M. and Perlman, I.: *Table of Isotopes.* New Jersey, Wiley.
6. Koch, R. C.: *Activation Analysis Handbook.* New York, Acad Pr.
7. Annual Reports on Analytical Atomic Spectroscopy 1971, 1972 vol. 1, 2. The Society for Analytical Chemistry.

CHAPTER 15

ALKALI METAL COMPLEXES AS PROBES FOR MEMBRANE TRANSPORT — A REQUIREMENT FOR MULTIMETHOD APPROACHES

David E. Fenton

The paracrystalline state seems the most suited to biological functions, as it combines the fluidity and diffusibility of liquids while preserving the possibilities of internal structure characteristic of crystalline solids.[*]

Joseph Needham

Introduction
Alkali Metals and the Cell
Selectivity Sequences
Structures of Alkali Metal Complexes
Conclusions

INTRODUCTION

THE RENAISSANCE of interest in complexes of the alkali metals may be considered as the product of a liaison between biology and chemistry. The requirement for information concerning the role of these metals in membrane transport processes stimulated efforts to simulate the conditions of transport and to attempt to understand the interaction of the metal with the environment provided by such systems.

Although it is possible to study the role of certain of these metals in physiological processes using radioactive tracer techniques, there are few other properties that can readily be used to investigate the environment of the metal. Nuclear magnetic resonance (nmr) techniques involving ^{23}Na

have been used for the direct study of sodium, but there has been little extension of the use of nmr techniques to other alkali metals. Consequently, the development of the use of isomorphous probe elements has evolved, together with the application of "multimethod techniques." This term is used to refer to the application of a wide range of techniques to solve a problem which, in other areas, may have been readily soluble either directly or by inference from an available model system. The absence of utilizable spectral parameters for the alkali metal cations contrasts with the transition metals where much information is available from spectral data.

This chapter is designed to show how a range of techniques has been used to elucidate the role that alkali metals play in systems that may simulate those concerning the transport of ions across membranes.

ALKALI METALS AND THE CELL

The majority of cells have a high internal potassium ion concentration and a low internal sodium ion concentration. This is in contrast to their surrounding fluids where the ratio is reversed. For example, in human

[*] Joseph Needham, *Biochemistry and Morphogenesis* (London, Cambridge, 1942), p. 661.

red blood cells the $K^+ : Na^+$ ratio is 105 : 10, while the ratio is 5 : 143 in blood plasma. When the cells are killed, or when their metabolism is reduced, the Na^+ level rises as the K^+ level falls.

The cell exercises a discriminatory control over the uptake of the alkali metal ions. If free diffusion of ions into and out of the cell could occur the concentration differences would not arise. It is the membrane of the cell which acts as a living barrier, controlling the diet of ions.

The energy that is required to maintain the process of "pumping" the ions across the membrane against the concentration gradient is provided by the hydrolysis of adenosine triphosphate (ATP) by enzymes known as Na^+/K^+ATP-ases. The actual mode of ion transfer has led to much conjecture and numerous theories. One concept concerns the active role of a "carrier" molecule, and it is proposed that the "carrier" molecule would complex reversibly with the alkali metal ions and transport them through the membrane system.

The architecture of the membrane is known, and not known. The chemical constituents are well established with phospholipids, proteins and cholesterol as the major constituents. Sialic acid, phosphatidic acid, glycolipids and inorganic ions, especially calcium, act as the minor constituents. The amount of constituent present varies with the type of membrane.

The actual organization of the membrane is less certain. Danielli and Davson proposed that the membrane consisted of a lipid bilayer coated on each side with protein. The advent of electron microscopy proved this hypothesis to be remarkably true, the membrane appearing in the electron micrograph as a pair of dense parallel lines with a more transparent area between them. More recently freeze-etch studies have indicated that the protein can

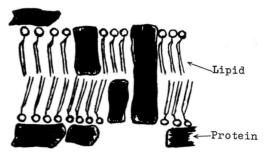

Figure 15-1. Schematic representation of the membrane.

penetrate into, and across, the lipid layer (Figure 15-1).

It is the requirement for information concerning the selective passage of cations

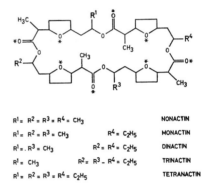

$R^1 = R^2 = R^3 = R^4 = CH_3$		NONACTIN
$R^1 = R^2 = R^3 = CH_3$	$R^4 = C_2H_5$	MONACTIN
$R^1 = R^3 = CH_3$	$R^2 = R^4 = C_2H_5$	DINACTIN
$R^1 = CH_3$	$R^2 = R^3 = R^4 = C_2H_5$	TRINACTIN
$R^1 = R^2 = R^3 = R^4 = C_2H_5$		TETRANACTIN

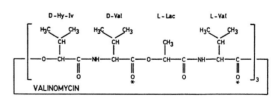

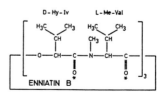

Figure 15-2(a). Antibiotics—Class I

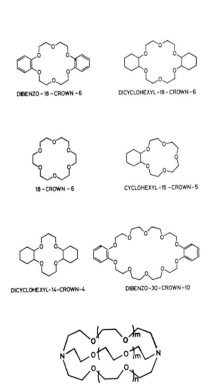

Figure 15-2(b). Antibiotics—Class II

DIBENZO-18-CROWN-6

DICYCLOHEXYL-18-CROWN-6

18-CROWN-6

CYCLOHEXYL-15-CROWN-5

DICYCLOHEXYL-14-CROWN-4

DIBENZO-30-CROWN-10

CRYPTATES

Figure 15-2(c). Synthetic polyethers

across such a structure that has led to the current interest in alkali metal complexes. In 1964, B. C. Pressman reported the phenomenon of antibiotic-induced ion transport in mitochondria. Certain streptomyces metabolites, such as valinomycin, the gramicidins, the enniatins, and the macrotetrolide actins were found to initiate the energy-linked accumulation of alkali metal cations. As it was possible to reverse the ion flow by interrupting the availability of energy, it was proposed that this was a form of active transport. A second class of antibiotics, nigericin and monensin, was found to reverse the effect of valinomycin, and these were believed to cause an inhibition of the ion pump.

The two classes of antibiotic differed in that the first group were all neutral molecules at physiological pH and so would transport ions as positively-charged species (cationic complexes), while the second group contained a carboxylic acid at physiological pH. This acid needed to be deprotonated prior to the antibiotic acting as a neutral carrier. These two classes of ioncarriers were termed "ionophorous agents" or "ionophores."*

In 1967, C. J. Pedersen published details of a class of synthetic polyethers which were capable of mimicking the discriminatory abilities of the antibiotics. A further class of selective agents, the "cryptates," a series of macrobicyclic amino ethers has

---

* The term *ionophore* was introduced by R. M. Fuoss in 1955, defining nomenclature for conductance. It was used to designate a species in which ions, and only ions, were present in the crystal lattice, e.g. NaCl. A substance whose molecular crystal lattice produced an electrolyte by reaction when dissolved in an appropriate solvent was termed an *ionogen*. Pressman used the term ionophorous agent to denote a species carrying ions across lipid barriers as lipid soluble complexes. It therefore would be better to adhere to this term rather than abbreviate it to ionophore.

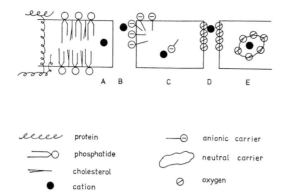

llll protein	—⊖ anionic carrier
⊐O phosphatide	⊂⊃ neutral carrier
= cholesterol	
● cation	⊘ oxygen

Figure 15-3. Modes of ion transport (after Eisenman)

been described by J. M. Lehn. Examples of the antibiotics and synthetic ethers are given in Figure 15-2.

Several modes of ion transfer across a membrane are illustrated schematically in Figure 15-3.

The mechanisms depicted are:

A. The permeation of a bare (or de-solvated) cation. This is unlikely as the lipid barrier has a low dielectric constant. A small leakage of cations could, however, occur.

B. The construction of a negatively charged pore, or canal, by the lipids through their head groups, or by a suitable anionic species could provide a means of transport through the membrane.

C. A lipophilic molecule capable of carrying the cation from one membrane surface to the other could exist. If negatively charged then the cation could hitchhike across the barrier as a neutral species. Nigericin and monesin are molecules of this type.

D. A cluster of neutral molecules with a suitably electron rich interior could form a neutral pore for transport.

E. A lipophilic, neutral molecule, such

as valinomycin, could transport an overall positively charged species across the membrane.

It is also possible that the protein constituent of the membrane could act in one of these capacities. Variations on these models may also exist; for example, would a carrier exist as an isolated species, or would a carrier relay system operate in which the cation is passed from molecule to molecule?

SELECTIVITY SEQUENCES

The selectivity sequences for simulated transport have been derived using a wide variety of techniques. Salt extraction by the antibiotics and models from aqueous solutions into organic solvents has been widely used to establish selectivities. Such experiments involve shaking an organic solvent phase containing the alleged carrier with aqueous solutions of alkali metal salts containing a lipid-soluble, colored anion. The intensity of color of the organic phase may then be monitored spectrophotometrically to determine the degree of extraction. Radioisotopes of the cation may also be used as probes, the migration of radiation into the organic phase being the measure of transport. The extraction of picrate, or dinitrophenolate, salts into *n*-hexane, dichloromethane or mixtures of solvents is generally used, as these solvents may be taken to represent the interior of the membrane di-electrically. The results obtained from these experiments can be compared with those designed to show the effects of the carriers on the electric-resistance properties of artificial phospholipid bilayers.

An interesting and effective modification of this procedure was the use of U-tubes to provide a transfer simulating system (Figure 15-4).

A semipermeable barrier ($CHCl_3$) was

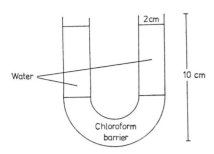

Figure 15-4. The U-tube experiment

interposed between two aqueous solutions and the appropriate antibiotic was dissolved in the chloroform layer. The appropriate salts were dissolved in the side-arms and the movement of salt through the barrier monitored—spectrophotometrically for the picrate ion and by pH meter for protons. Selectivity sequences were then compared for different antibiotics. Sequences determined using solvent extraction are given in Table 15-I.

The cyclic polyethers show a lower degree of activity than the antibiotics. This may be understood in terms of their having lower partition coefficients into the membrane phase, and in their restricted ability to encapsulate the alkali metal and shield it from solvent interaction in the cases studied.

The effects of the neutral antibiotics on the electrical properties of bilayers have

shown that the membrane resistance and membrane potential depend upon the nature of the carrier and cation and these effects follow the same orders as the selectivity sequences. The effect of several of the antibiotics on the electrical resistance of lipid bilayers prepared from lipids extracted from sheep red cell membranes showed that valinomycin, enniatin B and the actins showed a marked preference for potassium. Enniatin B however was less potent than valinomycin and the actins, and the synthetic crown polyethers were less potent than all of the antibiotics. The polyethers having five oxygen atoms showed marked sodium preference, while those with six oxygen atoms preferred potassium.

Stability constant measurements using vapor pressure osmometry, optical rotatory dispersion, conductivity measurements (using glass electrodes and liquid membrane electrodes), spectrophotometry, relaxation techniques and microcalorimetry have all contributed to establishing the sequences in Table 15-II. These have added quantitative weight to information retrieved from solvent extraction experiments.

Valinomycin and its analogues have so far shown the greatest specificity for potassium, and valinomycin has been used as a neutral carrier in liquid-membrane electrodes. In such an electrode the potassium over sodium specificity is of the order of 10^4. Experiments with frozen electrodes showed that this ratio dropped to approximately 2 for the valinomycin electrode. This is interpreted as supporting the view that valinomycin and similar antibiotics act as carriers, not pores, as on freezing the mobility of the molecule is lost, thus inhibiting transport.

Monitoring the antibiotic-mediated conductance of the bilayer has shown that the effectiveness of the presumed carrier is

TABLE 15-I

SELECTIVITY SEQUENCES FROM SOLVENT
EXTRACTION DATA

Valinomycin :	Rb > K > Cs > Na
Actins :	K > Rb > Cs > Na > Li
Antamanide :	Na > Li ~ K > Rb ~ Cs
Nigericin :	K > Rb > Na > Li
Monensin :	Na > K > Rb, Li
Dianemycin :	Na, K, Rb > Li
dibenzo-18-crown-6 :	K > Cs > Na > Li
dicyclohexyl-18-crown-6 :	K > Cs > Na > Li
benzo-15-crown-5 :	Na > K > Cs > Li

TABLE 15-II

LOG FORMATION CONSTANTS FOR ANTIBIOTIC-ALKALI
METAL COMPLEXES

Antibiotic	Log Formation Constant ($\log K_1$)					Soln.	$T°C$
	Li^+	Na^+	K^+	Rb^+	Cs^+		
Nonactin		2.3	4.1			EtOH	30°
		2.4	3.6	3.5	2.9	MeOH	30°
Enniatin B		3.1	3.6	3.6	3.3	EtOH	25°
		2.4	2.9			MeOH	25°
Valinomycin		0	6.3	6.4	5.8	EtOH	25°
		1.1	3.9			MeOH	25°
Beauveracin	2.0	2.5	3.5	3.5	3.5	EtOH	
Antamanide		3.4	2.4			EtOH	
Nigericin		4.4	5.2			MeOH	25°
Monensin		5.8	5.0			MeOH	25°

Compiled from W. E. Morf and W. Simon, *Helv Chim Acta*, 54 (1971), p. 2683.

also lost when the fluidity of the lipid bilayers is diminished on freezing. If gramicidin is present, however, the transport effects are unaltered and it may be proposed that this molecule provides a pore. It is also possible that both models could exist, perhaps simultaneously.

Antamanide is currently the only cyclic peptide system showing strong sodium over potassium selectivity. The actins show a strong preference for the ammonium ion over the alkali metals. For the synthetic models it has been established that the monocyclic systems are less selective than the encapsulating and bicyclic ligands. A dependence of stability constant upon hole size and on the number of available oxygen atoms was established for the cyclic polyethers (Table 15-III). It was also shown that the formation constants were higher in methanol than in water. This reflects the competition between complex formation and ion solvation, and is also observed for the antibiotics. A small cation has a strong electric field and so can attract both ligand and solvent, while the larger cations attract the ligand less and have lower solvation energies. The formation constants re-

TABLE 15-III

FORMATION CONSTANTS OF SOME CYCLIC POLYETHER COMPLEXES, LOG K_1

Cation		Na^+	K^+	Cs^+	Solvent
Ionic Diameter (Å)°		2.24	2.88	3.68	
Polyether (hole size in Å)					
Dicyclohexyl-14-crown-4	1.2-1.5	2.18	1.30		MeOH
Dibenzo-18-crown-6	2.6-3.2	4.36	5.00	3.55	MeOH
Dibenzo-30-crown-10		2.0	4.6		MeOH
Cyclohexyl-15-crown-5	1.8-2.2	3.71	3.58	2.78	MeOH
		0.3	0.6		
Dicyclohexyl-18-crown-6 (Å) .	2.6-3.2	4.08	6.01	4.61	MeOH
		1.5-1.85	2.18	1.25	H_2O

° The ionic diameters used throughout are those given by M. F. C. Ladd, *Theor Chim Acta*, Vol. 12 (1968), p. 333.

Compiled from H. K. Frensdorff, *J Am Chem Soc*, Vol. 93, (1971), p. 600.

TABLE 15-IV

FORMATION CONSTANTS FOR SOME CRYPTATE COMPLEXES, LOG K_1

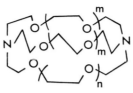

Cation	Li^+	Na^+	K^+	Rb^+	Cs^+
Ionic radius (Å)	0.86	1.12	1.44	1.58	1.84
Bicyclic Ligand (cavity size in Å)					
m = 0, n = 1 (0.8)	4.30	2.80	2.0	2.0	2.0
		2.7[a]†			
m = 1, n = 0 (1.15)	2.50	5.40	3.95	2.55	2.0
		9.0[a]			
m = n = 1 (1.4)	2.0	3.90	5.40	4.35	2.0
		9.0[a]			
m = 1, n = 2 (1.8)	2.0	2.0	2.2	2.05	2.20
		4.80[a]			
m = 2, n = 1 (2.1)	2.0	2.0	2.0	0.7	2.0
		2.80[a]			
m = n = 2 (2.4)	2.0	2.0	2.0	0.5	2.0

Compiled from J. M. Lehn and J. P. Sauvage, *JCS Chem Comm* (1971), p. 440.
† The values are for water solutions, except for *a* which are for methanol.

flect the balance of these properties as the smaller cations (Li^+, Na^+) would be strongly solvated whereas the larger cations would not attract the polyether. Ligand-ligand interactions, and the number of available oxygen atoms would modify the balance.

The cryptates exhibit a more cavity specific control as predicted from their cage-like nature (Table 15-IV). The cavity size is estimated from the radius of the sphere which would be included, without distorting the ligand, into the molecular cavity of space filling models of the ligand. The ligands can be envisaged as existing in exo-exo, endo-exo, and endo-endo forms, and the latter is used in estimating cavity size. Solvent dependency was again observed and it is possible to prepare complexes which are unstable in water in other solvents. For example the caesium complex of the cryptate with m = n = 1 may be isolated from chloroform solutions.

Further evidence for the efficiency of the cryptates in encapsulating metal ions is provided by the reaction with potassium fluoride, a high lattice energy salt not solubilized by the cyclic polyethers, and by the slow dissolution of barium sulphate in water in the presence of cryptate (iii).

STRUCTURES OF ALKALI METAL COMPLEXES
X-ray Crystallographic Studies

The establishing of the discriminatory powers of the ligands available for simulating transport leads to a requirement for information concerning the nature of the complex and of the immediate environment of the metal.

The first class of antibiotics as described by Pressman consist in the main of closed ring systems. It was first thought that they could encapsulate the hydrated alkali metal cation, but X-ray structural analysis has shown that, in the cases examined so far, the bare cation is encapsulated by the ligand. This occurs such that the metal is coordinated by six or eight oxygen atoms, and the external surface of the complex is lipophilic. In order that it may incorporate the metal ion the ligand must be flexible.

The second class of antibiotics, such as monensin and nigericin, may be formally written as linear systems. In the alkali metal salt the acid is wrapped round the cation, and the molecule is held cyclic by a hydrogen-bond between the deprotonated carboxylic acid group and an hydroxylic group. The metal coordination number ranges from five to seven, and the same criteria operate for incorporation as did for the neutral ligands. A survey of the available data for both systems is given in Table 15-V.

Valinomycin is a cyclic dodecadepsipeptide consisting of [D-valine, D-α-hydroxyvaleric acid, L-valine, L-lactic acid]$_3$. This provides a 36-atom ring with twelve carbonyl oxygen atoms. The structure of the potassium complex, determined as the tetrachloroaurate, can be seen to exist with the ligand folded round the metal in three sine waves (Figure 15-5b). The six ester carbonyl atoms point towards the center of the system where they produce an approximately octahedral environment for the potassium. Each peptide carbonyl is hydrogen bonded to the closest peptide = NH group to form a network of stabilizing hydrogen-bonds across the sine loops (Figure 15-5c). This may be used as a rule for such systems, i.e. cyclic dodecapeptides and cyclic dodecadepsipeptides. A further rule is that the side chains should not interfere with hydrogen bond formation. The importance of the hydrogen bonding may be illustrated by N-methylation of valinomycin when the selective activity of the antibiotic is lost.

The structure of uncomplexed valinomycin was found to be more elongated than

TABLE 15-V

	Ring Size	Coordination Number	Bond Lengths (Å)	Cation
a) Antibiotic (I)				
Valinomycin	36	6	2.8	K⁺
Enniatin B	18	6	2.6-2.8	K⁺
Nonactin	32	8	2.82 (furan)	K⁺
			2.77 (keto)	
			2.8 (furan)	Na⁺
			2.4 (keto)	
Tetranactin	32	8	2.94 (furan)	Rb⁺
			2.91 (keto)	(isomorphs)
			2.89 (furan)	K⁺
			2.79 (keto)	
			2.87 (furan)	K⁺
			2.77 (keto)	(isomorphs)
			2.82 (furan)	Na⁺
			2.43 (keto)	

	Carbon Atom Backbone	Coordination Number	Cation
b) Antibiotic (II)			
Nigericin	30	5	Ag⁺ (isomorphous K⁺, Na⁺)
Monensin	26	6	Ag⁺ (isomorphous K⁺, Tl⁺)
Dianemycin	30	7	Tl⁺, K⁺, Na⁺

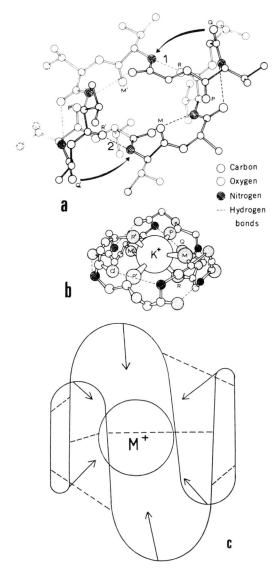

a

○ Carbon
○ Oxygen
● Nitrogen
--- Hydrogen bonds

b

c

Figure 15-5. The uncomplexed, (a), and K⁺ complexed, (b), conformations of valinomycin. The carbonyl oxygen atoms P. P′, M and M′ are those in the most exposed positions and they could begin complexing with K⁺ ions. The hydrogen bonds marked 1 and 2 could be disrupted, allowing atoms R and R′ to provide the remaining sixfold coordinate for the K⁺ ion. The rounding out of the molecule as the ion enters is accomplished by minor conformational changes which bring atom Q and Q′ into posi-

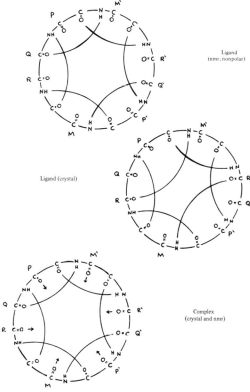

Figure 15-6. Primary structures for valinomycin showing the hydrogen bond patterns.

that of the complex (Figure 15-5a). The conformational difference involves the hydrogen bond pattern (Figure 15-6), two of the bonds being in 13-membered rings. Two free carbonyls (M) are directed towards the center of the ring, four are directed outward, two parallel to the axis of the

tion to complete the intramolecular hydrogen bonding of the K⁺ complexed ion. [From Duax, W. L., Hauptman, H., Weeks, C.M., and Morton, D. A.: *Science, 176*:911, 1972. Reproduced with permission.] (c) General architecture of a cyclodecapeptide folded to produce a cavity lined with six amide carbonyl donors (arrows). The remaining amide groups form hydrogen bonds (broken lines) to stabilise the skeleton. [After Gisin and Merrifield.]

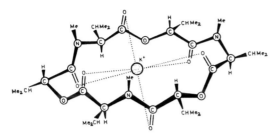

Figure 15-7. The structure of the K+-enniatin B complex. [From Hassall, C. H. and Thomas, W. A.: *Chemistry in Britain*, 146, 1971. Reproduced with permission.]

ring (P) and two perpendicular to this axis (Q). A mechanism for metal ion incorporation may be proposed in which the metal interacts with a pair of types M and

P, the ion could then disrupt the hydrogen bonding pattern and draw in the oxygen atoms of type Q. This would lead to encapsulation and the formation of new hydrogen bonds.

Enniatin B is a cyclic hexadepsipeptide with an 18-atom ring. The structure of its complex with potassium iodide (Figure 15-7) reflects the smaller ring size in that the degree of encapsulation is restricted. The ring provides six carbonyl oxygen atoms which surround the potassium with approximately octahedral geometry, and the side chains point outwards to provide lipophilicity. The molecule is flatter than the valinomycin-KAuCl$_4$ complex and resembles a disc. The stacking of the molecules in columns in the

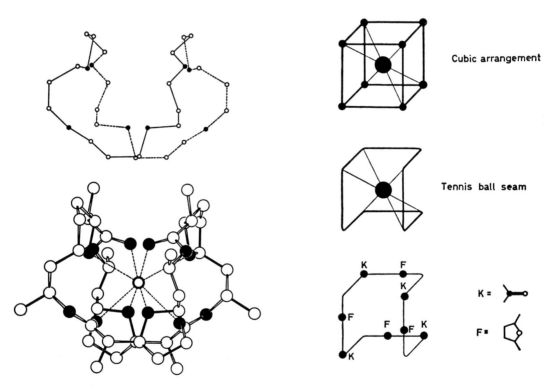

Figure 15-8. Representations of the conformation of the potassium-nonactin complex. [From Kilbourn, B. T., Dunitz, J. D., Pioda, L. A. R. and Simon, W.: *J Mol Biol*, *30*:559, 1967 and Diebler, H., Eigen, M., Ilgenfritz, G., Maass, G. and Winkler, R.: *Pure Appl Chem.*, 20:93, 1969. Reproduced with permission.]

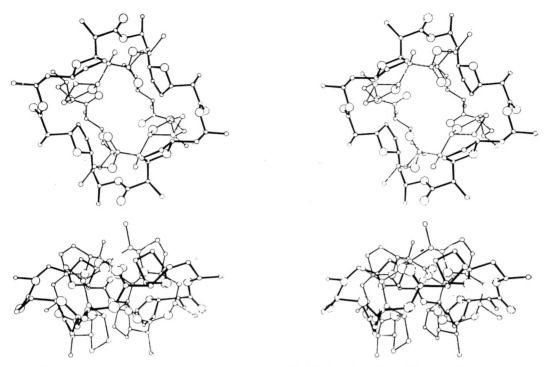

Figure 15-9. Stereographic view of nonactin (solid lines) and its K⁺ complex super-imposed. Top view looking down the twofold axis, bottom view with the twofold axis vertical. [From Dobler, M.: *Helv Chim Acta*, 52:1371, 1972. Reproduced with permission.]

crystal suggests that a similar association in the membrane could provide a pore, or tunnel, through which ions could pass.

Nonactin is a member of a family of macrocyclic tetralactones, the macrotetrolide actins. It has a 32-atom ring and in the K⁺NCS⁻ complex the ligand backbone winds round the metal in a way that has been described as resembling the seam of a tennis ball (Figure 15-8). The four oxygen atoms of the furan rings and four from the carbonyls interact with the metal to give nearly cubic coordination geometry, and the lipophilic groups are turned to the outside of the molecule.

The conformation of the free ligand is similar to that of the potassium complex as both have approximately S_4 symmetry. However, the free molecule is much flatter than the complexed molecule (Figure 15-9). If this molecule is regarded as a torus, then approach of the hydrated metal ion may be envisaged as first inserting into the central cavity with formation of hydrogen bonds between the water molecules and the oxygen atoms of nonactin. The complexation of the metal may then occur by a stepwise process in which removal of successive water molecules is coupled to conformational changes in the ligand as encapsulation occurs.

The structure of the nonactin-NaNCS complex does not have approximately equal metal-oxygen bond lengths. The carbonyl oxygens approach the sodium more closely than do the furan oxygen atoms (2.4Å cf. 2.8Å).

The second class of antibiotics have also been investigated crystallographically, but only dianemycin has been determined as an alkali metal salt. The other acids have been determined as their silver, or thallium, salts and these cations may isomorphously replace the alkali metal cation thus acting as a probe for the alkali metal. The ionic diameters for Tl^I and Ag^I are 3.08Å and 2.54Å compared with K^+, 2.88Å and Rb^+, 3.16Å.

Monensin is one of the polyalcohol, polyether monocarboxylic acids which form complexes with alkali metals. The dihydrated silver salt of monensin is dimorphic; one form is isomorphous with the sodium salt, and the second, which was determined, is isomorphous with the potassium and thallium salts. The carboxylate group was not found attached to the metal but forming hydrogen bonds to lock the anion into

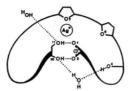 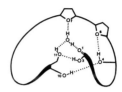

Figure 15-11. Schematic comparison of the hydrogen bond patterns in the Ag^+-monesin salt and in the free acid. [From Lutz, W. K., Winkler, F. K., and Dunitz, J. D.: *Helv Chim Acta*, 54:1103, 1971. Reproduced with permission.]

a macrocyclic form, the metal ion being encapsulated (Figure 15-10). The six-fold coordination of the silver was irregular and did not involve the water molecules present.

The free ligand was also found to be cyclic, and the central cavity contained a water molecule which aided the locking of the ligand into the cyclic form (Figure 15-11). This suggested that the water molecule could act as one of the molecules in the alkali metal solvation sphere during the initial complexation step. Subsequent removal of the solvation sphere leading to encapsulation could then take place accompanied by a minimum of ligand reorganization.

Similar structures have also been found for salts of nigericin and dianemycin.

The silver salt of nigericin was found to be isomorphous with the sodium salt, and for dianemycin the sodium, potassium and thallium salts were isomorphous. These provide five and seven coordinating oxygen atoms respectively and comparison with monensin shows that while the monensin-Ag^+ complex is relatively rigid and inflexible, the nigericin system is capable of adapting to larger cavity sizes and dianemycin is quite flexible and adaptable. This is reflected in the selectivities of the systems as dianemycin appears to have no prefer-

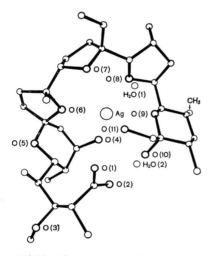

Figure 15-10. The structure of the Ag^+-monensin salt. [From Agtarap, A., Chamberlin, J. W., Pinkerton, M. and Steinrauf, L. K.: *J Amer Chem Soc*, 89:5737, 1967. Reproduced with permission.]

ence for Rb⁺, K⁺ and Na⁺, while nigericin prefers K⁺ to N⁺ and monensin prefers Na⁺ to K⁺.

The rigors of peptide synthesis are such that it is useful to have a more accessible and experimentally pliant system as a model. In 1967 C. J. Pedersen produced an intensive survey of a group of macrocyclic polyethers* (Figure 15-2). These compounds were found to mimic the neutral ionophorous agents and consolidation of some of the architectural requirements of carriers has come from X-ray structural determinations of these compounds.

The polyethers, depending on ring size and number of donor atoms, were envisaged as forming 1 : 1, "saturnine," species with the cation present, or 2 : 1 sandwich species. A 3 : 2 club sandwich was also envisaged but no confirmatory structural evidence has yet been presented.

The 1 : 1 complexes between dibenzo-18-crown-6 and RbNCS, and NaBr are basically saturnine. In the former, the rubidium cation lies below the plane of the six oxygen atoms of the polyether and the overall resemblance is to an inverted umbrella (Figure 15-12). The determination of this structure was complicated by a mixed metal site occupancy of 55% Rb and 45%

---

* Pedersen trivialized the I.U.P.A.C. nomenclature for his compounds as illustrated in the following structure (128):

(128)

is benzo-15-crown-5. There is one benzene ring, a 15-membered ring with five donor oxygen atoms, and the word crown suggests the shape of the ring. The problem with this system is that it does not give details of any isomerization present, as in cyclohexyl-15-crown-5, or of the nature of the ethylene bridges.

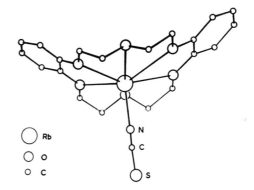

Figure 15-12. The structure of RbNCS (dibenzo-18-crown-6). [From Bright, D. and Truter, M. R.: *J Chem Soc, B*, 1544, 1970. Reproduced with permission.]

Na, the sodium cation was observed to lie closer to the oxygen ring than the rubidium. A second complication was the presence of uncomplexed dibenzo-18-crown-6, the conformation of which is different to that in the complex. In the NaBr complex two different types of sodium complex exist (Figure 15-13), due to the presence of

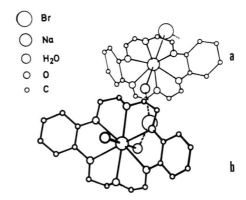

Figure 15-13. The structure of NaBr (dibenzo-18-crown-6). The broken lines indicate hydrogen bonds between water molecules and bromide ions. Complex b is dihydrated and complex a has the sodium cation displaced towards the bromide ion. [From Bush, M. A. and Truter, M. R.: *J Chem Soc, Chem Comm*, 1439, 1970. Reproduced with permission.]

water molecules. One type has a symmetrical sodium environment of two axial water molecules and the polyether and the cation sits almost in the plane of the oxygen atoms from the polyether. The second type has an asymmetric situation with one water molecule and one bromide anion sitting either side of the polyether ring with the sodium displaced toward the bromide ion. In all of these examples the metal sits at a center of optimal electron density.

The effect of ion size is more marked with benzo-15-crown-5. Here there is a 1 : 1 complex with NaI in which the sodium is pulled away from the plane of the oxygen atoms by an accompanying water molecule. With KI a 2 : 1 species is formed in which the two benzo-15-crown-5 molecules "encapsulate" the metal (Figure 15-14). The formation of 2 : 1 complexes has also been observed for the disc-like en-

niatins. The formation of the 2 : 1 species suggests that a 30-membered ring system— or a 10-oxygen atom system—should encapsulate the metal rather like nonactin. This is found in the 1 : 1 complex of dibenzo-30-crown-10 and KI. The resemblance to the potassium-nonactin complex is shown in Figure 15-15. The crystalline form of the uncomplexed molecule shows it to be a long closed loop which on undergoing conformational change encapsulates the metal.

One point of contrast between the structures of the cyclic polyether complexes and those for the antibiotics is that the interaction of dibenzo-24-crown-8 and KNCS gives gives a 1 : 2 complex. In this the two metal ions sit on either side of the almost flattened polyether ring (Figure 15-16). This suggests the feasibility of multimetal carriers. Other examples of 1 : 2 complexes have

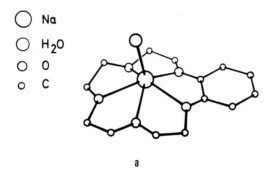

○ Na
○ H₂O
○ O
○ C

a

Figure 15-14

Figure 15-14. a) is the cation of NaI (benzo-15-crown-5), H_2O; b) is the cation of KI (benzo-15-crown-5)₂. [From Bush, M. A. and Truter, M. R.: *J Chem Soc, Chem Comm*, 1439, 1970 and Mallinson, P. R. and Truter, M. R.: *J Chem Soc, Perkin II*, 1818, 1972. Reproduced with permission.]

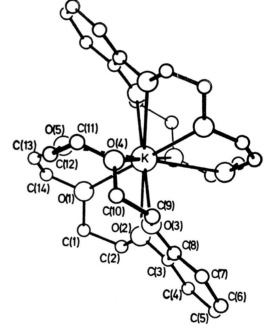

b

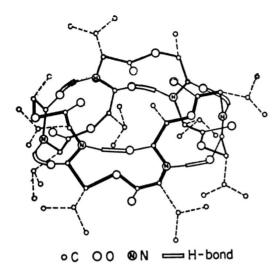

○ C   O O   ⊗ N   ▭▭ H-bond

Figure 15-20. The conformation of valinomycin in non-polar solvents. [From Ivanov, V. T. and Ovchinnikov, Yu: *Conformational Analysis,* ed. Chiordoglu, G. Academic Press (London), 111, 1971. Reproduced with permission.]

is excluded. The conformation of the non-polar form of enniatin B was shown to have the three structurally identical sub-units arrayed in different ways spatially. A further conformer existed in polar solvents. On complexation one would anticipate that with ions of varying size differences would be observed for the cavity size. This size would be determined by the orientation of the carbonyl groups which depend upon the cation radius. The nmr spectrum of (tri-N-desmethyl) enniatin B with Li^+, Na^+, K^+, Cs^+ showed a monotonous increase in the [3]NH-CH coupling constant, as the ring opened out to accommodate the cation.

Nonactin has been studied by ^{1}H nmr in anhydrous acetone-d_6 and acetone-d_6-water mixtures. The nonactin ring undergoes conformational changes in complexation as is evidenced by salt-induced chemical shifts for the nonactin protons. Na^+, K^+ and Cs^+ complexes are formed and the binding

constants are interesting. The respective values are 7×10^4, 7×10^4 and 1×10^4 in dry acetone, and 210, 2×10^4 and 400 in wet acetone. In wet acetone the ion must first be stripped of its hydration shell making the K^+ complex formation favored. In dry acetone the lack of selectivity may be caused by the similarity of the keto-M^+ interactions and binding energies in both the nonactin-complexes and the solvent. The conformational changes observed through chemical shifts were similar for each ion, this has been corroborated by ^{13}C nmr studies.

Studies with ^{13}C nmr have also been used to show that conformational changes occur upon complexation of potassium with valinomycin, nonactin and dicyclohexyl-18-crown-6, however little information concerning the species in solution was retrieved.

For the cyclic polyethers and cryptates there is little information on the nature of the species in solution. Optical spectroscopy studies show that for the cyclic polyethers separated ion pairs are formed on mixing equimolar quantities of fluorenyl-sodium with 18-crown-6 species in tetrahydrofuran. Contact and separated ion pairs are found for the 15-crown-5 species. A solvent effect in ^{1}H nmr is shown for the interaction of di(methylbenzo)-18-crown-6 with alkali metal-fluorenyl salts. In THF the sequence was found to be $Na^+ \gg K^+ > Cs^+ > Li^+$ whereas in oxetane the order was $K^+ > Na^+$. These sequences contrast with those found in water and methanol, $K^+ > Cs^+ > Na^+ > Li^+$. It is suggested that strong solvation by oxetane, water and methanol, compared with a weaker solvation by THF is the reason for this juxtaposition.

Studies on the cryptates involving ^{1}H nmr show that the metal sits inside the molecular cavity. This is evidenced in the thallium complex with cryptate(iii) where the spin-spin coupling of [203, 205]Tl nuclei to all of the protons of the ligand occurs.

Techniques involving [23]Na nmr have been used, but have yielded little or no information on the nature of the complexes. A very small amount of covalent interaction between the cation and the ligand oxygens has been proposed. Fast exchange between free and complexed Na^+ was shown to occur for valinomycin, enniatin B, the actins, dicyclohexyl-18-crown-6 and nigericin, but not for monensin. The kinetics of association and dissociation of valinomycin and K^+, studied by nmr, as a function of solvent showed that the reactions were immeasurably slow in $CDCl_3$ but rapid in the more polar 80% CH_3OH-20% $CDCl_3$. This latter rate was found to be close to the magnitude of the turnover number for valinomycin catalyzed transport, and supports the mobile carrier mechanism.

The kinetics of complexation for the carrier complexes have been studied by various relaxation techniques. An extremely high rate for the recombination of nigericin with Na^+ has been found, and a much slower rate observed for the actins and valinomycin. The rates for these systems, ($\sim 10^7$ $M^{-1}sec^{-1}$), are slower than that for a diffusion controlled process (3×10^9 $M^{-1}sec^{-1}$). The high rate is surprising as during any encounter between the carrier and solvated metal ion requires substitution of the solvation sheath. Each step must occur within 10^{-9} seconds, and stepwise ligation of the metal take place. High rates are necessary for carrier action as the metal ion to be transported must be moved across the membrane as quickly as possible. In the case of valinomycin, it has been shown that the rate of complexation corresponds to the rate of the slowest interconversion between conformers of the antibiotic alone, this means that the rate limiting step of complex formation is a ligand conformational change. The application of spectroscopic techniques to the study of cyclic depsipeptides in the presence of phospholipid (lecithin) vesicles has shown that while valinomycin is located towards the hydrophobic interior of the bilayer membrane, enniatin B is predominantly located at the membrane. The complexation of these membrane-bound antibiotics to alkali metal cations occurred at the membrane surface.

CONCLUSIONS

It must first be emphasized that none of the molecules referred to as ionophorous agents have yet been found endogenous to membranes, and they are therefore considered as models for transport by carrier or pore mechanisms. A cyclic peptide has been isolated from beef heart mitochondria which shares many of the properties of the ionophorous agents. However, no structural data has been made available.

The mode of action of transport may not involve a pore or carrier mechanism, or it may involve both. The ionophorous agents described here render alkali metal cations lipid soluble, and the structural data can be interpreted as supporting a pore hypothesis by virtue of the stacking of the molecules. The evidence presented from solution studies and freezing studies using ion selective electrodes, and the electrical properties of bilayers suggests that valinomycin and nonactin may act as carriers. This is supported by studies of the transport of K^+ in the presence of labelled nonactin across a bulk lipid membrane. The movement of K^+ was accompanied by equal movement of antibiotic as a 1 : 1 complex, together with an exchange of ligand. This would suggest a carrier relay process but nmr studies have suggested that the neutral antibiotics act as free carriers. The bulk lipid membrane is much thicker than a natural membrane and it is possible that free mobile carriage could occur in the latter.

The detection of ionophorous agent-cation complexes on phospholipid membranes has been observed using fluorescent probes. It has been suggested that gramicidin might form complexes by helix formation, or that it leads to the production of aqueous pores through the membrane rather than acting as a carrier. A further possible pore-former is alamethicin, a cyclic peptide of high molecular weight.

The mode of action may also come from general properties of the alkali metals and their capacity for binding to different anionic or neutral sites existing within the membrane. Conformational changes of the constituents close to these sites could provide a means of transferring the ions from site to site.

The structure of the protein subtilisin Novo shows two sites for ion chelation. At ion site 1 Tl^+ and K^+ can be bound, while at ion site 2 these ions are not bound. Na^+ may bind at both sites. It is possible that specific sites for alkali metals may exist in the membrane protein.

If a carrier, or pore, mechanism does exist and if it is based on the ionophorous agents the requirements of the carrier may be summarized as follows:

1. The ligand must be flexible in order to facilitate stepwise desolvation and subsequent encapsulation of the alkali metal cation. Too high an activation energy would be required to strip off all solvent molecules simultaneously.

2. The architecture of the ligand must be such that the exterior of the complex is lipophilic. Where hydrogen bonding plays an important role in stabilizing the structure modifications may not be made if they disrupt the bonding pattern. The requirement for a lipophilic exterior is evidenced by comparing nonactin $-K^+$ with dibenzo-30-crown-10-K^+ where although the latter has a higher stability constant, the former has a higher biological activity.

3. A best-fit situation is required with respect to the cavity size and the diameter of the incoming ligand. This need not occur at the minimum cavity size as steric fixation of the ligand at that point may provide inhibiting ligand-ligand repulsions. The maximization of the difference between the energy of formation of the complex and the solvation energy of the cation should be attained.

The precise structure of the membrane remains the key factor in elucidating the nature of transport. The use of models, and the accompanying oversimplification of the problems inherent in the natural system, can only, and then hopefully, give us pointers. The proposition that the paracrystalline, or liquid crystalline, state would be the most suited to biological functions as it combines the fluidities and diffusibilities of liquids while preserving the more structured form of the crystal, combined with the more recent concept that the protein present in the membrane consists of individual globules embedded in a two-dimensional viscous fluid mosaic of phospholipid, suggests that experiments on the behavior of alkali metal ions and their complexes in such media might yield information to bridge the gap between structures determined by X-ray crystallographic techniques and nmr spectroscopy.

It is always possible that when the structure of the membrane is determined, the work on carriers as we currently know it will be regarded as so much phlogiston.

If this is so then we must remember that the biologically orientated research into the chemistry of the alkali metals has expanded our knowledge of these metals and has shown that it consists of much more than just the preparation of caustic soda and the electrolysis of brine.

FURTHER READING

1. Williams, R. J. P.: The biochemistry of sodium, potassium, magnesium and calcium. *Quart Rev, 24:*331, 1970.

2. Frensdorff, H. K. and Pedersen, C. J.: Macrocyclic polyethers and their complexes. *Angew Chem, Intl Edn, 11:*16, 1972.

3. Pedersen, C. J. and Truter, M. R.: Cryptates. *Endeavour, 30:*142, 1971.

4. Symposium on biological and artificial membranes. *Fed Proc, 27:*1269, 1968.

5. Vasquez, D. (Ed.): *Molecular Mechanisms of Antibiotic Action on Protein Biosynthesis and Membranes.* Amsterdam, North Holland Publishing, 1972.

6. Ivanov, V. T. and Ovchinnikov, Yu. A.: Conformation of membrane-active cyclodepsipeptides. In Chiurdoglu, G. (Ed.): *Conformational Analysis—Scope and Present Limitations.* London, Acad Pr, 1971, p. 111.

7. Winkler, R.: Kinetics and mechanism of alkali ion complex formation in solution. In *Structure and Bonding.* 1972, vol. 10, p. 1. (Volume 16 of *Structure and Bonding* is devoted to alkali metal complexes.)

8. Singer, S. J. S. and Nicholson, G. L.: The fluid mosaic model of the structure of cell membranes. *Science, 175:*720, 1971.

9. Eisenmann, G., Szabo, G. and Ciani, S., McLaughlin and Krasne, S.: *Ion Binding and Ion Transport Produced by Neutral Lipid-Soluble Molecules* in *Progress in Surface and Membrane Science.* 1973, vol. 6, p. 139.

10. Chock, P. B. and Titus, E. O.: Alkali metal ion transport and biochemical activity. In *Progress in Inorganic and Nuclear Chemistry.* 1973, vol. 18, p. 287.

CHAPTER 16

OSCILLATING REACTIONS

Endre Körös

INTRODUCTION

THE MAJORITY OF *in vivo* reactions, both metabolic and genetic, occur in cycles, there being a wide range of periodicities (for example, the brain has *microsecond* oscillations, an aching tooth throbs every *second* or so, and the blood plasma concentrations of metals, our sleeping habits and our body temperature all have a *daily* cycle whereas the ovarian cycle is *monthly*). Such bodily cycles have two factors in common: (a) all involve bio-inorganic ions, and (b) all have built-in feedback mechanisms.

When compared with our knowledge of the thermodynamics and kinetics of chemical reactions, our knowledge of the mechanisms of chemical oscillations is meager. Not surprisingly, the subject has hardly been touched upon by bio-inorganic chemists but, as it is one of the fundamental aspects of life in humans, our reticence in researching this topic must be eliminated. The bio-inorganic chemistry of cell membranes has been overlooked by many researchers for far too long on the grounds of their complexity, and yet Chapter 15 has just demonstrated how bio-inorganic chemists have now commenced a study of this complex subject. So too, we must turn to researching oscillating reactions *in vivo*. Indeed the subject could well yield to the multimethod approach used on membranes (this chapter mentions the thermodynamics of Chapter 10, the kinetics of Chapter 11 and the bromide selective electrodes of Chapter 12).

The basic theoretical aspects of the subject have not had even occasional glances in textbooks of biochemistry, inorganic chemistry or medicine and so they are described in some length in the second section. But first of all, we ought to consider the thermodynamics and kinetics of a simple, closed, homogeneous system:

"In any closed homogeneous system at constant temperature and pressure all spontaneous chemical reactions must be accompanied by a decrease in the free energy of the system." The foregoing statement is a direct result of the second law of thermodynamics and is universally accepted. Another consequence of this statement is also accepted: A chemical system at constant temperature and pressure will approach equilibrium monotonically, i.e. the kinetics do not exhibit an oscillatory response. This means that oscillations about a final equilibrium state will not be observed in such a chemical system.

As a result of the rule of monotonically decreasing free energy in spontaneous

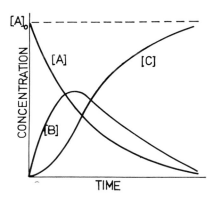

Figure 16-1. The variation of concentrations of A, B and C with time for a reaction of the type A → B → C.

chemical systems, the general observation is that the overall rate of the reaction decreases monotonically with time, and if there are any intermediates produced in the course of the reaction, then their concentration will either rapidly reach a steady state value or pass through a single maximum or minimum: In a system $A \xrightarrow{k_1} B \xrightarrow{k_2} C$ the variations with time of [A], [B] and [C] are shown schematically in Figure 16-1.

Oscillations in a closed homogeneous system were at one time considered to be thermodynamically impossible. However, one of the important results obtained from studying the thermodynamics of irreversible processes is that oscillating chemical reactions *are* thermodynamically possible around nonequilibrium stationary states.

Previous beliefs that oscillating chemical (and therefore *biochemical*) reactions could probably not take place in homogeneous systems and that they required coupling to diffusion and adsorption processes seem now to be unfounded. The mechanisms discussed below, and their general agreement with the experimentally observed oscillations, leave no doubt that such reactions can occur in a homogeneous phase.

THEORETICAL ASPECTS OF CHEMICAL OSCILLATION

Mathematical Models

Lotka formulated the following hypothetical reaction scheme

$$A + X \xrightarrow{k_1} 2\,X \qquad (130)$$
$$X + Y \xrightarrow{k_2} \text{byproducts} \qquad (131)$$
$$X + Y \xrightarrow{k_3} 2\,Y \qquad (132)$$
$$Y \xrightarrow{k_4} P \qquad (133)$$

and stated that this proposed set of chemical reactions (130-133) could generate sustained temporal oscillations in the concentrations of intermediates X and Y present during the net overall action A → P. Such behavior is possible only if the system is maintained far from equilibrium by the exchange of reactants and products.

If the system is closed, A cannot be maintained exactly constant because some portion of it is converted into products. However, if the concentration of A is sufficiently large, or if the system is not closed and A is continually supplied from the exterior, the changes in A may be small on the time scale used to monitor the changes in the concentrations of intermediates X and Y. Under these conditions

$$\frac{d\,[X]}{dt} = k_1[A][X] - (k_2 + k_3)[X][Y]$$
$$(134)$$

$$\frac{d[Y]}{dt} = -k_4[Y] + k_3[X][Y] \qquad (135)$$

An analysis of the kinetic equations shows that the system admits one nontrivial steady state and an infinity of undamped periodic solutions whose periods and amplitudes depend on the initial con-

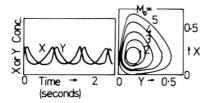

Figure 16-2. Analog computer solutions of the Lotka mechanism, showing typical kinetics and phase plane trajectories.

ditions. Equations (134 and 135) can be solved in closed form and yield semistable rotations around the singularity. Analog computer solutions are given in Figure 16-2.

The Lotka mechanism actually exhibits this peculiar behavior because it includes two autocatalytic reactions, but such autocatalysis is not essential for sustained oscillations. However, it is essential that at least one step in the mechanism involve feedback so that the product of the step affects its own net rate other than by mass reversibilities.

Four general types of chemical feedback loops can exist

I. $\longrightarrow X_1 \xrightarrow{\text{activation}} X_2 \cdots X_{n-1} \longrightarrow X_n \longrightarrow$

II. $\longrightarrow X_1 \xrightarrow{\text{inhibition}} X_2 \cdots X_{n-1} \longrightarrow X_n \longrightarrow$

III. $\longrightarrow X_1 \rightleftarrows X_2 \rightleftarrows \cdots X_{n-1} \xrightarrow{\text{activation}} \rightleftarrows X_n \longrightarrow$

IV. $\longrightarrow X_1 \rightleftarrows X_2 \rightleftarrows \cdots X_{n-1} \xrightarrow{\text{inhibition}} \rightleftarrows X_n \longrightarrow$

$$(136)$$

Schemes III and IV do not contain feedback unless the reaction steps in the sequence are reversible. All four types of feedback have been shown to produce sustained oscillations in hypothetical reaction schemes. No hypothetical reaction scheme with periodic solution is known which does not contain at least one of the four types of feedback.

Such feedback leads to nonlinearity in the differential equations that describe the mechanism and the system may exhibit all of the complexities resulting from nonlinear dynamic laws. The system must also be far from equilibrium and must be open if the oscillations are to be sustained. Chemical systems involving a number of specific types of feedback lead to nonlinear kinetic equations and oscillatory behavior.

In contrast, a system exhibiting limit cycle[*] behavior will approach a limiting periodic behavior defined only by the kinetic constants and independent of the initial concentrations of the oscillating intermediates.

A chemical mechanism (the Brusselator) exhibiting limit cycle is as follows

$$A \rightarrow X_1 \qquad (137)$$
$$2X_1 + X_2 \rightarrow 3X_1 \qquad (138)$$
$$B + X_1 \rightarrow D + X_2 \qquad (139)$$
$$X_1 \rightarrow E \qquad (140)$$

The net reaction is

$$A + B \rightarrow D + E \qquad (141)$$

This is the only chemical scheme that exhibits limit cycle behavior, has only two intermediate species, and develops complex spatial and temporal structures analogous to those exhibited in biological systems. A major objection to this scheme is that step (138) is third order with respect to the concentrations of the intermediates (X_1 and X_2).

---

[*] In the theory of two-dimensional bifurcation, a particularly important role is played by closed trajectories which obviously represent periodic motions. In a structurally stable system two closed trajectories are necessarily separated by a finite distance. These we call *limit cycles*.

Actual chemical processes which follow either the Lotka or Brusselator mechanisms are not known so far, i.e. we are not able to attribute chemical species to the letters A, X, Y.

Another approach to finding a mechanism which exhibits limit cycle behavior is to use the information actually obtained from an oscillating reaction. Thus, based on the experimental findings, and the proposed mechanism, of the Belousov-Zhabotinsky reaction the following general kinetic scheme has been suggested:

$$A + Y \rightarrow X \qquad (142)$$

$$X + Y \rightarrow P \qquad (143)$$

$$B + X \rightarrow 2X + Z \qquad (144)$$

$$2X \rightarrow Q \qquad (145)$$

$$Z \rightarrow Y \qquad (146)$$

which leads to the overall net reaction

$$A + 2B \rightarrow P + Q \qquad (147)$$

and which exhibits limit cycle behavior (Oregonator).

The model may be related to the mechanism of the Belousov-Zhabotinsky reaction by means of the key: $X \equiv HBrO_2$, $Y \equiv Br^-$, $Z \equiv Ce(IV)$, $A \equiv B \equiv BrO_3^-$.

This model is composed of five steps and involves three independent chemical intermediates (X, Y and Z). However, since the concentrations of two of the intermediates $HBrO_2$ and Br^- are stiffly coupled, a system of two independent variables can be generated such that it permits the already available theoretical developments to be applied to a system studied experimentally.

Thermodynamics

In an isolated system, a steady state other than that of equilibrium cannot exist, i.e. oscillations cannot take place around a transient state in the course of a system approaching equilibrium. However, beyond the domain of stability of steady states, i.e. far from equilibrium, in systems obeying nonlinear kinetic laws, stable, sustained, oscillations, of the limit cycle type bifurcating from an unstable steady state can occur. Thus sustained oscillations in an open system can be understood as supercritical phenomena. As far as the physical prerequisites for chemical oscillations (based upon our mathematical considerations) are concerned it can be concluded that, for the existence of stable sustained oscillations, positive or negative feedback, is necessary, and sometimes it is combined with cross-catalysis.

In a system that is closed to mass transfer, oscillations around a transient state remote from equilibria are also possible and, in fact, they have actually been observed experimentally. However, these oscillations are necessarily damped since, according to the second law of thermodynamics, the system will be continuously approaching equilibrium. In cases, having a reaction mixture in which some reagents are introduced initially as a large excess and in which the reaction rates are not too high, the oscillations can be treated as oscillations in open systems. Such systems closely approach the sustained oscillations observed in open systems. This phenomenon is observable with the Belousov-Zhabotinsky reaction in which the initial reaction mixture contains a large amount of malonic acid (*circa* 0.1-0.5 M) and potassium bromate (circa 0.1 M) and during one oscillation period (about 1 min.) only about 10^{-4} M of the reactants are consumed.

(For further details concerning the thermodynamics of oscillating reactions, the book by Glansdorff and Prigogine may be consulted).

OSCILLATIONS IN HOMOGENEOUS CHEMICAL SYSTEMS

Here we restrict our discussion to homogeneous chemical reactions and exclude the treatment of oscillating phenomena occurring in heterogeneous systems and of those driven by temperature changes (thermochemical oscillations).

There are only two chemical reactions which are worthy of mention:

A. The catalytic decomposition of hydrogen peroxide in sulphuric acid by iodic acid (Bray reaction). During the decomposition there are oscillations in the rate of O_2 evolution and in the concentrations of I_2, I^- and H^+. The maximum rate of O_2 evolution is synchronized with the phase of decreasing I_2 concentration. In spite of many studies of this reaction, the detailed mechanism of the oscillations is still an open question. It has been postulated that the reactions take place via a free radical chain mechanism, and that the oscillations are due to a quadratic branching of this chain. Computer calculations have revealed that a model containing a quadratic, branched, chain reaction can produce concentration oscillations with a waveform, very similar to those of the iodine concentration oscillations in Bray's reaction.

B. The other reacting chemical system is the catalytic oxidation of a variety of organic compounds that possess an active methylene group, e.g. malonic acid, bromomalonic acid, acetoacetic acid, oxaloacetic acid, citric acid, acetonedicarboxylic acid, 2,5-pentanedione with bromate in dilute sulphuric acid. The oxidation can be catalyzed by one-electron redox couples having a formal redox potential above 1 V; e.g. Ce^{4+}/Ce^{3+}, Mn^{3+}/Mn^{2+}, tris-1,10-phenanthroline-iron(II) / tris-1,10-phenanthroline-iron(III). This has been called the Belousov-Zhabotinsky-reaction.

The stoichiometry of the overall reaction (taking malonic acid as the organic compound) is:

$$2BrO_3^- + 3CH_2(COOH)_2 + 2H^+ \xrightarrow{\text{catalyst}} 2BrCH(COOH)_2 + 3CO_2 + 4H_2O \quad (148)$$

There are four principal reactions whose stoichiometries are given as follows:

$$BrO_3^- + 2Br^- + 3CH_2(COOH)_2 + 3H^+ \to 3BrCH(COOH)_2 + 3H_2O \quad (149)$$

$$BrO_3^- + 4Ce^{3+} + CH_2(COOH)_2 + 5H^+ \to BrCH(COOH)_2 + 4Ce^{4+} + 3H_2O \quad (150)$$

$$6Ce^{4+} + CH_2(COOH)_2 + 2H_2O \to 6Ce^{3+} + HCOOH + 2CO_2 + 6H^+ \quad (151)$$

$$6Ce^{4+} + BrCH(COOH)_2 + 2H_2O \to 6Ce^{3+} + 3CO_2 + 7H^+ + Br^- \quad (152)$$

The essential nonlinearity arises in reaction (150) whose underlying steps involve a first-order autocatalytic production of $HBrO_2$ along with its second-order decomposition. In addition, $HBrO_2$ participates in reaction (149). Depending upon the bromide concentration $HBrO_2$ may attain two different steady state concentrations: $[HBrO_2]_A$ and $[HBrO_2]_B$. The transition from $[HBrO_2]_A$ to $[HBrO_2]_B$ and *vice versa* occurs at a critical bromide concentration. Bromide inhibits reaction (150) and so there exists a well-defined threshold concentration switching reaction (150) on and off.

The reaction starts with a non-oscillatory period during which bromomalonic acid accumulates, and this is followed by the oscillations in the Ce^{4+}, Ce^{3+} and Br^- concentrations.

Typical potentiometric traces of $\log[Br^-]$ and $\log [Ce^{4+}]/[Ce^{3+}]$ vs time are shown in Figure 16-3.

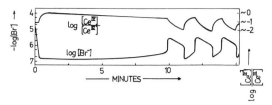

Figure 16-3. Potentiometric traces at 25°C of log [Br]⁻ and log [Ce^IV]/[Ce^III] versus time during the Belousov-Zhabotinsky reaction.

The detailed mechanism of this oscillating system which shows both temporal and spatial periodicities is rather sophisticated and the reader should consult the original papers.

The oscillating reactions are accompanied by heat evolution, and heat is produced in every phase of the oscillating reaction,* the rate of heat evolution, however, is periodic in character. At a certain stage in the process when changes are occurring rapidly in the system, i.e. the autocatalytic reactions are switched on, resulting in the transformation of large amounts of the reactant per unit time, the rate of heat evolution is high.

This periodicity in the rate of heat evolution is accompanied by a periodic rate of consumption of the reactants (BrO_3^-, malonic acid) and rate of formation of the end products, CO_2 and $BrCH(COOH)_2$.

In an oscillating system there are two types of concentration *vs* time functions, a periodic one—for the transient species, e.g. for Br⁻ and a nonmonotic one—for the nonrecycling species, e.g. for CO_2. The mode of synchronization is shown in Figure 16-4.

Oscillating reactions should be followed by the continuous recording of the concentrations of the oscillating species, e.g.

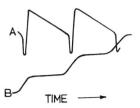

Figure 16-4. The change in bromide (a recycling species) concentration (curve A) and in carbon dioxide (a nonrecycling species) concentration or in temperature (curve B) during the Belousov-Zhabotinsky reaction.

in the Belousov-Zhabotinsky reaction the bromide concentration is traced with a bromide selective electrode, the cerium(IV) concentration is followed spectrophotometrically (or potentiometrically) and calorimetry is used to follow the temperature changes. To reveal the details of the mechanism of an oscillating reaction, the kinetics of the composite reactions, e.g. $Ce^{3+} + BrO_3$; $Ce^{4+} + BrCH(COOH)_2$ should be investigated. To this end, stopped-flow methods can be applied.

BIOCHEMICAL OSCILLATIONS

The present interest in chemical oscillations is due largely to the discovery of oscillating biochemical phenomena, which have been observed at both the metabolic and genetic, e.g. enzyme synthesis, levels.

The most completely investigated and well-understood example of oscillatory behavior in a biochemical system is that observed in the glycolytic pathway. In fact, all glycolytic metabolites have been observed to oscillate in concentrations. Glycolytic oscillations have been observed in suspensions of beef heart extracts and in tumor cells, but the oscillatory mechanism has been studied only in the yeast system. The source of the oscillation lies in the action of phosphofructokinase (PFK). This allosteric enzyme is activated by both

---

* It has been a rather widespread misbelief that during an oscillating reaction exothermic and endothermic steps follow each other.

its products ADP (via AMP) and fructose-1,6-diphosphate (FDP) and also by fructose-6-phosphate (F6P) which is one of its substrates. Oscillatory behavior is propagated along the glycolytic pathway by the coupling of the ATP system with all control points along the pathway.

Figure 16-5 illustrates the phase relationships of the glycolytic metabolites.

The role of metabolic oscillations is not clear. It can be supposed that the very complexity of biochemical control circuits, and the cooperative properties of certain enzymes involved therein, make oscillations inevitable. There is, however, another viewpoint which argues that metabolic oscillations constitute necessary elements for genetic and circadian oscillations. Further work is needed to elucidate the coupling between metabolic and circadian or genetic oscillators.

It is well established that certain biochemical reaction chains regulate the concentration of the metabolites involved

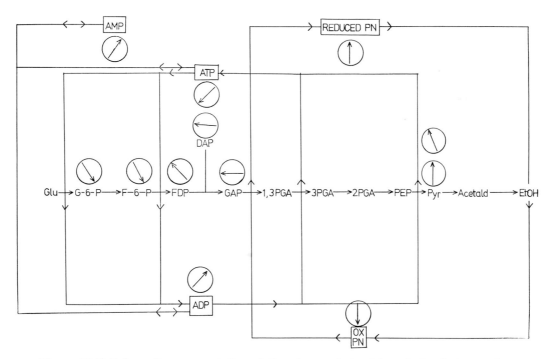

Figure 16-5. Schematic representation of the phase relationships of glycolytic metabolites. An upright arrow in the "clock" indicates a maximum concentration of a particular intermediate; an arrow pointing downward indicates a minimum concentration. Arrows pointed at intermediate angles indicate proportionally intermediate positions with regard to maximum and minimum concentrations. Abbreviations: Glu — glucose, AMP = adenosine-5'-monophosphate, ATP = adenosine-5'-triphosphate, PN = pyridine nucleotide, G-6-P = glucose-6-phosphate, F-6-P = fructose-6-phosphate, FDP = fructose-1,6-diphosphate, DAP = dihydroxy-acetone phosphate, GAP = glyceraldehyde-3-phosphate, 1,3PGA = 1,3-diphosphoglycerate, 3PGA = 3-phosphoglycerate, 2PGA = 2-phosphoglycerate, PEP = phosphoenol pyruvate, Pyr = pyruvate, Acetald = acetaldehyde, and EtOH = ethanol.

through various feedback mechanisms. One of the most common examples is that in which the product of a chain diminishes the activity of an enzyme synthesizing some distant precursor of the inhibitory product. An alternative type of feedback in biochemical reactions is that of repression (by the product of enzymatic reactions) of the genetic mechanism responsible for the synthesis of the enzyme. There are also biochemical reactions in which both the substrate and the product repress the enzyme activity. In all of these cases, chemical oscillations may be observable.

A few examples of oscillations of the genetic level are as follows: Periodicities have been observed in the rate of β-galactosidase synthesis in asynchronously growing cultures of *E. coli*, in the enzyme synthesis in synchronous populations of *Bacillus subtilis*, in the synthesis of lactate dehydrogenase in Chinese hamster cells, in the initiation of transcription of the tryptophan and the lactose operon, and in the activity of aspartate transcarbamylase and ornithine transcarbamylase in the absence of DNA synthesis.

THE ROLE OF OSCILLATIONS IN LARGE-SCALE BIOLOGICAL PHENOMENA

It is clearly beyond the scope of this chapter to discuss the role of oscillations in certain large-scale phenomena of biological interest. However, it is worth mentioning the rhythmic activity of the nervous system, i.e. 1) that certain parts of the brain respond in the form of damped or sustained oscillations to an external impulse-like stimulation, or without external stimulations self-oscillations arise in certain parts of the brain at characteristic frequencies; 2) that circadian clocks (the innate rhythms in living systems), 3) and

that the spatio-temporal control of various development processes, i.e. mechanisms which are responsible for switching genes on and off, all exhibit these phenomena.

The physiological role of supercellular oscillations is obvious. The basic properties of circadian rhythms, e.g. phase resetting, permit the organism to adapt in a flexible manner to the external conditions. Oscillations at the genetic level and during the process of development of the adult organism are essential for cellular division and for the temporal organization of the different functions during development.

FURTHER READING

1. Lotka, A. J.: Contribution to the theory of periodic reactions. *J Phys Chem, 14:*271, 1910.
2. Lotka, A. J.: Undamped oscillations derived from the law of mass action. *J Am Chem Soc, 42:*1595, 1920.
3. Gavalos, G. R.: *Non-linear Differential Equations of Chemically Reacting Systems.* Berlin, Springer-Verlag, 1968.
4. Glansdorff, P., and Prigogine, I.: *Thermodynamic Theory of Structure Stability and Fluctuations.* New York, Wiley, 1971.
5. Chance, B., Pye, E. K., Ghosh, A. K. and Hess, B. (Eds.): *Biological and Biochemical Oscillators,* New York and London, Academic Press, 1973.
6. Higgins, J.: *The Theory of Oscillating Reactions,* Ind. Eng. Chem., 595:19, 1967.
7. Nicolis, G. and Portnow, J.: Chemical oscillations. *Chem Revs,* 73:365, 1973.
8. Noyes, R. M., and Field, R. J.: Oscillatory chemical reactions. *Ann Rev Phys Chem,* 25:95, 1974.
9. Degn, H.: Oscillating chemical reactions in homogeneous phase. *Chem Edu,* 49:302, 1972.
10. Bray, W. C.: A periodic reaction in homogeneous solution and its relation to catalysis, *J Am Chem Soc, 43:*1262, 1921.
11. Degn, H.: Evidence of a branched chain reaction in the oscillating reaction of hydrogen peroxide, iodine and iodate, *Acta Chem Scand, 21:*1057, 1967.
12. Matsuzaki, I., Woodson, J. H. and Liebhafsky, H. A.: pH and temperature pulse during the

periodic decomposition of hydrogen peroxide. *Bull Chem Soc* (Japan), *43*:3317, 1970.

13. Field, R. J., Körös, E. and Noyes, R. M.: Oscillations in chemical systems. II. Thorough analysis of temporal oscillation in the bromate-cerium-malonic acid system. *J Am Chem Soc, 94*:8649, 1972.

14. Field, R. J., and Noyes, R. M.: Oscillations in chemical systems. IV. Limit cycle behavior in a model of a real chemical reaction. *J Phys Chem, 60*:1877, 1974.

15. Körös, E., Orban, M. and Nagy, Zs.: Periodic heat evolution during temporal chemical oscillations. *Nature Phys. Sci., 242*:30, 1973.

16. Körös, E., Burger, M., Friedrich, V.,

Ladányi, L., Nagy, Zs. and Orbán, M.: Chemistry of Belousov-type oscillating reactions. *Faraday Symposia Chem Soc, 9* (in press).

17. Hess, B. and Boiteaux, A.: Oscillatory phenomena in biochemistry. *Ann Rev Biochem, 40:* 271, 1971.

18. Walter, C.: *Chemical Oscillations. Enzyme Kinetics: Open and Closed Systems,* New York, Ronald, 1966, chapter 8.

19. Walker, M. D., and Williams, D. R., The chemical principles of chronotherapy as established from an *in vitro* model of circadian concentration rhythms, *Bio-inorganic Chemistry, 4*:117, 1975.

SECTION III
BIO-MEDICAL APPLICATIONS

CHAPTER 17

MEDICAL APPLICATIONS

Adrien Albert

BIOLOGICAL REACTIONS take place in an aqueous environment where evolutionary pressure has led to the selection of trace amounts of various elements for controlling biological reactions and equilibria. Deficiencies or excesses of metal ions cause most of the adverse biological effects. However, even nonmetals can be important, for example, a deficiency of iodine causes an endemic goiter, which the administration of iodine will quickly reverse. In biology and medicine, inorganic chemistry has already found many applications, and these are expected to increase rapidly with the current expansion of relevant chemical and biochemical knowledge.

In its applications to problems of biology and medicine, bio-inorganic chemistry is concerned mainly with cations. This is not to deny a certain importance to the nonmetals. In human medicine, boric acid, iodine, and sulphur (the latter as the element, or in the form of calcium sulphide) are used in dermatology, and sulphides are the most used depilatories. The fluoride anion is often added to a community's drinking water and to dentifrices. In agriculture, elemental sulphur is still much used as a fungicide, also selenous and boric acids to remedy any deficiency found in animals or plants (respectively). On the other hand, bromides and inorganic arsenic, which were before 1920 among the most frequently prescribed of all human medica-

ments, are no longer in use. The intake of nutritionally essential chlorine, sulphur, and phosphorus in food is usually sufficient to avoid any need for supplementation, except in pregnancy when extra phosphate is usually prescribed. The faith placed in phosphates as "tonics" by the medical profession for so many long years has now become transferred to the vitamins.

Cations, however, have *gained* in importance. Before referring to their many applications in medicine, a few words will be devoted to the effects of metal deficiency and excess; either extreme can endanger life, while intermediate concentrations are essential for its maintenance. In this account, medical and agricultural examples will be treated side-by-side; they have many features in common, such as a maximal concentration for optimal results, and the mutual antagonism of pairs of cations.

Until recognized as such, copper deficiency was the cause of many crop failures in the reclaimed areas of Holland and Denmark. Similarly, in farm animals, copper deficiency led to anaemia, and demyelination of the spinal cord causing ataxia in lambs. Excessive copper in the soil is deleterious to crops and, in sheep, causes haemolysis of the liver and eventual death. Cobalt, which is essential for animals but not for plants, manifests a different sequence of deficiency. Grass grown on cobalt-deficient soil is healthy, lush, and relished by sheep who, nevertheless, become anaemic, weak, and finally die through vitamin B_{12} deficiency. In man, as little as one microgram is a sufficient daily intake of this cobalt-containing vitamin; without it, pernicious anaemia sets in. The more common anaemias are caused by iron deficiency, and can be cured by oral doses of ferrous sulphate. Lack of iron in soil causes poor crops. Excess of iron in mammals can be poisonous, especially to children. Lack of zinc in the soil causes poor yields, but zinc deficiency (or excess) in mammals rarely occurs. Molybdenum in the soil is essential for the fixation of atmospheric nitrogen by root nodules. The aerial spraying of soil with copper, cobalt, iron, zinc, or molybdenum (after a deficiency has been indicated by soil analyses) has produced a tremendous improvement in pastures and crops.

Calcium deprivation is common in pregnant women and poorly nourished children, but responds to supplementation by this cation together with vitamin D. However, if this treatment is pushed to excess, symptoms of hypercalcification soon appear. Soils which differ greatly in their calcium content support different types of vegetation and control the hardness of the local water supply. A report in the *World Health Organization's Bulletin for 1972* showed that cardiovascular disease is statistically correlated with the softness of water and with the types of rock and soil underlying the area. Areas supplied with soft water usually have higher cardiovascular death rates; the exact cause is not known but trace elements are suspected. Potassium is the essential intracellular ion of all living matter, and animals require sodium extracellularly although most plants do not use it at all.

Many diseases of animals and plants are traceable to maladjustments in the *balance* between metals. Pastures which are rich in molybdenum cause signs of copper deficiency in sheep, relieved by feeding copper. Excess of zinc in the diet of rats causes an anaemia relieved by copper. Land in Holland which had been dressed with copper salts to prevent the common "reclamation disease" (a copper deficiency) began to produce crops with a marked deficiency of manganese, although the soil had a normal content of this metal. These antagonisms remind one of the simple antagonism (calcium versus potassium) which is responsible for regulating heart beat, as Ringer[4] showed in 1883. Again, calcium or magnesium are absorbed by the same biochemical path from the intestine, so that when the diet is poor in magnesium, more calcium is absorbed, and vice versa.

Metal dependent diseases are further discussed in Chapter 18. Illnesses caused by excess of a metal are not uncommon. Some were mentioned above. Others are caused by the use of unsuitable metal containers for food or drink, by industrial exposure to metallic fumes, and domestic exposure to toxic housepaints or fittings. Some people have a genetic error which makes them unable to cope with quite normal amounts of an essential metal, e.g. copper in Wilson's disease. These aspects are described more fully in Chapter 18, 19, and 20.

The therapeutic uses of metal cations are many and varied. First, there is direct replacement therapy, e.g. of K^+, Ca^{2+}, Fe^{2+} for humans and of the wider range of metals for crops and stock, as outlined above. Then there are direct nonphysiological uses such as gold in arthritis, magnesium and other salines as cathartics, silver salts in dental practice, and the extraordinarily effective use of silver sulphadiazine in healing severe burns. Zinc salts are still considered the ideal astringents in ophthalmology, and dermatology could not afford to give up the use of aluminium acetate to control eczema. Aluminium salts continue to be the preferred antiperspirants. The use of lead salts in human medicine has been discontinued, but astringent lotions containing lead acetate are still used in veterinary practice. The recent clinical trials of platinum-amine complexes in cancer seem promising.

Quite a different section of therapy is the devising and application of antidotes to deal with metallic poisoning whether arising from overdosage of therapeutic metals, from industrial emissions, or from suicidal intent. This aspect of therapy is dealt with fully in Chapter 20.

Another, and quite subtle, use of metals in therapy is to employ chelating agents as antibacterials and antifungals. One class of these agents (in particular they must be fairly liposoluble) can enter living cells where they usually do no harm. However, prior exposure of the agent (often accidentally) to metals of variable valence enables the metal to penetrate into the plasma membrane (or even into the cytoplasm) where the combination (of agent and metal) causes great damage. The relevant metals are iron for bacteria and copper for fungi. The site of damage is thought to be lipoic acid, the essential coenzyme for pyruvic acid metabolism, and the injury seems to be effected through an oxidative chain reaction. This use of chelating agents (so different from their use as antidotes) is known as "augmentation of the toxic effect of a metal by chelation."

This augmentation was first observed in Australia in 1953 when the lethal effect of oxine (8-hydroxyquinoline) on microorganisms was found to be suppressed when all traces of heavy metals were excluded.[5] Oxine and its halogenated derivatives are much used in dermatology and in the management of dysentery, but they are not suitable for systemic use. Pyrithione (2-mercaptopyridine-*N*-oxide, or "Omadine") acts similarly to oxine and is used in dermatology, e.g. it is the most effective known remedy for dandruff. Salts of dimethyldithiocarbamic acid, which are among the most widely used agents to prevent fungal attack of crops, also act similarly to oxine.

Finally, there are those chelating agents whose mode of action is not yet fully known, but seems to depend, at least in part, on metal binding. Foremost among these agents are the tetracycline antibiotics which are a sheet-anchor for systemic antibacterial chemotherapy. Many successful antitubercular drugs belong here, e.g. isoniazid, thiacetazone, ethambutol, and also some antiviral drugs, such as methisazone. Among the body's hormones, thyroxine, histamine, noradrenaline, and adrenaline show strong chelating properties that seem to affect their distribution and storage, and possibly their action too. Chapter 19 deals more fully with the use of metal ions and their complexes in therapy.

It is clear that the bio-medical applications of inorganic substances are steadily increasing as a greater variety of ways for employing them becomes apparent. Better understanding of orbital and ligand-field theory is contributing the background knowledge for indicating new areas for exploration and coordinating apparently unrelated observations.

REFERENCES

1. Albert, A.: *Selective Toxicity*, 5th edition. London, Chapman and Hall, 1973, Chap. 10.
2. Perrin, D. D.: *Organic Complexing Reagents*. New York, Interscience-Wiley, 1964.
3. Williams, R. J. P.: The biochemistry of sodium, potassium, magnesium and calcium. *Quarterly Reviews*, London, 331, 1970.
4. Ringer, S., *J Physiol*, 4:29, 1883.
5. Albert, A., Gibson, M. and Rubbo, S.: *Brit J Exper Path*, 34:119, 1953.

CHAPTER 18

MISCOORDINATION IN A METAL-DEPENDENT DISEASE: WILSON'S DISEASE

Bibudhendra Sarkar

INTRODUCTION

THE RECENT SURGE of interest in bio-inorganic chemistry can perhaps be considered a major triumph in the application of inorganic chemistry to biomedical sciences. Consequently, a reader of this book would do well to understand what relevance inorganic chemistry has in biomedical problems such as in metal-dependent diseases.

Some metal-dependent diseases are listed in Table 18-I but a detailed discussion on each of these conditions is beyond the scope of this volume. The purpose of this chapter is to acquaint the reader with the bio-inorganic aspects of just one essential trace metal in normal physiological systems and the same in pathological conditions.

There are scores of metabolic reactions

TABLE 18-I
METAL-DEPENDENT DISEASES

Metal	Deficiency Disease	Diseases Associated With Metal Accumulation and Intoxication
Iron	Anemia	Hemochromatosis
Copper	Anemia	Wilson's disease
	Kinky-hair syndrome	
Zinc	Dwarfism	Metal-fume fever
	Gonadal failure	
Cobalt	Anemia	Heart failure
		Polycythemia
Manganese	Gonadal dysfunction	Ataxia
	Skeletal abnormalities	
Chromium	Abnormal glucose metabolism	
Selenium	Liver necrosis	"Blind staggers"
	White muscle disease	(cattle)
Lead		Anemia
		Encephalitis
		Peripheral neuritis
Cadmium		Nephritis
Mercury		Encephalitis
		Peripheral neuritis

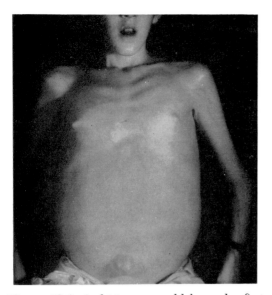

Figure 18-1. A thirteen-year-old boy who first came down with signs and symptoms of liver disease. He was later diagnosed to have Wilson's disease and at that time treatment was not available. The picture shows his thin shoulders and arms, and that he is severely malnourished. In contrast, he has a huge abdomen which is distended with large amounts of fluid. There are dilated veins over the abdominal wall and he has enlargement of the breasts, all of which are signs of his advanced liver disease.

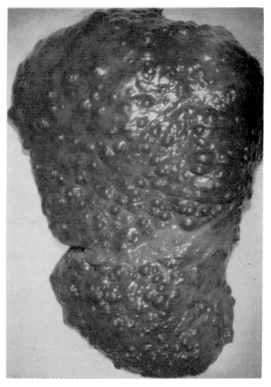

Figure 18-2. The boy in Figure 18-1 died one year later and at autopsy, his liver looked as it is shown in this picture. The liver is yellow instead of the normal brownish-red color. It is hob-nailed with multitudinous nodules measuring from a few millimeters to a centimeter of size. It is small and hard and shows the signs of advanced liver cirrhosis.

that maintain the normal body function which are related to metal ions. In a life system every metal has its own specific place. When something goes wrong, it reacts by coordinating with the wrong type of ligands which can be referred to as "miscoordination." This phenomenon is a complete departure from the normal physiological state and gives rise to many acute pathological and clinical conditions. There are many such examples of metal dependent diseases. Wilson's disease is one such example (Figures 18-1 and 18-2). The genetic defect underlying this condition leads to an excessive accumulation of coppor in the liver, kidney and brain which causes liver failure, to misfunction of the kidney and to various neurological abnormalities and death if unrecognized or untreated. From a bio-inorganic chemist's viewpoint, this disease poses many important challenges. The challenge concerns not only being able to find a better form of treatment for this ailment, but also knowing what has gone wrong in the pathways of copper metabolism which leads to the Wilson's disease conditions. There are two excellent reviews on this subject written recently by Sass-Kortsak,[1] and Evans,[2] which are highly recommended to the interested readers.

NORMAL COORDINATION OF COPPER IN HUMANS

Copper in Tissues and Biological Fluids. Copper Content

The copper content of the normal human adult is widely quoted as 50 to 120 mg. The concentration of copper in biological fluids of man is shown in Table 18-II. Copper is distributed throughout the body and certain organs have higher copper concentration than others. The liver, brain, heart and kidney, in decreasing order, contain the highest concentration of copper. Intermediate copper concentrations are found in lung, intestine and spleen. Endocrine glands, muscle and bone have the lowest concentration of copper. Because of the large mass of muscle and bone, these tissues contain approximately 50 percent of the total body copper, hepatic copper accounts for about 10 percent of the total. The copper levels of blood are around 100 mg/100 ml. Red blood cells contain

TABLE 18-II

COPPER CONTENT OF BIOLOGICAL FLUIDS IN MAN

Fluid	Copper $\mu g/100\ ml \pm$ S.D.	
Blood		
Whole	98.0 $\pm$	13.0
Plasma	109.0 $\pm$	17.0
Erythrocytes	115.0 $\pm$	22.0
Saliva	31.7 $\pm$	15.1
Bile	5.6—205.0	
Cerebrospinal fluid	1.6 $\pm$	0.4
Urine	18.0 $\pm$	7.2 $\mu g/24$h

a similar concentration of copper. Bile contains variable amounts of copper since it is the main excretory route for copper. The copper concentration in the cerebrospinal fluid of adults is low. There is very little copper excreted in urine or in sweat.

The State of Copper in Tissues

The properties of the element copper are such that free ionic copper cannot exist in appreciable amounts in the living organism. Only in the stomach may the relatively high degree of acidity allow the presence of copper in the ionic form. In the rest of the body, copper is in a complex more or less tightly bound form with proteins, peptides, amino acids and probably other organic substances. When copper is added to serum in excess of 1 : 1 molar ratio with respect to albumin, in addition to albumin and transferrin several other proteins in serum will also bind some copper. There are a few proteins of which copper is an integral part. Copper in these proteins is part of the molecular structure and not in dissociation equilibrium with ionic copper in solution as is found in the case of albumin bound to copper. There is a characteristic ratio between moles of protein and associated copper which makes these proteins a special class of metal-protein complex and establish them as metalloproteins. Those regularly present in mammals are listed in Table 18-III. Many of these copper proteins are enzymes and the copper is part of their active site, while others have no known enzyme activity. As far as is known, none of these proteins function as a respiratory carrier as hemocyanin does in mollusks. The copper in some of these copper proteins accounts for a sizeable portion or all of the copper present in the organ compartment where they are found.

TABLE 18-III

COPPER PROTEINS IN MAMMALS

Protein	Source	Molecular Weight	Color	Cu Content %	Cu (gram atom/ mole)	Enzyme Activity	Physiological Role
Ceruloplasmin	Human plasma	160,000	Blue	0.32	8	+	Unknown
Superoxide dismutase	Erythrocytes Liver	34,000	Bluish green	0.35	2	+	Protecting cells from damaging effect of super-oxide radical
Cytochrome C	Heart	——	——	0.07	1	+	Electron transport
Tyrosinase	Melanoma, skin	33,000	Colorless	0.02	1	+	Phenol oxidase
Uricase	Tissues of mammals except primates	100,000	——	0.056	1	+	Urate oxidation
Monoamine oxidase	Vertebrate tissues	195,000	——	0.07	4	+	Oxidative deamination of catecholamines and others
β-Mercapto-pyruvate transsulfurase	Rat liver	10,000	——	0.82	1	+	Transsulfuration of β-mercapto-pyruvate

Copper Proteins

Ceruloplasmin. This copper protein present in plasma deserves special attention. Human plasma contains approximately 32 mg/100 ml of this protein. Ceruloplasmin is an a_2-globulin and contains eight atoms of copper per molecule. In addition to being a metalloprotein it is a glycoprotein containing 7 percent carbohydrate. The molecular weight reported for human ceruloplasmin ranges from 132,000 to 160,-000 daltons. Copper analyses by various methods yielded an average level of 0.32% w/w. Calculations based on a molecular weight of 160,000 give eight atoms of copper per molecule. However, it should be noted that only six copper atoms are present in this molecule if the molecular weight is taken as 132,000. The protein absorbs light in the visible region with an absorption maximum at 610 nm over the pH range 5.5 to 8.0. The number of aspartic and glutamic acid residues in this protein was found to be high compared with those of other residues. It is relatively low in cystine. It has seventy-one tyrosine residues and only thirty-nine residues of histidine per molecule.

Numerous studies have indicated that the copper in ceruloplasmin exists in both cupric and cuprous forms, and also that the copper atoms may not be bound in identical fashion. Both by EPR and magnetic susceptibility measurement it has been established that half of the copper atoms in ceruloplasmin are in the cupric form and the other half in the cuprous state. It has been reported that four of the eight atoms of copper in ceruloplasmin were exchangeable with ionic ^{64}Cu when incubated with ascorbic acid at low pH. Digestion of ceruloplasmin with chymotrypsin also resulted in the liberation of four copper atoms. Partial removal of copper from ceruloplasmin results in the loss of its characteristic blue color, and loss of its enzymatic activity. The blue color is dependent on the presence of cupric atoms. The nature of the binding of copper to ceruloplasmin is not clear.

Ceruloplasmin exhibits enzymatic oxidase activity towards several substrates at pH 5.4 to 5.9. The best substrate is p-phenylene-diamine or its dimethyl derivative. Other substrates include hydroquinone, cate-cholamine, adrenaline, noradrenaline and serotonin. Ferrous iron is oxidized to ferric iron in the presence of ceruloplasmin. Many substances have been shown to inhibit the oxidase activity of ceruloplasmin. They include inorganic anions, carboxylate anions, sulfhydryl compounds, sodium azide and L-DOPA.

Treatment of native ceruloplasmin with urea and 3-mercaptoethanol reveals that it is composed of subunits, however the number and nature of these subunits are still unclear. Ceruloplasmin, like many other proteins, exhibits polymorphism. But all forms of ceruloplasmin so far investigated have the same copper content and the same enzymatic property.

The ceruloplasmin level in the newborn human is less than 10 mg/100 ml serum. It rises to adult levels (30.4 ± 5 mg/100 ml) by two to four months of age and continues to rise to a peak at two to three years and then declines to adult levels by twelve years of age. Different pathological states affect the ceruloplasmin level. A pronounced deficiency of this serum protein is characteristic of Wilson's disease. Ceruloplasmin levels are also influenced by hormones. The levels increase in pregnancy and also in response to administration of estrogens. Ceruloplasmin is synthesized exclusively in the liver. It is not known how the copper is incorporated into ceruloplasmin within the liver cell.

A number of hypotheses have been formulated over the years for the biological function of ceruloplasmin. First, it was thought that it might have a copper transport function because it contains 95 percent of the copper in plasma. However, the finding that patients with Wilson's disease have only a small amount of ceruloplasmin in plasma was used as an argument against it since these patients accumulate copper in tissues without it. It was then proposed that ceruloplasmin plays a role in the maintenance of the zero copper balance in the body. However, insufficient ceruloplasmin copper is normally excreted in the bile to account for the maintenance of copper balance, and there is no evidence for the excretion of the protein via the intestinal wall or the kidney. Subsequently, it was proposed that ceruloplasmin transfers copper specifically to cytochrome C oxidase. However, no supporting evidence was produced for this hypothesis. Meanwhile, there have also been some suggestions that ceruloplasmin is a vestigial protein; it probably had some role in the past but lost its function at some stage of the evolutionary processes. A recent suggestion is that ceruloplasmin oxidizes ferrous ion to ferric state which then goes on to bind to apotransferrin in the blood plasma. This theory too has come under heavy attack. The ferrous ion itself can bind apotransferrin and can get oxidized when it is bound to the protein. Also, patients with Wilson's disease do not usually have anemia. The fact that ceruloplasmin has a moderate amount of oxidase activity towards polyamines and polyphenols has suggested that it may oxidize ascorbic acid. However, its enzymatic activity is so low *in vitro* that the physiological significance of this property remains uncertain. Presently there is still no conclusive evidence regarding the function of ceruloplasmin.

SUPEROXIDE DISMUTASE. The protein referred to here as superoxide dismutase has undergone an interesting history of nomenclature. More than thirty years ago hemocuprein from bovine erythrocytes and hepatocuprein from bovine liver were isolated. The two proteins had the same copper content, the same molecular weight, but different color. Hemocuprein was blue whereas hepatocuprein was almost colorless. Several years later another copper protein was isolated from horse liver, similar to bovine hepatocuprein in copper content and molecular weight, but which had a bluish-green color. Later, another protein, cerebrocuprein, was reported from ox and human brain which resembled human hepatocuprein in all respects. A few years later yet another copper protein, erythrocuprein was isolated from human erythrocytes. In spite of the similarities between all these human copper proteins, cerebrocuprein from brain, erythrocuprein from erythrocytes and hepatocuprein from liver, it was not until recently that Carrico and Deutsch[3] found these proteins to be identical. These authors suggested that the protein be designated cytocuprein to eliminate referring to the same protein by different names. However, in a latter publication, the same authors indicated that cytocuprein is not an appropriate name since the protein also contains 2 gram-atoms of zinc per mole. Recently McCord and Fridovich[4] found an enzyme in bovine erythrocytes which could catalyze the dismutation of superoxide following the reaction:

$$O_2^- + O_2^- + 2H^+ \rightarrow O_2 + H_2O_2$$

They reported that the enzyme contained two copper atoms per molecule, exhibited a pale greenish-blue color and contained 0.38% copper. Its molecular weight was between 32,600 and 33,000 daltons. They concluded that the enzyme was erythrocuprein.

McCord and Fridovich compared the superoxide dismutase activities of their protein with that of a sample of human erythrocuprein prepared by Carrico and Deutsch and found that the two proteins had similar specific activities. Finally, superoxide dimutase activity was detected in the preparations from bovine heart, brain and liver.

This copper protein isolated from various sources was found to have an isoelectric point of 4.74 to 4.76 and a sedimentation constant which varied from 2.79 s to 2.84 s. They were also found to have similar molecular weights of approximately 34,000 daltons. They contained 0.35% of copper or two copper atoms per molecule. Their amino acid compositions were almost identical. The proteins were relatively high in glycine, aspartic acid and glutamic acid. Peptide mapping of tryptic and chymotryptic preparations did not reveal significant differences between the proteins. The same conclusion was drawn from visible and ultraviolet spectrophotometric studies. A broad peak was demonstrated at 675 nm and a maximum of 265 nm with a shoulder at 258 nm. The EPR spectra from the three proteins were indistinguishable and indicated that the copper was almost entirely in the cupric state. Immunodiffusion reactions also revealed a single identity.

Superoxide dismutase may have a vital role in protecting the cell from the damaging effects of the superoxide radical. There is high quantity of superoxide radical generated through various enzymatic processes. The superoxide radical reduces cytochrome

C which indicates the necessity for removal of this anion to prevent fixation of cytochrome C in the reduced state. Thus superoxide dismutase prevents the accumulation of a potentially toxic radical.

CYTOCHROME C OXIDASE. Cytochrome C oxidase is the terminal oxidase in the respiratory oxidation system. Located in the mitochondrial membrane of highly differentiated cells, it provides energy for the cell by coupling electron transport through the cytochrome chain with the process of oxidative phosphorylation. It requires copper for catalytic activity. Each water-soluble monomer of cytochrome oxidase contains one molecule of heme and one atom each of iron and copper.

The tendency of cytochrome C oxidase to aggregate makes the determination of its molecular weight rather difficult. A full discussion of this and various other associated matters are included in an excellent review by Lemberg.[5] Because of the difficulty of obtaining purified cytochrome oxidase, the enzyme as well as its mechanism of action have been difficult to characterize. Electron paramagnetic resonance studies suggest that cytochrome C reduces the copper of cytochrome oxidase followed by the reduction of oxygen in a series of steps yielding oxygen radicals.

As the terminal enzyme in the oxidative phosphorylation process, cytochrome oxidase is essential in cellular metabolism. Severe copper deficiency results in decreased cytochrome oxidase activity with a concurrent loss of capacity for oxidative phosphorylation.

Cytochrome C oxidase appears during the fetal period. There is a low activity in the first six months of gestation followed by a continuous rise in brain, liver and heart, reaching adult values at birth. It is well established that cytochrome C oxidase activity is reduced in animals fed on a copper-deficient diet. It is interesting to note that in copper-deficient animals the tissue respiration appears unchanged despite the depressed activity of cytochrome C oxidase.

OTHER COPPER PROTEINS. There are a number of other copper proteins distributed throughout the mammalian body. Monoamine oxidase, for example, has been purified and crystallized from both bovine and porcine plasma. The enzyme has a molecular weight of 195,000 daltons and contains four atoms of copper per molecule. Monoamine oxidase catalyzes the oxidative deamination of a variety of monoamines with a stoichiometric formation of the corresponding aldehyde, ammonia and hydrogen peroxide. Several experiments have demonstrated the enzyme to be involved in maintaining the structural integrity of both vascular and bone tissue.

Tyrosinase, another copper containing enzyme is essential in the pigmentation process. A genetic absence of tyrosinase results in albinism, which is characterized by extreme sensitivity to light. In addition, achromotrichia has been observed in several copper deficient species.

The copper-proteins discussed above represent only a few well-known copper proteins isolated from biological systems.

Absorption of Copper

Copper absorption is limited to the stomach and intestine. It is transported from the mucosal side to the serosal side by an energy dependent mechanism. The rate of absorption of orally administered ^{64}Cu increases rapidly and reaches a peak 0.5 hr. after administration of the isotope. Thereafter, the rate of absorption decreases and eventually reaches a constant level. It has been demonstrated that the L-amino acids facilitate the intestinal absorption of copper. Amino acid transport across the intestinal mucosa is known to be an energy-dependent process. Thus the energy-linked mechanism for copper transport may represent copper that is transported as a copper amino acid complex.

It has recently been identified that there are two copper binding proteins in the intestinal mucosa. One fraction of copper is contained within a protein that has the physical properties of the enzyme superoxide dismutase. The second and larger fraction of copper is bound to a sulfhydryl-rich protein with a molecular weight of 10,000 daltons. The properties of these proteins are very similar to those of metallothionein. Metallothionein has a molecular weight of 10,000 daltons, binds cadmium, copper and zinc. Highly purified metallothionein contains twenty-six sulfhydryl groups and the protein binds metals by the formation of mercaptides. The intestinal metallothionein may have at least two functions. First, it probably has a passive role in copper absorption by providing binding sites within the intestinal mucosa to insure that an adequate supply of the metal is accepted from the dietary source and stored temporarily for subsequent absorption. Second, metallothionein may represent a mucosal block to protect against absorption of toxic levels of copper.

Transport Form of Copper in Blood

Human serum contains approximately 1 μg of copper per ml. Most of this is bound to the copper protein ceruloplasmin (Figure 18-3). Ceruloplasmin is not a transport protein for copper in the same sense as transferrin functions for iron. The copper bound to ceruloplasmin does not constitute the transport form of copper. About 5 to 10 percent of the copper in serum is bound

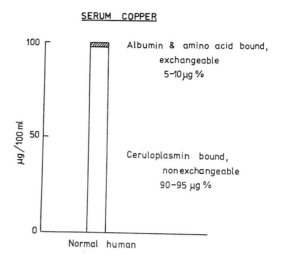

Figure 18-3. Distribution of copper in human blood serum.

to albumin. Albumin-copper is not a metalloprotein but this protein has one specific binding site for copper. The copper bound to albumin is in equilibrium with unbound copper or with copper bound by other ligands. There is ample evidence to suggest that the copper bound to albumin is the immediate transport form of copper.

The albumin molecule consists of a single polypeptide chain made up of approximately 600 amino acids and stabilized by about seventeen disulfide bonds. The copper-binding site of albumin is perhaps the best delineated metal binding site of a protein type in solution. Several groups have studied to characterize this site, but it was not until the determination of the amino acid sequence of the NH_2-terminal fragment of albumin that a comprehensive investigation was possible to identify the amino acid residues involved at the binding site. The copper-binding site of albumin has been proposed by Peters and Blumenstock[6] to involve the α-amino nitrogen, imidazole nitrogen from the histidine in position 3, and the two intervening pep-

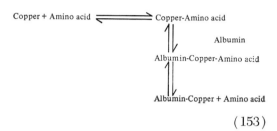

Figure 18-4. Structure of copper-binding site of human albumin.

tide bond nitrogens. The copper-binding site of human albumin is shown in Figure 18-4.

It has been found that the equilibration between albumin-bound copper in the blood and copper in tissues is extremely rapid, much more rapid than the equilibration of albumin between the intravascular and extravascular compartments. It became apparent that the rapid equilibration of copper between blood and tissues occurs via a small molecular ligand for copper which can effectively compete for the binding of copper with albumin. It was later established that these low molecular weight ligands are amino acids.

It was revealed that, of all the physiologically occurring amino acids in serum, L-histidine was the most important ligand for copper, followed by threonine, glutamine and some other amino acids. It was also found that a most unusual complex is present in human serum, a complex in which one atom of copper is bound to two different amino acids, such as L-histidine-copper-L-threonine. A small proportion of similar type of ternary complexes involving L-histidine and other amino acids, was also indicated. These amino acid-copper complexes have been separated and identified in the ultrafiltrates of normal human serum by Sarkar and Kruck.[7]

Most recent experiments have furnished evidence of the existence of an intermediary compound between albumin-copper complex and copper-amino acid complex. Main complex of this type was found to be albumin-copper-L-histidine. In view of these findings, Sarkar and coworkers[8, 9] proposed that the following equilibria are maintained in the physiological state:

$$\text{Copper} + \text{Amino acid} \rightleftharpoons \text{Copper-Amino acid}$$
$$\Updownarrow \quad \text{Albumin}$$
$$\text{Albumin-Copper-Amino acid}$$
$$\Updownarrow$$
$$\text{Albumin-Copper} + \text{Amino acid}$$

$$(153)$$

Further experiments have shown that the ternary complex is indeed rate limiting in the ligand exchange reaction. Consequently the mechanism shown above could play an important role in the exchange of copper between a macromolecule and a low molecular weight substance which in turn can readily be transported across the biological membrane.

Metabolism and Excretion of Copper

In the extrahepatic tissues, the metabolism of copper is confined mainly to the normal synthesis and degradation of copper-dependent enzymes, whereas in the liver, copper undergoes the metabolic interrelationship involved in the maintenance of copper homeostasis. Orally ingested radioactive copper appears rapidly in the blood indicating rapid absorption from the intestinal tract. It has been demonstrated that the radioactive copper was immediately bound to albumin after its entry into plasma and that the secondary rise in plasma radioactive copper was due to its incorporation into ceruloplasmin.

It is well documented now that bile is

the major pathway for excretion of copper. Bile contains both low and high molecular weight components. One fraction of copper is associated with amino acids and peptides and a second fraction is associated nonspecifically with high molecular weight molecules. The main route of excretion of copper from the body is via the feces. Urine contains extremely small amounts of copper (30 to 40 mg/day in man). Sweat contains only negligible quantities. There is very little copper in saliva and it is recirculated. Insignificant amounts are lost by menstruation.

MISCOORDINATION OF COPPER IN HUMANS: WILSON'S DISEASE

Genetic Aspects

Wilson's disease is one of the inborn errors of metabolism characterized by the classical triad of severe progressive liver disease (cirrhosis), a peculiar neurological syndrome and the presence of Kayser-Fleischer rings of the cornea. The disease is inherited in an autosomal-recessive fashion. There is a high incidence of consanguineous marriages among the parents. The disease is somewhat more prevalent in males. In one study, the frequency of the abnormal gene in the American population was calculated to be approximately 1:2,000 and the incidence of the disease 1:4,000,000. This may well be too conservative an estimate. It was suggested that the frequency of the abnormal gene may be higher in the population of southern Italy, among East European Jews and among Japanese. Patients with Wilson's disease have been reported in many countries and in individuals of European, Chinese, Japanese, Indian, Eskimo and Negro ancestry. Jewish patients from eastern Europe appear to have predominantly neurological symptoms with a later age of onset and

more slowly progressive character than is the tendency in other patients. The hepatic form of the disease with early onset and rapid deterioration is more prevalent among Oriental patients. These clinical differences may be influenced by genetic factors. However, the copper-rich diet of Oriental people may be a contributing factor to the earlier onset of the disease.

Clinical Features of Wilson's Disease

Kinnier Wilson, an eminent English neurologist, in his classical monograph,[10] first recognized the combination of neurological disease and cirrhosis of the liver as a disease entity which he called "Progressive Lenticular Degeneration." Patients generally fall into two broad categories. In one, the symptoms involve primarily the nervous system; in the other, primarily the liver. The neurological symptoms arise mainly from disorganization of the lenticular region of the brain. Tremor, like that seen in Parkinson's disease, rigidity, dysarthria and dysphagia appear. Spastic rigidity usually predominates in the neurological picture of younger patients, while progressively worsening tremor is common when the disease appears later in life. It has been suggested that the two types of symptoms have two different causes, with tremor arising from the toxic effect of copper and rigidity or dystonia from the effect of the second hepatic factor. Liver damage may be apparent in these forms from clinical signs and symptoms and from abnormalities in liver function. With continued refinement of biochemical tests for Wilson's disease the hepatic form of the disease is recognized as more common than was previously believed. The early hepatic symptoms resemble those of acute or chronic hepatitis. Postnecrotic cirrhosis follows. This type is sometimes referred to as the "abdominal form" of Wilson's disease.

The age of onset is generally during the first two decades of life. In general, the hepatic form of the disease predominates in younger patients and is rapidly progressive. The neurological form predominates in older patients and the disease may worsen slowly if untreated over a period of years. An important clinical feature is the Kayser-Fleischer rings. The rings are caused by the deposition of golden or greenish copper-containing pigment in the outer edge of the cornea. The rings, which are sometimes visible only by slit-lamp examination, are almost always present in symptomatic Wilson's disease and are absolute diagnostic criteria. However, they may be lacking in young patients with the hepatic form of the disease and their absence does not exclude the diagnosis.

The kidney is also frequently involved in Wilson's disease resulting in generalized amino aciduria and other signs of abnormal renal tubular function. These are usually not clinically obvious, although they have been known to cause bone lesions.

Psychiatric disturbances have been reported in many cases of Wilson's disease. It has been pointed out that the psychiatric abnormalities present have often been mistaken for specific psychiatric entities, such as schizophrenia.

Bone lesions and hemolytic crises infrequently accompany the disease. Copper concentration in the body apparently increases as the disease progresses. In one study liver-copper levels varied from four to forty times the normal mean level and brain-copper levels varied from seven to seventeen times the normal mean level. The increased copper content of the liver may be useful as a diagnostic feature in asymptomatic Wilson's disease. An increased amount of urinary copper in patients with the disease has been observed. Patients with symptomatic Wilson's disease usually excrete more than 100 μg copper per twenty-four hour urine compared with less than 50 μg per twenty-four hour urine in normal individuals. The level of urinary copper can vary considerably. Levels are not usually increased before clinical symptoms appear. The copper content in bile has been reported as normal. Fecal and serum-copper levels were found to be below normal levels in most patients with Wilson's disease. An important finding by Scheinberg and Gitlin[11] and independently by Bearn and Kunkel[12] showed the typical low blood levels of ceruloplasmin in Wilson's disease. In patients with Wilson's disease, the direct reacting copper fraction which is not bound to ceruloplasmin is increased. A low total serum-copper level is not a reliable feature of Wilson's disease. A moderately low ceruloplasmin level combined with an increased direct-reacting copper fraction can produce a normal serum-copper level in the patient.

There is no evidence, thus far, that patients with Wilson's disease have abnormalities of any other copper proteins. The enzyme tyrosinase must not be markedly decreased as in albinism because patients have normal pigmentation. Chromatographically defined copper proteins of the brain are the same in normal individuals and in patients, although in the latter, copper is also complexed to a variety of proteins which are not normally associated with copper. Excess copper in the liver of patients is bound to a protein in the mitochondrial fraction. The mean level of copper in erythrocytes was found to be similar in patients with Wilson's disease and in normal controls.

Copper Metabolism in Wilson's Disease

Metabolic abnormalities in Wilson's disease are listed in Table 18-IV. Comparison of response curves following an oral load-

TABLE 18-IV

METABOLIC ABNORMALITIES IN SYMPTOMATIC AND EARLY ASYMPTOMATIC
HOMOZYGOUS PATIENTS AND HETEROZYGOTES OF WILSON'S DISEASE

| | Homozygotes | | |
Abnormality	*Symptomatic*	*Asymptomatic*	*Heterozygotes*
Serum copper	Usually low	Usually low	Rarely low
Serum ceruloplasmin	Low	Low	Rarely low
Nonceruloplasmin-bound copper	High	Normal	Normal
Urinary copper	High	Normal	Normal
Copper in tissues	Elevated	Elevated	?
Handling of radiocopper			
Rate of clearance from blood	Slow	Normal	Normal
Incorporation into ceruloplasmin	None	None	Slow
Uptake in liver			
Initial rate	Slow	Normal	Normal
Extent	Low	Normal	Normal
Discharge from liver	Impaired	Impaired	Impaired
Excretion			
Urinary	High	Normal	Normal
Fecal	Low	Low	?
Other abnormalities			
Aminoaciduria	Present	Absent	Absent
Phosphaturia + low serum P	Present	Absent	Absent
Uricosuria + low serum urate	Present	Absent	Absent

ing dose of radioactive copper have shown an increased peak of radioactive copper and decreased rate of disappearance of radioactive copper from plasma or serum in patients with Wilson's disease when compared with control subjects. The higher peak of radioactivity has been interpreted to be a result of increased intestinal absorption in the patients with Wilson's disease. However, an equally valid explanation would be delayed clearance of radioactive copper from the plasma. In fact, when a loading dose of radioactive copper is administered intravenously, patients show a delayed fall in the radioactive copper level from their plasma in comparison with normal controls. Delayed clearance of radioactive copper from the serum, perhaps because of failure of an excretory pathway, is thus a feasible explanation for the results in patients.

The fall in the level of plasma radioactive copper after its oral or intravenous administration is followed in the normal individual by a secondary rise. It has been found that patients with Wilson's disease did not show this incorporation of labelled copper into the ceruloplasmin fraction. The failure of patients with Wilson's disease to incorporate labelled copper into ceruloplasmin is now well documented and has also been observed in patients with normal levels of ceruloplasmin. As the albumin-bound fraction of labelled copper disappears from the blood, labeled copper appears in the liver. The uptake of radioactive copper into the liver, the main organ in which concentration takes place, was found to be slower in patients with Wilson's disease than in control subjects. In the normal individual, the radioactive copper activity in the liver reaches a peak four to ten hours after administration, then gradually declines, while in the patient the activity rises more slowly and continues to rise apparently because there is no discharge of radioactive copper from the liver. This reduced ability of the liver to take up radioactive copper appears to be a cardinal feature of the disease mechanism.

With the high concentrations of copper in the liver of patients with Wilson's disease, an increased concentration of copper would be expected in the bile. It has been shown that rats whose livers have been loaded with copper prior to the administration of labeled copper have an increased excretion of radioactive copper in the bile. The ratio between bile and liver activity was similar in copper-loaded and control rats. The uptake of copper into the bile was increased in the copper-loaded rats. This is similar to the finding in early Wilson's disease. This evidence suggests that the low net uptake of radioactive copper found in symptomatic patients is the result of saturation of the normal copper-binding sites in the liver. Several studies have indicated a decreased fecal excretion of copper in patients with Wilson's disease. Since most of this copper originates from bile this finding combined with that outlined above provides strong evidence that biliary excretion of copper is impaired in Wilson's disease.

The urinary excretion of labelled copper is higher in patients than in controls. The fecal radioactive copper excretion is nevertheless considerably reduced, yet another evidence for a defect in biliary excretion of copper in the disease.

Pathogenesis

There is ample evidence that copper is present in excessive amounts on certain organs in individuals with Wilson's disease. When the copper deposits are removed by effective chelation therapy, the symptoms of the disease regress. Wilson's disease is apparently synonymous with copper intoxication. Evidence has been presented that in the cell, copper exerts a toxic effect on the mitochondria. It has also been shown that copper inhibits ATPase from the microsomal fraction of pigeon and rat brain homogenates. It has been suggested that copper exerts its toxic effects by inhibiting. ATPase and consequently the membrane transport of ions. The unanswered question is which of the above effects is more important, and indeed whether either one is the primary mechanism of the toxic action of copper.

The primary defect causing the copper intoxication is not yet known. A genetically determined defect in protein metabolism was proposed as a primary defect of Wilson's disease, which disordered copper metabolism as a secondary defect. It has been suggested that abnormal proteolytic activity in the tissues lead to the formation of protein or polypeptide residues with a high affinity for copper. Amino aciduria and peptiduria, consequences of proteolytic activity, were considered to be primarily biochemical manifestations of this disease. There is little support for this hypothesis. A delayed uptake of copper into the liver in advanced stages of Wilson's disease does not support the presence of abnormal substances with a high copper affinity. Furthermore, aminoaciduria is generally found as a late symptom while excess copper deposits are found even in completely asymptomatic patients.

The more favored hypothesis is that the basic defect lies in some aspect of copper metabolism. The ceruloplasmin deficiency when first recognized was proposed as the cause of Wilson's disease. It was proposed that this deficiency arises because of the lack of an enzyme converting ceruloplasmin from a precursor to final form. Also, a number of hypotheses involving various aspects of the transfer of copper have been proposed. Other hypotheses have included an unspecified block in the intracellular transport of copper, synthesis in Wilson's disease of an abnormal protein with a high affinity for copper, and the lack of enzymatic incorporation of copper into ceruloplasmin.

Treatment

While the exact nature of the genetic defect in Wilson's disease is still not known, it is almost universally accepted that the damage caused to vital organs (liver, brain, kidney), and thus the disease, is due to the toxic action exerted by excessive amounts of copper deposited primarily in these organs. If this is so, removal of this copper from the body, and in particular from these organs, should be beneficial. Treatment methods which can result in the production of a negative overall copper balance will lead to clinical improvement in symptomatic patients and, most importantly will also prevent the development of signs and symptoms of the disease in those patients who are lucky enough to be diagnosed in the presymptomatic phase of the disease.

To achieve significant "decoppering" of patients, agents are used which have the property of strongly binding to copper with the added requirement that they should be nontoxic and easily distributed in various body compartments. Also, the resultant complexes between copper and these agents should have an easy and rapid route of elimination from the body. Several such decoppering agents are available (Figure 18-5).

BAL(2,3-dimercaptopropanol) was originally used for the purpose of treatment of Wilson's disease with some success. EDTA (ethylenediaminetetraacetic acid) can also

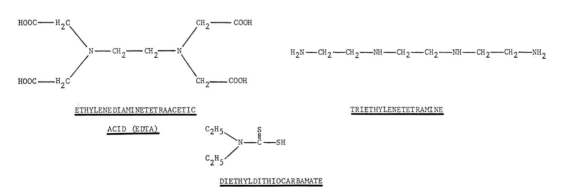

Figure 18-5. Decoppering agents used for treatment of Wilson's Disease.

be considered a decoppering agent. However, trials were not carried out over a period of time sufficient to demonstrate improvement of the disease in patients.

Orally administered sodium diethyldithiocarbamate has also been reported as a means whereby negative copper balance may be achieved in patients with Wilson's disease by increasing the fecal excretion of copper. Clinical improvement of neurological symptoms has been reported by the use of this agent.

The treatment of choice for Wilson's disease is oral administration of D-Penicillamine, proposed first by Walshe[13] and later confirmed by many others. It can be administered orally in capsules, at the usual recommended dose of 1g/day. This treatment is recommended to be continued all through the life of the patient. The immediate and most dramatic effect of the administration of D-Penicillamine is a marked increase in urinary copper excretion. In patients with Wilson's disease, in response to D-Penicillamine, urinary copper excretion at the start of treatment is always more than 1,000 μg/24-hr. and often as high as 4,000-5,000 μg/24-hr. With continuation of the treatment, the twenty-four hour urinary excretion gradually decreases so that after three to four months of treatment the excretion is often around 1,000 μg/day. This is not a sign of the patient becoming "refractory" to treatment, but is due rather to progressive reduction of abnormally high copper stores of these patients. At this point, particularly when the patient's clinical improvement is slow, the dose of D-Penicillamine may be increased to 1.5 to 2.0 or even 3.0 g/day but one must be mindful of the higher incidence of toxic reactions and untoward side effects which may more often be encountered on these higher doses.

Treatment with D-Penicillamine regularly leads to improvement in the clinical status of the patient. However, it is important to emphasize that improvement with this form of therapy is slow in terms of both clinical and laboratory manifestations of the disease. Also, the treatment may not always be successful. In very advanced cases patients may die with neurological deterioration or in hepatic failure during the first weeks or months of treatment. There are also undesirable side effects of treatment with D-Penicillamine. Recently, Walshe[14] has reported the administration of triethylenetetramine dihydrochloride which produced and maintained marked cupriuria and clinical improvement in patients with Wilson's disease who could not be treated with D-Penicillamine because of its nephrotoxicity. There are sensitivity reactions, bone marrow suppression, abnormalities in taste, formation of abnormal antibodies and dermatopathy due to D-Penicillamine. It also acts as an antimetabolite of Vitamin B_6 (pyridoxine). Recently it has become evident that D-Penicillamine may be reacting with one type of copper pool only. All these problems make it necessary to find other alternatives.

A new approach to this problem of removing copper has been initiated by Sarkar and coworkers[15-17] by designing suitable molecules which have the required specificity and affinity for copper. The specificity is an important criterion in building such molecules so that other metals in the body are not affected while removing copper. For this purpose advantage was taken of the type of copper-binding site that nature has created in albumin. The idea being that a small molecule is to be designed which is almost a replica of the human albumin-copper binding site but devoid of large extraneous sections of the molecule. It is expected that the designed molecule with its high affinity and specificity for copper will selectively pick up excess copper. The designed molecule-copper complex is small enough, compared to the big albumin molecule to go across the biological membrane to the kidneys and be excreted. With this type of compound it may even be possible to remove copper from pools which are normally untouched by presently available chelating agents. Several of such designed molecules have been synthesized which mimic the native copper-binding site of human albumin. However, these are still not ready for human use. Nevertheless, it is envisaged that this type of approach may be of tremendous value in the treatment of Wilson's disease.

REFERENCES

1. Sass-Kortsak, A.: Hepatolenticular degeneration (Kinnear Wilson's disease). Schwiegk (Editor), Berlin, Springer Verlag, 1974, In *Handbuch der Inneren Medizine.* 1973, vol. VII/1.

2. Evans, G. W.: Copper homeostasis in the mammalian system. *Physiol Revs,* 15:535-570, 1973.

3. Carrico, R. J. and Deutsch, H. F.: Isolation of human hepatocuprein and cerebrocuprein. Their identity with erythrocuprein. *J Biol Chem,* 244:6087-6093, 1969.

4. McCord, J. M. and Fridovich, I.: Superoxide dismutase. An enzymic function for erythrocuprein (hemocuprein). *J Biol Chem, 244:*6049-6055, 1969.

5. Lemberg, M. R.: Cytochrome oxidase. *Physiol Revs,* 49:48-221, 1969.

6. Peters, T., Jr. and Blumenstock, F. A.: Copper binding properties of bovine serum albumin and its amino-terminal peptide fragment. *J Biol Chem,* 242:1574-1578, 1967.

7. Sarkar, B. and Kruck, T. P. A.: Copper-amino acid complexes in human serum. In Peisach, J., Aisen, P. and Blumberg, W. (Eds.): *Biochemistry of Copper.* New York, Acad Pr, 1966.

8. Sarkar, B. and Wigfield, Y.: Evidence for albumin-Cu(II)-amino acid ternary complex. *Can J Biochem,* 46:601-607, 1968.

9. Lau, S. and Sarkar, B.: Ternary coordination complex between human serum albumin, copper(II) and L-histidine. *J Biol Chem,* 246: 5938-5943, 1971.

10. Wilson, S. A. K.: Progressive lenticular degeneration: A familial nervous disease associated with cirrhosis of the liver. *Brain, 34:* 295-509, 1912.

11. Scheinberg, I. H. and Gitlin, D.: Deficiency of ceruloplasmin in patients with hepatolenticular degeneration (Wilson's disease). *Science 116:*484-485, 1952.

12. Bearn, A. G. and Kunkel, H. G.: Biochemical abnormalities in Wilson's disease. *J Clin Invest, 31:*616, 1952.

13. Walshe, J. M.: Penicillamine, a new oral therapy for Wilson's disease. *Am J Med,* 21:487-495, 1956.

14. Walshe, J. M.: Copper chelation in patients with Wilson's disease: A comparison of penicillamine and triethylenetetramine dihydrochloride. *Quart J Med, 42:*441-452, 1973.

15. Sarkar, B.: *Molecular Design of the Metal-binding Site of Proteins.* Proceedings of the IXth International Congress of Biochemistry. Stockholm, Sweden, 1973.

16. Lau, S., Kruck, T. P. A. and Sarkar, B.: A peptide molecule mimicking the Copper(II) transport site of human serum albumin. *J Biol Chem,* 249:5878-5884, 1974.

17. Lao, S. and Sarkar, B.: Kinetic studies of copper(II)-exchange from L-histidine to human serum albumin and diglycyl-L-histidine, a peptide mimicking the copper(II)-transport site of albumin. *Can J Chem,* 53:710-715, 1975.

CHAPTER 19

THERAPEUTIC USES OF METALS, LIGANDS AND COMPLEXES

David R. Williams

Introduction
Metallotherapy
Ligand Therapy
Metal Complexes as Therapeuticals
Conclusion

INTRODUCTION

In 1828 FRIEDERICH WÖHLER (Figure 19-1) destroyed the myth that life was purely organic when he synthesized a biological molecule from an inorganic source. The twenty-five elements that are now known to be necessary for healthy human life are shown in Figure 19-2. In general, these elements follow the abundances of the elements in the earth's crust and in sea water (our distant ancestors evolved from aquatic life and the process of natural selection has removed organisms dependent upon less readily available elements) and for each element there is a cycle equivalent to the well known carbon or nitrogen cycles.

As more heavy elements are being used by our civilization the world is becoming more polluted. If we accept the view that evolution and adaption permit elements to traverse the scheme *poisons → tolerable impurities → useful elements → essential elements* it is understandable how the seven elements shown in parentheses in Figure 19-2 have just recently been shown to be "essential," e.g. dietary deficiencies produce animal growth rates as low as two-thirds the normal growth rates, because they oc-

Figure 19-1. Friederich Wöhler. (Photograph courtesy of Read Chemistry Museum, University of St. Andrews, Scotland.)

cupy periodic table positions adjacent to the eighteen essential elements whose biochemical properties are well known. For

334

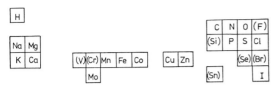

Figure 19-2. Elements that are essential for healthy human life.

similar reasons, it appears that boron, germanium and nickel have already started out along this evolutionary trail from "poison" to "essential" elements.[1]

It is now known that 0.7 percent of the atoms in the human body are metals and this comprises 3 percent of the total body weight. However, the preceding chapters have shown that life is dependent upon these elements far more than these figures suggest.

Paralleling the growth of bio-inorganic knowledge there is an overriding awareness of the usefulness of our discoveries in treating diseases. Diseases occur when excesses or deficiencies of *in vivo* metals appear, when metal pollutants enter the body (for example, cadmium, lead or mercury), or when poisons or viruses enter into the metal-ligand competition.

Chapter 18 has described the biochemistry of one metal-dependent disease in some depth and Chapter 20 deals specifically with *toxic* metal ions and their therapy. This present chapter describes the uses of metals, ligands and metal complexes in treating disease in general and, for illustrative purposes and to restrict the chapter to a reasonable length, refers to the cancer group of diseases in particular.

Cancer[2]

Cancer rates among the top three causes of deaths in the West (one in five deaths are attributed to cancer) and in spite of press reports of amazing new cures, this rating is likely to increase as better cures are found for other fatal diseases. Roe has defined cancer as "a disease of multicellular organisms which is characterized by the seemingly uncontrolled multiplication and spread within the organism of apparently abnormal forms of the organism's own cells." This term "cancer" actually embodies hundreds of different types of neoplastic diseases ranging from localized skin cancers to whole body leukaemias with representative cure rates as high as 95 percent or as low as 0 percent. However, the cure rates for most cancers fall between these extremes.

Carcinogens may be defined as substances that are capable of producing tumors in any test species by any route and at any dose level. This term includes quite inert materials such as gold, silver, sodium chloride, or plastics which can cause cancer by localized irritation, but in general it refers to the more widely recognized carcinogens summarized in Table 19-I. Carcinogens first come to light during epidemiological studies, and then the carcinogenicity of the suspect chemicals is measured using animal experiments. Difficulties that arise include (a) the extrapolation between one animal species and another, (b) the choice of the age of the animals (for example, newborns have less capabilities of

TABLE 19-I

CLASSIFICATION OF AGENTS KNOWN TO CAUSE CANCER

Chemical	Physical
Aromatic hydrocarbons and amines	Ionizing radiation
Aromatic heterocyclic ring compounds	Ultraviolet radiation burns
4-Nitroquinoline oxide	Burns
Nitrosamines	*Other*
Azo compounds	Chromosomal abnormalities
Alkylating agents	Viruses
Urethanes	
Polymers	
Metals	

developing immunological rejection reactions than older animals), and (c) the classification of identified carcinogenic chemicals into "initiators" or "promoters."

Our ideas concerning the molecular mechanism of cancer induction are still embodied mainly in tentative theories. Because cancer cells cannot be turned back into the forms of their parent normal cells, the following hypotheses are presently in vogue: (a) carcinogens may be metabolized into more active reagents, (b) the hereditary mechanism of the cell is then involved either by the carcinogen reacting with nuclear DNA (for example, sulphur mustard is thought to increase the cross-linking between juxtapositioned DNA strands) or by similar but more extensive reactions with cytoplasmic proteins. The significance of these carcinogen-protein interactions is still vague, but one view (the protein deletion hypothesis) is that these proteins usually suppress some chromosomes present in the normal cell; but when a carcinogen is bound to the proteins, the repression is not present and new cell characteristics then arise. There are many variations of these two mechanisms. The carcinogen so far may also be called an "initiator." However, there are some twenty more stages in the mechanism of our one abnormal cell appearing as a malignant tumor. Reactants required for these subsequent stages are called "promoters," or "co-carcinogens" (for example, cigarette smoke). Substances that provide initiators and promotors may be called "complete carcinogens." Two points worthy of note are that the speed and direction of cancer development are promoter dependent and that promoters themselves cannot cause cancer. Having been formed, the primary tumor metastasizes into secondary tumors either by direct contact and invasion or by cells from the primary tumor entering the blood and lymph streams and being carried to other parts of the body.

The popular view of cancer is of a painful, fast-growing lump which spreads to the surrounding tissues. In fact, the majority of cancers are at first present without a lump, and, as newly formed malignant tissue does not contain nerve endings, pain need not be felt until pressure is built up in some normal tissue. As far as the speed of growth is concerned, tumors usually grow more slowly than normal tissues can grow; however, the normal tissues are under homeostatic control and, except during childhood, do not demonstrate their maximum growth rates. Thus, the first symptoms of cancer are usually imitations of more innocent disorders (for example, a persistent cough). Clinical symptoms include the cancer tending to form at a source of irritation, a suppressed bone marrow activity, and a lower intercell adhesion. Microscopic analysis of tissue samples is the safest method of differentiating between benign and malignant growths because cancer cells can be perceived to differ from normal cells in their infinite range of shapes, sizes, and structures.

Once having diagnosed cancer in the primary locus and checked for growths in suspected secondary sites (for example, the lungs) by tissue sample analysis, a pattern of treatment is selected from chemotherapy, radiotherapy, and surgery. The radiotherapist's beam is used to kill the tumor and some surrounding tissues in an effort to prevent the spread of the disease. *Radiotherapy* is most effective in the presence of a host reaction, i.e. the effects of the radiation and the antibodies synergistically aid each other. *Surgery* is limited by the impossibility of removing vital organs until organ transplants become easier. *Chemotherapy* is discussed in subsequent sections of this chapter.

The Evolution of Modern Therapeuticals

The art of healing is many thousands of years old; the science of healing is comparatively modern.[3] Catherine in Shaw's "Arms and the Man" claimed that she could trace her ancestors back for almost twenty years! *Modern* pharmaceuticals have an equally brief ancestry.

Table 19-II summarizes the history of drugs and it can be seen that the scientific attack on disease occurred only during the last thirty or forty years. However, Paracelsus (1493-1541) was the true father of modern metallotherapy. "All things can be poisons, for there is nothing without poisonous qualities. It is only the dose which makes a thing a poison." He introduced heavy metals into therapy in doses balanced on that narrow plateau between their toxic effect and their maximum curative properties. Nevertheless, bio-inorganic progress was painfully slow until 1960 (during the intervening four centuries ferrous salts were used for the treatments of anaemias; sodium, calcium and magnesium salts were employed as intestinal treatments and there was some use of heavy metals for syphilis and trypanosomes).

At the beginning of this century organic drugs evolved as biproducts of industrial chemistry, the impact of these upon disease not being as extensive as one might assume. Breckon[3] points out that as recent as 1930, of the six *specific* drugs for fighting disease, three were relics of African and Hindu medicine, one originated as an old

wives remedy and only "Salvarsan" and vitamins were "modern."

Then, in 1935, came the birth of the sulphonamide era thus adding two additional elements, S, and N, to our basic three, C, H, O. The end of the second World War saw an expansion in industrial pharmaceutical research and a gradual transition from the situation where drugs were mainly discovered by academics to one where the majority of discoveries were made in industry. There are now some 30,000 drugs available, counting all the different formulations of the same chemical, some 8,000 being commonly prescribed. (Only 1 in every 5,000 newly synthesized drugs ever reaches the market!)

In the 1950's the antiviral properties of metal complexes were discovered and, as more cancers were found to be virus dependent, metallotherapy entered into the anticancer battle.[4]

There are four basic methods of finding a new drug:

(1) by isolating, analyzing and imitating natural products,

(2) by playing "molecular roulette" with existing drugs,

(3) by random screening of all chemicals, or

(4) by a theoretical attack upon a physiological basis.

The fourth is the most rewarding, but it must be stressed that it prerequires an understanding of biological processes occurring at the molecular level (as described in Chapter 18).

Thus, the time is opportune to synthesize new drugs involving *any* of the twenty-five essential elements shown in Figure 19-2, and others besides. To stress this point we will review some of the metals,

TABLE 19-II

HISTORY OF DRUGS

Date	Country	Drugs	Disease Treated	Notes
1600 B.C.	Egypt	Plants, minerals	All ailments	1500 B.C. iron, as rust, used to treat impotence (Melampus)
430 B.C.	Greece	Herbal remedies	All ailments	Drugs based upon case history of patient (Hippocrates)
2nd century	Asia Minor	Multiple herbal remedies	All ailments	Drugs based upon 4 "elements"—fire, air, water and earth (Galen)
16th century	Switzerland	As, Sb, Hg, Cd, Fe	Syphilis, anaemia	Condemnation of herbal mixtures; optimized drug doses (Paracelsus)
18th century	South America	Cinchona alkaloids	Malaria	Also digitalis isolated (Withering)
19th century	France, UK	Curare, etc.	Analgesics	Drugs first separated into ethical medicines (formula was published) and nonethical or proprietory (ingredients were a closely guarded secret). (Bernard et al.)
1899	Germany	Aspirin		C,H,O drugs as offshoots of the German dye industry (Bayer Co.)
1903-10	Germany	Veronal, Salvarsan	Insomnia, syphilis	(Ehrlich)
1920	Canada	Insulin	Diabetes	(Banting and Best)
1928	UK	Penicillin	Microbial infections	Unrecognized until 1940 (Fleming)
1935	Germany	Prontosil	Streptococci	S,N. The first sulphonamide (Domagk—first human trial was on his own daughter)
1943	UK	M and B 693	Pneumonia	Saved Sir W. Churchill's life and subsequently many thousands more since 1943.
1944-45	UK			Rapid expansion in drug industry (44,000 employed in UK)
1970	UK	many	Broad spectrum of diseases	(70,000 employed in UK)

ligands and complexes currently being used to treat disease.

METALLOTHERAPY

"Each organism has an optimal concentration of a given metal above and below which it does not thrive well."[5]

A. ALBERT

Metallotherapy is considered here in its broadest sense to encompass supplementing metal concentrations when bodily concentrations are deficient and also the lowering of metal concentrations when bodily contents are excessive.

Some metal-dependent diseases which necessitate the administration of a metal if the symptoms involve a metal deficiency or of a sequestering drug if the symptoms include an excessive accumulation of *in vivo* metal are listed in Table 19-III. Note that the converse situation may also apply when the *in vivo* concentrations of metals are monitored in order to suggest the presence of the diseases listed.

Supplementing the Essential Metals[6]

Although *sodium* and potassium have approximately equal abundances on the Earth's crust, plants have approximately ten times as much potassium as sodium therein. This leads to two types of supplementing being necessary: (1) the rapid uptake of potassium from soils necessitates using potassium salts as plant fertilizers, and (2) our bodies contain about *three* times as much potassium as sodium, so our diet needs supplementing with sodium chloride. Extensive sweating (for example, stokers or miners) increases the concentration of salt in the body, drinking water to slake their thirsts depletes the sodium chloride concentrations and "stoker's cramps" can occur (imbibing salt water is the well-known remedy).

Blood is isotonic with 0.15M sodium chloride. Administering hypotonic (hypertonic) solutions temporarily raises (lowers) the cerebrospinal pressure. Sodium chloride is also used to encourage water absorption by the tissues. Conversely, a sodium chloride-deficient diet is sometimes used to diminish the edematous increase in tissue fluids during nephritis. *Potassium* occurs so widely in foods that diet deficiencies never occur. Potassium salts are used as alkaline diuretics. Potassium poisoning may result from renal failure or from adrenal cortex malfunctions (Addison's disease).

Calcium is required in daily amounts of

TABLE 19-III

METAL-DEPENDENT DISEASES

Cadmium, lead, mercury, arsenic[*]	Metal poisoning
Calcium	Bone diseases, cataracts, gallstones, tetany
Cobalt	Pernicious anaemia
Copper	Lymphoma, myocardial infarction, pregnancy, rheumatoid arthritis, Wilson's disease, psoriasis, arterial lesions, infective hepatitis, leukaemia, cirrhosis, nephritis, Hodgkin's disease
Iron	Anaemia, haemolysis, pregnancy, mucosal irritation, diarrhoea, cardiac collapse
Lithium	Manic depression
Magnesium	Bone disease, calculi, gastrointestinal disorders
Sodium/Potassium	Addison's disease, gout, stoker's cramps
Zinc	Anaemia, cancer, dwarfism, hypogonadism, liver cirrhosis

[*] These poisons displace essential transition metals from metalloenzymes and so symptoms include imbalances in the concentrations of several metals.

0.5 to 1.0 g. Efficient calcium retention demands sufficient phosphate in the diet (and *vice versa*). If either is deficient neither is retained and rickets occurs. Injecting the required calcium into the blood stream has but a brief effect of hypercalcaemia. In fact, to increase blood calcium it is advisable to inject parathyroid extract which causes calcium to be yielded by the bones. Prolonged hypercalcaemia causes vomiting, diarrhoea, depressed central nervous system, coma and eventually death. Hypocalcaemia leads to excitability, tetany, and cramp. The calcium concentration is restored using calcium phosphate or calcium lactate (and sodium lactate and vitamin D to increase absorption). Intravenously, calcium is administered as the gluconate or, preferably, as the borogluconate which is more soluble and less irritant. *Magnesium* is only slowly absorbed in the intestine. The ion exerts an osmotic pressure which holds water in the intestine. Hence the use of magnesium salts, e.g. $MgSO_4 \cdot 7H_2O$—Epsom salts, as purgatives. Thus, oral administration of Mg salts does not increase plasma $[Mg^{2+}]$ (unless nephritis is present). Magnesium and calcium carbonates are also used as antacids.

Man does not suffer from *manganese* de-

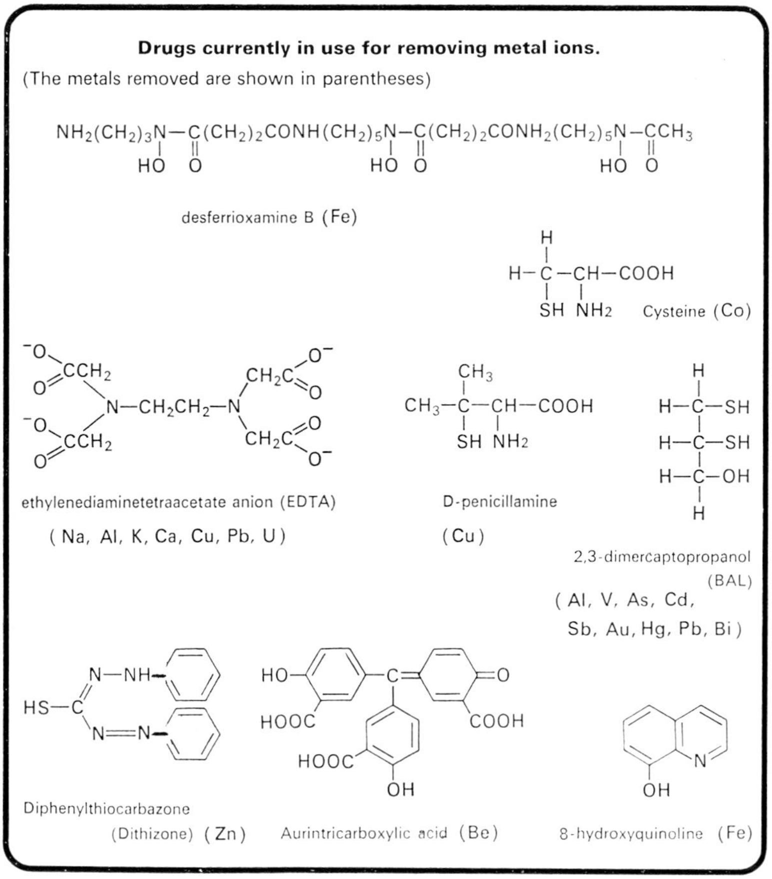

Drugs currently in use for removing metal ions.

(The metals removed are shown in parentheses)

$NH_2(CH_2)_3N-C(CH_2)_2CONH(CH_2)_5N-C(CH_2)_2CONH_2(CH_2)_5N-CCH_3$

desferrioxamine B (Fe)

Cysteine (Co)

ethylenediaminetetraacetate anion (EDTA)

(Na, Al, K, Ca, Cu, Pb, U)

D-penicillamine

(Cu)

2,3-dimercaptopropanol (BAL)

(Al, V, As, Cd, Sb, Au, Hg, Pb, Bi)

Diphenylthiocarbazone (Dithizone) (Zn)

Aurintricarboxylic acid (Be)

8-hydroxyquinoline (Fe)

Figure 19-3. Drugs currently in use for removing metal ions. The metals removed are shown in parentheses.

ficiencies. The only common drugs containing manganese are the sodium and potassium permanganates which are oxidizing disinfectants. *Iron* deficiencies are restored using ferrous salts (ferric salts are not absorbed in the small intestine) and these are often complexes of reducing ligands in order to maintain the iron in its +2 oxidation state, for example, carbohydrate, ascorbate or citrate complexes. *Copper* sulphate solutions have been used as emetics and antiseptics since the early days of Egyptian medicine. Pernicious anaemia is linked with *cobalt* deficiencies and this is corrected by taking liver extract or vitamin B_{12} (cyanocobalamin). *Zinc* sulphate is used to correct zinc deficiency, to promote wound healing, and it can also be used as a disinfectant and emetic. Zinc oxide, carbonate (calamine) and stearate are only slightly soluble and when applied as ointments are slowly converted to soluble salts which have this disinfectant action. Zinc oxide is sometimes used to thicken ointments into pastes. Zinc suspensions of insulin are used to prolong the sugar concentration buffering effects of the protein.

Reducing the Concentrations of Metals *In Vivo*

Drugs currently in use for sequestering and removing metal ions are shown in Figure 19-3. They complex through chelation

and possess some selectivity for the metal ions listed through the natures of the ligand donor atoms and of the metal ions being matched. An example is given in Table 19-IV. As one descends the table, increasing softness of donor atoms is paralleled by increasing softness of the acid removed. Each of these drugs is discussed in greater detail in Chapter 20.

Metals as Carcinogens

An enormous effort has been made to understand the mechanisms involved in the processes of chemically induced carcinogenesis.[7, 8] The field is replete with speculation but in the past few years interest has been gathering concerning the roles of metal ions in carcinogenesis. As evidence amasses it will be possible environmentally and dietetically to avoid those metals most deeply involved in carcinogenesis.

There are three factors involved (1) metals accumulate in one's body during ageing (especially aluminium, arsenic, barium, beryllium, cadmium, chromium, gold, lead, nickel, selenium, silicon, silver, strontium, tin, titanium and vanadium), (2) the distribution of some cancers has been correlated with a top-soil selenium, copper and zinc basis, (3) pure metals that have been reported to have actually caused cancer in test species are shown in Figure 19-4. These, in common with other widely

TABLE 19-IV

EARLIEST SEQUESTERING REAGENTS FOR THERAPEUTICALLY
TREATING METAL ION EXCESSES*

Ligand	Donor Atoms	Metals Removed	HSAB Classification
Desferrioxamine B	Several O	Fe(III)	Hard
EDTA	4 O, 2 N	Pb(II)/Co(II)	Borderline
D-Penicillamine	S, N, O	Cu(II)/Cu(I)	Borderline/soft
British antilewisite (BAL)	2 S	Arsenicals/Au(I)/Hg(II)/Hg(I)	Soft

* The chemical formulas are given in Figure 19-3. As one descends the table, increasing softness of donor atoms is paralleled by increasing softness of the metals removed.

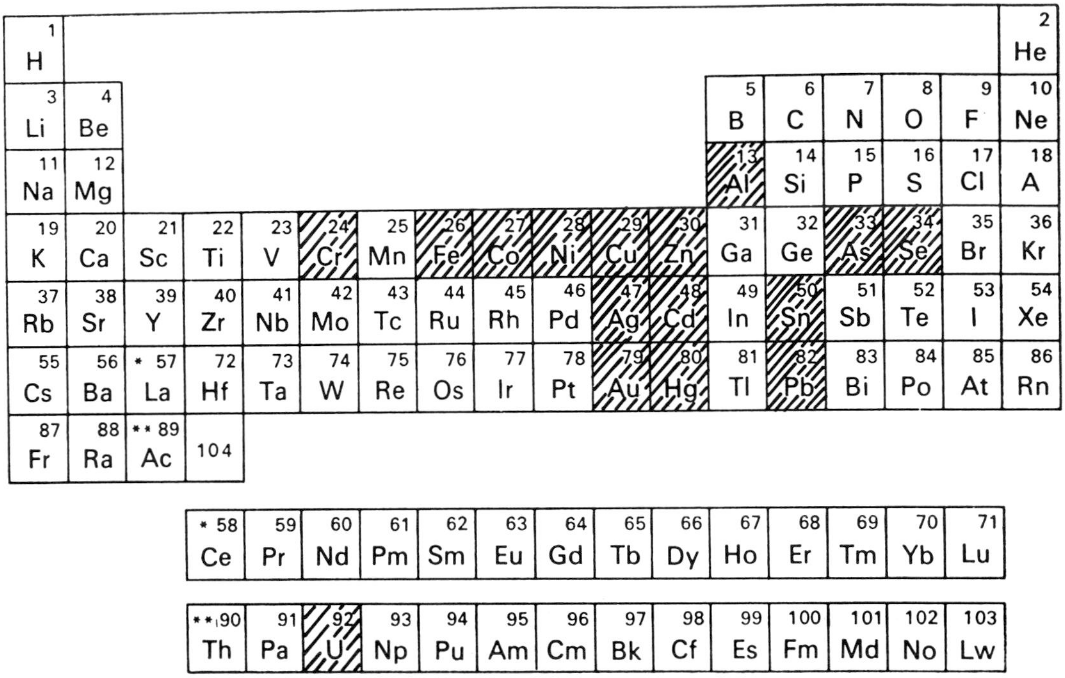

Figure 19-4. Shaded elements have produced cancer in at least one animal species.

used materials such as glucose, iron-dextran complexes, or even sodium chloride, sometimes have produced cancers following their subcutaneous implantation in test animals. During this process of carcinogenesis, if the metals dissolve in hsab hard solvents (for example, the blood) the higher oxidation states are likely, whereas in soft solvents (for example, lipids or some multiheaded enzymes) the lower oxidation states probably occur. However, at the present time, it appears that the technique employed for implantation (pieces, powders, or perforated sheets, etc.) is the important factor in determining whether the material will be carcinogenic. For the occasions when carcinogenesis does occur, Furst has suggested that the metals penetrate living cells and either advance or retard the kinetics of anabolic or catabolic enzymes by instigating a competition between the invading and normal metals. He

further suggested that viruses may aid cell penetration by these metals.

Metals of any oxidation state, whether complexed or not, if present as unstable isotopes emit ionizing radiations that can cause cross mutations and eventually cancer. A well-known group of radioprotective drugs (for example, cysteamine or mercaptoalkylamine) are specific copper-binding ligands that protect copper(I)-containing enzymes. Without such protection, irradiated mammals exude copper (and iron) from most organs.

The ions of metals and their complexes have been widely reported to have been involved in carcinogenesis. One of the first reports was that the excesses of ingested iron from iron cooking pots caused liver cancers. The process of ageing permits a wide range of soil and plant impurities to accumulate in various organs. For example, it has been found that the well-researched tobacco plant contains aluminium, barium, calcium, cobalt, copper, iron, lead, lithium, magnesium, manganese, molybdenum, nickel, sodium, strontium, tin, titanium, vanadium, and zinc.

Taking nickel as an example, the following compounds of nickel have been found to be carcinogenic: Ni_3S_2, NiO, $NiCO_3$, $Ni(CH_3COO)_2$, $Ni(cyclopentadiene)_2$ and Ni^0 (the pure metal).[7] In laboratory experiments, hamster carcinogenesis has been definitely shown to be zinc dependent, but the mechanism is still under investigation because, on the one hand, De Wys et al. report dietary zinc *deficiencies* as inhibiting cancer, while Poswillo and Cohen report zinc sulphate *excesses* as inhibiting cancer, some of this excess zinc appearing in the newly healed tissues. In humans, high zinc (and chromium) soil contents have been correlated with the regional incidences of stomach cancer.

Sometimes metal ions have the power of determining whether a carcinogen is active or not; for example, cyclohexylsulphamic acid and its sodium and calcium salts. Both salts are more than 98 percent ionized in aqueous solution at neutral pH. However, in animal experiments, under the most severe conditions, the sodium salt only manages to produce a mild self-limiting lesion, while only traces of the calcium salt produce progressive lesions. Analogously, in the crystalline state, it is usually the calcium salts that produce cancers (for example, calcium oxalate crystals produce bladder tumors by chronic irritations). All metals ions are inherently involved in the pH gradients that exist within the body and in the tendency of cancer cells to migrate from low pH toward neutral regions. Further, the pH within tumor masses is lower than in normal comparable tissues. This has been suggested as a means of powering the spread of cancer cells. It is im-

portant to remember that (a) the amount of metal hydroxy complexes present is pH dependent and, conversely, the type of metal determines the pH. This means that the free concentrations of metal ions are pH dependent also; (b) the distribution of the metal between the available complexes is pH dependent (an example is given in Figure 19-5). Here it is pertinent to realize that varying the pH changes the complexes present and that not all complexes can penetrate cell walls, for example, 8-hydroxyquinoline-Fe(III) complexes: The mono and bis forms are toxic and cannot penetrate cell walls; the tris form is nontoxic and can penetrate cells because it is uncharged; (c) the number of protons associated with a ligand is pH dependent and so proton ionizations from the active site of enzymes mean that enzyme catalysis is also pH dependent.

The metal dependence of carcinogenesis is not as easy to investigate as it may appear since (1) we are dealing with very small amounts of metal (for example, the

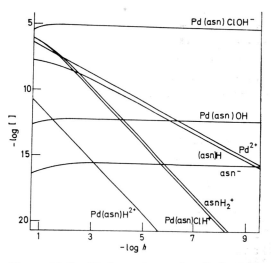

Figure 19-5. pH dependence of palladium-asparaginate-chloride systems. Total concentrations are—palladium = 10 μM, asparaginate = 10 μM, chloride = 150 mM.

essential transition metals are only present in minutest quantities in some tissues—only one atom in 10^7 being copper and in the body as a whole the total Cu : Mo : Co is in the ratio 1000 : 100 : 1),[8] and (2) paradoxes often arise where small amounts of trace metal may have one pharmacological effect but the *reversal* of these activities may occur if the concentration of the element rises beyond a certain value. Selenium is a typical element that illustrates how involved these matters can be:

Selenium at a level greater than 5μg (g of diet)$^{-1}$ is highly toxic causing "alkali disease" and "blind staggers" in animals. On the other hand, less than 3 μg (g of diet)$^{-1}$ improves growth and combats white muscle disease in sheep. Selenium deficiency causes the enzyme glutathione peroxidase to have reduced activity (it is thought that the Se may actually be an integral part of the enzyme). Nonprotein selenium is more *acutely* toxic than protein selenium while protein selenium is more *chronically* toxic than nonprotein selenium. Selenium is present in tuna fish and it protects the consumer against mercury poisoning (fortunately tuna fish accumulate Se alongside Hg!).

Selenium also appears to have a cancer protecting effect:[9]

1) Schamberger and Frost found that cancer mortalities in the twenty U.S. cities were inversely proportional to the selenium content of blood in blood banks.

2) Areas of Canada that grow selenium indicator plants have lower cancer mortaility rates.

3) Cancer mortality rates have been correlated with the selenium contents of grain and forage crops in all of the United States.

4) Plasma samples taken from cancer patients have lower than normal selenium contents.

Thus, the cancer protecting effect of selenium and its biological role as an antioxidant merit further investigations before a safe level of selenium can be recommended for diets.

Other metals being investigated along these lines are cadmium, copper, magnesium, molybdenum and zinc.

LIGAND THERAPY

Antibiotic Ligands[10, 11]

Group I metals usually form only poor complexing cations but some antibiotics complex with these metal ions very strongly, the usual binding order being $K^+ > Rb^+ > Cs^+ > Na^+ > Li^+$, an order which coincidentally is the same as is found for the permeability of membranes to these metal ions in the presence of antibiotics (Figure 19-6).

Model studies using cyclic polyethers have shown that for efficient active metal ion transport through a membrane:

1) The solvated metal ion requires as many solvent molecules as possible to be replaced by ligand atoms (if the atom is oxygen, the metal ion—O bond is predominantly ionic, the proportion of covalence increasing as O is replaced by N and S).

2) The ligand should form a cavity which completely envelops the metal ion and holds it through metal-oxygen induced dipoles.

3) The ligand needs to have a lipophilic exterior for membrane solubility, i.e. the overall effect is to solubilize ions in nonaqueous solvents.

4) The bonds within the ligand be sufficiently flexible to permit the metal ion to enter into its central cavity.

There are two groups of biologically produced antibiotics that illustrate these four principles:

1) The actins, valinomycins, gramicidins and enniatins (these compounds are fungal metabolites) are all *neutral, closed ring lig-*

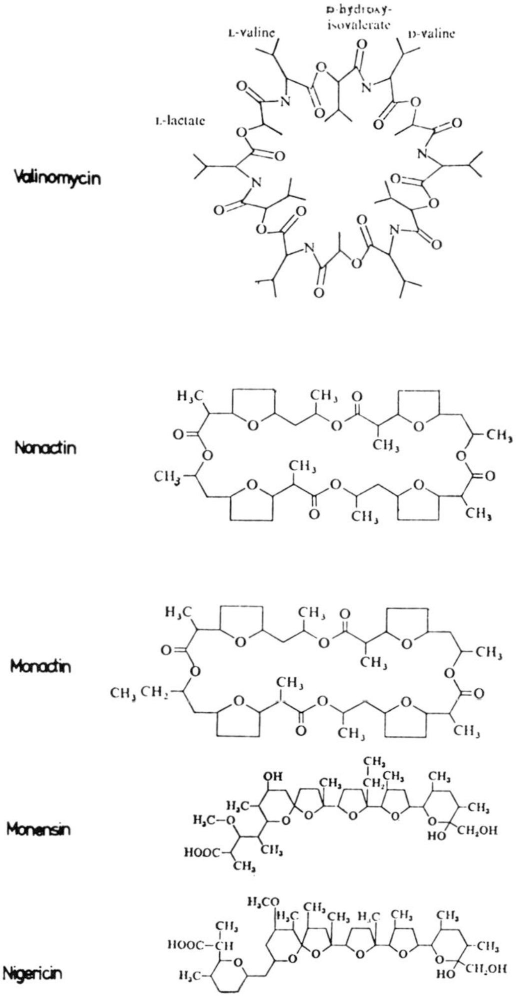

Figure 19-6. Three macrocyclic antibiotics capable of sequestering main group metal ions (chemically these ligands are known as "crown compounds"), and two macrotetralide antibiotics.

ands which through their lipid solubility increase the membrane permeabilities of K[+] and Rb[+] (Na[+] and Li[+] are not greatly affected).

2) Valinomycin antagonists such as monensin and nigericin contain carboxylate groups at physiological pHs and are linear chain molecules which (a) can cyclize through hydrogen bonding and (b) form

neutral complexes (RCOO[-]-Na[+]). These antibiotics increase the membrane permeability to protons and the efflux of alkali metal ions. For example, cells loaded with [86]Rb[+] show no significant Rb[+] loss on suspension in salt solutions until nigericin or monactin are added to the solution then the [86]Rb[+] exchanges rapidly for H[+], K[+] or Rb[+] and more slowly for Na[+] or Li[+]. This disruption of the normal ionic balance eventually destroys the cells involved.

Copper Sequestration in Wilson's Disease

S. A. K. Wilson's disease is a rare disease of the nervous system, the symptoms being progressive lenticular degeneration. Chemically the disease arises from an inherited disorder in the control mechanism governing one's copper metabolism, a ceruloplasmin deficiency, and this results in an excessive concentration of free copper in the liver, brain, tissues, accompanying neurological disorders, e.g. see Figure 19-7 and eventually the copper pressure causes hemolytic crises (red blood cells burst).[12]

There is no known cure for Wilson's disease, but the best care to date has employed the drug D-penicillamine. Associated medication problems are that the drug acts only slowly, and since it is not normally found in one's body it has no excretory mechanism and so tends to build up in the liver, i.e. it can be toxic.

Sarkar et al. have attempted to produce a replacement drug (a) that is built of amino acids which are normally found in man and thus can be reused or excreted, (b) was specific for copper leaving other metal ions largely untouched, and (c) would form a copper complex that was small enough to be excreted through the kidneys.[13]

The Toronto group took albumin, the copper transport protein in human blood, found it had a copper specific site con-

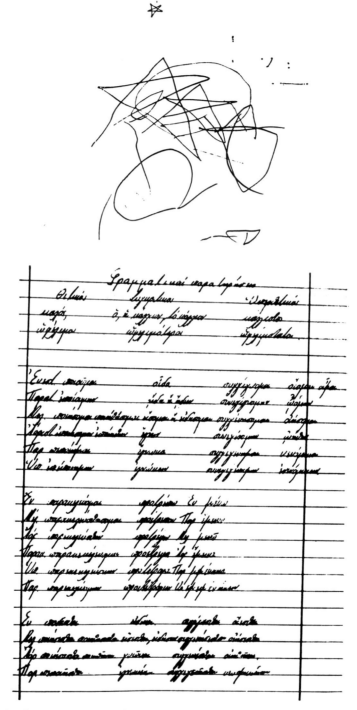

Figure 19-7. The handwriting of a patient with Wilson's disease (a) before (upper sample), and (b) after (lower), treatment with copper sequestering drugs.[12]

structed from three amino acids and copied this site by synthesizing the tripeptide glycylglycyl-L-histidine methyl ester. As expected this new drug promises to be more copper specific and less toxic than D penicillamine.

The lessons to be learned here are that other metal-dependent diseases might also be treated by mimicking their metal binding sites in metalloproteins. Also, by using amino acids, excesses of the drugs can easily be metabolized harmlessly by the body.

Thus, this new drug for treating a rare disease (one case per million of population) could well be remembered as a milestone establishing this principle of copying *in vivo* processes to treat other diseases and hence displacing other bio-inorganic ligands currently being used as drugs (see figure 19-3).

Ligands That Are Carcinogens

In this section we are referring to negative ligand therapy, i.e. ligands to be avoided. Paradoxically some ligands can

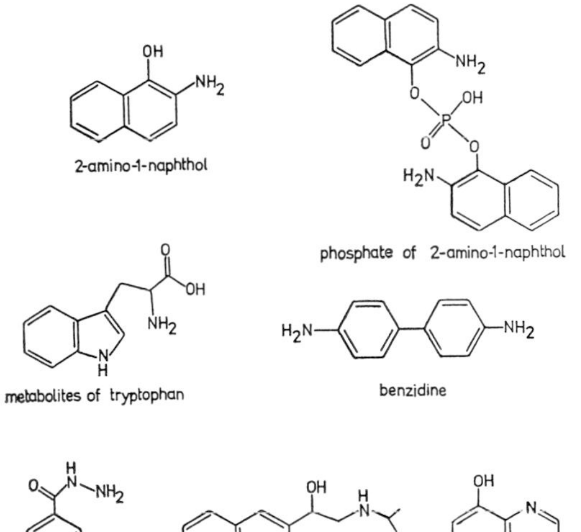

both cause and cure cancer depending upon the amounts and circumstances.

Some carcinogens are capable of acting directly as powerful ligands, either in an aqueous or in a nonaqueous environment, and others can react indirectly by undergoing metabolism into strong ligands. We shall discuss the subject under three headings: (a) the ligand donor atoms of known carcinogens, (b) is there an increase in the amount of ligands, i.e. a sequestering of metals, when cancer occurs?, and (c) correlations between ligand deficiencies and cancer.[2]

1) Ligand donor atoms of carcinogens. Figure 19-8 shows the most likely ligand donor atoms of known and suspected carcinogens. In general, the molecules may be described as (a) lipid soluble, and (b) consisting of coplanar rings. Furst has suggested that a third characteristic should be added, that of being (c) a metal binding compound or capable of being metabolized into one. Even if we restrict the choice of ions to those of the ten metals that are essential to life in humans, the range is broad enough to provide strongly complexing ions for all these ligands. Pairs of donor atoms that are separated by two or three other atoms are especially favored as they give rise to five- or six-membered chelate rings.

2) Increases in ligand and decreases in metal concentrations accompanying carcinogenesis. Normal liver cells have different homeostatic control mechanisms compared to malignant cells, e.g. animals fed on a tryptophan-rich diet usually adapt and develop extra tryptophan pyrolase in their livers and consequently the ligand excess is removed. However, the Morris 5123 experimental rat hepatoma does not do this, and so an excess of the amino acid accumulates in the animal. We must also remember that tumor cells have

Figure 19-8. Ligand donor groups of known and suspected carcinogens.

to compete with normal cells for essential nutrients and that some of these latter are metals or metal containing. It is conceivable that the metals can be won by the malignant cell by using more, or stronger, ligands than normal cells.

Holmberg has suggested yet another manner in which the ligands can be used by malignant cells to defeat normal cells. A variety of cancer cells that he studied excreted a toxin (a polypeptide of Gly, Cys, Glu, Arg, Val, Leu, Tyr, Ala; mol wt = 1900) that occurred at such a concentration that it did not affect the malignant cell but did poison the normal cell. It is thought that this toxin interferes with the S stage in the manufacture of DNA. The mechanism of poisoning by the toxin can be blocked by nucleosides such as deoxycytidine. Tumors are rich in deoxycytidine, and this is why they need so much higher concentrations of toxin before they are poisoned. On the other hand, normal cells have subprotective concentrations of deoxycytidine and so are selectively poisoned. Experiments have shown that the toxin, liberated from tumor cells when they die and break up, shortens the average lifetime of red blood cells by up to 20 percent.

3) Ligand deficiencies leading to cancer. It was shown in (2) that ligand excesses and carcinogenesis are interdependent. However, it also appears that deficiencies can also cause cancer. In general, starvation leads to (a) less waste products of metabolism, some of which might be carcinogenic, and (b) malnutrition which lowers one's body resistance to invasion by diseases in general. Choline- and methionine-deficient diets lead to cancer of the liver in rats, mice and chickens. However, treatment of nonmalignant psoriasis by using a tryptophan-deficient diet is also well known. Tryptophan is further coupled with

bladder cancer in that its metabolites (for example, 3-hydroxyanthranilic acid) cause tumors. Ligand deficiencies must also occur during the later stages of carcinogenesis when cell breakdown is reported to increase the blood copper concentration two- or threefold.

Ligand Therapy of Cancer

Chemotherapy of cancer has been practiced for seventy years but has remained fairly unsuccessful until the last two decades. Even now the treatment is not always successful, and the field is beset with innumerable difficulties; for example, only a proportion of cases respond to any one drug. Nevertheless, there have been fertile regions of discovery. Before discussing the range of drugs available, some of the problems associated with them will be mentioned. Frequently tumors have inadequate blood supplies, and so the drug has to be applied topically or injected right into the tumor rather than taken orally. Problems arise in defining the best kind of animal screening tests for potential new drugs. Yet another problem occurs when a tumor builds up a resistance to a drug, and so a different chemical has to be administered for a time. Finally, we should not underestimate the alarming side effects of these "anticancer" drugs. In fact, the drugs are less specific than the term implies and are really antigrowth reagents. Hence (a) slow-growing tumors do not respond well to them; (b) normal healthy rapidly multiplying parts of the body (for example, the bone marrow and stomach linings) are attacked by these drugs; and (c) in common with radiotherapy, chemotherapy depresses the number of lymphocytes, and so, even if cured of cancer, the patient is more likely to contract some other form of serious disease.

With only a few exceptions every chemotherapeutic agent that is effective against human or experimental neoplasms is a chelating agent or is metabolized into a chelating agent. These therapeuticals fall into four broad classifications.

1) The alkylating reagents are based upon nitrogen mustard, e.g. triethylenemelamine and myleran. They are sometimes called radiomimetic reagents as their biological effects resemble those of radiation. Hence, they are useful in treating leukaemias. Unfortunately, their high-antitumor activities are accompanied by high toxicities.

2) The antimetabolites have formulas very similar to a chemical that is essential for tumor growth. The tumor, mistakenly, builds its new cell using the administered antimetabolite and this hinders further growth. Possibly the most famous pair are aminopterin (antimetabolite of folic acid) and ethionine (antimetabolite of methionine).

3) Enzymes can also be used in treating cancers. For example, some cancer cells are asparagine-dependent, but cannot synthesize their own asparagine and so rely upon a supply from normal cells being passed to them *via* the blood stream. The enzyme asparaginase removes this blood-borne asparagine and so the malignant cell is starved of its supply.

4) Hormone treatment is the area of chemotherapy having the highest success rate to date. It is mainly used in the treatment of tumors of the breast, uterus, and prostate because these organs, being under hormonal control, support tumors that respond to the opposite hormones. Unfortunate side effects include masculinization in females and feminization in males. Researchers in chemotherapy must aim at making existing therapies more specific

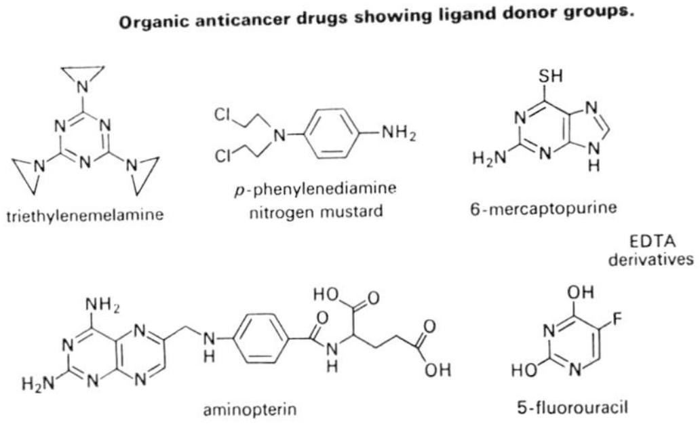

Figure 19-9. Organic anticancer drugs showing their ligand donor groups.

and searching for new families of therapeuticals.

Figure 19-9 demonstrates this concept that cancer drugs are usually viable ligands and illustrates the donor atoms probably involved in metal ion binding. Some of these drugs have an increased anticancer activity when administered as metal complexes. As with carcinogens, drugs incapable of being strong complexing ligands are usually metabolized into ligand species. Indeed, for nitrogen mustards it has been possible to correlate the rate at which the ligand (ethanolamine) was formed by metabolism with the extent of their therapeutic action, i.e. a N-C-C-Cl species was metabolized into the more powerful N-C-C-O or N-C-C-N ligands that form stable complexes with charged metal ions in aqueous solution.

Once having formed a ligand, the actual mechanism of anticancer activity is still open to speculation. In general, such hypotheses center around deactivating either carcinogenic metals or all enzymes necessary for rapid growth (both healthy and malignant).

In general, future cancer research is aimed (a) at discovering new anticancer and antiviral vaccines and new methods of early cancer detection, and (b) at improv-

Structure	Name	Use
(Tetracycline structure)	Tetracycline	Broad spectrum antibiotic
(Biallylamicol structure, $(C_2H_5)_2 \cdot N \cdot CH_2$, HO, C_3H_5, subscript 2)	"Biallylamicol"	Treatment of amoebic dysentry
(Hydrallazine structure, $NH \cdot NH_2$)	"Hydrallazine"	Reduces blood pressure
(Aspirin and phenacetin structures)	"Aspirin" and "phenacetin"	Analgesics
(Sulphanilamide structure, NH_2, SO_2NH_2)	"Sulphanilamide"	Bactericide
(Acetazolamide structure, $CH_3 \cdot C \cdot N$, SO_2NH_2)	"Acetazolamide"	Anticonvulsant
(Kojic acid structure, HO, CH_2OH)	Kojic acid	Antibiotic

FORMULA	NAME	USE

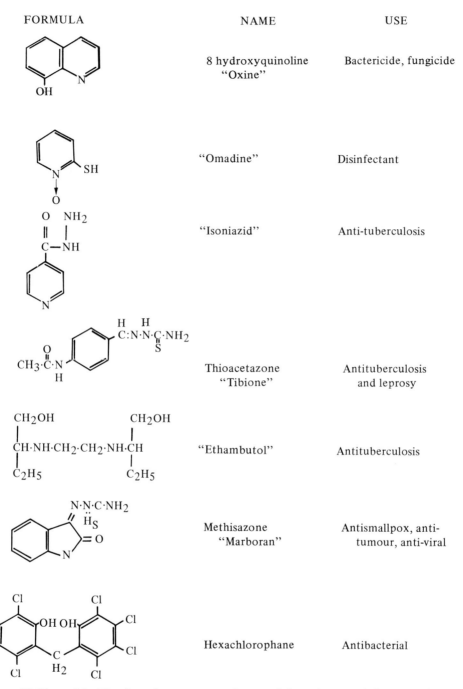

8 hydroxyquinoline "Oxine" — Bactericide, fungicide

"Omadine" — Disinfectant

"Isoniazid" — Anti-tuberculosis

Thioacetazone "Tibione" — Antituberculosis and leprosy

"Ethambutol" — Antituberculosis

Methisazone "Marboran" — Antismallpox, antitumour, anti-viral

Hexachlorophane — Antibacterial

Figure 19-10a and b. The formulae, names and uses of drugs having chelating abilities.

ing our attempts to remove environmental carcinogens, our knowledge of nucleic acid or protein carcinogen interactions, our matching of dose levels and spacings to tumor growth characteristics, and our synergistic combinations of treatments.

Other Chelating Ligands Used Therapeutically

Chelating drugs are of two general types —those acting *outside* and those acting *inside* cells. The major use of chelating drugs used outside cells is that of antidotes to metal poisoning. This aspect is described in detail in Chapter 20. In order for drugs to act inside cells they, or their metal complexes, must have some degree of lipophilicity in order to traverse the cytoplasmic membrane.

Albert[5] has reviewed the vast range of chelating ligands currently in use as drugs (examples are shown in Figure 19-10) and has described an effect whereby the drug is only effective when chelated to a trace metal:

Augmentation of Toxic Effects by Chelation

Some twenty years ago it was discovered that 8-hydroxyquinoline (oxine) destroyed microorganisms such as bacteria and funguses through chelating trace metals (not necessarily essential ones) in the support media. A lethal complex was being formed with one of the metal ions present. It is now known that many chelating drugs act against microorganisms through chelating any of the several metals present in trace amounts *in vivo*. The metals activating these drugs are usually those having variable valency properties (for example, copper is widely used in fungicides and iron in bactericides).

Extremely dilute solutions of oxine (*ca* 10^{-5}M) were found to be effective against staphylococci and streptococci, while higher concentrations (*ca* 10^{-3}M) were nonbactericidal. Such a situation occurred because it was the 1 : 1 or 2 : 1, but not the 3 : 1, oxine : metal ion complexes that were effective. Thus, the 10^{-3}M oxine solution originally quoted as nonactive can now be made so by elevating the iron in the support medium to approximately 10^{-3}M, i.e. causing the 1 : 1 complex to form.

Metals whose toxic effects cannot be augmented by chelation, for example cobalt, can protect bacteria against the effects of iron-oxine complexes through the cobalt successfully competing with the iron for the oxine.

The mechanisms through which oxine-metal complexes exhibit their activity are unknown, but it is postulated that atmospheric oxygen plus the iron or copper complexes oxidize mercapto compounds within the cell to produce hydrogen peroxide which in turn oxidizes more vital components such as lipoic acid. In fact, the antagonism observed with cobalt arises because the ion not only removes essential oxine but also breaks the chain reactions of these oxidative schemes.

Oxine, and many related compounds, have a similar antibacterial spectra of effectiveness as penicillin and are widely used against rashes and in controlling wound infections. However, they are not useful systemically as red cells deactivate oxine complexes.

METAL COMPLEXES AS THERAPEUTICALS

Inert, highly stable metal chelates have a considerable activity against microorganisms, fungi and viruses and they are also used in controlling the spread of neoplastic tissue. As germicides they are nonirritant to tissues and mucous cavities, at the concentrations used, and so are extensively used in controlling topical infections. However, when administered systemically several of them have a curariform effect.

The anticancer, antiviral and antibacterial drugs have several common characteristics: (1) in general, it is unknown whether it is the metal, the ligand or the complex (or two or three of these synergistically) that is/are the active component(s); (2) whatever the active factor in (1), it is presumed that it causes a more or less permanent dysfunction of an essential biological mechanism; (3) all these drugs have some degree of lipid solubility; (4) since many of the complexes are inert, it would seem prudent to correlate their drug activity with a variety of physical properties (note that metal complexes can be easily arranged in increasing size, charge, or lipophilicity series, etc. for shedding more light upon the mechanisms of the actions of these drugs); (5) both the antiviral, anticancer, and antibacterial complexes appear to prefer Fe and Rh or Pd and Pt, however this does not necessarily imply that there is anything in common with the mechanisms of drug activity, but rather that some metallotherapy schools have specialized in Pt complexes, while others have preferred Fe complexes.

As far as these mechanisms of antiviral and bacteriostatic activity are concerned, the "jury is still out." However, we shall describe three recent reports of new evidence.

Antibacterial Agents

Gillard et al.[14] have recently shed light upon the mechanisms of activity of metal complexes by screening a series of rhodium complexes against *E. coli,* strain B (examples are listed in Table 19-V). The ligands

TABLE 19-V

METAL COMPLEXES ACTIVE AT 50 μM CONCENTRATIONS AGAINST ESCHERICHI COLI

Rh(I)	[Rh(CO)$_2$Cl]$_2$
	[Rh(acetylacetonate)(CO)$_2$]
Rh(III)	*trans*[Rh(*N,N,N',N'*,-tetramethylethylene-diamine)$_2$Cl$_2$]Cl
	trans[Rh(pyridine)$_4$Cl$_2$]Cl

were often pyridines, or substituted pyridines, and *trans* halides. The lipophilicity of each complex was measured from partition coefficients and the ease of reduction of the Rh(III) complex to Rh(I) was determined polarographically. The most active complexes were those that were most easily reduced and, of these, those having high lipophilicities were best.

Although the antibacterial activity could be related to the ease with which Rh(I) may be generated, further experiments suggested that inhibition of bacterial growth is *not* directly proportional to the Rh(I) concentration in the growth medium, but rather to the amount of Rh(I) generated in the cellular membranes.

The suggested mechanism of anticancer activity of group VIIIB metal complexes is to be described later, but compared to anticancer complexes, the antibacterial agents do not need to be electrically neutral or to possess pairs of *cis* displaceable ligands. Thus, it appears that in spite of superficial similarities in their biological properties (for example, both platinum and rhodium bacteriostatic complexes cause cell filamentation) complexes of different metals might well invoke different mechanisms for arresting different diseases.

Antiviral Agents[4, 14]

A virus may be described as a core of nucleic acid encased in a layer of protein, lipid or polysaccharide. Viruses can only reproduce inside living cells where they take over the chemical and energetic supplies to the cell and use them to produce more viruses.

Viral attacks occur via absorption of the virus onto the cell (by electrostatic attraction), penetration through the cell membrane (by phagocytosis) and then the protein sheath ruptures liberating the viral nucleic acids into the cytoplasm whence they direct the synthesis of more viruses (each viral particle can produce as many as a thousand others). The new viruses are then released from the host cell and the cell soon dies.

The protein and nucleic acid portions of viruses are particularly good coordinators and thus cationic complexes have been introduced as antiviral agents. Their *modus operandi* is one of either (a) complexing the virus outside the cell and so reducing its activity, (b) occupying sites on the cell surface and so blocking virus absorption, or (c) preventing the viral replicative processes within the cell.

This metallotherapy approach has, so far, been reasonably successful especially when (1) the metal ion (a) is in the +2 oxidation state, and (b) is from groups VII and VIII of the transition series, (2) when the ligand is 1,10-phenanthroline (its organic rings confer liposolubility) or acetylacetone, and, (3) when the complex is cationic (Figure 19-11). Current trends are to replace these ligands with naturally occurring ligands which are less toxic when released from the metal complexes and so to increase the body's tolerance.

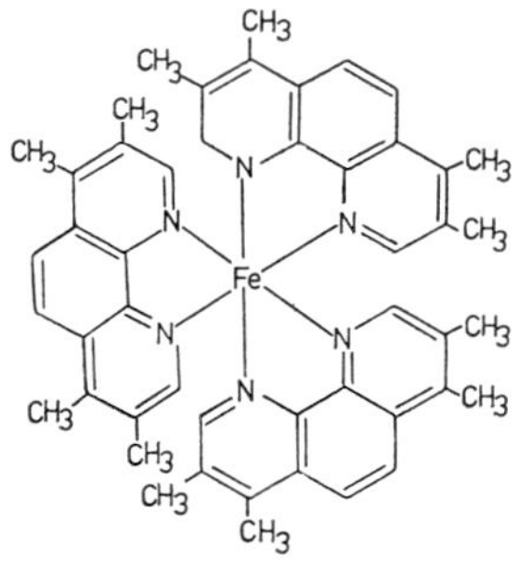

Mechanism Studies

Davidson and Rosenberg[15] have recently reported *cis*-diaquodiammineplatinum(II) as being an active antiviral agent *in vivo* and postulated that there are at least two

$$[\text{Fe}(3,4,7,8\text{-tetramethylphenanthroline})_3]^{++}$$

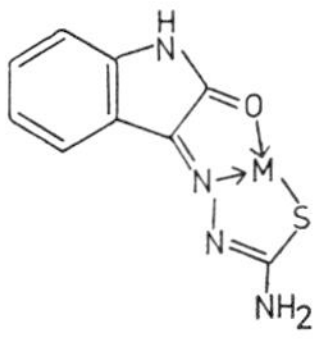

Isatin 3-thiosemicarbazone - metal (M) complex

Figure 19-11. Some antiviral metal complexes.

aspects of the mechanism of antiviral activity: (1) there is the interaction between the viruses host cell and the metal complex so that viral growth and propagation is inadequately supported; (2) there are direct reactions between virions and the metal complexes. Their researches have found that platinum complexes are only effective *in vivo* against viruses that require DNA replicative processes, e.g. Rous Sarcoma Virus and Fowl Pox Virus, and are ineffective against Newcastle Disease Virus which synthesizes its new RNA cores through RNA synthetase acting directly upon the RNA strands of the original infecting virions, DNA being unnecessary.

Virus inactivation is very rapid using *cis*Pt(NH$_3$)$_2$(H$_2$O)$_2{}^{2+}$, the virus being reduced by 60 percent within one minute and completely deactivated within thirty minutes. On the other hand, *cis*Pt(NH$_3$)$_2$Cl$_2$—an anticancer drug about to be described—is far less effective and these relative speeds have been explained by the water molecules of the former complex being rapidly replaceable, while the chlorides of the latter are replaced more slowly.

Oxford and Perrin[16] have recently re-examined the influence of the ligand donor atom and of the choice of central metal ion upon the effectiveness of these antiviral drugs. The *in vitro* screening of a logical sequence of ligands and metals was conducted by monitoring the extent of inhibition of RNA-dependent RNA polymerase activity associated with strains of influenza A and B viruses.

It was postulated that the antiviral complex might act inside the cell through the complex first dissociating to produce its component metal ion and ligands and then, (1) its ligands chelate an essential metal ion in the polymerase metalloenzyme, (2) the ligands oxidize —SH groups on this enzyme, or (3) an essential metalloenzyme, if it is activated by yet another metal such as magnesium or manganese, losing its supply of this activator metal when the administered ligands intercept and complex this metal ion *en route* to the metalloenzyme.

Oxford and Perrn's most promising results appear in Table 19-VI and suggest that, for the systems under investigation, postulate (1) is the most likely mechanism. The maximum concentration listed refers to that necessary to cause 50 percent inhibition in the enzyme reaction, i.e. the lower this figure, the more effective the drug.

Mass spectrographic analyses of the purified influenza B virus demonstrated that zinc was present (and occasionally copper). It was thus suggested that the RNA-dependent RNA-polymerase enzyme is a zinc metalloenzyme. This apparent requirement of the virus for zinc raises the possibility of chemotherapeutic control through two approaches:

1) During the synthesis of the enzyme depleting the level of free zinc ion in the medium could possibly control the rate of virus proliferation.

2) When the virus is active, administering a chelating agent could form a ternary drug-zinc-enzyme complex and inhibit the activity. In both instances the resulting deceleration of viral activity ought to give the body's defense mechanisms an improved chance of controlling the infection. Likely problems include the necessity of using a zinc chelating ligand that can penetrate the cell, the ambient zinc con-

TABLE 19-VI

ANTIVIRAL METAL COMPLEXES

Species Administered In Vitro	*Maximum Concentration for 50% Inhibition (μM)*
Disodium disulphonate salt of 4,7-diphenyl-1,9-dimethyl-1,10-phenanthroline (bathocuproine)	50
Disodium disulphonate salt of 2,9-dimethyl-1,10-phenanthroline (bathophenanthroline)	80
Copper(II) and Zinc(II) ions	60-80
1:1 and 2:1 Cu(I) and 1:1 Zn(II) complexes of bathocuproine and bathophenanthroline	10-20
2-acetylpyridine thiosemicarbazone	2
Cu(I), Zn(II), Ag(I) and Cd(II) complexes of 2-acetylpyridine thiosemicarbazone	1
N-methylisatin-3-thiosemicarbazone screened against influenza in mice and in tissue cultures and against polio virus in cell cultures	5

centration influence upon other zinc metalloenzymes present, and the toxicity of the ligands suggested.

An interesting observation is that Table 19-VI suggests that the antiviral effect of the complex is greater than the sum of the component effects of its component ligand(s) and metal ion. It has been suggested that not only do the complexes dissociate after being administered thus releasing ligands for control mechanisms (1) or (2) but also the metal ion liberated from the complex may inhibit at some other site on the polymerase enzyme, for example by a sulphydryl blocking action. Very recent tissue and animal experiments appear to exhibit similar trends (Table 19-VI).

Metal Complexes as Carcinostatic Agents

Many cancers have viruses associated with them and some animal, and a few human cancers are now believed to be caused by viruses. Hence, an anticancer drug may actually be an antiviral agent although the converse is not necessarily true. Compounds of the elements of group VIII in the d block have been particularly successful at destroying cancer cells: (a) Livingstone has reported a series of nickel, palladium, and platinum dialkyldithiophosphates which are capable of reducing mice tumors to 69 percent; (b) Rosenberg has reviewed a range of platinum complexes that are capable of being bacteriocidal if negatively charged (for example, hexachloroplatinate(IV) anion), of blocking the division but not the growth of a cell if neutral (for example, *cis*-tetrachlorodiammineplatinum(IV)), and of being antitumor and lysogenic if *cis* (for example, *cis*-dichlorodiammineplatinum(II)).[17] The complexes produced increased survival rates,

complete cures, and future immunity to tumors introduced by transplants, carcinogens, or viruses. Further, they had promising synergistic effects when combined with other drugs, and any cytotoxic damage produced in normal tissue was reversible. Their mode of action appears to be through forming *intra*strand links in a DNA chain; (c) Williams has added amino acids to these drugs to aid the anticancer specificity and cell membrane penetration. Examples of these anticancer drugs are shown in Figure 19-12[2]; (d) Connors, Ross, Tobe, et al. have recently prepared a number of platinum complexes which are thirty to forty times as effective as *cis*Pt(NH$_3$)$_2$Cl$_2$ at causing regressions in the PC6 mouse plasma cell tumor, e.g. the cyclopentylamine compound shown in Figure 19-12; (e) Furst has listed a wide range of lead, cobalt, cadmium, copper and bismuth complexes which are carcinostatic to specific tumors, e.g. lead arsenate, vitamin B$_{12}$, bismuth edtate.[8]

Before we can theorize concerning the mode of action of anticancer chelates we ought to have more information concerning physiological pH equilibria effects. Pending this information, several mechanisms have been postulated concerning the mode of action of metal chelates.

The simplest theory concerns Rosenberg's platinum complex (*cis*Pt(NH$_3$)$_2$Cl$_2$) and is that the site of its action is inside the cancer cells at the nuclear DNA. It is suggested that the complex selectively inhibits the synthesis of new DNA polymers. The

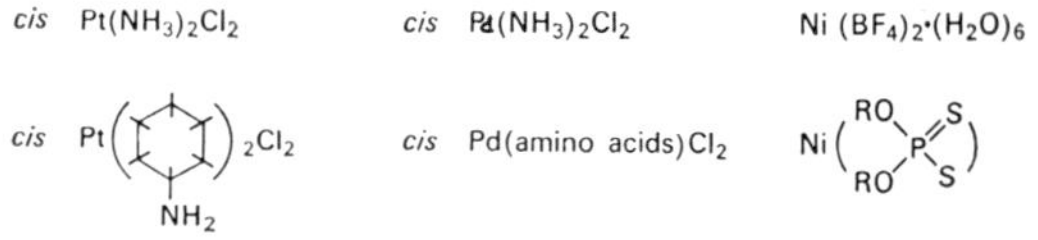

Figure 19-12. Anticancer metal complexes.

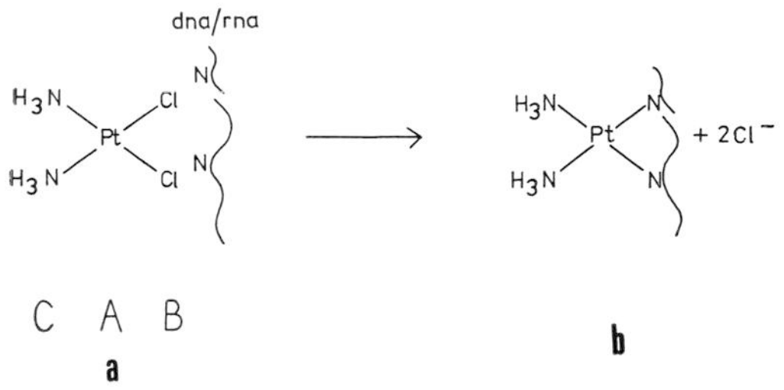

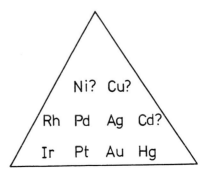

Figure 19-13. a ⟶ b is the suggested mode of action of *cis* Pt(NH₃)₂Cl₂. a also shows a convenient division of the metal complex into the three component parts, A, B, and C described in the text.

anticancer molecule contains Pt(II) and two *cis* Cl⁻ ions and it is suggested that these latter are lost to form a base—Pt—base cross link between nitrogens of two DNA purines or pyrimidines (Figure 19-13).

There are other theories, but whichever hypothesis eventually matures into a proven theory of anticancer activity for a particular type of cancer, the common factor appears to be a base nitrogen—metal ion—base nitrogen inter or intra-chain link, the nitrogens being fixed at a certain distance apart according to the three dimensional structure of DNA.

Chemical Design of Anticancer Metallotherapeutics[2, 18]

We shall mention some of the principles involved in the design of anticancer complexes in three overlapping parts (A) the choice of central metal ion, (B) the choice of the displaceable ligands, i.e. those that are removed when the metal-DNA bonds are formed), and (C) the choice of "carrier" ligand(s), i.e. the ligand(s) that occupies the remainder of the coordination sites on the metal ion and remain permanently attached throughout the anticancer action (Figure 19-13).

Choice of Central Metal Ion

It may appear that we have most of the metals and oxidation states of the periodic table to choose from but, in practice, our choice is somewhat more limited.

1) If our new drug is to be superior to existing C,H,O drugs we need to introduce wider bonding possibilities by invoking the d orbitals of the transition series rather than just the 2s and 2p orbitals of carbon.

2) To cause a cellular lesion, the DNA-metal bonds ought to be covalent rather than ionic so this effectively rules out the metals of groups I and II, in the s block.

(3) The HSAB approach to bonding suggests that metal-Cl and metal-purine N bonds (the latter necessarily being stronger than the former) can be formed by HSAB soft metal ions which lie in the Ahrland and Chatt triangle based upon group VIIIB of the transition series (class *b* metal ion acceptors) (Figure 19-14).

(4) The metal ions from the area of the periodic table that we have now focussed upon form square planar, octahedral, or sometimes square pyramidal coordination bonds. The interchange between a displaceable ligand and a DNA nitrogen usually occurs via aquation and an S_N2 mechanism. Square planar com-

Figure 19-14. The Ahrland, Chatt and Davies triangle.

plexes in particular lend themselves well to forming the coordinate intermediate since the apical positions offer convenient attacking points for incoming H_2O. These vacant positions and the available orbitals render the attack rapid.

Choice of Displacement Ligands

This section refers to the two monodentate ligands that are replaced at the DNA by basic nitrogens. In Rosenberg's active anticancer complexes these ligands have been two chlorides and Gillard has had some success using pyridines. The following two main properties are desirable:

1) It is clear from the active complexes discovered to date that *cis* displaceable ligands are necessary.

2) Drug activity is very dependent upon the exchange rate (a fast rate is necessary in order to overtake the tumor growth). It may be shown by the HSAB approach that for $Pt(NH)_3)_2{}^{2+}$ Cl^- is convenient but if we move to other metal complexes that are softer it may be necessary to use a softer pair of displaceable ligands, for example, I^- or Br^-, and *vice versa* for harder metal complexes. Indeed the high $[Cl^-]$ environment inside the cell could be a good justification for changing the anion released.

Choice of Carrier Ligands

Carrier ligands ought to be selected according to the following considerations:

1) The carrier ligand has the power of determining whether or not the proposed drug will be able to penetrate the membranes enclosing cells. Amino acids are particularly useful in this respect as not only do cells have active transport processes for most of them but also, healthy cells tolerate them well and have means of using, or of disposing of, the amino acids

after therapy. Additionally, some leukaemias rely upon a unidirectional flow of a supply of an amino acid into the malignant cell in order to multiply, e.g. lymphatic leukaemic cells require an external source of L asparagine.

2) The bond strengths of the carrier ligand-metal ion bond are also important. Firm ligand attachment is necessary and chelation usually facilitates this. When only monodentate ligands are used one risks a *cis* complex racemising into the thermodynamically more stable *trans* configuration. Bidentate ligands can only span *cis* positions and so remove this possibility of a drug deteriorating during storage. An L amino acid anion would seem to be an ideal choice of bidentate ligand for this task of conformation locking, but D forms ought not to be completely ruled out as there are instances of D amino acids forming better anticancer therapeuticals than the L variety.

Other Metal Complexes Used Therapeutically[6]

Several salts of nonessential metals are used as alimentary canal irritants and as disinfectants. Potassium, or ammonium, *aluminium* sulphate is an astringent, an antiseptic and an emetic. Kaolin (an aluminium silicate) is an insoluble powder having extensive absorptive powers and so it is used to absorb poisons from the intestine and to stop diarrhoea. The hydroxide of aluminium is also used as an antacid and absorbent.

Potassium *arsenite* is a treatment for malaria and rashes and it has also been used against chronic myelogenous leukaemia. Pentavalent organic arsenic compounds are administered to treat trypanosomes and are readily reduced to the trivalent forms *in vivo*.

Antimony compounds are employed against parasitic diseases endemic in the tropics (antimony potassium tartrate, lithium antimony thiomalate, antimony sodium thioglycollate, and sodium p-aminophenylstibonate is used against leishmania). The toxicity of antimony compound therapy is discussed in Chapter 20.

Mercury *bismuth* carbonate, bismuth salicylate or bismuth sodium tartrate are used in treating syphilis. Tripotassium dicitrato bismuthate is used in treating mouth, gastric and duodenal ulcers.

Gold is administered as Sanocrysin (sodium aurothiosulphate) intravenously for fighting tuberculosis or as Myocrisin (sodium aurothiomalate) for treating rheumatoid arthritis. The toxicity of gold compound therapy is discussed in Chapter 20.

Mercury has been used medicinally since the twelfth century. The effectiveness of mercury therapy depends upon the ease with which ionization occurs in the mercuric salts of chloride, cyanide, oxide and in mercurous chloride. These compounds along with several organic mercurials are employed as disinfectants and as diuretics. The toxicity of mercury compounds is discussed in Chapter 20.

Silver nitrate solution is used to treat warts and weak solutions of silver, aluminium, copper, zinc, iron and lead are all used as astringents or to produce vomiting.

CONCLUDING REMARKS

Anyone who attempts to predict future developments, especially in science, is likely to get his fingers burned. Nevertheless, we believe that future therapies are to be discovered through progress in three directions. Above all, we must avoid being lured into the structural empiricism that was a characteristic of the first half of this century. Instead we must progress through more education, more basic research and more epidemiological studies.

In *education* there is a need to remind scientists of the general dangers of the ligands and metals with which they work (deaths from some cancers are 25 percent more prevalent among chemists than among other professional men) and we should be reintroducing the analytical chemistries of the lighter, *in vivo* essential, elements into undergraduate chemistry course ($\pm$ a few percent is adequate to detect the presence of metal dependent diseases). From an interdisciplinary viewpoint, inorganic chemists ought to report their researches in such a manner that biochemists and pharmacologists can understand their relevance and also to show more interest in biological and polluting ligands, metal ions and complexes when selecting new research topics, i.e. synergistic research is needed.

At the *basic research* level, no single laboratory can ever expect to design, synthesize, screen and market a new drug and so the problems of communication and collaboration are important. As mentioned in Chapter 10, there are three levels of bio-inorganic research: (1) whole-body metal-ion interactions. Naturally the simpler species are the first to be examined, for example, nitrogen fixation bacteria; (2) metalloenzyme, including models of, studies, and (3) simple *in vivo* ligand-metal-ion investigations followed by computer simulation of the *in vivo* process. Current examples are the needs under (3) to establish the exact composition of the complex through which a drug acts (protonation or hydrolysis may modify the ingested compound), and under (1) to quantify the *in vivo* competition between the essential metals on the one hand and polluting metals on the other, e.g. there is an intimate relationship between cadmium and zinc metabolism. In cancer research more than any other field it is essential to have a planned, reasoned, approach because it may take up to a decade for a new drug to progress from the research laboratory into the armamentarium of the medical profession.

Epidemiological studies can assist (a) by highlighting new metal poisons before the problem reaches crisis level, and (b) suggest new links between the incidence of disease and the environment. Thus, the patterns of future researches may change as additional metals or complexes are found to be involved in different diseases.

REFERENCES

1. Frieden, E.: *Sci Am*, 227:52, 1972.
2. Williams, D. R.: *Chem Revs*, 72:230, 1972; *Inorg Chim Acta Revs*, 6:123, 1973.
3. Breckon, W.: *The Drug Makers*. London, Eyre Methuen, 1972.
4. Dwyer, F. P. and Mellor, D. P. (Eds.): *Chelating Agents and Metal Chelates*. London, Acad. Pr, 1964.
5. Albert, A.: *Aust J Sci, 30*:1, 1967-8; *Selective Toxicity*, 5th ed. London, Chapman, 1973.
6. Gaddum, J. H.: *Pharmacology*, 7th ed. Fair Lawn, Oxford U A, 1972.
7. Furst, A.: *Geol Soc Am*, Memoir No. *123*:109, 1971.
8. Furst, A.: *Chemistry of Chelation in Cancer*. Springfield, Thomas, 1963.
9. Schrauzer, G. N. and Rhead, W. J.: *Experientia*, 1069, 1971.
10. Christensen, J. J., Hill, J. O. and Izatt, R. N.: *Science, 174*:459, 1971.
11. Hughes, M. N.: *The Inorganic Chemistry of Biological Processes*. London, Wiley, 1972.
12. Peisach, J., Aisen, P. and Blumberg, W. E. (Eds): *The Biochemistry of Copper*. New York, Acad Pr, 1966.
13. Sarkar, B., Kruck, T. and Appleton, D.: *Chemistry in Canada*, Oct. 25, 1972.
14. Williams, D. R.: *The Metals of Life*. London, Van Nostrand, 1971.
15. Proceedings of Second International Symposium on Platinum Coordination Complexes in Cancer Chemotherapy, Oxford, 1973.
16. Oxford, J. S. and Perrin, D. D.: *Gen Virol, 18*:11, 1973; 23:59, 1974.
17. Rosenberg, B.: *Plat Met Rev, 15*(2):42, 1971.
18. Thomson, A. J., Williams, R. J. P. and Reslova, S.: *Structure and Bonding*. 1972, vol. 11, p. 1.

CHAPTER 20

METAL-INDUCED TOXICITY AND CHELATION THERAPY

D. D. Perrin and R. P. Agarwal

INTRODUCTION

THE INTEGRITY, maintenance and functioning of the mammalian body depend on the presence of sodium, potassium, magnesium, calcium and traces of other metals including iron, cobalt, copper, zinc, manganese and molybdenum. However, particularly as a result of developments in industrial technology, many people are exposed to the effects of other metal ions and to excessive concentrations of those that are essential. To illustrate the claim that man has become a potent geological force it is enough to reflect on the widespread exploitation of mineral deposits, the enormous consumption of fossil fuels, the massive amounts of garbage disposed of annually in the United States, and the continuing dispersal of lead compounds in dust from the exhausts of petrol-burning engines.

Because of their interactions with binding groups on proteins, heavy metal ions are potent enzyme inhibitors exerting toxic effects on living systems. Metal ions vary in their relative reactivities and the extent of binding to different types of groups, so that the numbers and types of sites occupied by particular kinds of metal ions will depend on many factors.

Most of the heavy metal ions are readily polarizable and the strengths of their binding to atoms in complex-forming groups lie in the sequence $S > N > O$. Thus mercury, silver and arsenic are well known for their strong bonding to sulphydryl groups. For the alkaline earth cations the sequence is $O > N > S$, and bonding to carboxyl and phosphate groups is relatively more important. Some metals such as zinc and lead are intermediate in character. The implications of differences in complex-forming abilities of metal ions with different ligands is discussed more fully elsewhere in this chapter. The ubiquitous occurrence of chloride and phosphate ions in biological systems also has important consequences, particularly in lead and mercury poisoning, because of the formation of insoluble lead phosphate and of only slightly dissociated mercuric chloride.

The actions of toxic metal ions on organisms are dependent on rates of absorption, distribution, deposition and excretion. Initial reactions tend to be with cell surfaces and cell membranes, followed by penetration into the cell, giving secondary changes in cell physiology. A distinction must also be made between acute and chronic poisoning, particularly as this is reflected in the distribution of toxic cations within the body. While chronic poisoning frequently leads to deposition in particular tissues such as lead in bone, or excessive copper in liver, brain and kidney, acute poisoning usually results in excessive concentrations of the metal in the gastrointestinal tract, the blood and the soft tissues. These distribution patterns are important when treatment with chelating agents is required, particularly if there is any possibility of the metal chelate being translocated to a more vital area. Thus the treatment of lead poisoning with 2,3-dimercaptopropanol is attended with the hazard of deposition of lead in the brain, with resulting damage.

It is better to avoid metal ion toxicities than to treat them, and a great deal can be achieved by preventive industrial medicine if dangers are realized. For example, the risk of lead poisoning would be greatly diminished if titanium dioxide-based paints were used instead of lead-based ones, and glazed earthenware was not used to store apple juice, cider, lime juice or other acidic or chelate-containing liquids.

The pattern of distribution of a toxic metal in the body tends to be that the most slowly established deposits also take longest to remove, so that, in general, the sooner chelation therapy is applied after metal poisoning is diagnosed, the more effective it will be. The technique chosen for the administration of a chelating agent will depend on its chemical properties and the ease with which it can pass across tissue boundaries; the route may be oral, subcutaneous, intraperitoneal, intravenous or by inhalation, important considerations being the maximum permissible ligand concentration in the blood and the necessary duration of the treatment.

Although there is currently a general community awareness of the dangers associated with lead and methyl mercury in the environment, hazards associated with metal ions have been recognized throughout historical times. More than two thousand years ago, Nicander described the effects of lead poisoning in humans. Pliny the Elder (A.D. 23-79) knew there were "most pernicious" effects of inhaling mercury while Ramazzini, an eighteenth century Italian physician, reported that from mercury mines came "the most cruel bane of all that deals death and destruction to miners." Distillation of rum in leaded stills was forbidden by the Massachusetts Bay Colony in 1723 to prevent "dry gripes," an intestinal condition. In the nineteenth century, nonferrous smelters in the Swansea Valley, Wales, produced copper, zinc and lead by processes emitting very large quantities of metal-rich sulphurous smokes and dusts which rendered derelict tracts of land near the smelters. In 1900 some thousands of people in Manchester were poisoned by drinking beer contaminated with arsenic. These examples, and others discussed below, serve to emphasize man's susceptibility to toxic metal concentrations.

MAJOR POLLUTANTS

Lead, mercury and cadmium are jointly grouped under this heading because upwards of 200,000 tons of lead is deposited on the earth annually from internal combustion engines, enough alkyl mercury compounds are present in some lakes that for many years to come the fish in them will be unsafe to eat, and the toxicity of cadmium has been abundantly demonstrated along the Jintsu river in Japan. In recent times the lead level in the surface waters of the northern oceans has been trebled. With the recognition of the danger, precautions have been taken to diminish the risk of further alkyl mercury poisoning, but lead poisoning is a continuing hazard and the possibility also remains that the incidence of cadmium poisoning will increase as the world consumption of cadmium rises. It is also possible that cadmium may have a widespread deleterious effect as an antagonist of zinc.

Cadmium

Cadmium accompanies zinc to the extent of about 0.5% in many of its ores and is obtained as a byproduct of its manufacture. About 5,000 tons of cadmium is used each year, mainly in plating, pigments, alkaline batteries, and metallurgy. Production has greatly increased since 1940, particularly in the United States. Although it has been known for over a century that cadmium is severely toxic to man, prior to 1940 there had been few reported cases of cadmium poisoning. Since that time there has been a dramatic increase.

Acute and chronic cadmium poisoning differ somewhat in their etiology. In acute poisoning, symptoms such as nausea, salivation, vomiting, diarrhoea, abdominal pains and myalgia may develop anything from fifteen minutes to five hours after consumption of cadmium-containing food or beverages, such as might result from the action of fruit juices on cadmium plated containers, and as little as 10 mg cadmium can cause serious symptoms. The toxicity of cadmium taken by mouth is partly offset by the vomiting it frequently induces. Chronic cadmium poisoning due to long-term, low-level dosage is primarily an industrial hazard but is best known as the Itai Itai disease of residents along the Jintsu river of northwestern Japan. The region is near an old disused nonferrous metal mine and over the years several hundred residents have died in agony from this disease (as its name implies), another of its characteristics being the brittleness of victims' bones. Cadmium deposition tends to be cumulative in the kidney and average levels of cadmium in ashed kidneys have been reported as 0.6, 0.3 and 0.1 percent for inhabitants of Japan, United States and England. The liver also stores cadmium at about one-tenth of these levels.

Another source of cadmium poisoning is by inhalation of toxic fumes, commonly of the oxide, during industrial operations. This leads to loss of smell, respiratory difficulties, weight loss and impairment of kidney, liver and blood-forming functions. In severe cases death frequently results. Toxic levels of cadmium in the atmosphere range from 1 to 30 mg Cd/m^3.

There is evidence suggesting that cadmium may be involved more extensively in some pathological conditions. Thus there is a good correlation between high renal cadmium levels and hypertension; the latter may reflect the extent to which cadmium has damaged the kidney. The cadmium content of public water supplies in the United States varies widely, from 0.0001 to 0.01 p.p.m., and similar wide variations are found in levels in foods. However, correlations have been claimed between regional cadmium levels and cardiovascular disease. Cadmium chloride, injected subcutaneously, is a potent carcinogen for the production of interstitial cell tumors, although this effect can be diminished by injections of zinc salts. Subcutaneous injection of cadmium chloride into animals also leads to irreversible testicular damage which can, however, be prevented by zinc or by complexing agents which include thiols, cysteine and dimercaptopropanol.

It is suggested that cadmium competes with zinc at essential reaction sites on enzymes, and the isolation from kidney and liver of a cadmium-containing metalloprotein, metallothionein, raises the possibility that this complex serves as a detoxification mechanism to bind cadmium and prevent it from inactivating otherwise sensitive enzymes. The renal cortex is known to be a specific site for cadmium deposition, possibly in the proximal convoluted tubules of the nephron.

Lead

Lead is widely distributed in the lithosphere, mainly as its sulphide galena, from which about 3.5 million tons of lead is produced annually. Of this, almost 10 percent is burned as lead alkyls in combustion engines. In Britain alone during 1970 about 10,000 tons of lead from this source was added to the environment. Large amounts are also used in accumulators, paints, lead pipes, solder and low-melting alloys. With industrial development there has been a steady accumulation of lead wastes in urban areas, resulting in a greater exposure of the public to the possibility of lead poisoning. Plumbism or saturnism has been known for at least 2,000 years as an insidious disease, difficult to diagnose and often unrecognized. Usually there is excessive uptake of lead for some months or longer before clinical symptoms appear. Again, these are initially mild and nonspecific, comprising general malaise, decreased appetite, irritability, clumsiness, headache, abdominal pain and vomiting. With continued exposure there can be a rapid progression to coma, convulsions and death, especially in infants.

The people at risk fall into several well defined categories:

1) Young children, especially in dilapidated areas, who consume flakes of leaded paint, plaster or putty. It has been estimated that 225,000 children in the United States may have abnormally high blood lead concentrations, with attendant risks of brain damage, mental retardation and kidney damage.

2) Those who eat or drink from improperly lead-glazed earthenware, especially if the liquid is acidic or contains chelating agents (such as citric acid in fruit juices).

3) Prolonged drinkers of illicit whisky distilled in solder-jointed metal equipment.

4) Regular consumers of rain water collected on roofs painted with lead-based paints or of "soft" water that has passed through lead pipes, especially after standing.

5) People exposed to prolonged and intense petrol combustion fumes, e.g. traffic policemen in large cities.

6) Foundry workers handling lead and lead alloys.

Unusual cases of lead poisoning have included the use of lead nipple shields by nursing mothers and inhalation of fumes from burning accumulator cases.

The normal daily intake of lead is about 0.3 mg, of which 90 percent passes through the gut unabsorbed. The respiratory intake is between 5 and 50 μg lead daily. The lead content of the soft tissues is relatively stable, but the concentration of lead in the bone increases steadily with age, probably by isomorphous replacement of some of the hydroxyapatite calcium by lead ion which has a comparable crystal radius. Under conditions of increased metabolic activity or of hypophosphataemia there can be a mobilization of lead from such bone deposits, so that lead poisoning may be superimposed on the initial disorder.

If the dietary intake exceeds 1 mg Pb/day, the rate of storage in bone is too slow and the blood lead level rises, producing metabolic, functional and clinical responses. These abate when the lead intake falls. Severe poisoning can damage the central nervous system and the blood-brain barrier, giving rise to severe cerebral edema with resulting destruction of brain tissue. This edema is associated with excessive permeability of the capillaries, so that they leak. Lead causes anaemia, encephalopathy, myelopathy and neuropathy. General symptoms include headaches, dizziness, seizures, coma, mental deterioration, colic and dense lead bands across the metaphyseal ends of the long bones. A feature that is diagnostic of lead poisoning is basophilic stippling of the red blood cells which is due to the retention of cytoplasmic constituents of red cell precursors, including mitochondria.

The insolubility of lead phosphate (log K_{so} for $Pb_3(PO_4)_2$ is -42) and lead chloride (log K_{so} for PbOHCl is -13.7) limits the concentrations of free lead ions that can be reached under physiological conditions and these insoluble salts may act as tissue reservoirs, thereby prolonging exposure of the organism to lead, so that it may be several months or years before blood lead levels return to normal. Lead exerts its best known effects because of its powerful combination with sulphydryl groups in certain enzymes. Thus it inhibits the biosynthesis of haem, the iron-containing constituent of haemoglobin and the cytochromes, acting specifically on the conversion of δ-aminolaevulinic acid (ALA) by ALA dehydrase in the cytoplasm to porphobilinogen and on the final formation by mitochondria of haem from protoporphyrin IX and iron, in both of which steps the enzymes depend for their activity on free sulphydryl groups. Interference by lead in haem synthesis has been demonstrated in tissue preparations of kidney, brain, liver and red cells. Lead may also be implicated

in the failure of mitochondria to form ALA from succinyl coenzyme A and glycine and in the nonconversion of coproporphyrinogen III to protoporphyrin IX. In lead poisoning, ALA is excreted in the urine along with coproporphyrin which, with protoporphyrin, accumulates in the red cells. Detection of these components is important in diagnosing lead poisoning and the diagnosis would be confirmed if lead levels in the blood exceeded about 60 μg/ 100 ml.

As a consequence of the disturbance of blood metabolism, there is a decrease in the life span and number of red cells, and in their haemoglobin content. The resulting anaemia is accompanied by hyperactivity of corpuscle-producing tissues, so that the blood contains increased numbers of immature red cells, reticulocytes and basophilic stippled cells. The anaemia is reversible if lead intake ceases.

Acute lead poisoning leads to impairment of kidney function. At the electron microscopic level, dense lead-protein inclusions are found in the nuclei and there are lead-containing granules in the mitochondria of kidney cells, including the proximal renal tubules. Uptake of lead in this way appears to be energy-dependent and to compete with other energy requiring processes. This metabolic disturbance may be the origin of the Fanconi syndrome seen in severe acute plumbism, in which there is increased loss of amino acids, glucose and phosphate in the urine because of impairment of the resorptive ability of tubular cells. Chronic lead poisoning commonly leads to scarring and shrinking of kidney tissues, with chronic nephritis and early death.

Another frequent effect of chronic lead poisoning is damage to the mitochondria of the Schwann cells, which synthesize the myelin sheath of nerve fiber, resulting in demyelination with impairment of nerve impulses to motor nerves of the extremities. In man this results in peripheral neuritis.

The extensively used petroleum additive tetraethyl lead, $Pb(C_2H_5)_4$, is highly toxic, but poisoning is rare because stringent precautions are observed in its manufacture and bulk handling. It is readily absorbed through the skin, and the respiratory and digestive tracts. Because of its lipid solubility it is concentrated in brain, body fat and liver, where it slowly breaks down with the release of lead ions. Typical lead alkyl poisoning is accompanied by mental disorder and differs significantly from inorganic lead poisoning. Intravenous injection of $CaNa_2EDTA$ gives slow improvement, the treatment probably being ineffective until the lead tetraethyl is degraded.

Poisoning directly due to lead alkyls in petrol is not a serious hazard for the general public, except possibly when petrol is used for degreasing and cleaning. The position is less certain regarding the inorganic lead salts—chlorobromide, sulphate, carbonate, phosphate—resulting from the burning of lead alkyls in petrol, partly because increased urban petrol consumption has been accompanied by a reduction in the use of coal as a heating fuel. Nevertheless, with the lead content of city dust at around 1 percent it is clear that a potential hazard exists. The possibility must also be considered that present lead levels in man, although below conventional clinical thresholds, may be high enough to produce adverse behavioral effects in at least some members of the community.

Mercury

Mercury is a rare element, making up less than 0.03 parts per million of the earth's crust. Professor D. Dyrssen has estimated that mercury is present in sea water at a mean concentration of 0.15 μg/l so that, the volume of the seas being 1.4×10^{21} l, the oceans comprise a reservoir containing 210 million tons of mercury. The normal atmospheric level is about 3 ng Hg/m^3, giving an atmospheric total of 12,000 tons. Mercury occurs, mainly as the red sulphide, cinnabar, in localized areas, a major source being the Almadén mine in Spain which has operated since Roman times and which currently produces about 15 percent of the world's output of 9,000 tons a year. About 50 percent of this output is lost each year in industrial and agricultural usage. Thus, there is no danger of a significant overall increase in the mercury level of the ocean, but serious damage can result in local areas.

This occurred, for example, at Minimata, Japan, in 1953 when neurological illness leading to fifty-two deaths followed the eating of fish and crustaceans contaminated with mercury discharged in the effluent of a chemical factory. Sea birds and domestic cats were similarly affected. In 1964 there was a similar outbreak at Niigata, on Honshu. Numerous fatal cases of mercury poisoning have been reported from Iraq, Pakistan, Guatemala and New Mexico following the consumption of grain seed treated with mercurial fungicide (methyl mercury dicyanodiamide). In some cases, poisoning followed the eating of animals that had been fed on the treated seed. Similarly, a decline in numbers of seed-eating birds, followed by decreases in their predators (kestrels, peregrine falcons and harriers), was observed in Sweden after the use of alkyl mercury compounds to treat seed grain. More recently, in 1970, high levels of mercury were found in tissues of seed-eating birds in Alberta and Saskatchewan. The use of alkyl mercury compounds in agriculture is now banned in many countries, including Sweden and the United States.

Much publicity has been given to the finding of high levels of mercury in fish, initially in the Baltic Sea, later in the Great Lakes area (particularly Lake St. Clair and Lake Erie), and more generally in large predatory fish such as swordfish, shark and tuna from a wide range of locations. In lakes and other enclosed bodies of water the high levels of mercury (up to 7 ppm in fish from Lake Erie) are due to industrial contamination, e.g. from chlor-alkali plants. With predatory fish, however, there is progressive accumulation of mercury along a food chain beginning with marine algae and passing through the series plankton, small fish, large fish. Fish-eating humans also belong to this chain, and up to 144 ppm mercury were found in the kidneys of the Minamata victims. Levels of up to 50 ppm (wet weight) were present in affected fish. The crab is a scavenger, so that crab meat may also be high in mercury: a level of 24 ppm wet weight was reported from Minamata. The United States Food and Drug Administration has laid down 0.5 ppm as an arbitrary safe limit for mercury in fish.

Three types of mercury poisoning can be distinguished, namely mercury vapor, inorganic mercury and alkyl mercury. Liquid mercury is reputedly not toxic, but it has an appreciable vapor pressure (0.01 mm at 50°, 0.3 mm at 100°, 2.8 mm at 150°). Inhalation of the vapor can be injurious, either acutely, when it causes bronchial irritation and destruction of lung tissue, or, more commonly, poisoning is chronic and leads to tremor, inflammation of the gums and general irritability. Mercury is readily absorbed into the lungs where it has a half-life of about seven hours before it is oxidized by oxygen and appears

as ionic mercury in the blood. On prolonged exposure a steady state is reached in which the daily mercury intake is balanced by the rate of excretion, with most of the mercury being located in the kidney. Because of the toxic effects of mercury vapor, workers in the Almadén mine are allowed to work for only eight days each month. Occupations marginally at risk from this source include thermometer fillers and laboratory technicians.

Soluble inorganic mercury salts are highly toxic. In excess, mercuric chloride causes corrosion of the intestinal tract and suppression of urine, resulting from kidney failure that ultimately leads to death. Use of mercuric nitrate in the felting process resulted in mercury poisoning becoming an occupational hazard of hatters, symptoms including tremors, loss of teeth, difficulty in walking and mental disability. The mental effects probably gave rise to the expression "as mad as a hatter" and also account for the use of the term "hatter" for the solitary, acutely shy gold miners on the Californian and New Zealand goldfields who recovered their gold by distillation of amalgam in crude stills. Poisoning from purely inorganic mercurial sources is rare, but might include munition workers and, at one time, British policemen who used a special mercury-containing fingerprinting compound. Nor does there seem to be a significant hazard in the use of amalgam fillings in dentistry.

Mercury ions have a very high affinity for thiol groups and as the latter are widely spread throughout proteins almost all proteins can bind mercury to some extent. Hence, almost all proteins in the body are potential targets for mercury poisoning. The protein metallothionein which accounts for most of the protein-bound fraction of mercury in the kidney may have an important role in protecting kidney tissue from damage by inorganic mercury.

Much the greatest danger from mercury is now recognized to be in its alkyl derivatives, as was shown dramatically at Minamata. Alkyl mercurials, especially the methyl and ethyl compounds, are far more toxic than mercury, itself. Using methyl-mercury-203 they have been shown to have a half-life in the body of ten weeks. They are rapidly taken up by erythrocytes and can readily pass the blood-brain barrier, a protective system that bars most toxins, and decrease the number of neurones in the cerebrum and in the granular cells of the cerebellum. This leads, as a direct consequence, to ataxic gait, convulsions, numbness of mouth and limbs, constriction of the visual field, difficulty in speaking and deafness. Injury to the brain cells is permanent, possibly because of the covalent character of the mercury-alkyl bond and also the strong affinity of mercury for sulphur in protein sulphydryl groups. Mercury bound to proteins in cell membranes may alter distribution of ions, change electrical potentials and interfere with the passage of nutrients across these membranes. Other possible effects include disturbance of function of mitochondria and lysosomes, as well as producing abnormalities of chromosomes. Unlike inorganic mercury compounds, alkyl mercurials can pass the placental barrier and affect the fetus, leading to cerebral palsy and congenital retardation, and can also damage bone marrow, lymph nodes, nerve fibers, liver and kidney.

Inorganic and aryl mercurials do not pass the blood-brain barrier, so that injury from them is less severe and much more nearly reversible. In fact, phenyl mercurials are comparatively nontoxic.

The main ecological hazard posed by mercury is in effluent discharge into lakes and estuaries. Anaerobic microorganisms in lake and stream beds can convert inorganic mercury and aryl mercurials into highly toxic methyl and dimethyl mercury, and the operation of a normal biological cycle may lead to dangerous levels persisting for many years even after pollution ceases. Similar processes can occur in estuarine mud, but progressive dilution by marine circulation would be much more rapid. Alkyl mercurials can arise from other sources; for example, poisoning by treated grain in Iraq was due to breakdown of ethyl mercury *p*-toluene sulphonanilide. Gas chromatographic procedures enable a distinction to be made between alkyl mercurial and inorganic mercurial contents of foodstuffs. They have been used to show that methyl mercury is the predominant form of mercury in Swedish fish, but the corresponding information does not appear to be available for pelagic fish. At present, a joint commission of the Food and Agriculture Organization and the World Health Organization suggests an upper limit of 0.05 ppm for mercury in foods other than fish and shellfish.

ESSENTIAL TRACE METAL IONS THAT ARE TOXIC IN EXCESS

It is almost a law of nature that anything taken into the body in excess is harmful, no matter how important it may be at lower concentrations. Although ions of metals like cobalt, copper, iron, zinc and manganese are essential for the human body, they become hazardous when present in excess. The undesirable accumulation may be the result of a metabolic disorder, as in Wilson's disease where a deficiency of the blood protein caeruloplasmin leads to excessively high levels of exchangeable serum copper(II) and to deposition of copper(II) in the brain and liver. More commonly, toxicity results from excessive exposure to sources of the trace metal, either in mines or industry, and to the accidental intake of concentrated materials, such as iron tablets by children.

Cobalt

Insufficient levels of cobalt in the diet of ruminants was shown, originally in Australia and New Zealand, to give rise to wasting diseases, known locally as enzootic marasmus and bush sickness. Trace amounts of cobalt-containing vitamin B_{12}, cobalamin, are essential for the synthesis of haemoglobin by mammals. Although the human body cannot make vitamin B_{12}, and must rely on microorganisms for its production, large therapeutic doses (up to 100 mg cobalt chloride daily) are used in the treatment of refractory anaemia, particularly when it is associated with renal failure. The mode of action of the cobalt is unknown, and undesirable side effects have been reported. Absorption of cobalt is diminished after a meal, varies greatly with quantity and type of dietary protein, pH of the stomach and duodenum, and is increased if an iron deficiency exists.

During the period from 1965 to 1966, cobalt sulphate added to beer in Quebec, Canada, at a concentration of 3 mg/gallon to improve the stability of the foam led to cardiomyopathy in heavy beer drinkers. (Similar cases were reported from Omaha, Nebraska and Leuven, Belgium.) Typically the first symptoms were gastrointestinal, beginning with severe anorexia, and these were followed by progressive signs of congestive heart failure, together with a characteristic cyanosis of the face and trunk. Of fifty cases examined, twenty died. At the time the cause of the outbreak was not known, but histological examination of

thyroid glands showed a pseudocarcinomatous change in the acinar cells and little or no colloid: These changes had previously been described following the administration of cobalt salts. Another frequent finding of polycythaemia also suggested cobalt. The level of cobalt in the beer meant that the daily intake was appreciably less than is commonly administered medicinally, but the beer apparently facilitated the uptake of the cobalt. A similar potentiation by beer was found in Manchester in 1900 when 2,000 cases of poisoning by arsenic in beer occurred. Subsequently, a case was reported of an industrial worker who died following exposure for four years to cobalt; post mortem examination showed the same type of heart failure, and analyses revealed levels of cobalt exceeding 1 ppm (dry basis) for heart, lung, liver and kidney.

In vitro, cobalt stimulates certain enzymes and inhibits others, probably by competition with magnesium and calcium ions needed for enzymatic activity.

Action on cell metabolism is an inhibition of keto acid oxidation by forming an oxygen-sensitive cobalt chelate with the dithiol form of lipoic acid, thereby blocking the Krebs cycle and aerobic cellular oxidation, and breaking the terminal electron transport mechanism. Enzymic inhibition may be combined with partial replacement of calcium on sites where it is needed for muscular contraction, leading to myocardial insufficiency because of defective utilization of high energy phosphates.

The biological effects of cobalt depend very much on its concentration, and are also modified by sulphydryl-containing amino acids which form complexes with cobalt. Small daily doses of cobalt, administered either orally or parenterally to animals, leads to polycythaemia. High levels of cobalt fed to intact animals result in histo-

toxic hypoxia and to a characteristic myocardial toxicity with pericardial edema. The
severe damage to the cardiac muscle cells
includes the disappearance of the myo-
fibrils, aggregation of mitochondria, fatty
degeneration, and final disappearance of
the cells, and fibrosis. Most of the short-
term effects of excessive intake of cobalt
are seen to be due to a generalized decrease in energy production. Implantation
of cobalt powder has been reported to
cause malignant tumors in muscles.

Copper

Major roles for copper in physiological
processes comprise requirements for the
formation of blood and (through oxidase
activity) of adequate crosslinking of elastin
in aortas and major blood vessels, and its
presence in many oxidative enzymes. Effects of copper deficiency, especially in
domestic animals, are numerous and the
consequences of a congenital lack of the
copper protein caeruloplasmin is described
in Chapter 18. Uptake and utilization of
copper is also sensitive to interactions of
other dietary constituents including zinc,
molybdenum and sulphate ion.

However, copper is a relatively toxic element with a wide range in species sensitivity. The main depot for storage of copper
is the liver, typical concentrations in ppm
dry matter being 600 for sheep, 200 for cattle and 40 for humans. Chronic copper
poisoning in sheep has been reported, particularly following ingestion of herbage in
areas such as orchards and vineyards previously sprayed with copper compounds, on
pastures high in copper (above 10 ppm
dry matter) or normal in copper but very
low in molybdenum, and as a result of liver
damage from eating *Heliotropum euro-
paeum*. Dosage of 125 to 500 mg copper

daily to sheep is usually fatal within one
to eight months. Copper poisoning in man
is unlikely, except as an industrial hazard
of copper miners and processers, although
an increased incidence might be expected
following the widespread use of copper for
water pipes. However, acute copper sulphate poisoning, both accidentally and with
suicidal intent, is common in New Delhi
where, in 1961, it accounted for one-third of
all poisoning cases admitted to a hospital.

In chronic copper poisoning of sheep,
copper concentrates in the liver without
apparent ill effect and the blood copper
level is normal, until the binding sites in
the liver are saturated. Toxicity is then
manifested as a necrotic hepatitis or cirrhosis and as an acute haemolytic crisis
following the release of excessive copper
to the circulating red cells, with severe
liver dysfunction as shown, among other
things, by an abrupt reduction in brom-
sulphthalein clearance. The haematocrit
level drops sharply, there is haemoglobin
in serum and urine, and methaemoglobinaemia and a profound hyperbilirubinaemia
are found. The approach of this crisis is
predictable three to eight weeks earlier by
moderate rises in serum glutamic oxaloacetate transaminase (SGOT) and lactic
dehydrogenase (LDH). At the crisis there
is a sudden massive increase in SGOT and
LDH, as well as in serum glutamic pyruvic
transaminase and arginase.

Chronic copper poisoning is almost unknown in man, but the predominant clinical
features of acute copper poisoning in humans are a marked haemolysis six to
twenty-four hours after exposure to copper, and associated jaundice, methaemoglobinaemia, hypoglycaemia, vomiting, diarrhoea, abdominal pain and liver tenderness.

Recently, it was pointed out that copper

poisoning is a potential hazard of hae-modialysis. Tap water passing through cop-per plumbing in Australian hospitals was found to contain up to 1 ppm copper in the dialysis fluid of haemodialysis units, and this was actively taken up by plasma proteins and red cells, with subsequent storage in the liver. Passage of 250 1 of such a fluid would add possibly 250 mg cop-per, as against a daily dietary intake of 2 to 3 mg. Analysis of the copper content of dialysis liquid would appear to be a desirable routine procedure. In some cases, acute copper poisoning has followed hae-modialysis treatment, but in these instances they were due to the action of acidic water from partially exhausted deionizers acting on copper components of an arti-ficial kidney machine.

Iron

The central position of iron in mam-malian physiology is apparent from its presence in haemoglobin, the cytochromes and flavoprotein enzymes, so that deficiency of dietary iron affects the formation of hae-moglobin and inhibits some of the iron-dependent enzymic processes. To correct such deficiencies, and in the treatment of anaemias, it has become customary to ad-minister tablets of soluble ferrous salts. Ac-cidental excessive consumption of these led to acute iron poisoning in a child in 1947, and by 1961 iron compounds were recog-nized as being, next to aspirin, the com-monest cause of poisoning of children in England; in the United States iron prepara-tions and vitamins ranked fourth as poisons of children under five years of age. A dis-tressing feature was the mortality rate around 50 percent, but this was subse-quently decreased dramatically by appro-priate therapy.

In acute iron poisoning the first clinical effects follow the highly corrosive action of iron salts on the gastrointestinal tract, with haemorrhage and necrosis of the stom-ach and intestinal mucosa, giving vomiting, pallor, shock, haematemesis, circulatory collapse and coma. Serum iron levels of 30 ppm are common. Collapse and coma may follow an apparently spontaneous improve-ment. Initial treatment includes the induce-ment of vomiting and gastric lavage with sodium bicarbonate or, better, disodium phosphate to form insoluble iron compounds which are removed in the lavage or ex-creted in the faeces. This may be followed by chelation therapy, particularly if there is hypotension, shock or coma.

Chronic conditions also exist in which iron is deposited in tissues and organs of the body. These are classified as haemo-chromatosis (where there is hepatic cir-rhosis and increased iron deposits in many organs including liver and pancreas) and haemosiderosis (where there is iron dep-osition but no cirrhosis). Haemochroma-tosis may be primary (genetic or idio-pathic) or secondary, the latter, like hae-mosiderosis, arising from malnutrition, treatment of anaemia or excessive intake of oral iron. Excessive oral iron may be as a result of medication, but in the case of Bantu haemochromatosis it arises from the use of iron cooking pots in preparing a carbohydrate-rich gruel. Although iron de-posits in some organs such as kidney may be anatomically striking they do not cause dysfunction. However, deposits in the liver tend to produce reactive fibrosis and, oc-casionally, clinical features of hepatic dys-function. Cardiac iron deposits can cause congestive heart failure. Also, prolonged treatment by blood transfusion of anaemia due to thalassaemia major can lead to iron deposition with impaired organ function and pathological changes resembling idiopathic haemochromatosis.

Manganese

Although manganese is an essential element for plants and animals it is known so far to be involved in only a small number of enzymic reactions, including the formation of glucosamine-serine linkages in the synthesis of cartilage mucopolysaccharides, the action of pyruvate carboxylase and the utilization of glucose. Liver arginase has been claimed to contain manganese and a manganese-porphyrin compound has been reported to be present in erythrocytes. In the form of its salts, manganese is not appreciably toxic when taken by mouth, diets containing 0.1 to 0.2% manganese having no effect when fed to rats; higher levels interfered with phosphate retention. High levels of manganese added to iron-deficient diets further depressed the iron uptake and utilization by rabbits, piglets and lambs.

Of greater interest is the chronic manganese poisoning found as an occupational hazard among miners in Chile. A case has also been reported in a crane driver operating above furnaces of molten manganese steel. Oxide dust enters the lungs and the gastrointestinal tract, giving rise to a disease *(locura manganica)* which resembles schizophrenia. This is followed by a permanently crippling neurological disorder clinically similar to Parkinson's disease, and showing tremor and impaired muscular coordination. The main site of the lesion is in the extrapyramidal motor system and is associated with decreased brain dopamine levels and disturbance of norepinephrine metabolism, arising from selective damage to dopaminergic neurones. Some of the neurological signs can be alleviated with L-dopa, the precursor of dopamine, but the patients regress to their initial condition when therapy ceases. Workers in the dry battery industry who are exposed for a long time to dust high in manganese dioxide may also develop the same condition.

Molybdenum

Molybdenum is involved in biological processes as part of the prosthetic group of the flavoprotein enzymes xanthine oxidase and nitrite reductase, and its effect on legume growth by stimulating the activity of nitrogen-fixing organisms is well known. The toxic effects of molybdenum on animals arise from a sensitive interrelation of dietary levels of copper, molybdenum and sulphate. For many years cattle grazed on certain areas in Somerset contracted a disease, teart, which was characterized by severe scouring and loss of condition. Similar effects were found in parts of California and New Zealand. In the teart area, soils contained up to 0.01% molybdenum, while pastures grown on them had molybdenum levels of 20 to 100 ppm dry matter as against 1 to 3 ppm in normal pasture. Treatment with copper sulphate, by mouth or by injection, quickly controlled the disease.

This copper-molybdenum antagonism is also shown by the ability of molybdenum to counteract the chronic copper poisoning of sheep in Australia, and by the tendency of pastures low in molybdenum to favor accumulation of copper in the tissues of sheep. A high level of inorganic sulphate in the diets of sheep, cattle and rats leads to a lower molybdenum retention in the tissues, by reducing interstitial absorption and increasing urinary excretion. Alleviation of molybdenum toxicity in the sheep is not helped by malonate, citrate, phosphate or a number of other anions, but diets rich in methionine or cystine can be beneficial because of oxidation to sulphate in the body.

Recent studies suggest that copper does not affect the intestinal absorption of molybdenum, but a copper : molybdenum complex is formed which inhibits copper utilization, so that caeruloplasmin activity is sharply reduced. This may also arise from impaired caeruloplasmin synthesis by the liver, resulting from the apparent interference by molybdenum in the transfer of copper from the blood to the liver. Other effects of molybdenum toxicity have been shown in rats to be a decrease in alkaline phosphatase activity in the liver, with increases in the kidney and intestine, and a depression of liver sulphide oxidase activity.

Selenium

Probably because it is the least abundant and the most toxic element known to be essential for mammals, the toxicity of selenium was established a long time before it was found that selenium, in traces, was also an essential element for chickens, lambs, calves and rats. Thus, adequate dietary intakes of selenium prevent liver necrosis in rats, exudative diathesis in chickens and muscular dystrophy (white muscle disease) in lambs and calves. Treatment of white muscle disease by injecting soluble selenium salts has caused some accidental deaths in livestock.

Two diseases of livestock occurring in the midwestern United States and known as "blind staggers" and "alkali disease" have been shown to be due to acute and chronic selenium poisoning, respectively. Although its cause was not known, a similar disease was reported by Marco Polo near the city of Su-chan in Tangut and was also described in Colombia in 1560. In the arid Midwest, certain plants, particularly of the *Astragalus* species, require selenium for growth, and they accumulate it in such concentrations as to constitute a hazard to grazing stock.

Cattle, pigs and horses eating seleniferous plants lose hair and hooves, probably because of displacement of sulphur in keratin by selenium. Other effects include serious hepatic cirrhosis and heart injury. Ingestion of water-soluble selenium compounds leads to focal necrosis of the liver, haemorrhagic areas in the viscera and general weakness. Selenium toxicity leads to loss of fertility and congenital malformation. This teratogenic effect is consistent with its reported carcinogenic activity under certain conditions of diet (the feeding of seleniferous grain) and environment, but there is no evidence that selenium in seleniferous areas has produced any tumors in humans or livestock. In fact there is the possibility that incorporation of selenium into cystine might inhibit the development of tumors and leukaemia.

Toxicity of selenium in humans is rare, being found only in persons living in seleniferous areas and consuming local food. Effects include discolored and decayed teeth, yellow skin color, chronic arthritis, atrophic brittle nails, edema, gastrointestinal disorders and a "garlic" breath.

Selenium can be incorporated into at least six sulphur-containing amino acids, and hence in chronic selenium poisoning it accumulates in hair, hoof and other parts of the organism that are rich in sulphur. For the same reason, the presence of sulphur derivatives in the diet reduces selenium toxicity. In a way not yet understood, methionine protects against moderately high levels of selenium in rats provided adequate levels of vitamin E or certain fat-soluble antioxidants are also present.

Zinc

Toxic effects due to zinc are rare, but a high intake in the diets of animals can exert an antagonistic effect on other minerals, leading to secondary deficiencies of copper, iron and calcium, and resulting in anaemia and osteoporosis. High oral dosage in humans can produce vomiting and diarrhoea, the emetic dose being 220 to 450 mg. This is a transitory condition and may result from the preparation of acidic food in galvanized containers. A more serious hazard is "metal fume fever" or "brass founder's ague" which arises from exposure to zinc oxide fumes and which may occur wherever zinc oxide is a product, e.g. in zinc smelting, brass founding, the manufacture of zinc oxide or zinc powder and in the welding of galvanized iron. The condition is accompanied by pulmonary distress, fever and chills. The effect is not cumulative; the disease is self-limiting and it is clearly more desirable to prevent the disease than to treat it. The same general remarks apply to the inhalation of magnesium oxide fumes. However, a similar condition arising from exposure to cadmium oxide fumes may produce a severe reaction and, possibly, death.

LESS COMMON TOXIC METAL IONS

Technological developments have led to an increasingly wide range of metals coming into extensive use, and to changes in the applications of more familiar metals. Examples include the use of beryllium in light-weight and heat-resistant alloys and, until recently, as phosphors, of thallium compounds as rat killers and hair removers, and of tellurium in the electronic industry.

Hypersensitivity reactions to metals are sometimes due to metals other than those that appear to be involved: recently allergic reactions produced by gold alloy rings were found to be due to nickel added as a hardener. It is probably true to say that all metals currently find some applications in industry or commerce, and that most of these are toxic if ingested or injected at sufficiently high levels. Topical application may also have effects, for example the production of skin cancers as a hypersensitivity reaction following the use of zirconium lactate in deodorants. However, to keep the present chapter to a reasonable size, only a limited selection can be discussed. The increasing use of atomic power and radio isotopes, and the continuation of atomic bomb testing has also raised the level of risk of contamination for people in general, posing unique problems associated with continued irradiation of living tissue.

Antimony

The main uses of antimony are in the manufacture of alloys, such as type metal, Britannia metal, pewter and antifriction metal, and of sodium and potassium antimonyl tartrate and organic antimony compounds, for use in dyeing and medicine. Antimony and its compounds are irritant poisons, causing vomiting and purging. Chronic poisoning is characterized by nausea, progressive weakness, and may lead to death from exhaustion. Inhalation of dust containing antimony or antimony sesquioxide is reported to be much more toxic, leading to cell injury, than is the ingestion of these materials. Although recorded cases of antimony poisoning in industry are rare, such an occupational hazard should not be ignored.

Contamination of food by antimony can arise from contact with vulcanized rubber and from alloys in food processing equipment, or from defective enamelware. Antimony is used in such ware to give a more opaque enamel, permitting a thinner application; in poor enamel, however, antimony may be leached out by acid foods or foods containing complexing agents.

Currently, more interest has been taken in the use of trivalent antimony compounds such as the antimonyl tartrates, antimony pyrocatechol sodium disulphonate and antimony dimercaptosuccinate to treat sufferers from schistosomiasis. These treatments are potentially hazardous because antimony accumulates in the body, especially in the liver, and a small proportion of the patients die from necrosis of the liver. During treatment, serum glutamic oxaloacetic transaminase and serum glutamic pyruvic transaminase increase significantly, apparently indicating hepatotoxicity. Heart muscle damage can also occur, so that abnormal cardiograms may be found during therapy, but recovery follows cessation of treatment. However, it may take several months before the elimination of antimony from the body is effectively complete.

2,3-Dimercaptopropanol is probably the most useful chelating agent for treating antimony poisoning.

Arsenic

In 1965, fourteen workers in a Swedish metal refining plant suffered poisoning by accidental inhalation of arsine, AsH_3, probably resulting from the action of water on aluminium arsenide. Although such poisoning is rare, mortality in affected cases was formerly high because rapid haemolysis and nephrotoxic effects led to kidney damage and anuria. The availability of dialysis equipment now enables affected patients to be given repeated haemodialysis until kidney function is restored.

However, the most common causes of arsenic poisoning are the accidental ingestion of pesticides and rodenticides containing arsenious oxide, Paris green or calcium or lead arsenate. Formerly, other hazards included arsenic compounds in dyes and wallpaper (where, under damp conditions, mold growth could lead to the liberation of volatile arsenic compounds). Arsenic poisoning, especially as a dermatitis, occasionally resulted from the use of organic arsenicals in prepenicillin therapies of venereal diseases.

After absorption, arsenic is deposited in the liver, spleen, kidneys and keratin-rich tissues. In acute cases, nausea, vomiting and diarrhoea occur suddenly and may be followed by collapse and death. Survivors develop, after some days or weeks, changes in skin (dermatitis) and fingernails (transverse white striae), and nerve damage becomes apparent as a tingling and numbness in the feet and hands, leading to weakness and muscular pain. Liver necrosis and brain damage may also occur. Chronic exposure leads to the same result but gastrointestinal effects may be absent. Because of the nonspecific nature of the disease the most reliable diagnosis is based on an arsenic content exceeding 0.1 mg in a twenty-four-hour urine sample. Recovery may take some months but is not always complete.

Arsenic acts by binding through sulphydryl groups of proteins and enzymes, and this is why it is stored in hair and other tissues. The peripheral neuropathy associated with arsenic intoxication is believed to be due to the interaction of arsenic with pyruvate oxidase, thus blocking the supply of energy from the breakdown of carbohydrates to the nerve cell; the ability of trivalent arsenic to act in this way has been shown in isolated enzyme systems. Prolonged treatment with 2,3-dimercaptopropanol, by intramuscular injection, chelates and removes arsenic from the body, and can actually reverse the inhibition of pyruvate kinase, but it does not lead to an accelerated recovery of sensory or motor function. For this reason, treatment should begin as soon as possible after a positive diagnosis of arsenic poisoning, if neurological effects are to be kept to a minimum.

While alkyl mercury compounds are much more poisonous than inorganic mercury salts, alkylation apparently suppresses the toxicity of arsenic. Levels of arsenic up to 200 ppm dry weight have been found, as $As(CH_3)_3$, in shrimp.

Beryllium

To meet the demands of modern technology, beryllium ore production has risen from 1,000 tons in 1938 to a current level of about 20,000 tons. Awareness of the toxicity of beryllium compounds developed quickly in the early 1940's when a number of young women engaged in fluorescent lamp manufacture suddenly developed a pneumonia-like condition with, in some cases, a fatal pulmonary edema. This was found to be due to the inhalation of beryllium phosphor powder. Since 1948 this material has not been used in fluorescent lamps, but applications of beryllium in other industries have increased so that people at risk include those engaged in the extraction of beryllium ores, those who live near factories concerned with refining beryllium, and those employed in making or machining the lightweight, hard and heat-resistant beryllium alloys, e.g. with aluminium, copper or zinc that are used in cathode ray tubes, X-ray machine windows, atomic energy units and aircraft. Beryllium is also used in ceramics.

Acute beryllium poisoning, nonspecific chemical pneumonitis, results primarily from inhalation exposure and thus primarily affects the lungs. However, beryllium is a systemic poison and other organs are affected, including the production of a skin hypersensitivity. Chronic beryllium poisoning is characterized by a cancer-like epithelioid cell granuloma found in the interstitial lung tissue, in other organs and in hilar lymph nodes. They are histologically indistinguishable from granulomas found in other diseases, including a type of tuberculosis, so that criteria for diagnosis depend on known exposure and a positive patch test, with confirmation being sought by chemical analysis of tissues and urine.

Beryllium can persist in the tissues for a long time. The length of exposure prior to the development of disease has been up to twenty-three years, and lesions may persist for over ten years. There is a high fatality rate and no satisfactory method of treatment.

Gold

A current method of treating rheumatoid arthritis is by the injection of sodium aurothiomalate and other sulphur-containing gold complexes. Many modes of action have been suggested, including the inhibition of responses to histamine and serotonin-releasing substance, the stabilization of lysosomes, retardation of the formation of disulphide bonds thereby preventing the formation of abnormal disulphide-linked proteins, and reaction with collagen. Long-term dosage with gold salts can lead to eczematous rashes, dermatitis and other undesirable reactions including aplastic anaemia, proteinuria and renal toxicity: The percentage of toxic reactions to injections is reported to lie between 5 and 62 percent. Acute reactions to gold tend to clear spontaneously when administration of gold ceases. If they fail to do so, D-penicillamine given orally appears to be suitable treatment.

Nickel

Nickel salts, ingested by mouth, are poorly absorbed and are relatively nontoxic. Thus, a diet containing 0.1% has been fed to rats without affecting their growth rates or reproduction. However, above a level of 0.07% nickel in the diet of chickens there was some depression of growth and impairment of nitrogen retention. High levels of nickel carbonate fed to calves produced nephritis.

The toxicity of nickel manifests itself in other ways. Allergic contact dermatitis from external response to metal is very commonly seen with nickel and, especially in the nickel alloys used in jewelry including nickel silver and some types of hardened gold. A case has been recorded of eczematous dermatitis induced by a nickel-containing orthopaedic metal implant. (Similar effects are sometimes produced by contact with chromate and mercury.) Possibly related to this physiological response, nickel and its compounds have been shown to be carcinogens, leading to a high incidence of lung cancer and nasal sinus cancer among people working in nickel mines and refineries. These lesions, which have been recognized as occupational hazards of nickel workers, may develop many years after exposure ceases. As with nickel, exposure to chromate dust can also lead to tumor production, and there is a substantial hazard of both types of cancer in men employed in dusty occupations such as furnace work in refineries. Subcutaneous or intramuscular injection of a suspension of nickel powder leads to the production of tumors.

Nickel carbonyl, $Ni(CO)_4$, boils at 43° and has a high vapor pressure at room temperature. It is highly toxic and may be encountered industrially in the separation of nickel from its ores, in the preparation of intermediates in organic synthesis, in electroplating and, inadvertently, by the action of carbon monoxide on active nickel. The initial symptoms following inhalation are nonspecific (headache, coughing, giddiness) and may disappear quickly. Severe symptoms develop insidiously, hours or days later, possibly terminating in delirium and convulsions. Chronic exposure to nickel carbonyl greatly increases the risk of developing carcinoma of the lungs and respiratory passages. Cases of recent exposure to nickel carbonyl have been treated orally with sodium diethyldithiocarbamate for some days, giving an increased output of nickel in the urine and minimizing delayed reaction; calcium disodium EDTA was reported to be ineffective.

In the light of these facts it has been suggested that heavy smoking of cigarettes could transfer significant amounts of nickel as the carbonyl compound to the lungs. It is also possibly undesirable to use nickel chelates for topical application because of the risk of carcinogenic activity.

Silver

The absorption of silver salts can lead to argyria, characterized by a permanent bluish or black discoloration of skin and tissues. A famous example was the "blue man" of Barnum and Bailey's Circus. Although the effect may be unsightly there seems to be little evidence of toxicity. High concentrations of silver ion can lead to skin irritation. However, with the lessened therapeutic application of silver compounds and the practice of preventive industrial medicine, both of these conditions have become much less common.

Recently, the treatment of extensive body burns with 0.5% silver nitrate dressings has been found to lead, sometimes, to methaemoglobinaemia (in which the iron is oxidized from the normal ferrous to the ferric form), with consequent decrease in the ease of oxygen transport by the blood. This effect is probably due to nitrite ion produced by nitrate-reducing bacteria, and not to silver.

Tellurium

Tellurium is used in the rubber, electronics and metallurgical industries, and workers exposed to tellurium oxide sludge or fumes develop a garlic odor of the breath, sweat and urine (due to dimethyl telluride), dryness of the mouth and skin, a metallic taste and somnolence. Toxicity is related to the formation of soluble tellurium compounds; elemental tellurium which is stored in the tissues is much less toxic. Successful treatment with BAL has been reported.

Fed as the tetrachloride, tellurium was acutely toxic to ducks at dietary levels of 0.05 and 0.1%, death occurring within two weeks from myocardial haemorrhage and necrosis. Necrotic lesions were also found in the brain and liver. Tellurium compounds can cross the blood-brain and placental barriers, so that elemental tellurium fed to gestating mammals can result in hydrocephaly in the offspring. In weanling rats on diets containing elemental tellurium there was degeneration of Schwann cells and a demyelination that led to paralysis of the hind legs. Older rats fed tellurium dioxide developed necrosis of the liver and degeneration of kidney tubule epithelium.

Thallium

The use of thallium acetate as a depilatory and of thallium sulphate as a rodent exterminator is potentially hazardous and has led to human toxicities and occasional deafness. Thus, the ingestion of 0.2 to 1 g of tasteless, odorless thallium sulphate can be lethal, and for this reason the use of thallium in rodenticides and depilatory agents is now restricted in the United States. Thallium oxide is an ingredient in some optical glass manufacture.

Thallium is readily absorbed through the skin, the gastrointestinal tract and parenterally, and it is firmly bound to most tissues, especially the liver, kidney and intestine, so that it is a cumulative poison and may persist in the body for weeks. It apparently becomes fixed on sulphydryl groups on mitochondrial membranes causing loss of cristae and an increase in size and density of mitochondrial granules. Inhibition of oxidative phosphorylation and interference with the metabolism of sulphur-containing amino acids has been demonstrated.

The symptoms of thallium poisoning are not initially specific. Acute thallium poisoning leads to nausea, vomiting, diarrhoea and gastrointestinal haemorrhage, giving place within ten days to pain and tremor. Disturbance of cerebral activity is common. In survivors, the most striking feature is severe loss of hair after about two to three weeks, probably as a consequence of thallium accumulation in hair follicles interfering with the formation of keratin. Recovery may be complete, or there may be residual effects such as tremor, optic atrophy and mental abnormalities. These effects are much more marked in cases of chronic thallium poisoning.

Diagnosis is difficult before hair-shedding occurs, but one possible indicator is the appearance of black pigment at the base of the hair several days after ingestion of thallium. If thallium poisoning is suspected early enough, gastric lavage with sodium iodide to precipitate insoluble thallous iodide may help prevent its absorption. Thallium poisoning has been successfully treated by giving patients activated charcoal (to increase the elimination of thallium in the feces), potassium chloride (to assist in mobilization of intracellular thallium) and dithizone (to chelate the thallium and increase its excretion in urine and feces). Haemodialysis may be helpful if there is kidney damage.

The earlier members of Group IIIb (gallium and indium) are not appreciably toxic when given by mouth. Thus, although injection of gallium salts causes renal injury and neuromuscular toxicity, levels of gallium salts up to at least 0.1% of the diet are not readily absorbed from the gastrointestinal tract.

Tin

There is little danger from inorganic tin compounds such as might be added to food by possible contamination from cans. This is because of the difficulty of introducing tin into the body in soluble form. However, in the possible industrial use of volatile and water-soluble organic tin compounds it must be remembered that on a molar basis tin in the form of its alkyl derivatives was found to be five times as toxic as beryllium. In the mid-1950's over 100 deaths in France followed the treatment of carbuncles by oral dosage of diethyl tin diiodide dissolved in linoleic acid ("Stalinon"). In both acute and chronic poisoning, triethyl tin salts produce muscular weakness and affect mainly the central nervous system, its most serious action in the brain being to decrease the rate of oxidation of pyruvate derived from glucose. In experimental animals it produced cerebral edema and splitting of the myelin sheaths around nerve fibers. Acute poisoning can lead to convulsions and death. Tetraalkyl tin is not initially very toxic, but becomes so after partial degrada-

tion in the body. Medication with BAL antagonized the effects of dialkyl tin compounds, but did not influence the response to trialkyl tin. A naturally occurring polyhydroxyphenyl-chromone, silymarin, from a South African plant, *Silybum marianum*, is reported to prevent functional deterioration of the central nervous system induced by triethyl tin sulphate poisoning, probably because of its chelating ability.

Radioactive Metals

Hazards in the use of radium were pointed out dramatically in 1924 by the observation that young women who painted dials with luminous paint, and who "tipped" their brushes with their lips, were liable to develop osteomyelitis of the jaw, leading to malignancies in the bone and bone marrow. Similar damage throughout the body can result from the uptake of ^{90}Sr, which is a common constituent of radioactive fallout, ^{140}Ba and Pu. In these cases, the cations exchange with bone calcium so that they can be retained for long periods in the body.

The main radioelements produced in nuclear explosions are fission products of uranium and plutonium, together with bomb constructional materials that have been activated by neutron irradiation. Fallout materials that may be absorbed via the food chain or by inhalation is limited among the metals to ^{90}Sr, ^{89}Sr and ^{137}Cs as having long enough half-lives and sufficient biological retention times to be potentially dangerous.

Otherwise, accidental intake of radioactive materials is likely to be mainly by inhalation, which may involve either the deposition of insoluble dusts in the lung passages or the uptake of soluble dusts which are rapidly absorbed into the blood stream. Thus, a biological half-life of 500 days has been found for the retention of PuO_2 particles in the mouse, and up to 140 days for radium sulphate in man. Sometimes there is translocation of insoluble particles from the lungs to lymph nodes where they are retained for a long time: Uranium oxide particles are localized in this way.

The uptake of radioisotopes by particular tissues or organs is familiar from the medical use of ^{131}I to assess thyroid function and from the example of strontium and barium retention by bone. Although a general discussion of radiation damage is outside the scope of the present account, it can be seen that such localization increases the risk of radiation damage to particular areas of the body and makes it more difficult to remove the radioactive material from the body. There may also be a well-defined distribution sequence. When ^{59}Fe enters the blood plasma it is rapidly cleared to bone marrow and it then subsequently appears in circulating erythrocytes, finally coming to equilibrium with various metabolic and storage pools.

The main problem in treating radioisotope toxicities lies in the need for rapid and effectively complete removal of the active material because the toxic effects of radioactive metals on living organisms are the result of the emitted ionizing radiation.

For example, the maximum amount of plutonium in the body should not exceed 1.5 μg, making it one of the most toxic materials known to man. Accidental contamination of workers involved in the manufacture or use of Pu can result in levels substantially higher than this.

Treatment of body surfaces to remove any Pu that may be present is straightforward and involves copious washing with water, detergents and chelating agents such as DTPA or EDTA. If soluble Pu salts have been swallowed or inhaled, injection and oral administration of Zn-DTPA, $CaNa_2EDTA$ or $CaNa_3$ diethylenetriaminepentaacetate may lower the body's burden of Pu sufficiently by excretion in the urine. Similar injections may be advantageous following the flesh deposition of ^{239}Pu. Early treatment is essential if immobilization of Pu in bone is to be kept to a minimum. Where insoluble Pu-containing material is driven into tissue or inhaled a much graver problem exists. So long as ^{239}Pu is present, irrespective of its chemical form, radiation damage due to its emitted α-particles will continue, with the risk of malignancy developing in bone, liver, lungs, lymph nodes and tissue around the Pu. If Pu contamination is severe, surgical excision may be the only effective treatment.

There appears to be little risk of radiation damage from uranium; earlier reports of lung cancer among uranium miners were probably due to prolonged inhalation of traces of radon.

THE USE OF CHELATING AGENTS TO TREAT METAL POISONING

Most of the biological effects of toxic metal ions are probably due to the binding of these metals to metabolically important groups, often with the displacement of an essential trace metal ion. Hence, there is a good chance that, given a more powerful complexing agent, an undesirable metal ion might be able to be removed from the site at which it is producing a "biochemical lesion." The situation is complicated by the tendency of some heavy metals to form "robust" low-spin complexes that do not readily undergo reversible dissociation but, in principle, factors underlying the design or selection of suitable chelating agents for therapeutic purposes are probably better understood than for any other group of synthetic drugs. Nevertheless, there may frequently be unexpected side effects as shown, for example, by the much greater toxicity of L-penicillamine than of its D-isomer, or of trisaminoethylamine relative to triethylenetetramine. Nor is successful treatment to be expected in cases of pulmonary disease arising from inhalation of insoluble metallic oxides which tend to accumulate in lymph nodes and may lead to cancerous growths. Again, the lack of selectivity of most chelating agents for metal ions can result in induced deficiencies of essential metal ions such as zinc, especially if treatment is prolonged: This difficulty can be overcome by administering supplementary metal ions to restore balances. Some metal ions, such as those of aluminium and titanium, are so extensively hydrolysed under physiological

conditions that they are not readily absorbed from the digestive tract and hence are not toxic, but in many other cases physiological conditions limit the rate of absorption of an inorganic species into the human body and its distribution among the tissues of the body.

The chelating agents currently of choice in dealing with metal toxicities are summarized in Table 20-I.

Choice of Ligand

Chelation therapy, then, is concerned with the use of appropriate chelating agents for the controlled removal of undesirable metal ions, whether these are adventitious or essential cations present in excess. Modern chemical theory and extensive compilations of stability constants provide a sound basis for choosing likely ligands which will greatly diminish the concentration of the free metal ion and

TABLE 20-I

SELECTED LIGANDS FOR TREATMENT OF METAL POISONING

Metal	Suggested Ligand
Antimony	BAL
Arsenic	BAL
Beryllium	—
Cadmium	—
Cobalt	CaNa$_2$EDTA
Copper	Penicillamine, triethylenetetramine, BAL
Gold	Penicillamine, BAL
Iron	Desferrioxamine
Lead	(Acute) CaNa$_2$EDTA, (chronic) penicillamine
Manganese	(Acute) CaNa$_2$EDTA (?)
Mercury	(Inorganic) BAL, N-Acetylpenicillamine; (alkyl) thiol resin (?)
Molybdenum	—
Nickel	(As Ni(CO)$_4$) sodium diethyldithiocarbamate
Plutonium	CaNa$_3$DTPA
Selenium	—
Silver	—
Tellurium	BAL
Thallium	Dithizone
Tin	(Dialkyls) BAL, (trialkyls) silymarin
Zinc	CaNa$_2$EDTA

Dashes indicate no suitable chelation therapy currently available.

favor competitive displacement from its biological binding sites. An essential, but not sufficient, requirement is that the metal complex which is to be excreted should have a high conditional stability constant. A useful qualitative guide is given by the concept of "hard" and "soft" ligands and metal ions, with the strongest complexes being formed between pairs having comparable "hardness." Some of the more common cations and ligands are classified in this way in Table 20-II. Another important factor is the stereochemistry of a metal ion in its complexes. For many metal ions the preferred structures are octahedral, but in other cases square planar, tetrahedral or linear complexes are involved. For example, the resulting square planar configuration favors complex formation of copper(II) with triethylenetetramine rather than with trisaminoethylamine (where this structure is not possible).

The five chelating agents most commonly used as such in medicine fall into clearly defined types, reflecting the natures of cations they are intended to remove. Depending on the properties of the chelating agents, including water solubility, resistance to metabolic degradation, and ease of transfer across membranes, and the speed with which they must act, administration may be oral, subcutaneous, intraperitoneal, intravenous or by inhalation. Usually, the initial result of treatment is the rapid excretion as the metal chelate of much of the toxic metal present in the blood and the soft tissues. Subsequently there is a much slower rate of removal dependent on the rate at which the metal redistributes itself within the body. For efficient metal ion removal, the ligand concentrations must be high relative to the metal.

2,3-Dimercaptopropanol. The arsenical poison gas Lewisite, $ClCH = CHAsCl_2$, reacts with essential $-SH$ groups of enzymes such as succinoxidase and pyruvic oxidase, thereby preventing their action.

TABLE 20-II

"HARD" AND "SOFT" METAL IONS AND LIGANDS

Hard												
Li	Be											
Na	Mg	Al										
K	Ca	Sc	Ti	V	Cr(III)	Mn(II)	Fe(III)	Co(III)	Ga	As		
Rb	Sr	Y	Zr	Mo					In			
Cs	Ba	La	Hf									
			Th									
			U, and rare earths									
H_2O	OH^-	$RCOO^-$	PO_4^{3-}	SO_4^{2-}	CO_3^{2-}	NO_3^-	ROH	RO^-	R_2O	F^-	Cl^-	

Intermediate						
	Fe(II)	Co(II)	Ni	Cu(II)	Zn	
	Ru(II)	Rh(III)		Sn(II)	Sb(III)	
	Os(II)	Ir(III)		Pb(II)	Bi	
NH_3	RNH_2 N_2H_4	Br^-	N_3^-	NO_2^-	SO_3^{2-}	pyridine aniline

Soft					
		Cu(I)			
	Pd(II)	Ag(I)	Cd		Te(IV)
	Pt(II, IV)	Au(I)	Hg(I, II)	Ti(I, III)	
R_2S	RSH RS^-	SCN^-	$S_2O_3^{2-}$	R_3P $(RO)_3P$	I^- CN^-

As an antidote to this type of poisoning, the oily liquid 2,3-dimercaptopropanol (Dimercaprol, British Anti-Lewisite, BAL),

$$
\begin{array}{c}
CH_2SH \\
| \\
CHSH \\
| \\
CH_2OH
\end{array}
\qquad (153)
$$

was developed during World War II. It combines with trivalent arsenicals to form stable chelates of the type

$$
\begin{array}{c}
CH_2S \\
| \quad \diagdown \\
\qquad As - R \\
| \quad \diagup \\
CHS \\
| \\
CH_2OH
\end{array}
\qquad (154)
$$

liberating essential —SH groups that are otherwise blocked. Subsequently, BAL was also found to be effective against poisoning by antimony, bismuth, gold and mercury. It is also sometimes used in acute lead poisoning.

Unfortunately, there are difficulties associated with the use of BAL. It has a very unpleasant odor which permeates the tissues of patients, it is easily oxidized and is rapidly degraded in the body if it is not chelated, and it is unstable in aqueous solution so that it has to be used as a 5% solution in arachis oil containing 10% benzyl benzoate. It is injected intramuscularly, along with procaine. This treatment frequently leads to vomiting, nausea and pain. However, solution in oil has the advantages of slower absorption, more pro-

longed action and less danger of exceeding toxic dose levels. BAL can be applied as an ointment because it is absorbed through the skin; it has been used in this way to treat dermatitis due to arsenic and gold, as well as chromium ulcers of the hands in chrome-plating factories.

Dimercaptopropanol has been used in the treatment of acute copper poisoning and of Wilson's disease, but it is doubtful if its ability to chelate copper more strongly than D-penicillamine offsets its disadvantages. In an endeavor to overcome these, many other potentially useful dithiols have been synthesized, including unithiol (2,3-dimercaptopropanesulphonic acid) which has improved solubility in water. Disodium dimercaptosuccinate is less toxic and has been injected to treat arsenic and mercury poisoning; it has also been suggested for use as a detoxicant for cadmium.

Similar complexing agents in this group include dihydrothioctic acid, proposed as an alternative to BAL for treating arsenical poisoning, and β-mercaptoethylamine, injected intravenously, for treating thallium poisoning. Many sulphur-containing ligands are possible antidotes for mercury(II) poisoning, including potassium methyl- and ethylxanthates, sodium diethyldithiocarbamate, cysteine, β-mercaptoethylamine and thioctic acid.

D-Penicillamine. The use of D-penicillamine (Distamine®, Cuprimine®) as a che-

$$
\begin{array}{c}
\quad\quad CH_3 \quad H \\
\quad\quad | \quad\quad | \\
CH_3 - C - C - COOH \\
\quad\quad | \quad\quad | \\
\quad\quad SH \quad NH_2
\end{array}
\qquad (155)
$$

lating agent to promote the removal of copper from patients with Wilson's disease followed the observation that penicillamine was more stable in the body than was cysteine. The nonphysiological D-isomer is used clinically, not the natural L-isomer derived from penicillin, because it is less toxic. (L-Penicillamine has an antipyridoxine effect.) It has the advantages of being water-soluble and stable, can be given by mouth, is tolerated in larger doses than is BAL, and is excreted rapidly. Patients with Wilson's disease have been maintained in good health for some years by penicillamine therapy. After the initial treatment with CaNa$_2$EDTA to remove the bulk of the lead present in the soft tissues, penicillamine is suitable for the prolonged administration necessary to lower the lead burden in the bodies of chronically lead poisoned people in consequence of the slow redistribution of lead from less accessible stores.

N-Acetylpenicillamine has been proposed for use in treating mercury poisoning.

EDTA and Its Derivatives

The aminopolycarboxylic acids that comprise this group are much less selective in their metal-chelating tendencies than are sulphur- and nitrogen-types of ligands. They are not readily absorbed when taken by mouth, and hence are normally administered by intravenous injection. Nor are they readily metabolized, but removal via the kidney is rapid. Injection of free EDTA or its sodium salts is potentially hazardous

$$^-OOC-CH_2 \quad\quad\quad CH_2-COO^-$$
$$\underset{^-OOC-CH_2}{\overset{}{\diagdown}} \overset{+}{HN}-CH_2-CH_2-\overset{+}{NH} \diagup \underset{CH_2-COO^-}{\overset{}{}}$$

$$(156)$$

because of the risk of depleting blood calcium levels enough to induce tetany. This difficulty can be avoided by using CaNa$_2$

EDTA which does not disturb body calcium levels but readily exchanges its calcium for lead or other heavy metal ions, forming water-soluble complexes that are readily excreted. Intravenous injection of CaNa$_2$EDTA in man up to 3 g/day is usually without untoward effects, but it should not be used if renal damage is present.

Lead poisoning is treated by injecting CaNa$_2$EDTA and, sometimes, BAL, so that there is a rapid depletion of high lead levels in soft tissues and adverse metabolic effects of lead are suppressed. The combined use of these two chelates can reduce substantially the mortality rate from acute lead encephalopathy in children. The lead chelates are excreted through the kidney and liver. No therapy can restore any brain damage or scarring of kidney tissue caused by the lead.

Use of EDTA salts has been shown to prevent cobalt intoxication in animals, raising the possibility that this might be a useful treatment in cases of cobalt poisoning. Similarly, although EDTA therapy is ineffective in patients with chronic manganese intoxication, it might well be valuable in the initial stages of manganese poisoning.

The slight extent to which EDTA is absorbed from the gut makes it useful in decreasing the availability of metal ions present in the gastrointestinal tract. Addition of EDTA to the diet has masked magnesium and calcium in the rumen, thereby preventing bloat in ruminants. It has overcome growth depression in turkeys arising from high uptakes of manganese, zinc and copper, and prevented the toxic effects of vanadium in a diet fed to chickens. As its sodium salt, EDTA has been used to chelate calcium in treating lime burns of the cornea.

Attempts to improve the chelation possibilities of EDTA have included the synthesis of related compounds such as cyclohexane-1,2-diaminotetraacetic acid (CDTA), triethylenetetraminehexaacetic acid (TTHA) and diethylenetriaminepentaacetic acid (DTPA). Claims as to their relative effec-

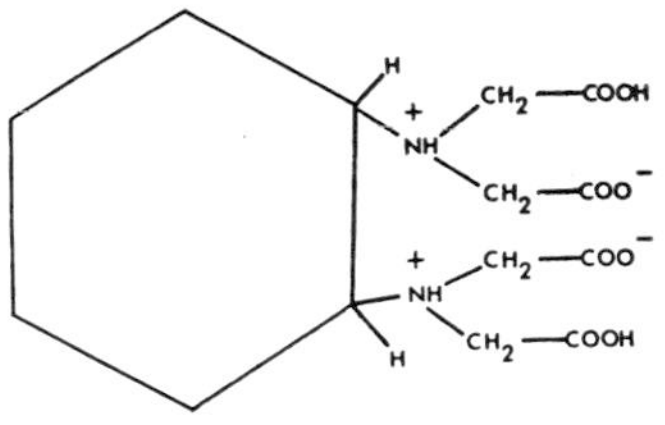

(CDTA)

(157)

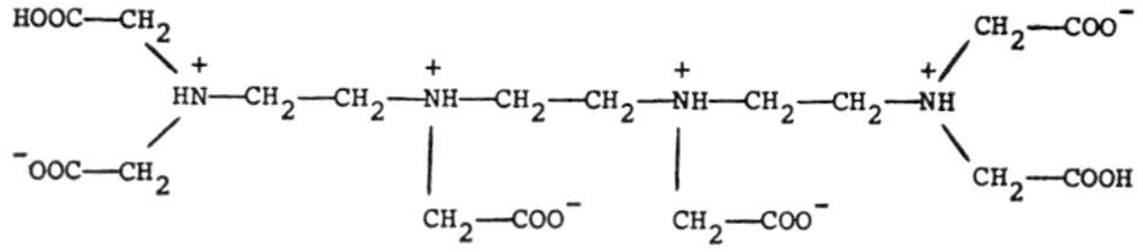

(TTHA)

(158)

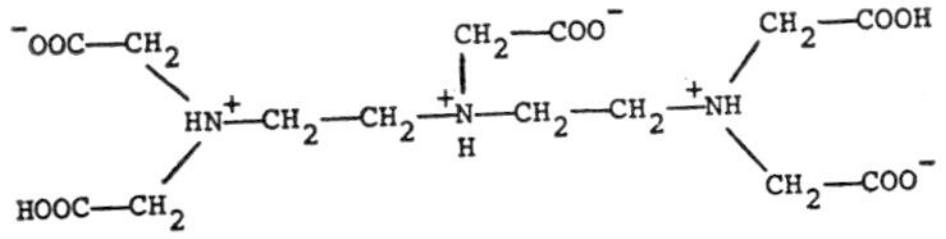

(DTPA)

(159)

tiveness are often conflicting but CaNa$_3$ DTPA is reported to be better than CaNa$_2$ EDTA for intravenous injection to bind iron in acute ferrous sulphate poisoning and, perhaps, for the removal of lead.

Differences in the relative stability constant values of different metal ions enables EDTA-type ligands to be used for the selective removal of toxic metal ions. Thus the Mg-EDTA complex might be used to lower hypertension by decreasing the blood calcium level by exchanging some of the calcium for magnesium present in the Mg-EDTA complex. This reagent would also

remove other metals such as copper and zinc. Injection of the Ca-EDTA complex would not affect blood calcium or magnesium levels, but would still remove copper and zinc. Hence the use of Zn- and Mn-EDTA complexes, injected into the blood of rats, facilitated the excretion of radio-isotopes of zinc and manganese. Sometimes the differences in binding abilities for different metal ions is not great enough to be useful. This is why DTPA, although better than EDTA, is not able satisfactorily to remove radioactive strontium from bone.

Like EDTA, DTPA has been used in diets to diminish the uptake of copper from the alimentary canal.

Desferrioxamine

Ferrioxamines are iron chelates occurring in bacteria; they are believed to be concerned in the enzymatic incorporation of iron into the porphyrin skeleton. Desferrioxamines, derived from them by removal of the iron by treatment with acid, have very strong chelating ability for iron(III) and only a weak affinity for most other cations. They are readily soluble in water, but are only poorly absorbed from the digestive tract. One of these compounds, Desferrioxamine B (Deferoxamine®, Desferal®) is used as its methanesulphonate to regulate

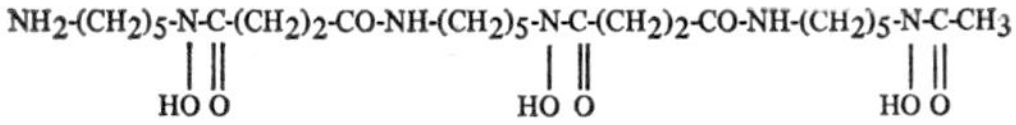

NH$_2$-(CH$_2$)$_5$-N-C-(CH$_2$)$_2$-CO-NH-(CH$_2$)$_5$-N-C-(CH$_2$)$_2$-CO-NH-(CH$_2$)$_5$-N-C-CH$_3$

|| || ||
HO O   HO O   HO O

(160)

dietary iron uptake because, taken orally, it largely blocks the absorption of iron from the alimentary canal. This is the basis of one form of control of idiopathic haemochromatosis, a disease in which excessive amounts of iron are absorbed and stored in the liver, spleen and other organs, but if depletion of these stores is desired a

and parenteral injection is the recommended treatment.

Dithizone

The stability constants of most thallium (I) complexes are rather small, so that the use of BAL or EDTA salts in the treatment of thallium poisoning is ineffective, except at dose levels producing toxic reactions.

(161)

long course of intramuscular injections of desferrioxamine is required. The alternative, and more rapid, treatment is venesection.

Desferrioxamine can attack depot iron in the body and also, to some extent, transport iron but not porphyrin iron. Intramuscular injection or intravenous drip of desferrioxamine depletes iron reserves in the body, slow addition of the ligand being important to avoid the accumulation of toxic concentrations of ferrioxamine by allowing adequate time for excretion via the kidney.

Of the various therapies suggested for the treatment of acute iron poisoning, chelating agents have the greatest theoretical promise for detoxifying absorbed iron and removing it from the body by excretion in the urine. While EDTA and DTPA have been found not to reduce mortality in serious cases, the use of desferrioxamine has undoubtedly led to a great improvement in prognosis. Cases of acute iron poisoning should not be given desferrioxamine by mouth if there is appreciable iron still in the gastrointestinal tract because the resulting ferrioxamine may be absorbed in sufficient amounts to be toxic. Gastric lavage

Although treatment with dithizone does not increase the urinary excretion of thallium it does lead to clinical improvement (and possibly to increased removal of thallium in the feces). A similar removal mechanism is suggested to explain the mode of action of Prussion blue (ferric cyanoferrate) which, given by mouth, can impair the uptake by rats of thallium from the intestine and may be suitable as an antidote immediately following poisoning and as an ion-exchanger to interrupt the enteral circulation of thallium.

Ion-exchange Resins

A similar interruption of the enterohepatic circulation of a toxic metal species by the use of an ion-exchange resin has been suggested as a means of depleting the body burden of methyl mercury. Rats dosed with methylmercuric chloride and then fed a diet containing a nonabsorbable polythiolated chelating resin showed an increased removal of mercury from the body. The therapeutic value of such treatment has not yet been evaluated but it is known that 80 percent of the methylmercury in the bile of mercury-poisoned rats is present as a low-molecular-weight thiol complex which is probably formed by a dipeptide.

Limitations and Dangers, Including Translocation

In some instances, damage done by toxic metal ions is irreversible, so that chelation therapy can, at best, only ameliorate the condition. Examples include brain damage due to lead, mercury or manganese. Thus, crippled exminers gradually clear the manganese they have absorbed during their work, without any improvement of their neurological conditions.

The chelating agent may also be toxic or result in undesirable translocation of metal complexes. In thallium poisoning, treatment with sodium diethyldithiocarbamate increases the excretion rate of thallium but its use may lead to deterioration in general condition and the appearance of central nervous system symptoms. This may be related to the lipid solubility of the thallium diethyldithiocarbamate chelate permitting easier penetration of the central nervous system than for thallium ion and leading to disturbances in cerebral function. This is consistent with clinical observation of patients treated in this way and with the finding of increased levels of thallium in the brains of experimental rats. Similarly, the intraperitoneal injection of sodium diethyldithiocarbamate into mice significantly influenced the distribution of copper in mice and increased its toxicity. The copper content of the brain was increased, probably because of the lipophilic character of the copper-diethyldithiocarbamate complex.

A similar risk is attendant on the use of dimercaptopropanol in treating chronic poisoning with organic mercurial compounds because it may result in increased mercury concentration in brain, liver and muscle. Fairly high levels of dimercaptopropanol

are needed if it is to be effectively in removing thallium; better methods should be sought. Dimercaptopropanol is also unsatisfactory for removing cadmium from the body because its cadmium complex causes kidney damage.

In a small proportion of cases, undesirable side reactions occur in patients with Wilson's disease when they are maintained for a long time on penicillamine therapy, necessitating its discontinuation. Such cases can quite satisfactorily be switched to treatment with triethylenetetramine provided care is taken to ensure that this material is not contaminated with trisaminoethylamine.

Dangers arising from the absorption of ferrioxamine or iron-EDTA from the gut in acute iron poisoning has already been mentioned, and a similar risk is attendant upon the oral administration of penicillamine or $CaNa_2EDTA$ in acute lead poisoning.

The difficulty of achieving selectivity in metal ion removal is well known, and there is the continuing problem that in prolonged treatment with chelating agents deficiencies of essential metal ions might be induced. Thus, differences in the stability constants of zinc and cadmium complexes of currently available ligands are not great enough for suitable treatment of cadmium poisoning by depletion of body cadmium deposits without simultaneous derangement of body zinc reserves. This difficulty could be avoided by administering the zinc complex of a ligand for which the cadmium complex was several powers of ten more stable than the zinc complex.

Computer Evaluation of the Biological Effectiveness of Ligands

A high stability constant is not a sufficient guarantee that a chelating agent will be effective under biological conditions. Thus, although HgEDTA²⁻ has a higher constant than ZnEDTA²⁻, administration of EDTA will remove zinc but not mercury from the body. Many dynamic equilibria exist in the mixtures of metal ions and ligands that are present in living tissues, and to investigate the distribution of metal ions among ligands in such systems, computer programs such as COMICS and HALTAFALL have been developed.

Input information for COMICS comprises the total concentrations of the different metal ions and complexing agents, the pH of the solution and the relevant equilibrium constants (pK_a and stability constant values). This approach has been used to compute the distribution of copper(II) and zinc(II) ions among seventeen amino acids present in human blood plasma, taking account of 159 equilibria.

To illustrate the use of such a computer program, equilibrium concentrations have been computed for the system lead/zinc/histidine/penicillamine using concentrations of zinc and histidine near to those in blood plasma and stability constants taken from the literature. Typical results are shown in Figure 20-1, from which it can be seen that the penicillamine effectively competes with histidine for lead but not for zinc. Mixed-ligand complexes of lead have not been allowed for, but their effect would be to increase further the extent to which lead is complexed.

One of the exciting possibilities of this approach is in the potential ability to model biological systems, particularly if the model is extended to make it a multicompartmental one, and to assess in advance of clinical trials, whether or not any suggested therapeutant would be able under physiological conditions to compete successfully with the ligands already present for the

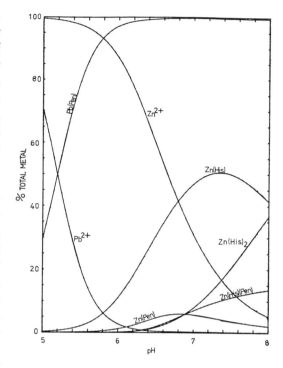

Figure 20-1. pH-Dependence of the computed distribution of zinc(II) and lead(II) in the presence of histidine (His) and D-penicillamine (Pen). Total zinc = 5.10^{-5}M, total lead = 1.10^{-6}M, total histidine = 1.10^{-4}M, total D-penicillamine = 1.10^{-5}M.

metal ion it was desired to remove. This approach has already been used to show that oxidized penicillamine and triethylenetetramine are able to remove copper from a plasma amino acids, copper, zinc mixture, in agreement with their use in treating Wilson's disease. The successful exploitation of this kind of study will require a great proliferation in the amount of available stability constant data, including a major increase in the data for mixed-ligand complexes.

Chemical aspects of metal complex formation are now well enough understood to raise the possibility of synthesizing ligands that form metal complexes with higher stability constants and with a better selectivity for particular kinds of metal ions. Other requirements if a chelating agent is to be of use in medicine are that it must be nontoxic, that it can form a soluble metal chelate that is less toxic than the metal and that can be excreted via the urine or bile, and, preferably, that it is active when given by mouth. Differences in metal-binding abilities among the polypeptides may well make them an attractive group of chelating agents for further study.

REFERENCES

1. Chenoweth, M. B.: Clinical uses of metal-binding drugs. *Clin Pharmacol Ther,* 9:365, 1968.
2. Chisolm, J. J.: Chelating agents in the treatment of lead intoxication in childhood. *J Pediatr,* 73:1, 1968.
3. Chisolm, J. J.: Lead poisoning. *Sci Am,* 224 (2):15, 1971.
4. Flick, D. F., Kraybill, H. F. and Dimitroff, J. M.: Toxic Effects of Cadmium: A Review. *Environ Res,* 4:71, 1971.
5. Goldwater, L. J.: Mercury in the environment. *Sci Am,* 224(5):15, 1971.
6. Goodman, L. S. and Gilman, A. (Eds.): Heavy metals and heavy-metal antagonists. In *The Pharmacological Basis of Therapeutics,* 4th ed. London, Macmillan, 1971, chap 45 and 46.
7. Hallman, P. S., Perrin, D. D. and Watt, A. E.: The computed distribution of copper(II) and zinc(II) ions among seventeen amino acids present in human blood plasma. *Biochem J, 121*:549, 1971.
8. Nelson, N. et al.: Hazards of mercury. Special Report to the Secretary, Pesticide Advisory Committee. Washington, D.C. United States Department of Health, Education and Welfare. *Environ Res,* 4:1, 1971.
9. Passow, H., Rothstein, A. and Clarkson, T. W.: The general pharmacology of the heavy metals. *Pharmacol Revs, 13*:185, 1961.
10. Perrin, D. D.: Biological applications. In *Masking and Demasking of Chemical Reactions.* New York, Wiley-Interscience, 1970, chapter 13.

CONCLUSIONS

David R. Williams

INTRODUCTION

As PORTRAYED in the aforegoing chapters, the fate of bio-inorganic chemistry is somewhat precariously poised between the disciplines of biochemistry, inorganic chemistry and medicine. It is the task of this chapter to suggest directions in which this balanced situation may lean.

Some apologies seem in order here: (1) Our introduction to bio-inorganic chemistry may not be acceptable to the conventionally minded reader; we have not protracted the introduction by mentioning the "particle-in-a-box origin of theories of bonding" nor have we introduced metallotherapy of cancer from a "DNA into protein sequence translation" discussion. These omissions have been intentional: As an analogy we may note that it is quite exciting and rewarding to master the technique of weighing without, in the first instance, needing to understand the principles of gravitation. (2) Some topics have been omitted partly because these topics have not yet reached a degree of maturity such that a simple survey at the level of the present book is capable of achieving anything constructive. (3) The reader will probably find gaps in the correlations attempted in the previous chapters. If these gaps promote further research or interest then our efforts have been worthwhile.

THE CHALLENGES

Bio-inorganic chemistry offers a challenge to many disciplines: Scientists have a particular involvement in the health and future of *Homo sapiens*. Everyone must age and, of course, life is finite. Nevertheless, *biochemists* may well be able to control the pattern of aging and so the quality of life may well be metal *ion* or *complex* controllable just as the *pure* metals copper, silver and gold have controlled it for many centuries. *Inorganic chemists* can either condemn the human race to pollution or they can design more specific ligands as drugs. It has been said that more *medical researchers* have lost their reputations in cancer research than in any other subject. On the other hand, Section III has demonstrated that bio-inorganic drugs have already made significant contributions to the reputations of some cancer research workers. We do not claim that there is a bio-inorganic solution to all our current problems but, at least, the subject is helpful in framing the pertinent questions and in providing a discussion forum for medical researchers to air their problems and for inorganic chemists and biochemists to exhibit their techniques and instruments that may conceivably be necessary for the solution of these problems.

THE FUTURE

We have talked a great deal about recently discovered bio-inorganic facts and suggested that these may be the beginning of a fascinating future. What exactly does this future hold in store for us? First of all, let me point out that the concluding paragraphs to most of the previous chapters suggest directions of what seems to be the more immediate future. But what of the long-term future? In 1968, I wrote: "The (long-term) future will only be bright if we can delete our present ignorance of the connections between chemical structure and biological activity." This book has shown that in the intervening half decade some of this ignorance has indeed been removed. However, it still requires clairvoyance to predict possible future trends. To record one's predictions in print so that they are readable many years hence is foolhardy (people with predictive abilities ought to write Patent Applications rather than books!) and yet, unless one is committed to one's future destination one is likely to finish up some place else.

From what has been written in previous chapters it should be clear that bio-inorganic chemistry is full of surprises and so simple extrapolation of tendencies observed in established results may not completely foretell the future progress of the subject. It ought also to be clear that the applications of bio-inorganic knowledge may be responsible for profound social and medical changes (for example, the relationships between lithium and manic depression, platinum and cancer therapy, and zinc and wound healing). To insist on progressing in these directions will require far more than advanced instrumentation and gigantic finances; it needs that characteristic which is the most important of all the virtues—courage.

In the next decade bio-inorganic chemistry will develop under two, often conflicting, influences—the *internal* and *external* pressures for bio-inorganic research.

Internally, a far deeper understanding of biochemistry, inorganic chemistry and medicine is needed for complete maturity of the subject. We must guard against any one area developing more rapidly than the others lest the subject assumes a biased character (for example, it would be wrong for it to only be understood by, and be taught by, inorganic chemists). This termination of the interdisciplinary dialogue could be fatal. Instead, each topic ought to develop in parallel formations through purely descriptive phases (as exemplified by the majority of this book), to generalizations and theories, and eventually to a quantitative phase (for example, the computer calculation of the exact doses necessary to correct some metabolic malfunctioning of an organ).

Externally, social pressures for bio-inorganic knowledge will undoubtedly distort the pace of development of particular topics as agricultural, environmental, industrial, medical and military needs yield their respective pressures (for example, molybdenum in sheep ranching, or strontium 90 in the atmosphere). These external demands are likely to increase but we sincerely hope that they will not sway our attempts to complete the development of our internal knowledge of the basic principles of the subject.

For those completing their scientific training at the end of a current degree course, it may have escaped their notice that it is still possible to take the subject further upon a purely amateur basis without the relaxation of standards implied by the term "amateur." Increasing leisure hours provide more time for observer-experimenters to explore and contemplate such topics as ecology, the environment, pollution, the fortunes of research accidents, or the human angles behind new discoveries. Eventually, science in general and bio-inorganic chemistry in particular may be important among the aesthetic and cultural interests of our next generation of students.

TELEOLOGICAL SUGGESTIONS

The teleology of bio-inorganic chemistry has four main goals:

(1) We have unravelled many of the structures of metal complexes that occur *in vivo*. Now we must strive to place mechanistic explanations into the science. This will not be an easy task since the concepts of bio-inorganic mechanisms are many dimensions more complicated than those of the structures of the components of the mechanism. To take a structural analogy, it has been more than fifty years since the word "chelate" was first applied to metal complexes and yet only recently has it become commonly used in pharmacology. Let us hope that modern computers and interdisciplinary dialogues can abbreviate this timespan as far as bio-inorganic mechanisms are concerned!

(2) As stated earlier, one of the ends we must have in sight is an enhancement of the quantitative side of the subject. For example, most blanket antiviral therapies kill both the viruses and their hosts. Quantitative medication, rather than molecular roulette, is more likely to seize upon the differences in toxicological thresholds between virus and host to the detriment of the virus.

(3) Both goals (1) and (2) necessitate

the production of improved models of biological systems. The first stage will be a more realistic model of blood plasma. This must incorporate multiphase equilibria in which the relative volumes of the phases are specified, distribution coefficients across phase boundaries, and dynamic equilibria involving all phases (membranes, solids, liquids, gases, and active sites); for example, the effects of diet, environment, exercise and diurnal variations upon the kinetics and equilibria involved in blood plasma. Subsequent stages could eventually lead to the ultimate synthesis—a computer model of man. If such a goal is realizable then surely no one can doubt its value in leading to better understandings of the life process and in diagnosing malfunctions of the life process. Perhaps Sillén has most nearly approached the truly global concept in his equilibrium model produced to study the constant compositions of the oceans and the atmosphere.

There are so many unanswered questions pleading for such models—How does man regulate pH, metal ion uptake and excretions and water content? Or, how critical are the essential twenty-five elements and their concentrations in the hydrosphere? (After all, man has been adapting to these elemental concentrations for millions of years, surely our society's recent random addition of various other elements to the ecosystem must affect our *in vivo* equilibria and kinetics.) Or, which trace elements should we add to the increasing volume of artificial foods?

(4) Bio-inorganic chemistry is too important a subject to be left to the chemists. It has taken centuries for chemistry departments to escape from the hangover of the "Vital Force" theory which divided the subject into separate, artificial compart-

ments labelled organic and inorganic. Above all, we must avoid creating yet another compartment. Perhaps this book has been rather iconoclastic in making the point that life is not just hydrogen, carbon, nitrogen and oxygen but many more elements besides. Nevertheless our eventual aim must be to be reassimilated into biochemical, inorganic and medical courses.

"The vineyard is indeed large, and the laborers are few; But that does not matter, provided that the wine is good."

INDEX

MAY 2 1997

Chem
QD
415
.I58